General data and fundamental constants

Quantity	Symbol	Value	Power of ten	Units
Speed of light	c	2.997 924 58	10^8	m s^{-1}
Elementary charge	e	1.602 177	10^{-19}	C
Faraday constant	$F = N_A e$	9.6485	10^4	C mol^{-1}
Boltzmann constant	k	1.380 66	10^{-23}	J K^{-1}
Gas constant	$R = N_A k$	8.314 51		J K^{-1} mol^{-1}
		8.314 51	10^{-2}	L bar K^{-1} mol^{-1}
		8.205 78	10^{-2}	L atm K^{-1} mol^{-1}
		6.2364	10	L Torr K^{-1} mol^{-1}
Planck constant	h	6.626 08	10^{-34}	J s
	$\hbar = h/2\pi$	1.054 59	10^{-34}	J s
Avogadro constant	N_A	6.022 14	10^{23}	mol^{-1}
Atomic mass unit	u	1.660 54	10^{-27}	kg
Mass				
electron	m_e	9.109 39	10^{-31}	kg
proton	m_p	1.672 62	10^{-27}	kg
neutron	m_n	1.674 93	10^{-27}	kg
Vacuum permittivity	ε_0	8.854 19	10^{-12}	J^{-1} C^2 m^{-1}
	$4\pi\varepsilon_0$	1.112 65	10^{-10}	J^{-1} C^2 m^{-1}
Magneton				
Bohr	$\mu_B = e\hbar/2m_e c$	9.274 02	10^{-24}	J T^{-1}
nuclear	$\mu_N = e\hbar/2m_p c$	5.050 79	10^{-27}	J T^{-1}
Bohr radius	$a_0 = 4\pi\varepsilon_0 \hbar^2/m_e e^2$	5.291 77	10^{-11}	m
Rydberg constant	$R = m_e e^4/8h^3 c\varepsilon_0^2$	1.096 77	10^5	cm^{-1}
Standard acceleration of free fall	g	9.806 65		m s^{-2}

The Elements of
Physical Chemistry

The Elements of Physical Chemistry

SECOND EDITION

Peter Atkins

Professor of Chemistry and Fellow of
Lincoln College, Oxford

W. H. Freeman and Company
New York

Library of Congress Cataloging-in-Publication Data

Atkins, P. W. (Peter Willam), 1940–
    The elements of physical chemistry / P. W. Atkins. — 2nd ed.
      p. cm.
    Includes index.
    ISBN 0–7167–3077–4 (Hard Cover)
    1. Chemistry, Physical and theoretical.   I. Title.
  QD453.2.A87   1996b
  541—dc20   96–27901
                                           CIP

Published in the United States by W. H. Freeman and Company,
41 Madison Avenue, New York, NY 10010.
Published in the United Kingdom by Oxford University Press.

Printed in the United States of America

Third Printing 1998

Preface

The aim in this edition, as in the first, is to present an introduction to the essentials of physical chemistry without too much mathematics. I have had students of life sciences particularly in mind while preparing this revision, but the text should be useful to anyone wanting an introduction to the subject.

Where relatively lengthy or calculus-based derivations are appropriate, I have introduced them in *Justifications*. Where I judge it important to provide some background material—such as on classical physics or electromagnetic radiation—I have written a *Further Information* section. Because this background material may be relevant to more than one section of the text, I have collected it at the end of the book.

The second edition has been improved in a number of ways: I have added 'Strategy' sections to the *Worked Examples* to help readers collect their thoughts before embarking on a solution. The number and variety of end-of-chapter *Exercises* have also been increased. A number of these exercises have been given a more interesting context and in many cases a more biological slant.

I strongly believe that good illustrations help enormously, and I have put a great deal of effort into improving this aspect of the text by redrawing almost the entire art program and expanding it to more than 300 illustrations. There is much new material, some trimming of old, some lengthening of others, and a lot of detailed rearrangement and elucidation.

As always, there is a difficult line to tread between providing an account of the broad principles of the subject, which are largely timeless in so mature a field, and showing that the subject is still lively by importing new and currently fashionable topics. To broaden the appeal of the text, I have strengthened the spectroscopy sections, with more detailed accounts of rotational and Raman spectroscopy, and have taken magnetic resonance as far as a brief introduction to magnetic-resonance imaging. Other additions in the spectroscopy sections include circular dichroism and the spectroscopic analysis of mixtures. These changes will make the text more useful for students who need to know how techniques are applied.

All authors draw heavily on the generous-spirited contributions of colleagues, either formally involved or informal correspondents. I would particularly like to thank the translators of the first edition, who provided me with so much detailed commentary. I am also grateful to all those users who wrote to me with unsolicited comments and who have guided my thoughts for this revision. I hope they will continue to do so, and that in these pages they will see that I listen. The people who have been more formally involved in this edition are: J. Albright, Texas Christian University; B. Cleaver, University of Southampton; D. Goss, City University of New York; G. W. Gray, Selkirk College; P. Jagodzinski, West Virginia University; M. L. James, University of Northern Colorado; B. Johnson, University of Leeds; B. Joshi, State University of New York College at Geneseo; H. F. Leach, University of Edinburgh; P. Monk, Manchester Metropolitan University; W. Plachy, San Fransisco State University; S. Roser, University of Bath; B. Robinson, University of East Anglia; S. Schullery, Eastern Michigan University; P. D. Sullivan, Ohio University; and G. Wikanda, Umea University. I am deeply grateful to them and have learned a lot from their wise advice.

Oxford   P.W.A.
June 1996

Contents

Contents

Contents

States of matter and the properties of gases

Chemistry is the science of matter and the changes it can undergo. The branch of the subject called **physical chemistry** is concerned with the physical principles that underlie chemistry. It seeks to account for the structure of matter and the changes it undergoes in terms of fundamental concepts such as atoms, electrons, and energy. Physical chemistry provides the basic framework for all other branches of chemistry—for inorganic chemistry, organic chemistry, biochemistry, geochemistry, and engineering. It also provides the basis of modern methods of analysis, the determination of structure, and the elucidation of the manner by which chemical reactions occur. To do all this, it draws on two of the great foundations of modern science, thermodynamics and quantum mechanics. This text will introduce you to the central concepts of these two subjects and show how they are used in chemistry. However, their principles reach out into other disciplines, and through them you will come to have a sure foundation for understanding a wide range of modern science and its never ending—and highly successful—quest for new materials and new ways of looking at the world.

We begin our journey by thinking about matter in bulk. The broadest classification of bulk matter is into one of three **physical states**, namely gas, liquid, and solid. Later we shall see how this classification can be refined, but these three broad classes are a good starting point. The three states of matter can be recognized by their behaviour when they are enclosed in a container:

A **gas** is a fluid state of matter that fills the container it occupies.
A **liquid** is a fluid state of matter that possesses a well-defined surface and (in a gravitational field) fills the lower part of the container it occupies.
A **solid** retains its shape independent of the shape of the container it occupies.

One of the roles of physical chemistry is to establish the link between the properties of bulk matter and the behaviour of the particles—atoms, ions, or molecules—of which it is composed. A physical chemist tries to provide a clear *picture* of each state of matter, and then tries to show how its properties can be understood in terms of this insight. The existence of the three states of matter is a first illustration of this procedure, for each state is composed of particles with different freedom of movement. So, a physical chemist might

have in mind the following images when thinking about the three states of matter:

> A gas is composed of particles in continuous rapid, chaotic motion. A particle travels several (often many) diameters before colliding with another particle, and for most of the time the particles are so far apart that they interact with each other only very weakly. The overriding image is one of busy chaos.
>
> A liquid consists of particles that are in contact with each other, but are able to move past each other in a restricted manner. The particles are in a continuous state of motion, but travel only a fraction of a diameter before bumping into a neighbour. The overriding image is one of movement, but with molecules jostling one another.
>
> A solid consists of particles that are in contact with one another and unable to move past one another. Although the particles oscillate around an average location, they are essentially trapped in their initial positions. The overriding image is one of almost stationary, regularly arrayed particles.

The essential difference between the three states of matter comes down to the freedom of the particles to move past one another. If the particles are widely separated on average, there is hardly any restriction on their motion, and the substance is a gas. If the particles interact so strongly with each other that they are locked together, then the substance is a solid. If the particles have an intermediate mobility between these extremes, then the substance is a liquid. The melting of a solid and the vaporization of a liquid can be understood in terms of the progressive increase in the liberty of the particles as a sample is heated and the particles are able to move more vigorously.

In this chapter we shall see how to describe gases, the simplest of these three states of matter. However, much of the material we cover will in due course be seen to be useful in many parts of physical chemistry. For instance, the description of gases introduces the concepts of pressure and temperature, which are needed whenever we are discussing the behaviour of bulk matter. Moreover, although gases are simple—both to describe and in terms of their internal structure—they are of immense importance. We spend our whole lives surrounded by gas in the form of air, and the local variation in its properties is what we term 'weather'. To understand the atmospheres of this and other planets we need to understand gases. As we breathe, we pump gas in and out of our lungs, where it changes composition and temperature, and to understand the construction of our bodies and their response to anaesthetics we need to understand the properties of gases. Many industrial processes, the Haber process for the synthesis of ammonia, for instance, involve gases, and both the outcome of the reaction and the design of the reaction vessels depend on a knowledge of gases. A knowledge of gases turns out to be useful in all manner of surprising places. The interiors of stars, for example, although immensely dense and at first sight quite unlike any gas we meet on Earth, can be described by the laws that we shall develop here.

The description of states of matter

The term 'state' has many different meanings in chemistry, and it is important to keep them all in mind. We have already met one meaning in the expression 'the states of matter' and specifically 'the gaseous state'. Now we meet a second: by **state** we shall mean a particular condition of a sample of matter that is described in terms of the volume, pressure, temperature, and amount of substance present. (The precise meanings of these terms are described below.) So, 1 kg of hydrogen gas in a container of volume 10 L at a specified pressure and temperature is in a particular state. The same mass of gas in a container of volume 5 L is in a different state. Two samples of a given substance are in the same state only if they have the same mass, volume, pressure, and temperature.

To see more precisely what is involved in specifying the state of a substance, we need to define the terms we have used. One property—the volume—can be dealt with without any fuss: the **volume**, V, of a sample is a measure of the space it occupies. Thus, we write $V = 100$ cm^3 if the sample occupies 100 cm^3 of space. Some of the units used to express volume (including cubic metres, m^3; litres, L; millilitres, mL), and units and symbols in general, are reviewed in *Further information 2*.[†]

The other properties we have mentioned (pressure, temperature, and amount of substance) need more introduction, for even though they may be familiar from everyday life, they need to be defined carefully for use in science.

1.1 Pressure

By **pressure**, p, we mean force divided by the area on which the force is exerted:

$$\text{pressure} = \frac{\text{force}}{\text{area}}$$

When you are standing on ice, you generate a pressure on the ice as a result of the gravitational force pulling you towards the centre of the Earth, but the pressure will be quite low because the force is spread over the area equal to that of the soles of your shoes. However, if you stand on skates, then the area of the blades in contact with the ice is much smaller, so although the force is the same, the pressure is much greater (Fig. 1.1). The pressure may be so great, in fact, that it modifies the arrangement of water molecules at the surface of the ice, and hence allows you to slide smoothly over the surface.

Although the gravitational pull of the Earth on an object can result in a pressure, pressure can arise in other ways. For example, the impact of gas molecules on a surface gives rise to a force, and hence to a pressure. If an

[†]All the *Further information* sections are collected at the end of the book.

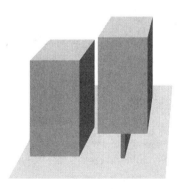

Fig. 1.1 These two blocks of matter have the same mass. They exert the same force on the surface on which they are standing, but the one on the right exerts a greater pressure because it exerts the force on a smaller area than the one on the left.

object is immersed in the gas, it experiences a pressure over its entire surface because molecules collide with it from all directions. In this way, the atmosphere exerts a pressure on all the objects in it. We are incessantly battered by molecules of gas in the atmosphere, and experience this battering as the atmospheric pressure. The pressure is greatest at sea level because the density of air, and hence the number of colliding molecules, is greatest there. The atmospheric pressure is very considerable: it is the same as would be exerted by loading 1 kg of lead (or any other material) on to a surface of area 1cm^2: we go through our lives under this heavy burden pressing on every square centimetre of our bodies. Some deep sea creatures are built to withstand even greater pressures: at 1000 m below sea level the pressure is 100 times greater than at the surface. Creatures and submarines that operate at these depths must withstand the equivalent of 100 kg of lead loaded on to each square centimetre of their surfaces. The pressure of the air in our lungs helps us withstand the relatively low but still substantial pressures that we experience.

If a gas is confined to a cylinder fitted with a movable piston that separates it from the atmosphere, then the position of the piston adjusts until the pressure of the gas inside the cylinder is equal to that exerted by the atmosphere. When the pressures on either side of a movable piston are the same, we say that the two regions on either side are in **mechanical equilibrium**. The pressure of the confined gas arises from the incessant impact of the particles: they batter the inside surface of the piston and counter the battering of the molecules in the atmosphere that is pressing on the outside surface of the piston (Fig. 1.2). So long as the piston is weightless (that is, so long as we can neglect any gravitational pull on it), the gas is in mechanical equilibrium with the atmosphere whatever the orientation of the piston and cylinder, because the external battering is the same in all directions.

Pressure is measured in the unit called the **pascal**, Pa:

$$1\,\text{Pa} = 1\,\text{kg}\,\text{m}^{-1}\,\text{s}^{-2}$$

The pressure of the atmosphere at sea level is about 10^5 Pa (100 kPa). This fact lets us imagine the magnitude of 1 Pa, for we have just seen that 1 kg of lead resting on 1 cm^2 on the surface of the Earth exerts about the same pressure as the atmosphere; so $1/10^5$ of that mass, or 0.01 g, will exert about 1 Pa (which shows that the pascal is rather a small unit of pressure). Some of the other units that are used to report pressures are summarized in Table 1.1. They include the following unit that we shall use frequently:

$$1\,\text{bar} = 10^5\,\text{Pa}$$

Table I.I Pressure units and conversion factors

SI unit: pascal (Pa)	$1\,\text{Pa} = 1\,\text{N}\,\text{m}^{-2}$
bar	$1\,\text{bar} = 10^5\,\text{Pa}$
atmosphere	$1\,\text{atm} = 101.325\,\text{kPa}$
torr*	$760\,\text{Torr} = 1\,\text{atm}$
	$1\,\text{Torr} = 133.32\,\text{Pa}$

*The name of the unit is torr; its symbol is Torr.

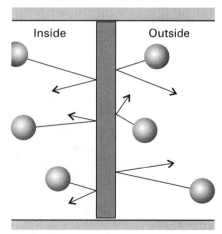

Inside   Outside

Fig. 1.2 A system is in mechanical equilibrium with its surroundings if it is separated from them by a movable wall and the external pressure is equal to the pressure of the gas in the system.

Atmospheric pressure at sea level is close to 1 bar. You should also be able to convert between the units in the table, because atmospheres (atm), torr (Torr), and millimetres of mercury (mmHg) are still widely used. The following example illustrates the procedure.

Example *Converting units*
Express a pressure of 1.000 bar in atmospheres.

Strategy
The general procedure for converting between units is to write the relation between them, namely

$$\text{units given} = \text{units required}$$

in the form of a conversion factor:

$$\frac{\text{units required}}{\text{units given}}$$

and then to write

$$\text{quantity in units required} = \text{quantity in units given} \times \frac{\text{units required}}{\text{units given}}$$

Solution
From Table 1.1 we have

$$1.013\,25\,\text{bar} = 1\,\text{atm}$$

The calculation then takes the form

$$\text{pressure in atm} = (1.000\,\text{bar}) \times \frac{1\,\text{atm}}{1.013\,25\,\text{bar}} = 0.9869\,\text{atm}$$

Notice how the units (bar in this case) cancel, just like numbers.

Exercise E1.1

Express a pressure of 723 Torr in kilopascals.

[*Answer*: 96.4 kPa]

Atmospheric pressure (a property that varies with altitude and the weather) is measured with a **barometer**, which was invented by Torricelli, a student of Galileo's. A barometer consists of an inverted tube of mercury that is sealed at its upper end and stands with its lower end in a bath of mercury. The height of the mercury in the tube is proportional to the atmospheric pressure (Fig. 1.3). The pressure at the foot of a column of incompressible liquid (a

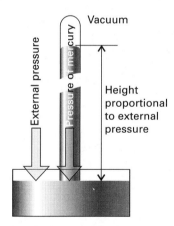

Fig. 1.3 The operation of a mercury barometer. The space above the mercury in the vertical tube is a vacuum, so no pressure is exerted on the top of the mercury column; however, the atmosphere exerts a pressure on the mercury in the reservoir, and pushes the column up the tube until the pressure exerted by the mercury column is equal to that exerted by the atmosphere. The height reached by the column is proportional to the external pressure, so the height can be used as a measure of this pressure.

5

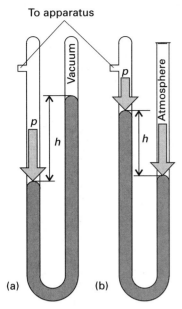

To apparatus

Vacuum

Atmosphere

p

h

p

p

h

(a)   (b)

Fig. 1.4 Two versions of a manometer used to measure the pressure of a sample of gas. (a) The height difference, h, of the two columns in the sealed-tube manometer is directly proportional to the pressure of the sample. (b) The difference in heights of the columns in the open-tube manometer is proportional to the difference in pressure between the sample and the atmosphere. In this case, the pressure of the sample is lower than that of the atmosphere.

good approximation for any liquid) is proportional to the height, h, of the column and the density, ρ (rho), of the liquid:

$$p = g\rho h \qquad (1)$$

where g is the acceleration of free fall, a measure of the Earth's gravitational pull on the liquid ($g = 9.81\,\mathrm{m\,s^{-2}}$ at sea level). This expression provides an easy means of relating the observed height of a barometer to the pressure in pascals. For example, the pressure at the foot of a column of mercury of height 760 mm and density $13.6\,\mathrm{g\,cm^{-3}}$ ($1.36 \times 10^4\,\mathrm{kg\,m^{-3}}$) is

$$p = (9.81\,\mathrm{m\,s^{-2}}) \times (1.36 \times 10^4\,\mathrm{kg\,m^{-3}}) \times (0.760\,\mathrm{m})$$
$$= 1.01 \times 10^5\,\mathrm{kg\,m^{-1}\,s^{-2}}$$

This pressure corresponds to 101 kPa (1.00 atm).

The pressure of a gas *inside* a container can be measured by a variety of different pressure gauges. The simplest is a **manometer**, which is a U-tube containing liquid (water is sometimes used), with one limb connected to the container and the other open to the atmosphere (Fig. 1.4). The difference in heights of the liquid in the two arms of the U-tube is proportional to the *difference* in pressure between the gas in the container and the external atmosphere. For example, according to eqn (1), a column of water of height 10.0 cm corresponds to a pressure

$$p = (9.81\,\mathrm{m\,s^{-2}}) \times (1.00 \times 10^3\,\mathrm{kg\,m^{-3}}) \times (0.100\,\mathrm{m}) = 0.981\,\mathrm{kPa}$$

Therefore, if the pressure of the atmosphere at the time of the experiment is 100.021 kPa, and the column of water is higher on the apparatus side of the manometer like that shown in Fig. 1.4(b), which indicates that the pressure is lower in the apparatus than outside, then the pressure in the apparatus is 100.021 kPa − 0.981 kPa = 99.040 kPa.

1.2 Temperature

The **temperature**, T, of a sample is a familiar concept in everyday life—as a measure of how 'hot' or 'cold' something is—but it is quite difficult to give the concept a precise definition. The temperature of an object is a property that determines in which direction energy will flow when it is in contact with another object: energy ('heat') flows from the higher temperature to the lower temperature. When the two bodies have the same temperature, there is no net flow of heat between them. In that case we say that the bodies are in **thermal equilibrium** (Fig. 1.5).

Temperatures in science are measured on either the Celsius scale or the Kelvin scale. On the **Celsius scale**, the freezing point of water corresponds to 0 °C and the boiling point corresponds to 100 °C. This scale is now in widespread everyday use. Temperatures on the Celsius scale will be denoted by the Greek letter θ (theta) throughout this text. However, it turns out to be much more convenient in many scientific applications to adopt the **Kelvin** scale and to express the temperature in kelvins, K. *Whenever we use T to denote a temperature, we shall mean a temperature on the Kelvin scale.* The Celsius and Kelvin scales are related by

$$T(\text{in kelvin}) = \theta(\text{in degrees Celsius}) + 273.15$$

That is, to obtain the temperature in kelvins, add 273.15 to the temperature in degrees Celsius. Thus, water freezes at 273 K and boils at 373 K; a warm day (25 °C) corresponds to 298 K.

A more sophisticated way of expressing the relation between T and θ, and one that we shall use in other contexts, is to regard the value of T as the product of a number and a unit (K), so that T/K is a pure number; for example, if $T = 298$ K, then $T/K = 298$. Likewise, $\theta/°C$ is also a pure number. For example, if $\theta = 25$ °C, then $\theta/°C = 25$. Then we can write the relation between the two scales as

$$T/K = \theta/°C + 273.15 \tag{2}$$

This expression is a relation between pure numbers.

Exercise E1.2

Use eqn (2) to express body temperature, 37 °C in kelvins.

[*Answer*: 310 K]

1.3 Amount of substance

The **mass**, m, of an object is a measure of the quantity of matter it contains. Thus, 2 kg of lead contains twice as much matter as 1 kg of lead and indeed twice as much matter as 1 kg of anything. On average, a man contains more matter than a woman. The SI unit of mass is the **kilogram** (kg), with 1 kg currently defined as the mass of a certain block of platinum-iridium alloy preserved at Sèvres, outside Paris. For typical laboratory-sized samples it is usually more convenient to use a smaller unit and to express mass in grams (g), where $1\,\text{kg} = 10^3$ g.

In chemistry, where we focus on the behaviour of atoms, it is usually more useful to know the number of atoms, molecules, or ions in a sample rather than the mass of the sample. Because even 10 g of water consists of about 10^{23} H_2O molecules, it is clearly appropriate to define a new unit that can be used to express such large numbers in a much simpler fashion. Chemists have therefore introduced the **mole** (mol; the name is derived, ironically, from the Latin word meaning 'massive heap') and is defined as follows:

1 mol of particles is equal to the number of atoms in exactly 12 g of carbon–12.

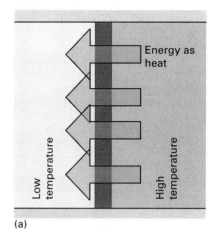

(a)

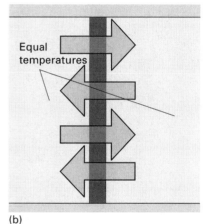

(b)

Fig. 1.5 The temperatures of two objects act as a signpost showing the direction in which energy will flow as heat through a thermally conducting wall: (a) heat always flows from high temperature to low temperature. (b) When the two objects have the same temperature, although there is still energy transfer in both directions, there is no net flow of energy.

In practice,

$$1 \, \text{mol} \approx 6.022 \, 14 \times 10^{23} \text{ particles}$$

For example, a sample of hydrogen gas that contains 6.022×10^{23} hydrogen molecules consists of $1.000 \, \text{mol} \, H_2$, and a sample of water that contains $1.2 \times 10^{24} (= 2.0 \times 6.022 \times 10^{23})$ water molecules consists of $2.0 \, \text{mol} \, H_2O$. A number of awe-inspiring illustrations of the number of particles in a mole have been devised; for instance, 1 mol of soft-drink cans would bury the surface of the Earth to a depth of 300 km.

Note that we must always specify the particles when using the unit mole, for that avoids any ambiguity. If, improperly, we said that a sample consisted of 1 mol of hydrogen, it would not be clear whether it consisted of 6×10^{23} hydrogen atoms $(1 \, \text{mol} \, H)$ or 6×10^{23} hydrogen molecules $(1 \, \text{mol} \, H_2)$. The important feature of the mole is that if we know that a sample of gas contains $1 \, \text{mol} \, O_2$ and another sample contains $1 \, \text{mol} \, N_2$, then we know that both samples contain the same number of molecules. The two samples would have different masses (different quantities of matter), but the numbers of molecules would be the same in each case.

Just as a kilogram is the unit used to report a certain physical property (the mass), the mole is the unit used when reporting the value of another physical property. In this case the property is called the **amount of substance**, n, of a sample. Thus, we can write $n = 1 \, \text{mol} \, H_2$, and say that the amount of hydrogen molecules in a sample is 1 mol. The term 'amount of substance', however, has not yet found wide acceptance among chemists, and in casual conversation you will often hear them speaking of 'the number of moles' in a sample. The term **chemical amount**, however, is becoming more widely used, and we shall often use it in this book.

There are various useful concepts that stem from the introduction of the concept of the chemical amount and its unit the mole. One is the **Avogadro constant**, N_A, the number of particles (of any kind) per mole of substance:

$$N_A = 6.022 \, 14 \times 10^{23} \, \text{mol}^{-1}$$

The Avogadro constant makes it very simple to convert from the number of particles N (a pure number) in a sample to the chemical amount n (in moles) it contains:

$$n = \frac{\text{number of particles}}{\text{number of particles per mole}} = \frac{N}{N_A} \tag{3}$$

For example, 8.8×10^{22} copper atoms corresponds to

$$n(\text{Cu}) = \frac{8.8 \times 10^{22}}{6.022 \, 14 \times 10^{23} \, \text{mol}^{-1}} = 0.15 \, \text{mol}$$

Notice how much easier it is to report the amount of Cu atoms present rather than their actual number.

Exercise E1.3

How many Xe atoms are present in a sample that contains 1.8 mol Xe?

[*Answer*: 1.1×10^{24} Xe]

The second very important concept is that of **molar mass**, M, the mass per mole of substance. For example, the molar mass of C atoms is $12.01\,\mathrm{g\,mol^{-1}}$ and that of H_2O molecules is $18.02\,\mathrm{g\,mol^{-1}}$.[†] When we refer to the molar mass of an element we shall always mean the mass per mole of its atoms. When we refer to the molar mass of a compound, we shall always mean the molar mass of its molecules or, in the case of ionic compounds, the mass per mole of its formula units. A **formula unit** is the physical entity corresponding to a specific chemical formula. For example, the formula unit for Na_2SO_4 consists of two Na^+ ions and one SO_4^{2-} ion.

The molar mass of an element is determined by mass spectrometric measurement of the mass of its atoms and then multiplying the mass of one atom by the Avogadro constant. Care has to be taken to allow for the isotopic composition of an element, and a suitably weighted mean of the masses of the atoms is used. The values obtained in this way are printed in the periodic table on the inside front cover. The molar mass of a compound is calculated by taking a sum of the molar masses of its constituent atoms.

The molar mass is used to convert from the mass of a sample (which we can measure) to the amount of substance present (which, in chemistry, we often need to know):

$$n = \frac{\text{mass of sample}}{\text{molar mass}} = \frac{m}{M} \tag{4}$$

For example, to find the amount of C atoms present in 21.5 g of carbon, given the molar mass of C is $12.01\,\mathrm{g\,mol^{-1}}$, we write

$$n(C) = \frac{21.5\,\mathrm{g}}{12.01\,\mathrm{g\,mol^{-1}}} = 1.79\,\mathrm{mol}$$

Exercise E1.4

What amount of H_2O molecules is present in 100 g of water?

[*Answer*: 5.55 mol]

[†]The former names for molar mass were atomic weight, for the mass per mole of atoms, and molecular weight, for the mass per mole of molecules; both terms are still far from dead.

Equations of state

Now we are at the stage where we can specify the state of any sample of substance by giving the values of the following properties:

V, the volume the sample occupies.
p, its pressure.
T, its temperature.
n, the amount of substance it contains.

However, an astonishing experimental fact of nature is that *these four quantities are not independent of one another*. For instance, we cannot arbitrarily choose to have a sample of $5.55 \, mol \, H_2O$ in a volume of $100 \, cm^3$ at $100 \, kPa$ and $500 \, K$: it is found *experimentally* that the state simply does not exist. If we select the amount, the volume, and the temperature, then we find that we have to accept a particular pressure. The same is true of all substances. This experimental generalization is summarized by saying the substance obeys an **equation of state**, an equation that relates one of the four properties to the other three.

The equations of state of most substances are not known, so in general we cannot write down the mathematical relation between the four properties that define a state. However, certain equations of state are known. In particular, the equation of state of low-pressure gases is known, and proves to be very simple and very useful. This is the equation we mentioned in the introduction, and which is used to describe the behaviour of gases taking part in reactions, the behaviour of the atmosphere, as a starting point for problems in chemical engineering, and even in the description of the structures of stars.

1.4 The perfect gas equation of state

The equations of states of gases were among the first results to be established in physical chemistry. The original experiments were carried out by Robert Boyle in the seventeenth century, and there was a resurgence in interest in the late seventeenth century when people began to fly in balloons. This technological progress demanded more knowledge about the response of gases to changes of pressure and temperature.

Boyle's and his successors' experiments led to the formulation of the following **perfect gas equation of state**:

$$pV = nRT \tag{5}$$

In this equation—which is probably the most important in the whole of physical chemistry—R is a constant with the same value for all gases. It is known as the **gas constant**, and has the value

$$R = 8.3145 \, kPa \, L \, K^{-1} \, mol^{-1}$$

This is the value to use when pressures are quoted in kilopascals and volumes in litres, L (1 L = 1 dm³): when other units are used for these variables, you should use one of the values of R given in Table 1.2.

The perfect gas equation of state—more briefly, the 'perfect gas law'—is so-called because it is an idealization of the equations of state that gases actually obey. Specifically, it is found that all gases obey the equation ever more closely as the pressure is reduced towards zero. That is, eqn (5) is an example of a **limiting law**, a law that is not obeyed exactly by any actual gas, but becomes increasingly valid as the pressure is reduced and is obeyed exactly in the limit of zero pressure.

A hypothetical fluid that obeys eqn (5) at any pressure is called a **perfect gas**. From what has just been said, an actual gas, which is termed a **real gas**, behaves more and more like a perfect gas as its pressure is reduced, and behaves exactly like a perfect gas when the pressure has been reduced to zero. In practice, normal atmospheric pressure at sea level ($p \approx 100$ kPa) is already low enough for most real gases to behave almost perfectly, and we shall always assume in this text that the gases we encounter behave like a perfect gas (and therefore that eqn (5) can be used to describe the relation between their volume, pressure, temperature, and amount). The reason why a real gas behaves differently from a perfect gas can be traced to the attractions and repulsions that exist between actual molecules and which are absent in a perfect gas; the origin of these interactions is described in Chapter 10.

The perfect gas law summarizes three sets of observations. One is **Boyle's law**, that at constant temperature the pressure of a fixed amount of gas is inversely proportional to its volume. Thus, when n and T are constant, the perfect gas law becomes $pV = $ constant, and hence

$$p \propto \frac{1}{V} \tag{6}$$

The graph obtained by plotting p against V is shown in Fig. 1.6. Each curve is called an **isotherm** because it shows the variation of a property (the pressure) at constant temperature. (*Iso* is Greek for equal and *therm* is derived from the Greek for heat.) Instead of plotting p against V, we can plot p against $1/V$, which should give a straight line (Fig. 1.7).[†] Equation (6) also shows that if we compress a fixed amount of gas at constant temperature into half its original volume, then its pressure will double.

The second observation summarized by eqn (5) is **Charles's law**, that the volume of a fixed amount of gas at constant pressure is proportional to the temperature, T. To verify that eqn (5) agrees with this observation, we first rearrange it into

$$V = \frac{nRT}{p}$$

and then note that when the amount n and the pressure p are both constant, we can write

$$V \propto T \tag{7}$$

[†]Mathematical techniques and graphs are reviewed in *Further information 1*.

Table 1.2 The gas constant in various units

$R =$	8.314 51 J K⁻¹ mol⁻¹
	8.314 51 kPa L K⁻¹ mol⁻¹
	8.205 78 ×10⁻² L atm K⁻¹ mol⁻¹
	62.364 L Torr K⁻¹ mol⁻¹
	1.987 22 cal K⁻¹ mol⁻¹

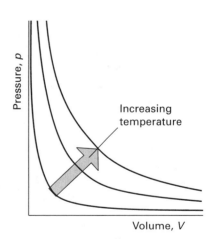

Fig. 1.6 The volume of a gas decreases as the pressure on it is increased. For a sample that obeys Boyle's law and that is kept at constant temperature, the graph showing the dependence is a hyperbola, as shown here. Each curve corresponds to a single temperature, and hence is an isotherm.

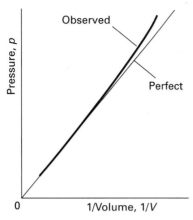

Fig. 1.7 A good test of Boyle's law is to plot p against $1/V$ (at constant temperature), when a straight line should be obtained. This diagram shows that the observed pressures (the heavy line) approaches a straight line as the volume is increased and the pressure reduced. A perfect gas would follow the straight line at all pressures; real gases obey Boyle's law in the limit of low pressures.

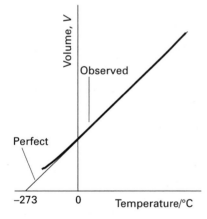

Fig. 1.8 This diagram illustrates the content and implications of Charles's law, which asserts that the volume occupied by a gas (at constant pressure) is proportional to the temperature. When plotted against Celsius temperatures (as here), all gases give lines that extrapolate to $V = 0$ at $-273\,°C$. This extrapolation suggests that $-273\,°C$ is the lowest attainable temperature.

as Charles's law requires. It follows that doubling the temperature (on the Kelvin scale, such as from 300 K to 600 K, corresponding to an increase from $27\,°C$ to $327\,°C$) doubles the volume, provided the pressure remains the same.

Charles's law indicates that there exists an **absolute zero** of temperature, a temperature below which it is impossible to cool an object. If the volume of a sample of gas is plotted against temperature, then the observations extrapolate to zero volume as T approaches zero (Fig. 1.8). Because the volume of a body cannot be negative, it follows that T cannot be reduced to below zero. We conclude that $T = 0$ is the lowest possible temperature. According to eqn (2), $T = 0$ corresponds to $\theta = -273.15\,°C$.

The third feature of eqn (5) that is consistent with observations is **Avogadro's principle**, which states that *at a given temperature and pressure, equal volumes of gas contain the same number of molecules*. That is, if a sample of air at 100 kPa occupies 1.00 L at 300 K, and a sample of pure carbon dioxide occupies the same volume under the same conditions, then we can infer that both samples contain the same number of molecules. Moreover, doubling the number of molecules results in a doubling of the volume of the sample, as long as its pressure and temperature remain constant. We can therefore write:

$$V \propto n \tag{8}$$

at constant pressure and temperature. This result follows easily from eqn (5) if we regard p and T as constants. Avogadro's suggestion is a principle rather than a law, because it is not a direct summary of experience (it is based on the model of a gas as a collection of molecules).

The **molar volume**, V_m, of any substance (not just a gas) is the volume it occupies per mole of molecules present in the sample:

$$V_m = \frac{V}{n} \tag{9}$$

Avogadro's principle implies that the molar volume of a gas should be the same for all gases at the same temperature and pressure. That this conclusion is almost true under normal conditions (normal atmospheric pressure of about 100 kPa and room temperature) can be confirmed by measuring out 1 mol of molecules of various gases and determining the volume they occupy. Some experimental data are given in Table 1.3.

1.5 Using the perfect gas law

The perfect gas law is used when we want to predict the pressure of a gas given its temperature, its chemical amount, and the volume it occupies. Then we rearrange it into

$$p = \frac{nRT}{V} \tag{10}$$

and substitute the data.

Example *Predicting the pressure of a sample of gas*
What pressure is exerted by 1.25 g of nitrogen gas in a flask of volume
250 mL at 20 °C?

Strategy
All we need do is to insert the data into eqn (10). However, to use this
formula, we need to know the amount of molecules (in moles) in the
sample, which we can obtain from the mass and the molar mass.
Convert the temperature to the Kelvin scale.

Solution
The amount of N_2 molecules (of molar mass $28.02 \, g \, mol^{-1}$) present is

$$n = \frac{mass}{molar \; mass} = \frac{1.25 \, g}{28.02 \, g \, mol^{-1}} = 4.46 \times 10^{-2} \, mol$$

We have rounded the result to the correct number of figures at this
stage. In practice, you would probably store the intermediate numbers in a memory register of your calculator, and do the rounding
step at the end of your calculation. The temperature of the sample
is

$$T/K = 20 + 273.15 = 293$$

That is, $T = 293 \, K$. Therefore, from eqn (10) we obtain

$$p = \frac{(4.46 \times 10^{-2} \, mol) \times (8.3145 \, kPa \, L \, K^{-1} \, mol^{-1}) \times (293 \, K)}{250 \times 10^{-3} \, L}$$

$$= \frac{(4.46 \times 10^{-2}) \times 8.3145 \times 293}{250 \times 10^{-3}} \, kPa = 435 \, kPa$$

Note that the units cancel like ordinary numbers.

Table 1.3 The molar volumes of
gases at 25°C and 1 bar

Gas	$V_m/L \, mol^{-1}$
Perfect gas	24.79
Ammonia	24.8
Argon	24.4
Carbon dioxide	24.6
Nitrogen	24.8
Oxygen	24.8
Hydrogen	24.8
Helium	24.8

Exercise E1.5

Calculate the pressure exerted by 1.22 g of carbon dioxide confined to
a flask of volume 500 mL at 37 °C.

[*Answer*: 143 kPa]

It is often the case, though, that we are given the pressure under one set of
conditions, and are asked to predict the pressure of the same sample under a
different set of conditions. In this case we use the law as follows. Suppose the

pressure is p_1 when the temperature is T_1 and the volume is V_1, then from eqn (5) we can write

$$\frac{p_1 V_1}{T_1} = nR$$

If the conditions are changed to T_2 and V_2, and the pressure changes to p_2 as a result, then we also know that

$$\frac{p_2 V_2}{T_2} = nR$$

The nR on the right of these two equations is the same in each case, because R is a constant and the amount of gas molecules has not changed. It follows that we can combine the two equations into a single equation:

$$\frac{p_1 V_1}{T_1} = \frac{p_2 V_2}{T_2} \tag{11}$$

This expression is known as the **combined gas equation**. It can easily be rearranged to give one unknown (such as p_2, for instance) in terms of the other variables.

Exercise E1.6

What is the final volume of a sample of gas that has been heated from $25\,^\circ\text{C}$ to $1000\,^\circ\text{C}$ and its pressure increased from $10.0\,\text{kPa}$ to $150.0\,\text{kPa}$, given that its initial volume was $15\,\text{mL}$?

[*Answer*: $4.3\,\text{mL}$]

The perfect gas law can also be used to calculate the molar volume of a perfect gas at any temperature and pressure. If eqn (5) is substituted into eqn (9), we find

$$V_\text{m} = \frac{V}{n} = \frac{(nRT/p)}{n} = \frac{RT}{p} \tag{12}$$

Chemists have found it convenient to report much of their data at a particular set of standard conditions. By **standard ambient temperature and pressure** (SATP) they mean a temperature of $25\,^\circ\text{C}$ (more precisely, $298.15\,\text{K}$) and a pressure of $100\,\text{kPa}$ (that is, 1 bar).[†] The **standard pressure** is normally denoted $p^\ominus$ with $p^\ominus = 1\,\text{bar}$ exactly. At SATP, the molar volume of a perfect

---

[†]An earlier set of standard conditions, which is still encountered, is 'standard temperature and pressure' (STP), namely $0\,^\circ\text{C}$ and 1 atm. The sign $^\circ$ is also widely used to denote the standard value of a property.

gas is $24.79 \, \text{L mol}^{-1}$, as can be verified by substituting the values of the temperature and pressure into eqn (12).

Mixtures of gases: partial pressures

Chemists are often concerned with mixtures of gases, such as when they are considering the properties of the atmosphere in meteorology, the composition of exhaled air in medicine, or the mixtures of hydrogen and nitrogen used in the industrial synthesis of ammonia. We need to be able to assess the contribution that each component of a gaseous mixture makes to the total pressure.

In the early nineteenth century, John Dalton carried out a series of experiments that led him to formulate the following law:

Dalton's law: The pressure exerted by a mixture of perfect gases is the sum of the pressures exerted by the individual gases occupying the same volume alone.

The contribution that a gas J makes to the total pressure is the **partial pressure**, p_J. It follows that

1. The total pressure of a mixture of gases is the sum of the partial pressures of all the gases present.
2. The partial pressure of each gas is equal to the pressure it would exert if it occupied the container alone.

Dalton's law is strictly valid only for mixtures of perfect gases (or for real gases at such low pressures that they are behaving perfectly), but under most conditions we shall encounter, gases can be considered to be perfect.

As an example, suppose that we were interested in the composition of inhaled and exhaled air, and we knew that a certain mass of carbon dioxide exerts a pressure of 5 kPa when present alone in a container, and that a certain mass of oxygen exerts 20 kPa when present alone in the same container at the same temperature. Then, when both gases are present in the container, the partial pressure of the carbon dioxide in the mixture is 5 kPa, that of the oxygen is 20 kPa, and the total pressure of the mixture is 25 kPa (Fig. 1.9).

The partial pressure of each component in a mixture can be calculated from the perfect gas equation of state once we know the amount of molecules of each gas. For example, suppose the mixture consisted of an amount n_A of A molecules and n_B of B molecules, then the partial pressures of A and B are each given by eqn (5) as

$$p_A = \frac{n_A RT}{V} \qquad p_B = \frac{n_B RT}{V} \qquad (13)$$

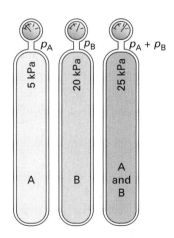

Fig. 1.9 The partial pressure p_A of a perfect gas A is the pressure that it would exert if it occupied a container alone; similarly, the partial pressure p_B of a perfect gas B is the pressure that it would exert if it occupied the same container alone. The total pressure p when both gases simultaneously occupy the container is the sum of their partial pressures.

15

According to Dalton's law, the total pressure of the mixture is the sum of these two partial pressures:

$$p = p_A + p_B \tag{14}$$

A similar calculation can be carried out for mixtures of three or more gases, with each partial pressure being given by a term like those in eqn (13).

The easiest way of setting up partial pressure calculations is to introduce the concept of mole fraction (we shall find this concept useful in several other contexts too):

> The **mole fraction**, x_J, of a species J is the chemical amount of J expressed as a fraction of the total amount of molecules present in the sample.

In a sample that consists of an amount n_A of a species A, n_B of a species B, and so on, the mole fraction of A present in the mixture is

$$x_A = \frac{n_A}{n_A + n_B + \cdots} \tag{15}$$

with similar expressions (with n_B, etc in the numerator) for the mole fractions of B, etc. For a **binary mixture**, one that consists of two species,

$$x_A = \frac{n_A}{n_A + n_B} \qquad x_B = \frac{n_B}{n_A + n_B} \qquad x_A + x_B = 1 \tag{16}$$

When only A is present, $x_A = 1$ and $x_B = 0$. When only B is present, $x_B = 1$ and $x_A = 0$. When both are present in the same amounts, $x_A = \frac{1}{2}$ and $x_B = \frac{1}{2}$.

Exercise E1.7

Calculate the mole fractions of N_2, O_2, and Ar in dry air at sea level, given that 100.0 g of air consists of 75.5 g of N_2, 23.2 g of O_2, and 1.3 g of Ar. [*Hint*: Begin by converting each mass to an amount in moles.]
[*Answer*: 0.780, 0.210, 0.009]

The advantage of introducing the mole fractions of the species present in a mixture is that each partial pressure p_J is related to the total pressure p by

$$p_J = x_J p \tag{17}$$

Therefore, if we know the mole fraction of the species and the total pressure of the mixture, we can immediately write down its partial pressure.

Justification

The total pressure p exerted by a sample that consists of a total amount n of molecules (of any kind, including a mixture of molecules) is related to the volume and temperature by the perfect gas equation of state, $pV = nRT$. It follows that

$$\frac{RT}{V} = \frac{p}{n}$$

Substitution of this expression into eqn (13) results in

$$p_J = n_J \times \frac{p}{n}$$

However, n_J/n is the mole fraction of J in the mixture. Therefore,

$$p_J = x_J \times p$$

as written above.

Example *Calculating partial pressures*

Many environmental and biophysical arguments depend on a knowledge of the composition of the atmosphere, and in particular on the partial pressures of its principal components. Given that the composition of a sample of mass 100 g of dry air at sea level is 75.5 g nitrogen, 23.2 g oxygen, and 1.3 g argon, what is the partial pressure of each component at 100 kPa total pressure?

Strategy

To use eqn (17), we need to know the mole fractions of the components. As the mole fractions were calculated in Exercise E1.7, all we need do is substitute the data into eqn (17).

Solution

From Exercise E1.7, we have $x(N_2) = 0.780$, $x(O_2) = 0.210$, and $x(Ar) = 0.009$. It then follows from eqn (17) that

$$p(N_2) = x(N_2)p = 0.780 \times (100\,\text{kPa}) = 78.0\,\text{kPa}$$

Similarly, for the other two components we find $p(O_2) = 21.0\,\text{kPa}$ and $p(Ar) = 0.9\,\text{kPa}$. The partial pressure of oxygen in air plays an important role in the aeration of water, to enable aquatic life to thrive, and in the absorption of oxygen by blood in our lungs (see Section 4.3).

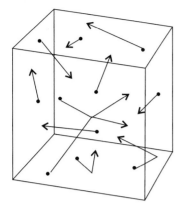

Fig. 1.10 The model used for discussing the molecular basis of the physical properties of a perfect gas. The pointlike molecules move randomly with a range of speeds and directions, both of which change when they collide with the walls or with other molecules.

Exercise E1.8

Calculate the partial pressures of a sample of gas consisting of 2.50 g of oxygen and 6.43 g of carbon dioxide with a total pressure of 88 kPa.

[*Answer*: 31 kPa, 57 kPa]

The kinetic theory of gases

We have remarked that a gas may be pictured as a collection of particles in continuous, chaotic motion (Fig. 1.10). Now we shall develop this model of the gaseous state of matter to see how it accounts for the perfect gas laws. As we have remarked, one of the most important functions of physical chemistry is to convert qualitative notions into quantitative statements that can be tested experimentally by making measurements and comparing the results with predictions. Indeed, an important component of science as a whole is its technique of proposing a qualitative model and then expressing that model mathematically. The kinetic theory of gases is an excellent example of this procedure: the model is very simple, and the quantitative prediction (the perfect gas law) is experimentally verifiable.

The **kinetic theory of gases** is based on three assumptions:

1. A gas consists of molecules in ceaseless random motion.
2. The size of the molecules is negligible in the sense that their diameters are much smaller than the average distance travelled between collisions.
3. The molecules do not interact, except during collisions.

The assumption that the molecules do not interact unless they are in contact (assumption 3) implies that their energy is independent of their separation. Because the total energy is the sum of the kinetic energy (the energy arising from motion) and the potential energy (the energy arising from position), and the latter is zero, the total energy of a sample of gas is the sum of the kinetic energies of all the molecules present in it. It follows that the faster the molecules travel (and hence the greater the average kinetic energy of the molecules), the greater the total energy of the gas. (The various contributions to the energy are reviewed in *Further information 3*.)

1.6 The pressure of a gas

The kinetic theory accounts for the steady pressure exerted by a gas in terms of the collisions the molecules make with the walls of the container. Each collision gives rise to a brief force on the wall, but as billions of collisions take place every second, the walls experience a virtually constant force, and hence the gas exerts a steady pressure (Fig. 1.11).

For the details of the calculation of the pressure exerted by the gas see *Further information 4*. Briefly, to calculate the pressure of a gas, we need to calculate the total force exerted by the molecules as they collide with a wall, and then divide that total force by the area of the wall. The outcome of the calculation is

$$p = \frac{nMc^2}{3V} \tag{18}$$

where M is the molar mass and c is the **root-mean-square speed** (r.m.s. speed) of the molecules. The latter is the square root of the average of the square of the speeds of all the molecules in the sample; for a sample consisting of N molecules with speeds $s_1, s_2, \cdots, s_N$,

$$c = \sqrt{\frac{s_1^2 + s_2^2 + \cdots + s_N^2}{N}} \tag{19}$$

The r.m.s. speed is quite close in value to another measure of molecular speed, the mean speed of the molecules; the mean speed is smaller by a factor of 0.92 for large numbers of molecules.

Exercise E1.9

Cars pass a point travelling at 45.00 (5), 47.00 (7), 50.00 (9), 53.00 (4), 57.00 (1) km h^{-1}, where the number of cars is given in parentheses. Calculate (a) the r.m.s speed and (b) the mean speed of the cars.

[*Answer*: (a) 49.06 km h^{-1}, (b) 48.96 km h^{-1}]

The kinetic theory gives an expression that already resembles the perfect gas equation of state, for eqn (18) can be rearranged to

$$pV = \tfrac{1}{3}nMc^2 \tag{20}$$

which is very similar in form to $pV = nRT$. This conclusion is a major success of the kinetic model, for the model implies an experimentally verified result.

1.7 The speeds of gas molecules

The ability of the kinetic theory to account for the form of the perfect gas law suggests that it is a valid model of perfect gas behaviour. That being so,

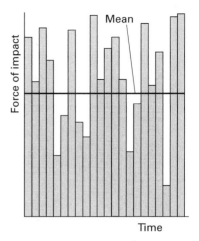

Fig. 1.11 The wall of a vessel containing a gas experiences collisions from the gas molecules. The impacts have different strengths (depending on the velocities of the molecules that strike the surface). This diagram illustrates the strengths of the impulses that a wall experiences in a very brief period (the entire interval shown could be as brief as 1 fs, or even less, depending on the pressure and temperature of the gas). The impulses vary so rapidly that the wall experiences an almost perfectly steady force, which is the average of the individual impacts.

we can take another major step. We shall suppose that the expression for pV derived from kinetic theory, eqn (20), is indeed the equation of state of a perfect gas. That being so, we can equate the expression on the right of the equation to nRT, which gives

$$\tfrac{1}{3}nMc^2 = nRT$$

and rearrange this expression to obtain a formula for calculating the r.m.s. speed of the gas molecules at any temperature:

$$c = \sqrt{\frac{3RT}{M}} \qquad (21)$$

Substitution of the molar mass of O_2 ($32\,\mathrm{g\,mol^{-1}}$) and a temperature corresponding to $25\,°C$ (that is, 298 K) gives an r.m.s. speed for these molecules of $482\,\mathrm{m\,s^{-1}}$. The same calculation for nitrogen molecules gives $515\,\mathrm{m\,s^{-1}}$. Both these values are not far off the speed of sound in air ($346\,\mathrm{m\,s^{-1}}$ at $25\,°C$); that is reasonable, because sound is a wave of pressure variation that is transmitted by the movement of molecules, and so the speed of propagation of a wave should be approximately the same as the average speed at which molecules can adjust their locations.

The important conclusion to draw from eqn (21) is that *the r.m.s. speed of molecules in a gas is proportional to the square root of the temperature.* Therefore, doubling the temperature (on the Kelvin scale) increases the r.m.s. speed of molecules by a factor of $\sqrt{2} = 1.4$. Cooling a sample of air from $25\,°C$ (298 K) to $0\,°C$ (273 K) reduces the original r.m.s. speed of the molecules by a factor of

$$\sqrt{\frac{273\,\mathrm{K}}{298\,\mathrm{K}}} = \sqrt{\frac{273}{298}} = 0.957$$

So, on a cold day, the molecules of the air are moving on average at about 4 per cent more slowly than on a warm day.

The Maxwell distribution of speeds

So far, we have dealt only with the *mean speeds* of molecules. Not all molecules, however, travel at the same speed: some are travelling more slowly than the average (until they collide, and get accelerated to a high speed, like the impact of a bat on a ball), and others may briefly be travelling at much higher speeds than the average. There is continual interchange of speeds among molecules as they undergo collisions, and at one moment a molecule may be travelling much faster than the average, but at the next moment it might be brought to almost a complete standstill when it collides with another. These collisions take place about every nanosecond ($1\,\mathrm{ns} = 10^{-9}\,\mathrm{s}$) or so in a gas under normal conditions.

The mathematical expression that tells us the fraction of molecules that have a particular speed is called the **distribution of molecular speeds**. Thus, the distribution might tell us that at $20\,°C$ a fraction 1.9×10^{-2} of all

O_2 molecules (19 in 1000) have a speed in the range between 300 and 310 m s^{-1}, that a fraction 2.1×10^{-2} (21 in 1000) have a speed in the range 400 to 410 m s^{-1}, and so on. The precise form of the distribution was worked out by James Clerk Maxwell towards the end of the nineteenth century, and is known as the **Maxwell distribution of speeds**. According to Maxwell, the fraction f of molecules that have a speed in a narrow range s to $s + \Delta s$ is

$$f = 4\pi \left(\frac{M}{2\pi RT} \right)^{3/2} s^2 e^{-Ms^2/2RT} \Delta s \qquad (22)$$

This is the formula used to calculate the figures quoted above. Although the formula looks quite complicated, its features can be picked out quite readily. First, the fraction in the range Δs increases in proportion to the width of the range: if at a given speed we increase the range (but still ensure that it is narrow), then the fraction in that range increases. Second, the fact that the expression includes a decaying exponential (a function of the form e^{-x}, with x proportional to s^2 in this case) implies that fraction of molecules with very high speeds will be very small because e^{-x} becomes very small indeed when x is large. Third, the factor s^2 that multiplies the exponential goes to zero as s goes to zero, so the fraction with very low speeds will also be very small.

The graph of the Maxwell distribution is very important, and is shown in Fig. 1.12. As we should expect from the form of eqn (22), we see that only a very few molecules in the sample have speeds much smaller than the average speed and, similarly, only very few molecules have speeds much greater than the average. However, the fraction with very high speeds increases sharply as the temperature is raised, as the tail of the distribution reaches up to higher speeds. This feature plays an important role in the rates of gas-phase chemical reactions, for (as we shall see in Section 7.5), the rate of a reaction in the gas phase depends on the energy with which two molecules crash together, which in turn depends on their speeds.

Another feature of the Maxwell distribution is shown in Fig. 1.13, where it is drawn for molecules with different molar masses. As can be seen, not only do heavy molecules have lower average speeds at a given temperature, but they also have a significantly narrower *spread* of speeds: most molecules will be found with speeds close to the average. In contrast, light molecules (such as H_2) have high average speeds and a wide spread of speeds: many molecules will be found travelling either much more slowly or much more quickly than the average. This feature plays an important role in determining the composition of planetary atmospheres, because it means that a significant fraction of light molecules travel at sufficiently high speeds to escape from the planet's gravitational attraction. The ability of light molecules to escape is one reason why hydrogen (molar mass 2.02 g mol^{-1}) and helium (4.00 g mol^{-1}) are very rare in the Earth's atmosphere.

Diffusion and effusion

The process by which the molecules of different substances mingle with each other is called **diffusion**. The atoms of two solids diffuse into each other when

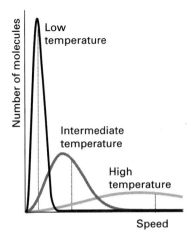

Fig. 1.12 The Maxwell distribution of speeds and its variation with the temperature. Note the broadening of the distribution and the shift of the r.m.s. speed (denoted by the locations of the vertical lines) to higher values as the temperature is increased.

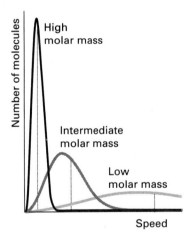

Fig. 1.13 The Maxwell distribution of speeds also depends on the molar mass of the molecules. Molecules of low molar mass have a broad spread of speeds, and a significant fraction may be found travelling much faster than the r.m.s. speed. The distribution is much narrower for heavy molecules, and most of them travel with speeds close to the r.m.s. value (denoted by the locations of the vertical lines).

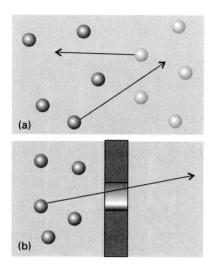

Fig. 1.14 (a) Diffusion is the spreading of the molecules of one substance into the region initially occupied by another species. Note that molecules of both substances move, and each substance diffuses into the other. (b) Effusion is the escape of molecules through a small hole in a confining wall.

the two solids are in contact, but the process is very slow. The diffusion of a solid through a liquid solvent is much faster, but mixing normally needs to be encouraged by stirring or shaking the solid in the liquid (the process is then no longer pure diffusion). Gaseous diffusion is much faster. It accounts for the largely uniform composition of the atmosphere, for if a gas is produced by a localized source (such as carbon dioxide from the respiration of animals, oxygen from photosynthesis by green plants, and pollutants from cars and industrial sources), then the molecules of gas will diffuse from the source and in due course be distributed throughout the atmosphere. (In practice, the process of mixing is accelerated by winds.) The process of **effusion** is the escape of a gas through a small hole, as in a puncture in an inflated balloon or tyre (Fig. 1.14).

The rates of diffusion and effusion of gases increase with increasing temperature, for both processes depend on the motion of molecules, and molecular speeds increase with temperature. The rates also decrease with increasing molar mass, for average molecular speeds decrease with increasing molar mass. The dependence on molar mass, however, is simple only in the case of effusion, for then only a single substance is in motion, not the several gases that may be involved in diffusion.

The experimental observations on the dependence of the rate of effusion of a gas on its molar mass are summarized by a law proposed by Thomas Graham in 1833:

> **Graham's law of effusion**: At a given pressure and temperature, the rate of effusion of a gas is inversely proportional to the square root of its molar mass.

For example, the rates at which hydrogen (molar mass $2.02\,\mathrm{g\,mol^{-1}}$) and carbon dioxide ($44.01\,\mathrm{g\,mol^{-1}}$) effuse under the same conditions of pressure and temperature are in the ratio

$$\frac{\text{rate of effusion of H}_2}{\text{rate of effusion of CO}_2} = \sqrt{\frac{44.01\,\mathrm{g\,mol^{-1}}}{2.02\,\mathrm{g\,mol^{-1}}}} = \sqrt{\frac{44.01}{2.02}} = 4.67$$

The different rates of effusion through a porous barrier are employed in the separation of uranium-235 from the more abundant and less useful uranium-238 in the processing of nuclear fuel. The process depends on the formation of uranium hexafluoride, a volatile solid. However, because the ratio of the molar masses of $^{238}\mathrm{UF}_6$ and $^{235}\mathrm{UF}_6$ is only 1.008, the ratio of the rates of effusion is only $\sqrt{1.008} = 1.004$. As a consequence, thousands of successive effusion stages are required to achieve a significant separation. The rate of effusion of gases was once used to determine the molar masses of gases and volatile liquids by comparison of the rate of effusion of the gas with that of a gas of known molar mass. However, there are now much more precise methods available (mass spectrometry, for instance).

The form of Graham's law can be explained by referring to the expression for the r.m.s. speed of molecules of a gas, eqn (21), which is inversely proportional to the square root of the molar mass. Because the rate of effusion through a hole in a container is proportional to the rate at which molecules pass through the hole, it follows that the rate should be inversely proportional to $\sqrt{M}$, which is in accord with Graham's law.

1.8 Molecular collisions

The average distance that a molecule travels between collisions is called its **mean free path**, λ (lambda, Fig. 1.15). The mean free path in a liquid is less than the diameter of the molecules, for molecules in liquids meet a neighbour even if they move only a fraction of a diameter. However, in gases, the mean free paths of molecules can be several hundred molecular diameters. The **collision frequency**, z, is the average number of collisions per second made by one molecule. It follows that the inverse of the collision frequency, $1/z$, is the average time that a molecule spends in flight between two collisions: as we shall see, this average time is typically about 1 ns at 1 atm and room temperature.

Speed is distance travelled divided by the time needed for the journey. Therefore, the r.m.s. speed, c, can be identified with the average length of the flight of a molecule between collisions (that is, with the mean free path, λ) divided by the average length of time needed for the flight (that is, divided by $1/z$). It follows that the mean free path and the collision frequency are related by

$$c = \lambda z \qquad (23)$$

Therefore, if we can calculate either λ or z, then we can find the other from this equation.

To find expressions for λ and z we need a slightly more elaborate version of the kinetic theory of gases. The kinetic theory supposes that the molecules are effectively pointlike; however, to obtain collisions, we need to assume that two 'points' score a hit whenever they come within a certain range d of each other, where d is the diameter of the molecules of the gas (Fig. 1.16). When this feature is added to the model, it turns out that the mean free path and the collision frequency depend on the quantity $\sigma = \pi d^2$. This area is called the **collision cross-section** of the molecules. The collision cross-section can be thought of as the target area that a molecule presents to an incoming projectile. In terms of this quantity,

$$\lambda = \frac{RT}{\sqrt{2}N_A\sigma p} \qquad z = \frac{\sqrt{2}N_A\sigma cp}{RT} \qquad (24)$$

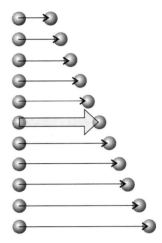

Fig. 1.15 In a gas, a molecule traces a chaotic path—a random walk—that changes direction on each collision; each step is over a different distance. The average of the distances that any molecule travels between collisions is the mean free path, denoted here by the broad arrow.

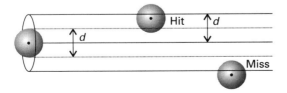

Fig. 1.16 To calculate features of the gas that are related to collisions (such as the mean free path and the collision frequency), a point is regarded as being surrounded by a sphere of diameter d. A molecule will hit another molecule if the latter lies within a cylinder of radius d.

States of matter and the properties of gases

Table 1.4 Collision cross sections of atoms and molecules

Species	σ/nm^2
Ar	0.36
C_2H_4	0.64
C_6H_6	0.88
CH_4	0.46
Cl_2	0.93
CO_2	0.52
H_2	0.27
He	0.21
N_2	0.43
Ne	0.24
O_2	0.40
SO_2	0.58

$1\,\text{nm}^2 = 10^{-18}\,\text{m}^2$

The collision cross-sections of some common atoms and molecules are given in Table 1.4. For example, the mean free path of O_2 molecules in a sample of oxygen at SATP ($25\,^\circ$C, 1 bar) is

$$\lambda = \frac{(8.3145\,\text{J K}^{-1}\,\text{mol}^{-1}) \times (298\,\text{K})}{\sqrt{2} \times (6.022 \times 10^{23}\,\text{mol}^{-1}) \times (0.40 \times 10^{-18}\,\text{m}^2) \times (1.00 \times 10^5\,\text{Pa})}$$
$$= 7.3 \times 10^{-8}\,\text{m}$$

or 73 nm. There are two points to note about this calculation. First, we have used R in its 'fundamental' SI unit form: this form is usually appropriate in calculations in the kinetic theory. Second, we have made use of the relation $1\,\text{J} = 1\,\text{Pa m}^3$; see *Further information 2*. Under the same conditions, the collision frequency is $6.2 \times 10^9\,\text{s}^{-1}$ (6.2 billion collisions per second).

In physical chemistry (as in any of the physical sciences), it is important to develop the skill of interpreting the content of equations. The important features of eqn (24) are as follows. First, because $\lambda \propto 1/p$, it shows that

1. The mean free path decreases as the pressure increases.

This decrease is a result of the increase in the number of molecules present in a given volume as the pressure is increased, so each molecule travels a shorter distance before it collides with a neighbour. For example, the mean free path of an O_2 molecule decreases from 72 nm to 36 nm when the pressure is increased from 1.0 bar to 2.0 bar. Second, because $\lambda \propto 1/\sigma$,

2. The mean free path is shorter for molecules with large collision cross-sections.

For instance, the collision cross-section of a benzene molecule ($0.88\,\text{nm}^2$) is about four times greater than that of a helium atom ($0.21\,\text{nm}^2$), and at the same pressure and temperature its mean free path four times shorter. A third point is that because $z \propto p$,

3. The collision frequency increases with the pressure of the gas.

This dependence follows from the fact that, provided the temperature is the same, the molecules take less time to travel to its neighbour in a denser, higher-pressure gas. For example, although the collision frequency for an O_2 molecule in oxygen gas at SATP is $6.2 \times 10^9\,\text{s}^{-1}$, at 2.0 bar and the same temperature the collision frequency is doubled, to $1.2 \times 10^{10}\,\text{s}^{-1}$. Finally, because eqn (24) shows that $z \propto c$, and we know that $c \propto 1/\sqrt{M}$,

4. Providing their collision cross-sections are the same, heavy molecules have lower collision frequencies than light molecules.

Heavy molecules travel more slowly on average than light molecules do (at the same temperature), so they collide with other molecules less frequently.

Real gases

So far, everything we have said applies to perfect gases, ones in which the molecules are so far apart on average that they move completely independently of one another. In terms of the quantities introduced in the previous section, a perfect gas is a gas for which the mean free path, λ, of the molecules in the sample (the distance they travel between collisions) is much greater than d (the separation at which they are regarded as being in contact). This condition is written $\lambda \gg d$. As a result of this large average separation, a perfect gas is a gas in which the only contribution to the energy comes from the kinetic energy of the motion of the molecules and there is no contribution to the total energy from the potential energy arising from the interaction of the molecules with one another. However, in fact all molecules do interact with one another provided they are close enough together, so the 'kinetic energy only' model is only an approximation.

1.9 Intermolecular interactions

There are two contributions to the interaction between molecules. At relatively long distances (a few molecular diameters), two molecules attract each other. This **intermolecular attraction** is responsible for the condensation of gases into liquids at low temperatures. At low enough temperatures the molecules of a gas have insufficient kinetic energy to escape from each other's attraction and they stick together. Second, although molecules attract one another when they are a few diameters apart, as soon as they come into contact they repel one another. This **intermolecular repulsion** is responsible for the fact that liquids and solids have a definite bulk and do not collapse to an infinitesimal point.

Intermolecular interactions give rise to a potential energy that contributes to the total energy of a gas. Because attractions correspond to a lowering of total energy as molecules get closer together, they contribute a *negative* potential energy: they reduce the total energy as the molecules approach. On the other hand, intermolecular repulsions correspond to a raising of the total energy as the molecules squash together. This increase in energy is represented by a *positive* potential energy. The general form of the variation of the potential energy is illustrated in Fig. 1.17. At large separations the energy-lowering interactions are attractive, but at short distances the energy-raising repulsions dominate.

The effect of the intermolecular interactions shows up in the bulk properties of a gas. For example, the isotherms of real gases have shapes that differ from those given by Boyle's law, particularly at high pressures. A set of experimental isotherms for carbon dioxide are shown in Fig. 1.18. They should be compared with the perfect-gas isotherms shown in Fig. 1.6. Although the experimental isotherms resemble the perfect-gas isotherms at high temperatures (and at low pressures, off the scale on the right of the

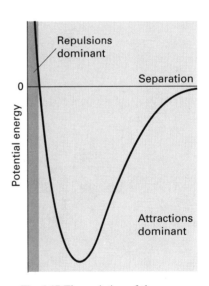

Fig. 1.17 The variation of the potential energy of two molecules with their separation. High positive potential energy (at very small separations) indicates that the interactions between them are strongly repulsive at these distances. At intermediate separations, where the potential energy is negative, the attractive interactions dominate. At large separations (on the right) the potential energy is zero and there is no interaction between the molecules.

25

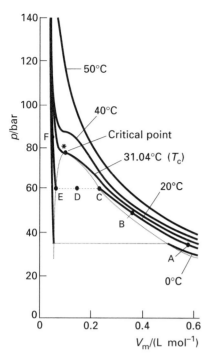

Fig. 1.18 The experimental isotherms of carbon dioxide at several temperatures. The critical isotherm is at 31.04 °C.

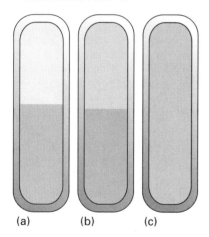

Fig. 1.19 When a liquid is heated in a sealed container, the density of the vapour phase increases and that of the liquid phase decreases. There comes a stage at which the two densities are equal and the interface between the two fluids disappears. This disappearance occurs at the critical temperature. The container needs to be strong: the critical temperature of water is at 373 °C and the vapour pressure is then 218 atm.

graph), there are very striking differences between the two at temperatures below about 50 °C and at pressures above about 1 bar.

1.10 The critical temperature

The shape of the isotherm at 20 °C—two rising curves joined by a horizontal dashed line—indicates that a liquid forms when pressure is applied to carbon dioxide at this temperature. To understand this interpretation, consider the significance of the points A to F marked on the isotherm. At point A the sample is a gas. As the volume of the sample is reduced to B by pressing in a piston, the pressure increases broadly in agreement with Boyle's law, and the increase continues until point C is reached. At this stage, we find that the piston can be pushed in without any further increase in pressure, through D to E. After E has been reached, the reduction in volume from E to F corresponds to a very steep increase in pressure. This variation of pressure with volume is exactly what you would expect if the gas at C condenses to a compact liquid at E. Indeed, if we could see the sample we would see it begin to condense to a liquid at C and the condensation would be complete when the piston was pushed in to E: the piston would then be resting on the surface of the liquid. The subsequent reduction in volume, from E to F, corresponds to the very high pressure needed to compress a liquid into a smaller volume. In terms of intermolecular interactions, the step from C to E corresponds to the molecules being so close on average that they attract each other and cohere into a liquid. The step from E to F represents the effect of trying to force the molecules even closer together when they are already in contact, and hence trying to overcome the strong repulsive interactions between them.

If we could look inside the container at point D, we would see a liquid separated from the remaining gas by a sharp surface (Fig. 1.19). At a slightly higher temperature (at 30 °C , for instance), a liquid forms, but it needs a higher pressure to produce it. It might be difficult to make out the surface because the remaining gas is at such a high pressure that its density is similar to that of the liquid. At the special temperature of 31.04 °C (304.19 K) the gaseous state appears to transform continuously into the condensed state, and at no stage is there a visible surface between the two physical states. At this temperature, which is called the **critical temperature**, T_c, and at all higher temperatures, a single phase fills the container at all stages of the compression and there is no separation of a liquid from the gas. Above T_c, there is no separation of phases whatever pressure is applied. We have to conclude that *a gas cannot be condensed to a liquid by the application of pressure alone unless the temperature is lower than the critical temperature*. The critical temperatures of some common gases are listed in Table 1.5. The data there imply, for example, that no degree of compression of nitrogen gas will result in the formation of a liquid unless the temperature is below 126 K (−147 °C).

The dense fluid obtained by compressing a gas when its temperature is higher than its critical temperature is not a true liquid, but it behaves like a liquid in many respects (it has a similar density, for instance); despite its density, it is not strictly a liquid because it never possesses a surface that separates it from a vapour phase. Nor is it much like a gas, because it is so dense. It is an example of a **supercritical fluid**. Supercritical fluids are currently being used as solvents; for example, supercritical carbon dioxide is used to extract caffeine in the manufacture of decaffeinated coffee, where (unlike organic solvents) it does not result in the formation of an unpleasant and possibly toxic residue.

1.11 Real equations of state

According to the perfect gas equation of state,

$$\frac{pV_m}{RT} = \frac{pV}{nRT} = 1 \tag{25}$$

where V_m is the molar volume of the gas. This expression suggests that pV_m/RT should be equal to 1 for all gases that behave perfectly. However, when the value of pV_m/RT is measured for actual gases, it is found to differ from 1, and it varies with pressure as shown in Fig. 1.20. The deviation of pV_m/RT from 1 arises from the intermolecular interactions in the gas. At low pressures, some gases (methane, ethane, and ammonia, for instance) show a decrease below 1. The value $pV_m/RT < 1$ means that for a given temperature and pressure, the molar volume of the gas is less than that expected for a perfect gas. The reason for the reduction in molar volume can be traced to the dominant attractive interactions between the molecules, which tend to draw the molecules together and hence reduce the space they occupy.

The experimental value of pV_m/RT rises above 1 at high pressures whatever the identity of the gas, and for some gases (hydrogen in the illustration) its value is greater than 1 at all pressures. When $pV_m/RT > 1$, the molar volume is greater than that expected for a perfect gas of the same temperature and pressure. This behaviour can be traced to the dominant repulsive forces. These forces tend to drive the molecules apart when they are forced to be close together at high pressures. For hydrogen, the attractive interactions are so weak that the repulsive interactions dominate even at low pressures.

The virial equation of state

The departure of pV_m/RT from its 'perfect' value of 1 can be used to construct an *empirical* (observation based) equation of state. To do so, we suppose that the perfect gas equation of state is only the first term of a lengthier expression, and write

$$\frac{pV_m}{RT} = 1 + \frac{B}{V_m} + \frac{C}{V_m^2} + \cdots \tag{26}$$

This technique, of taking a limiting law and supposing that it is the first term of a more complicated expression, is quite common in physical chemistry: the limiting law is the first approximation to the true expression, whatever that may be, and the additional terms take into account the effects that the limiting law ignores. Equation (26) is called the **virial**

Table 1.5 The critical temperatures of gases

	Critical temperature/°C
Noble gases	
He	−268 (5.2 K)
Ne	−229
Ar	−123
Kr	−64
Xe	17
Halogens	
Cl_2	144
Br_2	311
Small inorganic species	
H_2	−240
O_2	−118
H_2O	374
N_2	−147
NH_3	132
CO_2	31
Organic compounds	
CH_4	−83
CCl_4	283
C_6H_6	289

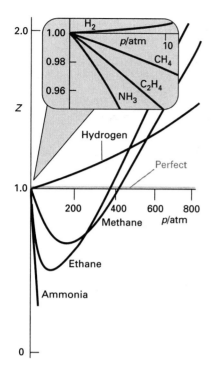

Fig. 1.20 The variation of $Z = pV_m/RT$ with pressure for several gases at $0\,°C$. A perfect gas has $Z = 1$ at all pressures. Of the gases shown, hydrogen shows positive deviations at all pressures (at this temperature); all the other gases show negative deviations initially but positive deviations at high pressures. The negative deviations are a result of the attractive interactions between molecules and the positive deviations are a result of the repulsive interactions.

equation of state.[†] The coefficients B, C, etc., which vary from gas to gas and have to be determined experimentally in each case, are called the **virial coefficients**. From the graphs in Fig. 1.20, it follows that, for the temperature to which the data apply, B must be positive for hydrogen (so that the right hand side of eqn (26) is greater than 1) but negative for methane, ethane, and ammonia (so that the right hand side of eqn (26) is less than 1). However, regardless of the sign of B, the term C/V_m^2 becomes large at high pressures—when V_m^2 is very small—and the right hand side of eqn (26) becomes greater than 1, as in the graphs in Fig. 1.20. The values of virial coefficients for many gases are known from measurements of pV_m/RT over a range of pressures and fitting the data to eqn (26) by varying B and C until a good match is obtained.

A final point about the virial equation concerns its form at very low pressures, when the molar volume is very large. In that case, the terms B/V_m and C/V_m^2 are both very small, and to a good approximation the gas is described by the perfect gas equation of state:

$$\frac{pV_m}{RT} \to 1 \qquad \text{as } V_m \to \infty \text{ when } p \to 0 \qquad (27)$$

The van der Waals equation of state

Although it is the most reliable equation of state (because it is a fit of observed pressures and molar volumes), the virial equation does not give us much immediate insight into the behaviour of gases and their condensation to liquids. The **van der Waals equation**, which was proposed in 1873 by the Dutch physicist Johannes van der Waals, is only an approximate equation of state, but it has the advantage of showing how the intermolecular interactions contribute to the deviations of a gas from the predictions of the perfect gas law. We can view the van der Waals equation as another example of taking a soundly based qualitative idea and building up a mathematical expression that can be tested quantitatively. To do so, we recognize that a real gas differs from a perfect gas in that the molecules interact with one another. The interaction is attractive when the molecules are within a few molecular diameters of each other, but is strongly repulsive as soon as they are in contact.

The presence of the repulsive interaction implies that two molecules cannot come closer than a certain distance. Therefore, instead of being free to travel anywhere in a volume V, the actual volume in which the molecules can travel is reduced to an extent that is proportional to the number of molecules present and the volume they each exclude. We can therefore model the effect of the repulsive, volume-excluding forces by changing V in the perfect gas equation to $V - nb$, where b is the proportionality constant between the reduction in volume and the amount of molecules present in the container. With this modification, the perfect gas equation of state changes to

$$p = \frac{nRT}{V - nb}$$

[†]The word 'virial' comes from the Latin word for force, and it reflects the fact that intermolecular forces are now significant.

This equation of state (it is not yet the full van der Waals equation) should describe a gas in which repulsions are important. Note that when the pressure is low, the volume is large compared with the volume excluded by the molecules (which we write $V \gg nb$), the nb can be ignored in the denominator, and the equation reduces to the perfect gas equation of state.

The effect of the attractive interactions between the molecules is to reduce the pressure that the gas exerts on the walls of the container. We can estimate how the reduction in pressure depends on n and V by noting that the attraction experienced by a given molecule is proportional to the molar concentration, n/V, of molecules in the container. Because the attractions slow the molecules down, the molecules strike the walls less frequently *and* with a lower impact. The reduction in pressure is therefore proportional to the *square* of the molar concentration. The argument behind this remark can be expressed mathematically as follows:

reduction in pressure $\propto$ (reduction in frequency of impacts)$\times$

(reduction in average strength of impacts)

$\propto$ (molecular attraction) $\times$ (molecular attraction)

$\propto$ (molar concentration) $\times$ (molar concentration)

$$\propto \frac{n}{V} \times \frac{n}{V}$$

If the constant of proportionality is written as a, we can write

$$\text{reduction in pressure} = a\left(\frac{n}{V}\right)^2$$

It follows that the equation of state allowing for both repulsions and attractions is

$$p = \frac{nRT}{V - nb} - a\left(\frac{n}{V}\right)^2 \tag{28}$$

This expression is the **van der Waals equation of state**. To show the resemblance of this equation to the perfect gas equation $pV = nRT$ it is sometimes rearranged into

$$\left\{p + a\left(\frac{n}{V}\right)^2\right\}\{V - nb\} = nRT \tag{29}$$

We have built the van der Waals equation using physical arguments about the volumes of molecules and the effects of forces between them. It can be derived in other ways, but the present method has the advantage that it shows how to derive the form of an equation out of general ideas. The derivation also has the advantage of keeping imprecise the significance of the **van der Waals parameters** a and b: they are much better regarded as empirical parameters than as precisely defined molecular properties.

We can judge the reliability of the van der Waals equation by comparing the isotherms it predicts, which are shown in Fig. 1.21, with the

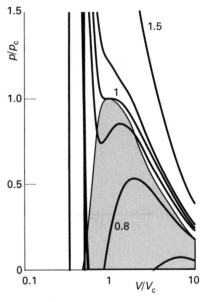

Fig. 1.21 Isotherms calculated by using the van der Waals equation of state. The axes are labelled with the reduced pressure, p/p_c, and reduced volume, $V/V_{m,c}$, where $p_c = a/27b^2$ and $V_{m,c} = 3b$. The individual isotherms are labelled with the reduced temperature, T/T_c, where $T_c = 8a/27Rb$. The isotherm labelled 1 is the critical isotherm (the isotherm at the critical temperature).

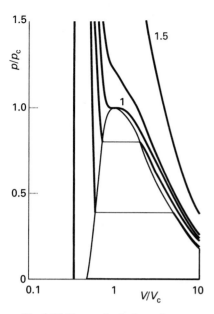

Fig. 1.22 The unphysical van der Waals loops (where volume increases as pressure is increased) are eliminated by drawing straight lines that divide the loops into areas of equal size. With this procedure, the isotherms strongly resemble the observed isotherms.

experimental isotherms already shown in Fig. 1.18. Apart from the wave-like shape below the critical temperature they do resemble experimental isotherms quite well. The waves, which are called **van der Waals' loops**, are unrealistic because they suggest that under some conditions an increase of pressure results in an increase of volume (notice that in some places the isotherms rise as V increases). The wave-like regions are therefore trimmed away and replaced by horizontal lines (Fig. 1.22). The van der Waals coefficients are found by fitting the calculated curves to the experimental, and the values for some gases are listed in Table 1.6.

We should note that perfect-gas isotherms are obtained from the van der Waals equations at high temperatures and large molar volumes. To confirm this remark, we need to note that when the temperature is high, RT may be so large that the first term in eqn (28) greatly exceeds the second. Furthermore, as we have already seen, if the molar volume is large (in the sense $V_m \gg b$, so that $V \gg nb$), then the denominator $V - nb$ can be replaced by V. Hence, under these conditions (of high temperature and large molar volume) the equation reduces to $p = nRT/V$, the perfect gas equation.

The van der Waals equation of state is an interesting demonstration of the conversion of physical concepts into a quantitatively testable theory. The equation also provides yardsticks to judge when a gas can be treated as perfect (when V_m is large compared with both a/RT and b). It is also useful in practice for estimating the pressure of a gas that cannot be treated as perfect.

1.12 The liquefaction of gases

A gas may be liquefied by cooling it below its boiling point at the pressure of the experiment. For example, chlorine at 1 atm can be liquefied by cooling it to below $-34\,°C$ in a bath cooled with dry ice (solid carbon dioxide). For gases with very low boiling points (such as oxygen and nitrogen, at $-183\,°C$ and $-186\,°C$, respectively), such a simple technique is not practicable unless an even colder bath is available.

One alternative and widely used commercial technique makes use of the forces that act between molecules. We saw earlier that the r.m.s. speed of molecules in a gas is proportional to the square root of the temperature (eqn (21)). It follows that if there is a way of reducing the r.m.s. speed of the molecules in a gas, then that is equivalent to cooling the gas. If the speed of the molecules can be reduced to the point that neighbours can capture each other by their intermolecular attractions, then the cooled gas will condense to a liquid.

To slow the gas molecules, we make use of an effect similar to that seen when a ball is thrown into the air: as it rises it slows on account of the gravitational attraction of the Earth. Molecules attract each other, as we have seen (the attraction is not gravitational, but the effect is the same), and if we can cause them to move apart from each other, like a ball rising from a

Table 1.6 Van der Waals constants of gases

	a/L^2 atm mol^{-2}	b/L mol^{-1}
Air	1.4	0.039
Ammonia	4.17	0.037
Argon	1.35	0.032
Carbon dioxide	3.59	0.043
Ethane	5.49	0.064
Ethene	4.47	0.057
Helium	0.034	0.024
Hydrogen	0.244	0.027
Nitrogen	1.39	0.039
Oxygen	1.36	0.032
Xenon	4.19	0.051

planet, then they should slow. It is very easy to move molecules apart from each other: we allow the gas to expand, which increases the average separation of the molecules. To cool a gas, therefore, we allow it to expand without allowing any heat to enter from outside. As it does so, the molecules climb apart struggling against the attraction of all the other molecules, and as they move apart, they travel more slowly. Because the molecules now travel more slowly on average, the gas is now cooler than before the expansion. This process of cooling by expansion is called the **Joule–Thomson effect**, because it was first observed by James Joule (whose name is commemorated in the unit of energy) and William Thomson (who later became Lord Kelvin). The procedure works only for real gases in which the attractive interactions are dominant, because the molecules have to climb apart against the attractive force in order for them to travel more slowly.

In practice, the gas is allowed to expand several times by recirculating it through a device called a **Linde refrigerator** (Fig. 1.23). On each successive expansion the gas becomes cooler, and as it flows past the incoming gas, the latter is cooled further. After several successive expansions, the gas becomes so cold that it condenses to a liquid.

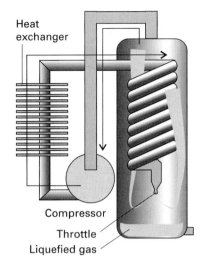

Fig. 1.23 The principle of the Linde refrigerator is shown in this diagram. The gas is recirculated and cools the gas that is about to undergo expansion through the throttle. The expanding gas cools still further. Eventually, liquefied gas drips from the throttle.

EXERCISES

Treat all gases as perfect unless instructed otherwise.

1.1 Express (a) 110 kPa in Torr, (b) 0.997 bar in atmospheres, (c) 2.15×10^4 Pa in atmospheres, (d) 723 Torr in pascals.

1.2 What pressure is exerted by a sample of nitrogen gas of mass 2.045 g in a container of volume 2.00 L at 21°C?

1.3 A sample of neon of mass 255 mg occupies 3.00 L at 122 K. What pressure does it exert?

1.4 Much to everone's surprise, nitrogen monoxide (NO) has been found to act as a neurotransmitter. To prepare to study its effect, a sample was collected in a container of volume 250.0 mL. At 19.5°C its pressure is found to be 24.5 kPa. What amount (in moles) of NO has been collected?

1.5 A domestic water-carbonating kit uses steel cylinders of carbon dioxide of volume 250 mL. They weigh 1.04 kg when full and 0.74 kg when empty. What is the pressure of gas in the cylinder at 20°C?

1.6 The effect of high pressure on organisms, including humans, is studied to gain information about deep-sea diving and anaesthesia. A sample of air occupies 1.00 L at 25°C and 1.00 atm. What pressure is needed to compress it to 100 cm^3 at this temperature?

1.7 You are warned not to dispose of pressurized cans by throwing them on to a fire. The gas in an aerosol container exerts a pressure of 125 kPa at 18°C. The container is thrown on a fire, and its temperature rises to 700°C. What is the pressure at this temperature?

1.8 Until we find an economical way of extracting oxygen from sea-water or lunar rocks, we have to carry it with us to inhospitable places, and do so in compressed form in tanks. A sample of oxygen at 101 kPa is compressed at constant temperature from 7.20 L to 4.21 L. Calculate the final pressure of the gas.

1.9 To what temperature must a sample of helium gas be cooled from 22.2°C to reduce its volume from 1.00 L to 100 cm^3?

1.10 Hot-air balloons gain their lift from the lowering of density of air that occurs when the air in the envelope is heated. To what temperature should you heat a sample of air, initially at 340 K, to increase its volume by 14 per cent?

1.11 At sea level, where the pressure was 104 kPa and the temperature 21.1 °C, a certain mass of air occupied 2.0 m^3. To what volume will the region expand when it has risen to an altitude where the pressure and temperature are (a) 52 kPa, -5.0 °C, (b) 880 Pa, -52.0 °C?

1.12 A gas mixture being used to simulate the atmosphere of another planet consists of 320 mg of methane, 175 mg of argon, and 225 mg of nitrogen. The partial pressure of nitrogen at 300 K is 15.2 kPa. Calculate (a) the volume and (b) the total pressure of the mixture.

1.13 A determination of the density of a gas or vapour can provide a quick estimate of its molar mass even though for practical work mass spectrometry is far more precise. The density of a gaseous compound was found to be 1.23 g L^{-1} at 330 K and 25.5 kPa. What is the molar mass of the compound?

1.14 In an experiment to measure the molar mass of a gas, 250 cm^3 of the gas was confined in a glass vessel. The pressure was 152 Torr at 298 K and the mass of the gas was 33.5 mg. What is the molar mass of the gas?

1.15 Calculate the pressure exerted by 1.0 mol C$_2$H$_6$ behaving as (a) a perfect gas, (b) a van der Waals gas when it is confined under the following conditions: (i) at 273.15 K in 22.414 L, (ii) at 1000 K in 100 cm^3. Use the data in Table 1.6.

1.16 A vessel of volume 22.4 L contains 2.0 mol H$_2$ and 1.0 mol N$_2$ at 273.15 K. Calculate (a) their partial pressures and (b) the total pressure.

1.17 A diving bell has an air space of 3.0 m^3 when on the deck of a boat. What is the volume of the air space when the bell has been lowered to a depth of 50 m? Take the mean density of sea water to be 1.025 g cm^{-3} and assume that the temperature is the same as on the surface.

1.18 What pressure difference must be generated across the length of a 15 cm vertical drinking straw in order to drink a water-like liquid of density 1.0 g cm^{-3}?

1.19 A meterological balloon had a radius of 1.0 m when released at sea level at 20 °C and expanded to a radius of 3.0 m when it had risen to its maximum altitude where the temperature was -20 °C. What is the pressure inside the balloon at that altitude?

1.20 How reliable is the perfect gas law in comparison with the van derWaals equation? Calculate the difference in pressure of 10.00 g of carbon dioxide confined to a container of volume 100 cm^3 at 25.0 °C between treating it as a perfect gas and a van der Waals gas.

1.21 Express the van der Waals equation of state as a virial expansion in powers of $1/V_m$ and obtain expressions for B and C in terms of the parameters a and b. The expansion you will need is

$$\frac{1}{1-x} = 1 + x + x^2 + \cdots$$

Measurements on argon gave $B = -21.7\,\text{cm}^3\,\text{mol}^{-1}$ and $C = 1200\,\text{cm}^6\,\text{mol}^{-2}$ for the virial coefficients at 273 K. What are the values of a and b in the corresponding van der Waals equation of state?

1.22 Show that there is a temperature at which the second virial coefficient, B, is zero for a van der Waals gas, and calculate its value for carbon dioxide. *Hint*: Use the expression for B derived in the preceding exercise.

1.23 The composition of planetary atmospheres is determined in part by the speeds of the molecules of the constituent gases, because the faster moving molecules can reach escape velocity and leave the planet. Calculate the mean speed of (a) He atoms, (b) CH$_4$ molecules at (i) 77 K, (ii) 298 K, (iii) 1000 K.

1.24 At what pressure does the mean free path of argon at 25 °C become comparable to the diameter of a spherical vessel of volume 1.0 L that contains it? Take $\sigma = 0.36\,\text{nm}^2$.

1.25 At what pressure does the mean free path of argon at 25 °C become comparable to 10 times the diameters of the atoms themselves? Take $\sigma = 0.36\,\text{nm}^2$.

1.26 When we are studying the photochemical processes that can occur in the upper atmosphere, we need to know how often atoms and molecules collide. At an altitude of 20 km the temperature is 217 K and the pressure 0.050 atm. What is the mean free path of N$_2$ molecules? Take $\sigma = 0.43\,\text{nm}^2$.

1.27 How many collisions does a single Ar atom make in 1.0 s when the temperature is 25 °C and the pressure is (a) 10 bar, (b) 100 kPa, (c) 1.0 Pa?

1.28 Calculate the total number of collisions per second in 1.0 L of argon under the same conditions as in Exercise 1.24.

1.29 How many collisions per second does an N_2 molecule make at an altitude of 20 km? (See Exercise 1.26 for data.)

1.30 The spread of pollutants through the atmosphere is governed partly by the effects of winds but also by the natural tendency of molecules to spread. The latter depends on how far a molecule can travel before colliding with another molecule. Calculate the mean free path of diatomic molecules in air using $\sigma = 0.43$ nm^2 at 25 °C and (a) 10 bar, (b) 103 kPa, (c) 1 Pa.

1.31 Use the Maxwell distribution of speeds to estimate the fraction of N_2 molecules at 500 K that have speeds in the range 290 to 300 m s^{-1}.

1.32 How does the mean free path in a sample of a gas vary with temperature in a constant-volume container?

Thermodynamics: the First Law

The branch of physical chemistry known as **thermodynamics** is concerned with the study of the transformations of energy, and in particular the transformation of energy from heat into work and vice versa. That concern might seem remote from chemistry; indeed, thermodynamics was originally formulated by physicists and engineers interested in the efficiency of steam engines. However, thermodynamics has proved to be of immense importance in chemistry. Not only does it deal with the energy output of chemical reactions but it also helps to answer questions that lie right at the heart of everyday chemistry, such as why reactions reach equilibrium, what their composition is at equilibrium, and how reactions in electrochemical (and biological) cells can be used to generate electricity.

Classical thermodynamics, the thermodynamics developed during the nineteenth century, stands aloof from any models of the internal constitution of matter: you could develop and use thermodynamics without ever mentioning atoms and molecules. However, the subject is greatly enriched by acknowledging that atoms and molecule do exist, and interpreting thermodynamic properties and relations in terms of them. Wherever it is appropriate, we shall cross back and forth between thermodynamics, which provides useful relations between observable properties of bulk matter, and the properties of atoms and molecules, which are ultimately responsible for these bulk properties. The theory of the connection between atomic and bulk thermodynamic properties is called **statistical thermodynamics**.

Almost every argument and explanation in chemistry boils down to a consideration of some aspect of a single property: the **energy**. We shall find that energy determines what molecules may form, what reactions may occur, how fast they may occur, and (with a refinement in our conception of energy) in which direction a reaction has a tendency to occur. Energy is central to chemistry, yet it is extraordinarily difficult to give a satisfactory account of what energy is. It is easy to give a bland definition:

Energy is the capacity to do work.

(We shall explain the scientific meaning of work shortly.) This definition implies that a raised weight has more energy than one on the ground because the former has a greater capacity to do work: it can do work as it falls to the level of the lower weight. The definition also implies that a gas at high

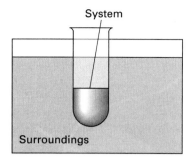

Fig. 2.1 The sample is the system of interest; the rest of the world is its surroundings. The surroundings are where observations are made on the system. They can often be modelled (as here) by a large water bath.

temperature has more energy than the same gas at a low temperature: the hot gas has a higher pressure and can do more work in driving out a piston.

The conservation of energy

People struggled for centuries to create energy from nothing, for they believed that if they could create energy, then they could produce work (and wealth) endlessly. However, without exception, despite strenuous efforts (many of which degenerated into deceit), they failed. As a result of their failed efforts, we have come to recognize that energy can be neither created nor destroyed. This property of energy is called the **conservation of energy**. The conservation of energy is of great importance in chemistry. Most chemical reactions release energy or absorb it as they occur, so according to the conservation of energy, we can be confident that all such changes must involve only the *conversion* of energy from one form to another or its transfer from place to place, not its creation or annihilation.

In this first section we see how to take into account the various forms in which energy can be transferred into or out of a sample of matter. To do so, we must make a distinction between a system and its surrounding. A **system** is the part of the world in which we have a special interest. The **surroundings** are where we make our observations. Thus, a system may be a stoppered flask containing a reaction mixture; the surroundings may be a constant-temperature bath in which the flask is immersed (Fig. 2.1). We also need to distinguish three types of system (Fig. 2.2):

1. An **open system** is a system that can exchange matter with its surroundings.
2. A **closed system** is a system that cannot exchange matter with its surroundings.
3. An **isolated system** is a system that can exchange neither matter nor energy with its surroundings.

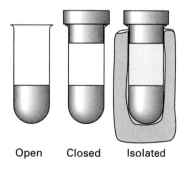

Open   Closed   Isolated

Fig. 2.2 A system is *open* if it can exchange matter and energy with its surroundings, *closed* if it can exchange energy but not matter, and *isolated* if it can exchange neither energy nor matter.

An example of an open system is a flask that is not stoppered and to which various substances can be added. A biochemical cell is an open system because nutrients and waste can pass through the cell wall. An example of a closed system is a stoppered flask: energy can be exchanged with the contents of the flask because the walls may be able to conduct heat. An example of an isolated system is a sealed flask that is thermally, mechanically, and electrically insulated from its surroundings. A system that is thermally insulated from its surroundings (like water in a vacuum flask) is called an **adiabatic system** (from the Greek words for 'not passing through'). For the rest of this chapter we consider closed and isolated systems. Such systems are very good approximations to many of the actual systems that we meet in practice, and by focusing on them the discussion is greatly simplified. For example, a sealed flask and the water bath in which it is immersed jointly make up a very good approximation to an isolated system.

2.1 Work and heat

There are only two ways in which the energy of a closed system can be changed: by transferring energy as work and by transferring energy as heat:

Work is a transfer of energy that can be used to change the height of a weight somewhere in the surroundings.
Heat is a transfer of energy as a result of a temperature difference between the system and the surroundings.

For instance, consider a chemical reaction that produce gases, such as the thermal decomposition of calcium carbonate

$$CaCO_3(s) \xrightarrow{\Delta} CaO(s) + CO_2(g)$$

(The symbol Δ signifies an elevated temperature; this reaction proceeds at about 800 °C.) We suppose that it takes place inside a cylinder fitted with a piston so that the gas produced drives out the piston and raises a weight in the surroundings (Fig. 2.3). In this case, the system has transferred energy to the surroundings because a weight has been raised in the surroundings and they (the surroundings) are now capable of doing more work than before the weight was raised. Alternatively, suppose we carry out a typical acid–base neutralization, such as

$$HCl(aq) + NaOH(aq) \longrightarrow NaCl(aq) + H_2O(l)$$

in a flask immersed in an ice bath (Fig. 2.4). No weight is raised in the surroundings (so the reaction does no work), but some of the ice melts. We have to conclude that energy has left the system (because ice must be supplied with energy if it is to melt) and has travelled from the system to the ice bath. Because the latter has a lower temperature than the system, we conclude that energy has left the system as heat.

A process that releases heat into the surroundings is called **exothermic**. A process that absorbs heat from the surroundings is called **endothermic**. All combustions are exothermic. Endothermic reactions are much less common, but one spectacular example is the reaction that takes place between ammonium thiocyanate, NH_4SCN, and barium hydroxide octahydrate, $Ba(OH)_2 \cdot 8H_2O$, when the two solids are ground together:

$$2\,NH_4SCN(s) + Ba(OH)_2 \cdot 8H_2O(s)$$
$$\longrightarrow Ba(SCN)_2(aq) + 10\,H_2O(l) + 2\,NH_3(g)$$

The reaction releases the water of hydration from the barium hydroxide, and is so endothermic that the resulting solution freezes, as does any water that happens to condense on the outside of the flask. The endothermic dissolution of ammonium nitrate in water is the basis of the instant cold-packs that are included in some first-aid kits. They consist of a plastic envelope containing water dyed blue (for psychological reasons) and a

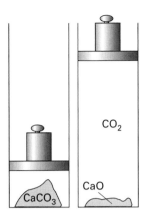

Fig. 2.3 When calcium carbonate decomposes, the carbon dioxide produced must push back the surrounding atmosphere (represented by the weight resting on the piston), and hence must do work on its surroundings. This is an example of energy leaving a system as work.

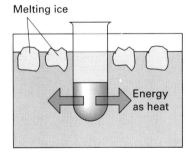

Fig. 2.4 Energy may also leave a system as heat. The loss into the surroundings can be detected by noting whether ice melts as the process proceeds. Note that in this case and in Fig. 2.3 the distinction between energy transfer by work or by heat is made by observations in the surroundings (the raising of a weight or the melting of ice).

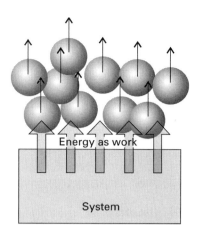

Fig. 2.5 Work is transfer of energy that causes or utilizes uniform motion of atoms in the surroundings. For example, when a weight is raised, all the atoms of the weight (shown magnified) move in unison in the same direction.

small tube of ammonium nitrate, which is broken when the pack is to be used.

The clue to the molecular nature of work comes from thinking about the motion of a weight in terms of its component atoms. When a weight is raised, all its atoms move in the same direction (upwards). This observation suggests that *work is the transfer of energy that achieves or utilizes uniform motion in the surroundings* (Fig. 2.5). Whenever you think of work being done, you can always think of it in terms of uniform motion of some kind. Electrical work, for instance, corresponds to electrons being pushed in the same direction through a circuit. Mechanical work corresponds to atoms being pushed in the same direction against an opposing force.

Now consider heat. When energy is transferred to an ice bath and some of the ice melts, the H_2O molecules in the ice oscillate more rapidly around their positions and those at the surface may escape into the surrounding liquid. The key point is that the motion stimulated by the arrival of energy from the system is disorderly, not uniform as in the case of work. This observation reveals the molecular nature of heat: *heat is the transfer of energy that achieves or utilizes disorderly motion in the surroundings* (Fig. 2.6). A fuel burning, for example, generates disorderly molecular motion in its vicinity, and so it releases energy as heat. The nuclear reactions in the Sun shake the atoms in their immediate surroundings into disorderly oscillation, and so their energy is released as heat.

It is an interesting social point that the molecular difference between work and heat correlates with the chronological order of their application. The liberation of heat by fire is a relatively unsophisticated procedure because the energy emerges in a disordered fashion from the burning fuel. It was developed—stumbled upon—early in the history of civilization. The generation of work by a burning fuel relies on a carefully controlled transfer of energy so that myriads of molecules move in unison. It came thousands of years later, with the development of the steam engine.

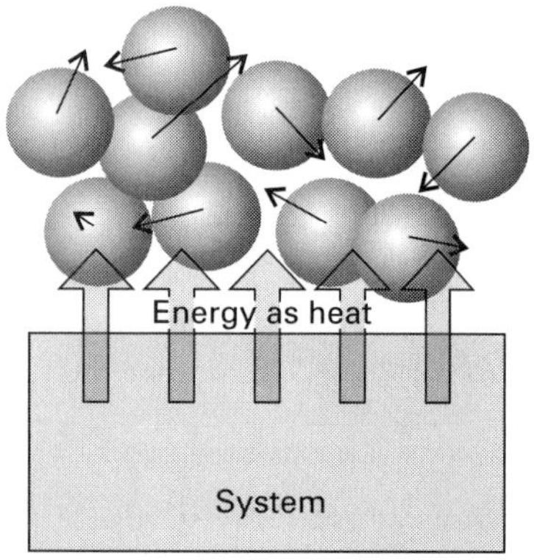

2.2 The measurement of work

When we want to calculate the work done by a given process, we have to use the definition given in physics that work is equal to the product of distance and the force opposing the motion:

$$\text{work} = \text{distance} \times \text{opposing force}$$

It follows that moving through a long distance against a strong opposing force (like climbing a high mountain) requires a lot of work. If the force is the gravitational attraction of the Earth on a mass m, the force opposing raising the mass vertically is mg where g is the acceleration of free fall ($9.81\ \text{m s}^{-2}$). Therefore, the work needed to raise a mass m through a height h on the surface of the Earth is

$$\text{work} = h \times mg = mgh \qquad (1)$$

Fig. 2.6 Heat is the transfer of energy that causes or utilizes chaotic motion in the surroundings. When energy leaves the system (the shaded region), it generates chaotic motion in the surroundings (shown magnified).

For example, raising a book like this one (of mass about 1.0 kg) from the floor to the table 75 cm above requires

$$\text{work} = (1.0\,\text{kg}) \times (9.81\,\text{m s}^{-2}) \times (0.75\,\text{m}) = 7.4\,\text{kg m}^2\,\text{s}^{-2}$$

The unit used to report energy is the joule (J), which is named after James Joule, the Manchester brewer who made a detailed study of heat and work in the nineteenth century:

$$1\,\text{J} = 1\,\text{kg m}^2\,\text{s}^{-2}$$

Therefore, the work we have just calculated would be reported as 7.4 J. Each beat of the human heart does work equal to about 1 J, so about 100 kJ of energy is expended each day driving the blood around one's body.

Exercise E2.1

A useful relation between joules and pascals is $1\,\text{J} = 1\,\text{Pa m}^3$. Confirm this relation from the definition of pascals in *Further information 2*.

$$[1\,\text{Pa} \times 1\,\text{m}^3 = 1\,\text{kg m}^{-1}\,\text{s}^{-2} \times 1\,\text{m}^3 = 1\,\text{kg m}^2\,\text{s}^{-2} = 1\,\text{J}]$$

In chemistry, a very important type of work is **expansion work**, the work done when a system expands against an opposing pressure. The thermal decomposition of calcium carbonate illustrated in Fig. 2.3 is an example of expansion work, where the opposing pressure is provided by the downward force of the weight. It is quite easy to calculate the work that a system does as it expands through a volume ΔV by considering a piston of area A moving out through a distance h (Fig. 2.7). The force opposing the expansion is the constant external pressure p_{ex} multiplied by the area of the piston (because pressure is force divided by area, so force is pressure times area). The work done is therefore

$$\begin{aligned}\text{work done by system} &= \text{distance} \times \text{opposing force} \\ &= h \times (p_{ex} \times A) = p_{ex} \times (h \times A) \\ &= p_{ex} \times \Delta V\end{aligned}$$

The last line follows from the fact that hA is the volume of the cylinder swept out by the piston as the gas expands, so we can write $hA = \Delta V$. That is, for expansion work,

$$\text{work done by system} = p_{ex}\Delta V \qquad (2)$$

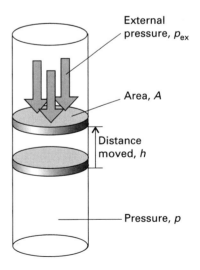

Fig. 2.7 When a piston of area A moves out through a distance h, it sweeps out a volume $\Delta V = Ah$. The external pressure p_{ex} opposes the expansion with a force $p_{ex}A$.

39

Exercise E2.2

Calculate the work done by a system in which a reaction results in the formation of 1.0 mol of gas molecules at 25 °C and 100 kPa. [*Hint*. The increase in volume will be 25 L under these conditions (if the gas is treated as perfect); note the relation between pascals and joules established in Exercise E2.1.]

[*Answer*: 2.5 kJ]

A very important type of expansion work is a special case of the one we have just described. Suppose we were interested in the *maximum* work that we can extract from a gas as it expands isothermally (at constant temperature). According to eqn (2), the system does maximum work when the external pressure has its maximum value because then the force opposing the expansion is greatest and most effort is required to push the piston out. However, that external pressure cannot be greater than the pressure of the gas inside the system, for otherwise the external pressure would compress the gas instead of allowing it to expand. Therefore, *maximum work is obtained when the external pressure is only infinitesimally less than the pressure of the gas in the system.* In effect, the two pressures are the same. In Chapter 1 we called this balance of pressures a state of mechanical equilibrium. Therefore, we can conclude that *a system in mechanical equilibrium does maximum expansion work.*

There is another way of expressing the condition for obtaining maximum work. Because the external pressure is infinitesimally less than the pressure of the gas, the piston moves outwards. However, suppose we increase the external pressure so that it became infinitesimally greater than the pressure of the gas; now the piston moves inwards. That is, *when a system is in a state of mechanical equilibrium, infinitesimal changes in the pressure can result in opposite directions of change.* A change that can be reversed by an infinitesimal change in a parameter—in this case, the pressure—is said to be **reversible**. In everyday life 'reversible' means a process that can be reversed; in thermodynamics it has a stronger meaning—it means that a process can be reversed by an *infinitesimal* modification in some property (such as the pressure).

We can summarize this discussion by the following remarks:

1. A system does maximum expansion work when the external pressure is equal to that of the system ($p_{ex} = p$).
2. A system does maximum expansion work when it is in mechanical equilibrium with its surroundings.
3. A system does maximum expansion work when it is changing reversibly.

All three statements are equivalent, but they reflect different degrees of sophistication in the way the point is expressed.

The maximum work obtainable from the isothermal expansion of a gas from an initial volume V_i to a final volume V_f cannot be written down simply by replacing p_{ex} in eqn (2) by p (the pressure of the gas in the cylinder) because, as the piston moves out and the volume of the cylinder increases, the pressure inside the system falls (by Boyle's law). Therefore, at each stage of the expansion, the external pressure must be reduced slightly to match the slightly lower pressure of the gas. The calculation of the work available from the system under such circumstances can be done by using calculus, and the result is

$$\text{work done by system} = nRT \ln \frac{V_f}{V_i} \qquad (3)$$

where n is the amount of substance of gas and T is the temperature (which is maintained constant, for example by immersing the cylinder in a water bath).

Justification

Although we cannot replace p_{ex} by p in the overall change (because p changes during the expansion), we can think of the expansion as taking place in an infinite number of infinitesimally small steps. During each one of these steps, the external pressure is set equal to the current pressure of the gas (Fig. 2.8). For an infinitesimal increase in volume dV when the pressure inside and out is p, we can write

$$\text{work done by system} = p\,dV$$

(For a note on the use of calculus, see *Further information 1*.) Now we need to take into account the fact that p changes as the gas expands. According to the perfect gas equation of state, the pressure is related to the volume by

$$p = \frac{nRT}{V}$$

That is, when the volume is V, the external pressure must be set equal to nRT/V. Therefore, the *total* work of expansion as

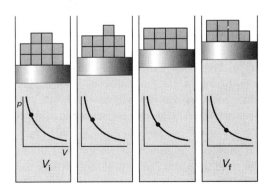

Fig. 2.8 For a gas to expand reversibly, the external pressure must be adjusted to match the internal pressure at each stage of the expansion. This matching is represented in this illustration by gradually unloading weights from the piston as the piston is raised and the internal pressure falls. The procedure results in the extraction of the maximum possible work of expansion.

41

the system expands from an initial volume V_i to a final volume V_f is the sum (integral) of all the infinitesimal contributions:

$$\text{work done by system} = \int_{V_i}^{V_f} p\,dV = nRT \int_{V_i}^{V_f} \frac{dV}{V}$$

$$= nRT \ln \frac{V_f}{V_i} \qquad \left(\text{because} \int \frac{dx}{x} = \ln x\right)$$

Exercise E2.3

Calculate the work done when $1.0\,\text{mol Ar(g)}$ that is confined in a cylinder of volume $1.0\,\text{L}$ at $25\,^{\circ}\text{C}$ expands isothermally and reversibly to $2.0\,\text{L}$.

[*Answer*: $1.7\,\text{kJ}$]

The internal energy

We need some way of keeping track of the energy changes in a system as energy is transferred in and out as work or heat. This is the role of the property called the **internal energy**, U, of the system. The internal energy is an accounting device, like a country's gold reserves for monitoring transactions with the outside world (the surroundings).

When a system releases $10\,\text{kJ}$ of energy to the surroundings as work (that is, when $10\,\text{kJ}$ of work is done in the surroundings by raising a weight), the internal energy of the system decreases by $10\,\text{kJ}$, and we write $\Delta U = -10\,\text{kJ}$. The minus sign signifies the reduction in internal energy that has occurred. If the system releases $20\,\text{kJ}$ of energy as heat, then the internal energy decreases by $20\,\text{kJ}$, and we write $\Delta U = -20\,\text{kJ}$. If the system releases $10\,\text{kJ}$ as work *and* $20\,\text{kJ}$ as heat (as in an inefficient internal combustion engine), the internal energy falls by a total of $30\,\text{kJ}$, and we write $\Delta U = -30\,\text{kJ}$. On the other hand, if we do $10\,\text{kJ}$ of work on the system, for instance, by winding a spring it contains, or pushing in a piston to compress a gas (Fig. 2.9), then the internal energy of the system increases by $10\,\text{kJ}$, and we write $\Delta U = +10\,\text{kJ}$. Likewise, if we supply $20\,\text{kJ}$ of energy as heat, then the internal energy increases by $20\,\text{kJ}$, and we write $\Delta U = +20\,\text{kJ}$. Notice that ΔU *always* carries a sign explicitly, even if it is positive: we never write $\Delta U = 20\,\text{kJ}$ but always $+20\,\text{kJ}$.

The internal energy is the sum of all the kinetic and potential contributions to the energy of all the atoms, ions, and molecules in the system: it is the grand total energy of the system. When energy is transferred into the system by

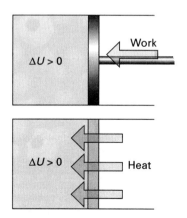

Fig. 2.9 When work is done on a system, its internal energy rises. The internal energy also rises when energy is transferred as heat.

heating it or doing work on it, the increased energy is stored in the increased kinetic and potential energies of the molecules (in a gas, the molecules move faster). Likewise, when energy is lost, it is given up by the molecules as they lose kinetic or potential energy (in a gas, the molecules move more slowly). In practice, we do not know and cannot measure this total energy, because it includes the kinetic and potential energies of all the electrons and all the components of the atomic nuclei. Nevertheless, there is no problem with dealing with the *changes* in internal energy, because we can determine those by monitoring the energy supplied or lost as heat or work. All practical applications of thermodynamics deal only with ΔU, not with U itself.

2.3 Notation

The preceding remarks can be summarized by introducing the following notation:

w is the energy supplied *to* the system as work.
q is the energy supplied *to* the system as heat.

The word 'to' in these definitions has a lot of significance, because it implies a sign convention. A positive value of w or q means that energy is indeed supplied *to* the system; however, if the value of w or q is negative, then that signifies that energy has left the system. These conventions mean that

$w = +10\,$kJ if 10 kJ of energy is supplied to the system as work.

$w = -10\,$kJ if 10 kJ of energy is lost from the system as work.

$q = +10\,$kJ if 10 kJ of energy is supplied to the system as heat.

$q = -10\,$kJ if 10 kJ of energy is lost from the system as heat.

In terms of this notation, the change in internal energy when energy is transferred as heat or work can be expressed succinctly as

$$\Delta U = w + q \qquad (4)$$

Example *Calculating the change in internal energy*
Nutritionists are interested in the use of energy by the human body, and we can consider our own body as a thermodynamic 'system'. Calorimeters have been constructed that can accommodate a person to measure (non-destructively!) their net energy output. Suppose in the course of an experiment someone does 622 kJ of work on an exercise bicycle and loses 82 kJ of energy as heat. What is the change in internal energy of the person?

Strategy
This example is an exercise in keeping track of signs correctly. When energy is lost from the system, w or q is negative. When energy is gained by the system, w or q is positive.

Solution

To take note of the signs we write $w = -622\,\text{kJ}$ and $q = -82\,\text{kJ}$. Then eqn (4) gives us

$$\Delta U = w + q = (-622\,\text{kJ}) + (-82\,\text{kJ}) = -704\,\text{kJ}$$

We see that the person's internal energy falls by $704\,\text{kJ}$.

Exercise E2.4

An electric battery is charged by supplying $250\,\text{kJ}$ of energy to it as electrical work, but in the process it loses $25\,\text{kJ}$ of energy as heat to the surroundings. What is the change in internal energy of the battery?

[*Answer*: $+225\,\text{kJ}$]

Keeping track of signs can seem a burden in thermodynamics, but it is straightforward if you always keep in mind what is actually happening physically. For example, the expressions we have written for expansion work can now be written in terms of w and the appropriate sign:

Expansion (of any substance) against a constant external pressure:

$$w = -p_{ex}\Delta V \qquad (\Delta V = V_f - V_i) \tag{5}$$

Reversible, isothermal expansion of a perfect gas:

$$w = -nRT\,\ln\frac{V_f}{V_i} \tag{6}$$

In each case, the negative sign signifies that energy leaves the system as work when it expands. Thus, when the system expands against a constant external pressure, ΔV is positive and so $-p_{ex}\Delta V$ is negative: this sign tells us that the system has *lost* energy as work. Likewise, in the reversible expansion, the ratio V_f/V_i is greater than 1 (because V_f is larger than V_i in an expansion), and as the logarithm of a number larger than 1 is positive it follows that w is negative. Once again, the sign tells us that energy has left the system as work.

Equation (4) expresses the fact that *work and heat are equivalent ways of changing the internal energy of a system*. That is, it is immaterial whether we supply a given quantity of energy to a system as heat, work, or a combination of the two: the change in internal energy is the same in each case. Moreover, because there is nothing distinctive about the energy once it has been transferred into the system, any internal energy that has been supplied as work (for instance) can be withdrawn as heat. The system is like a bank that can accept deposits and make payments in either of two currencies (as work or heat), but stores its reserves as the thermodynamic equivalent of gold (that is, as internal energy).

An important characteristic of the internal energy is that it is a state function:

A **state function** is a physical property that depends only on the current state of the system and is independent of the path by which that state was reached.

A state function is very much like altitude: each point on the surface of the Earth can be specified by quoting its latitude and longitude, and (on land areas, at least) there is a unique property, the altitude, that has a fixed value at that point. The altitude at any location is independent of the path that we may have taken to arrive there, so altitude is a state property that is determined by the two variables latitude and longitude (Fig. 2.10).

The internal energy, being the sum of all the kinetic and potential energies that the atoms in a system happen to have at a particular instant, is a property of the state of the system. If we were to change the temperature of the system, then change the pressure, then adjust the temperature and pressure back to their original values, the internal energy would return to its original value too.

More generally, the fact that U is a state function implies that *a change ΔU in the internal energy between two states of a system is independent of the path between them*. For example, if we had a sample of gas which we compressed to a certain pressure and then cooled to a certain temperature, the change in internal energy would have a particular value. If, on the other hand, we changed the temperature and then the pressure, but ensured that the two final values were the same as in the first experiment, then the overall change in internal energy would be exactly the same as before. This path independence of the value of ΔU is of the greatest importance in chemistry, as we shall soon see.

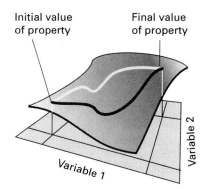

Fig. 2.10 The curved sheet shows how a property (for example, the altitude) varies as two coordinates (for example, latitude and longitude) are changed. The altitude is a state property, because it depends only on the current state of the system. The change in the value of a state property is independent of the path between the two states. For example, the difference in altitude between the initial and final states shown in the diagram is the same whatever path (as depicted by the dark and light lines) is used to travel between them. The internal energy is a state property in the same sense, but now the variables include the pressure and temperature of the system.

2.4 The First Law

Finally, we can move to the climax of this part of the chapter. Suppose we now consider an isolated system, one that is insulated mechanically and thermally from its surroundings. Then because the system can neither do work nor supply heat, it follows that its internal energy cannot change. That is,

The internal energy of an isolated system is constant.

This statement is the **First Law of thermodynamics**. It is obviously closely related to the law of conservation of energy, but it should be remembered that the internal energy is expressed in terms of work and heat; so the First Law is a statement that concerns *thermodynamics*, the transformation of energy into different forms (including heat), not just mechanics (which does not deal with the concept of heat).

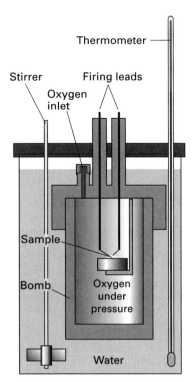

Fig. 2.11 A constant-volume bomb calorimeter. The 'bomb' is the central, sturdy vessel, which is strong enough to withstand high pressures. The calorimeter is the entire assembly shown here. In order to ensure that no heat escapes into the surroundings, the calorimeter may be immersed in a water bath with a temperature that is continuously adjusted to that of the calorimeter at each stage of the combustion.

The measurement of ΔU

The definition of ΔU in terms of w and q points to a very simple method for measuring the change in internal energy of a system when a reaction takes place. We have seen already that the work done by a system when it pushes against a fixed external pressure is proportional to the change in volume. Therefore, if we seal the reaction into a container that cannot change its volume, the system can do no expansion work and we can set $w = 0$. Then eqn (4) simplifies to

$$\Delta U = q \text{ at constant volume} \tag{7}$$

It follows that, to measure a change in internal energy, we need only measure the heat supplied to or absorbed by a system that cannot change its volume.

The apparatus used in practice is called a **bomb calorimeter** (Fig. 2.11; the name comes from 'calor', the Latin word for heat). It consists of a sturdy container, in which the reaction occurs, a thermometer, and a surrounding water bath. The entire assembly is insulated from the rest of the world, so overall it forms an isolated system. The first step is to *calibrate* the calorimeter (the entire assembly—the reaction vessel and the water bath) by comparing the observed change in temperature with a change in temperature brought about by a given quantity of heat. One procedure is to heat the calorimeter electrically by passing a known current for a measured time through a heater, and record the increase in temperature. The heat supplied by the heater is given by

$$\text{heat supplied to system} = I\mathcal{V}t \tag{8}$$

where I is the current in amperes (A), $\mathcal{V}$ is the potential of the supply in volts (V), and t is the time in seconds for which the current flows. For example, if a current of 1.23 A from a 12.0 V source flows for 123 s, then the heat supplied is

$$\text{heat supplied to system} = (1.23\,\text{A}) \times (12.0\,\text{V}) \times (123\,\text{s}) = 1.82\,\text{kJ}$$

(We have used the relation 1 A V s = 1 J, which follows from the definition of the electrical units.) If we observe that the temperature of the calorimeter rises by 4.47 °C, then the **calorimeter constant**, C, the ratio of the heat supplied to the increase in temperature, is

$$C = \frac{\text{heat supplied}}{\text{increase in temperature}} = \frac{1.82\,\text{kJ}}{4.47\,°\text{C}} = 0.407\,\text{kJ}\,(°\text{C})^{-1}$$

If in an experiment on an unknown reaction we measure a temperature rise of 3.22 °C in the same calorimeter, then we could conclude that the heat output of the reaction is

$$\text{heat output} = C \times \text{increase in temperature}$$
$$= (0.407\,\text{kJ}\,(°\text{C})^{-1}) \times (3.22\,°\text{C}) = 1.31\,\text{kJ}$$

The value of q obtained in this way is identified with the change in internal energy of the system. An alternative procedure is to carry out a reaction of known heat output. One common procedure is the combustion of benzoic acid, which has an accurately known heat output of $3227\,kJ\,mol^{-1}$. Similar procedures apply when a reaction releases heat or absorbs heat: the temperature of a calorimeter rises if heat is released and it falls if heat is absorbed.

The heat capacity

When heat is transferred to a system at constant volume, its internal energy rises. In most cases, the temperature of the system also rises.[†] This suggests that there may be a relation between the change in temperature of a system and the change in its internal energy, and that we could monitor the latter simply by monitoring the temperature. In practice, experiments have shown that, for small temperature increases, the rise in temperature is proportional to the quantity of heat supplied, so twice as much heat is required to raise the temperature of a sample of water by 10 K than is required to raise it by 5 K.

If the increase in temperature is written ΔT, the observed approximate proportionality between temperature increase and heat q supplied means that we can write $\Delta T \propto q$. It is conventional to denote the constant of proportionality as $1/C$; then

$$\Delta T = \frac{q}{C} \tag{9}$$

The constant C is called the **heat capacity**. With heat measured in joules and the temperature rise in kelvins, the units of C are joules per kelvin, $J\,K^{-1}$. Because ΔT is inversely proportional to the heat capacity, a system with a large heat capacity undergoes only a small increase in temperature for a given input of heat.

The heat capacity of a substance depends on the size of the sample and it is common to report values as the **specific heat capacity**, the heat capacity divided by the mass of the sample (typically, in joules per kelvin per gram, $J\,K^{-1}\,g^{-1}$) or the **molar heat capacity**, the heat capacity divided by the amount of substance (in joules per kelvin per mole, $J\,K^{-1}\,mol^{-1}$). The value of the molar heat capacity of water, $75\,J\,K^{-1}\,mol^{-1}$, for example, shows that the increase in temperature of 100 g of water (5.55 mol H_2O) when 1.0 kJ of heat is supplied is approximately

$$\Delta T = \frac{1.0 \times 10^3\,J}{(5.55\,mol) \times (75\,J\,K^{-1}\,mol^{-1})} = +2.4\,K$$

It turns out that the numerical value of the heat capacity depends on whether the sample is maintained at the same volume (like a gas contained in a rigid vessel) or if it is free to expand against a constant external pressure (as when a solid or liquid is heated in the atmosphere or a gas is heated in a cylinder equipped with a movable piston). The two heat capacities are called,

[†]The temperature does not rise if a phase change is taking place, such as when water is boiling.

respectively, the **constant-volume heat capacity**, C_V (for a system that cannot expand) and the **constant-pressure heat capacity**, C_p (for a system that is free to expand). The numerical values of the two types of heat capacity are very similar for solids and liquids (neither of which change volume very much with temperature, even if they are free to), but they differ appreciably for gases. For perfect gases, the two molar heat capacities ($C_{V,m}$ and $C_{p,m}$, the subscript m normally denotes a molar quantity) are related by

$$C_{p,m} = C_{V,m} + R \qquad (10)$$

This expression tells us that the constant-pressure heat capacity of a gas is larger than its constant-volume heat capacity. The difference arises because some of the energy supplied as heat at constant pressure is used to drive back the atmosphere as the system expands. Because some energy is used to drive back the atmosphere, less is available to raise the temperature of the system, so a given input of energy as heat brings about a smaller rise in temperature than for the same sample at constant volume. That smaller increase corresponds to a larger heat capacity. Some typical values are given in Table 2.1.

Exercise E2.5

The temperature of 15.0 g of water vapour increases by 8.61 °C when 0.241 kJ of heat is supplied at constant pressure. What is its molar constant-volume heat capacity if it behaves as a perfect gas?

[*Answer*: 25.3 J K^{-1} mol^{-1}]

Table 2.1 Heat capacities of some materials

Substance		Specific heat capacity $C_{p,s}$/J K^{-1} g^{-1}	Molar heat capacity* $C_{p,m}$/J K^{-1} mol^{-1}
Air		1.01	29
Benzene, C_6H_6		1.05	136.1
Brass		0.37	
Copper, Cu		0.38	
Ethanol, C_2H_5OH		2.42	111.5
Glass (Pyrex)		0.78	
Granite		0.80	
Marble		0.84	
Polyethylene		2.3	
Stainless steel		0.51	
Water, H_2O	solid	2.03	37
	liquid	4.18	75.29
	vapour	2.01	33.58

* Molar heat capacities are given only for air and well defined, pure substances. See also Appendix 1.

Next, we make the connection between the change in temperature and the change in internal energy. We have already seen that at constant volume $\Delta U = q$. This relation can be combined with the expression defining the heat capacity, which can be rearranged to $q = C\Delta T$, to obtain

$$\Delta U = C_V \Delta T \qquad \text{at constant } V \qquad (11)$$

For example, when 100 g of liquid water (5.55 mol H_2O) is heated through 5.0 K in a fixed-volume container, its internal energy increases by

$$\Delta U = (5.55\,\text{mol}) \times (75\,\text{J K}^{-1}\,\text{mol}^{-1}) \times (5.0\,\text{K}) = +2.1\,\text{kJ}$$

The enthalpy

Much of chemistry takes place in vessels that are open to the atmosphere and subjected to constant pressure as distinct from being maintained at constant volume in a sturdy, sealed vessel. In general, when a change takes place in a system open to the atmosphere, the volume of the system changes. For example, the thermal decomposition of 1.0 mol $CaCO_3$ at 1 bar results in an increase in volume of 89 L at 800 °C on account of the carbon dioxide gas produced. To create this large volume for the carbon dioxide to occupy, the surrounding atmosphere must be pushed back. That increase in volume means that the system must perform expansion work. Therefore, although a certain quantity of heat may be supplied to bring about the endothermic decomposition, the increase in internal energy of the system is not equal to the energy supplied as heat because some energy has been used to do work of expansion (Fig. 2.12). In other words, because the volume has increased, some of the heat supplied to the system has leaked back into the surroundings as work.

Another example is the oxidation of a fat, such as tristearin, to carbon dioxide in the body. The overall reaction is

$$2\,C_{57}H_{110}O_6(\text{aq}) + 163\,O_2(\text{g}) \longrightarrow 114\,CO_2(\text{g}) + 110\,H_2O(\text{l})$$

In this exothermic reaction there is a net *decrease* in volume equivalent to the elimination of 49 mol of gas molecules. The decrease in volume is about 600 mL at 25 °C for the consumption of 1 g of the fat. Because the volume occupied by the substances decreases in the course of the reaction, the atmosphere does work *on* the system as the reaction proceeds. That is, energy is transferred as work from the surroundings to the system as it contracts. There is therefore more energy available to be released back into the surroundings as heat. For this reaction, the energy released as heat will be greater than the decrease in the internal energy of the system.

The complication of having to take into account the work of expansion when a reaction (or any other process occurs) can be avoided by introducing a property that is equal to the heat transfer at constant pressure. This

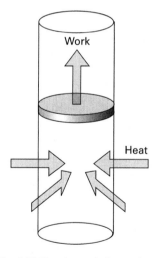

Fig. 2.12 The change in internal energy of a system that is free to expand or contract is not equal to the energy supplied as heat because some energy may escape back into the surroundings as work. However, the change in enthalpy of the system under these conditions is equal to the energy supplied as heat.

property is called the 'enthalpy': it will be at the centre of our attention throughout the rest of the chapter and will recur throughout the book.

2.5 The definition of enthalpy

The **enthalpy**, H, of a system is defined so that

$$\Delta H = q \qquad \text{at constant } p \qquad (12)$$

That is, if 10 kJ of heat is supplied to the system that is free to change its volume at constant pressure, then the enthalpy increases by 10 kJ and we write $\Delta H = +10$ kJ. On the other hand, if the reaction is exothermic and releases 10 kJ of heat when it occurs, then we write $\Delta H = -10$ kJ. For the particular case of the combustion of tristearin mentioned above, in which 90 kJ of energy is released as heat, we would write $\Delta H = -90$ kJ. Note that an *endothermic reaction corresponds to an increase in enthalpy of the system* because energy enters the system as heat and its enthalpy increases (hence the positive sign). On the other hand, *an exothermic process corresponds to a decrease in enthalpy of the system* because energy leaves the system as heat and so its enthalpy falls (hence the negative sign). All combustion reactions (including the controlled combustions that contribute to respiration) are exothermic and are accompanied by a reduction in the enthalpy of the reaction system. These relations are consistent with the choice of the name 'enthalpy', which is derived from the Greek words meaning 'heat inside': the 'heat inside' the system is increased if the process is endothermic and absorbs heat from the surroundings; it is decreased if the process is exothermic and releases heat into the surroundings.[†]

We have seen that the internal energy of a system rises as the temperature is increased. The same is true of the enthalpy, which also increases when the temperature is raised. For example, the enthalpy of 100 g of water is greater at 80 °C than at 20 °C. The change can be measured simply by monitoring the energy that must be supplied as heat to raise the temperature through 60 °C when the sample is open to the atmosphere (or subjected to some other constant pressure); it is found that $\Delta H \approx +25$ kJ in this instance.

The relation between the change in enthalpy and the change in temperature can be found by combining the statement that $\Delta H = q$ at constant pressure with the definition of the constant-pressure heat capacity C_p (the expression $q = C_p \Delta T$):

$$\Delta H = C_p \Delta T \qquad \text{at constant } p \qquad (13)$$

For example, when the temperature of 100 g of water (5.55 mol H_2O) is raised from 20 °C to 80 °C (so $\Delta T = +60$ K) at constant pressure, the enthalpy of the sample changes by

$$\Delta H = (5.55 \, \text{mol}) \times (75.29 \, \text{J K}^{-1} \, \text{mol}^{-1}) \times (60 \, \text{K}) = +25 \, \text{kJ}$$

[†]But heat does not actually 'exist' inside: only energy exists in a system; heat is a means of recovering that energy or increasing it. Heat is energy in transit, not a form in which energy is stored.

There is a very simple relation between the enthalpy and the internal energy of a system:

$$H = U + pV \tag{14}$$

Therefore, to calculate the enthalpy of a system, we simply add the product of the pressure and its volume to the value of its internal energy.

Justification

The consistency of eqns (12) and (14) can be verified quite easily by considering a system that is open to the atmosphere, so that its pressure p is constant and equal to the external pressure p_{ex}. Initially, the enthalpy is

$$H_i = U_i + pV_i$$

After a reaction has occurred at constant pressure, the enthalpy is

$$H_f = U_f + pV_f$$

The change in enthalpy is therefore the difference between these two quantities, and is

$$H_f - H_i = U_f - U_i + p(V_f - V_i)$$

or

$$\Delta H = \Delta U + p\Delta V$$

However, we know that the change in internal energy is given by eqn (4) with $w = -p_{ex}\Delta V$, and substituting that expression into this one gives

$$\Delta H = (-p_{ex}\Delta V + q) + p\Delta V$$

However, $p_{ex} = p$ (because the system and the surroundings have the same pressure), so

$$\Delta H = (-p\Delta V + q) + p\Delta V = q$$

This equation is the same as eqn (12). We can therefore conclude, that if the enthalpy is defined as in eqn (14), then the change in enthalpy is equal to the heat absorbed at constant pressure.

The importance of the relation between H and U given in eqn (14) is that it shows that *the enthalpy is a state function*. This conclusion follows from the fact that U, p, and V are all state functions, and so H must be too. That the enthalpy is a state function, and hence has a value that is independent of how the state was reached, is of the greatest importance in chemistry, as we shall see.

2.6 The enthalpy of physical change

The rest of this chapter is an extended illustration of the role of enthalpy in chemistry. Here we consider the enthalpy changes accompanying two types of physical change: the conversion of one bulk phase (such as a liquid) to another (such as a vapour), and the conversion of individual atoms and molecules to their ions or other fragments.

Bulk change

The vaporization of a liquid, such as the conversion of liquid water to water vapour when a pool of water evaporates at 20 °C or a kettle boils at 100 °C is an endothermic process, because heat must be supplied to bring about the change. At a molecular level, molecules are being driven apart from the grip exerted on them by the attractive interaction of their neighbours, and this process requires an input of energy. One of the body's strategies for maintaining its temperature at about 37 °C is to use the endothermic character of the vaporization of water, for the evaporation of perspiration requires heat and withdraws it from the skin.

The heat that must be supplied at constant pressure per mole of molecules that are vaporized is called the **enthalpy of vaporization** of the liquid, and is denoted $\Delta_{vap}H$ (Table 2.2).[†] For example, the heat required to vaporize 1 mol H_2O from the liquid at 25 °C is 44 kJ, so $\Delta_{vap}H = +44\,kJ\,mol^{-1}$.

Table 2.2 Standard enthalpies of physical change*

Substance	Formula	Freezing point T_f/K	$\Delta_{fus}H^{\ominus}/$ kJ mol^{-1}	Boiling point T_b/K	$\Delta_{vap}H^{\ominus}/$ kJ mol^{-1}
Acetone	CH_3COCH_3	177.8	5.72	329.4	29.1
Ammonia	NH_3	195.3	5.65	239.7	23.4
Argon	Ar	83.8	1.2	87.3	6.5
Benzene	C_6H_6	278.7	9.87	353.3	30.8
Ethanol	C_2H_5OH	158.7	4.60	351.5	43.5
Helium	He	3.5	0.02	4.22	0.08
Mercury	Hg	234.3	2.292	629.7	59.30
Methane	CH_4	90.7	0.94	111.7	8.2
Methanol	CH_3OH	175.5	3.16	337.2	35.3
Water	H_2O	273.2	6.01	373.2	40.7

* Values correspond to the transition temperature. For values at 25°C, use the data in Appendix 1.

[†]The attachment of the subscript vap to the Δ is an agreed international convention; however, the older convention in which the subscript is attached to the H, as in ΔH_{vap}, is still widely used.

Alternatively, we can report the same information by writing the **thermochemical equation**

$$H_2O(l) \longrightarrow H_2O(g) \qquad \Delta H = +44\,kJ$$

A thermochemical equation shows the enthalpy change that accompanies the conversion of an amount of reactant equal to its stoichiometric coefficient in the accompanying chemical equation (in this case, $1\,mol\,H_2O$). If the stoichiometric numbers in the chemical equation are multiplied through by 2, then the thermochemical equation would be written

$$2\,H_2O(l) \longrightarrow 2\,H_2O(g) \qquad \Delta H = +88\,kJ$$

This equation signifies that 88 kJ of heat is required to vaporize $2\,mol\,H_2O$.

Example *Determining the enthalpy of vaporization of a liquid*
Ethanol, C_2H_5OH, is brought to the boil at a pressure of 1 atm. When an electric current of 0.682 A from a 12.0 V supply is passed through a heating coil for 500 s, it is found that 4.33 g of ethanol is vaporized. What is the enthalpy of vaporization of ethanol at its boiling point?

Strategy
Because the heat is supplied at constant pressure, the heat supplied, q, can be identified with the change in enthalpy of the ethanol when it vaporizes. We need to calculate the heat supplied and the amount of ethanol molecules vaporized. Then the enthalpy of vaporization is the heat supplied divided by the amount. The heat supplied is given by the formula quoted in eqn (8): $q = IVt$. Recall that $1\,A\,V\,s = 1\,J$. The amount of ethanol molecules can be determined by dividing the mass of ethanol vaporized by the molar mass.

Solution
The energy supplied as heat is

$$q = IVt = (0.682\,A) \times (12.0\,V) \times (500\,s)$$
$$= 0.682 \times 12.0 \times 500\,J = 4.09\,kJ$$

Therefore, for the quoted extent of vaporization,

$$\Delta H = +4.09\,kJ$$

The amount of ethanol molecules (of molar mass $46.07\,g\,mol^{-1}$) vaporized is

$$n(C_2H_5OH) = \frac{4.33\,g}{46.07\,g\,mol^{-1}} = 0.0940\,mol$$

It follows that

$$\Delta_{vap}H = \frac{\Delta H}{n} = \frac{4.09\,kJ}{0.0940\,mol} = +43.5\,kJ\,mol^{-1}$$

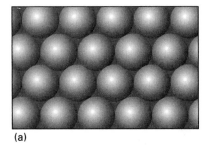

(a)

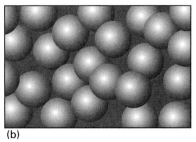

(b)

Fig. 2.13 When a solid (a) melts to a liquid (b), the molecules separate from one another only slightly, the intermolecular interactions are reduced only slightly, and there is only a small change in enthalpy. When a liquid vaporizes (not shown), the molecules are separated by a considerable distance, the intermolecular forces are reduced almost to zero, and the change in enthalpy is much greater.

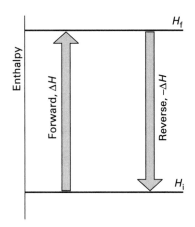

Fig. 2.14 The enthalpy change accompanying a reverse process is the negative of the enthalpy change for the forward process.

Exercise E2.6

In a similar experiment to the above, it was found that 1.36 g of boiling benzene, C_6H_6, is vaporized when a current of 0.835 A from a 12.0 V source is passed for 53.5 s. What is the enthalpy of vaporization of benzene at its boiling point?

[*Answer*: $+30.8\,\text{kJ mol}^{-1}$]

There are some striking differences in enthalpies of vaporization: although the value for water is $+44\,\text{kJ mol}^{-1}$, that for methane, CH_4, at its boiling point is only $+8\,\text{kJ mol}^{-1}$. Even allowing for the fact that vaporization is taking place at different temperatures, the difference between the enthalpies of vaporization signifies that water molecules are held together in the bulk liquid much more tightly than methane molecules are in liquid methane. (We shall see in Chapter 10 that the interaction responsible for the low volatility of water is the hydrogen bond.) The high enthalpy of vaporization of water has profound ecological consequences, for it is partly responsible for the survival of the oceans and the generally low humidity of the atmosphere. If only a small amount of heat had to be supplied to vaporize the oceans, the atmosphere would be much more heavily saturated with water vapour than is in fact the case.

Another common phase transition is **fusion**, or melting, as when ice melts to water or iron becomes molten. The enthalpy per mole of molecules that accompanies fusion is called the **enthalpy of fusion**, $\Delta_{fus}H$. Its value for water at $0\,°C$ is $+6.01\,\text{kJ mol}^{-1}$, which signifies that 6.01 kJ of energy is needed to melt $1\,\text{mol}\,H_2O(s)$ at $0\,°C$. Notice that the enthalpy of fusion of water is much less than its enthalpy of vaporization. In the latter transition, the molecules become completely separated from each other, whereas when a solid melts, the molecules are merely loosened without separating completely (Fig. 2.13).

The reverse of vaporization is **condensation** and the reverse of fusion is **freezing**. The enthalpy changes are, respectively, the negative of the enthalpies of vaporization and fusion, because the heat that is supplied to vaporize or melt the substance is released when it condenses or freezes. It is always the case that *the enthalpy change of a reverse transition is the negative of the enthalpy change of the forward transition* (under the same conditions of temperature and pressure):

$$H_2O(s) \longrightarrow H_2O(l) \qquad \Delta H = +6.01\,\text{kJ}$$
$$H_2O(l) \longrightarrow H_2O(s) \qquad \Delta H = -6.01\,\text{kJ}$$

This relation follows from the fact that H is a state property, and so it must return to the same value if a forward change is followed by the reverse of that change (Fig. 2.14). The high enthalpy of vaporization of water $(+44\,\text{kJ mol}^{-1})$, signifying a strongly endothermic process, implies that the condensation of water ($-44\,\text{kJ}\,\text{mol}^{-1}$) is a strongly exothermic process. That exothermicity is the origin of the ability of steam to scald severely, because the energy is passed on to the skin.

The direct conversion of a solid to a vapour is called **sublimation**, the reverse process is called **vapour deposition**. Sublimation can be observed on a cold, frosty morning, when frost vanishes as vapour without first melting. The frost itself forms by vapour deposition from cold, damp air. The vaporization of solid carbon dioxide ('dry ice') is another example of sublimation. The molar enthalpy change accompanying sublimation is called the **enthalpy of sublimation**, $\Delta_{\text{sub}}H$. Because enthalpy is a state property, the same change in enthalpy must be obtained both in the *direct* conversion of solid to vapour and in the *indirect* conversion, in which the solid first melts to the liquid and then that liquid vaporizes (Fig. 2.15):

$$\Delta_{\text{sub}}H = \Delta_{\text{fus}}H + \Delta_{\text{vap}}H \qquad (15)$$

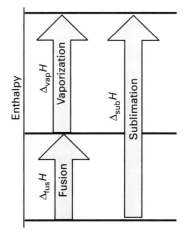

Fig. 2.15 The enthalpy of sublimation at a given temperature is the sum of the enthalpies of fusion and vaporization at that temperature. In general, the enthalpy change of an overall process is the sum of the enthalpy changes for the possibly hypothetical steps into which it may be divided.

Exercise E2.7

Calculate the enthalpy of sublimation of ice at $0\,^{\circ}\text{C}$ from its enthalpy of fusion at $0\,^{\circ}\text{C}$ ($+6.01\,\text{kJ}\,\text{mol}^{-1}$) and the enthalpy of vaporization of water at $0\,^{\circ}\text{C}$ ($+45.07\,\text{kJ}\,\text{mol}^{-1}$). Note that, as in this exercise, we must use the values of the enthalpy changes at the same temperature.

[*Answer*: $+51.08\,\text{kJ}\,\text{mol}^{-1}$]

So far we have used the fact that enthalpy is a state property in two different ways:

1. Because enthalpy is a state property, the enthalpy change in a reverse process is the negative of the enthalpy change in the forward process between the same two states:

$$\Delta H(\text{reverse}) = -\Delta H(\text{forward})$$

2. Because enthalpy is a state function, the enthalpy change in a direct route between two states is equal to the sum of the enthalpy changes for a sequence of changes between the same two states:

$$\Delta H(\text{indirect route}) = \Delta H(\text{direct route})$$

We shall build on these relations in the more complicated cases we shall now consider.

Atomic and molecular change

One group of enthalpy changes that we shall employ quite often in the following pages are those accompanying changes to individual atoms and molecules. Among the most important of these processes is the **enthalpy of ionization**, $\Delta_{ion}H$, the molar enthalpy change accompanying the removal of an electron from a gas-phase atom (or ion). For example, because

$$H(g) \longrightarrow H^+(g) + e^-(g) \qquad \Delta H = +1312\,kJ$$

the enthalpy of ionization of hydrogen atoms is reported as $+1312\,kJ\,mol^{-1}$. This value signifies that 1312 kJ of heat must be supplied to ionize 1 mol H. Ionization of neutral atoms is endothermic in all cases, so all enthalpies of ionization of neutral atoms are positive (Table 2.3).

It is often convenient to consider a succession of ionizations, such as the conversion of magnesium atoms to Mg^+ ions, and then the ionization of these Mg^+ ions to Mg^{2+} ions, and so on. The successive molar enthalpy changes are called, respectively, the **first ionization enthalpy**, the **second**

Table 2.3 First and second (and some higher) ionization enthalpies of the elements in kilojoules per mole ($kJ\,mol^{-1}$).

H							He
1310							2370
							5250
Li	Be	B	C	N	O	F	Ne
519	900	799	1090	1400	1310	1680	2080
7300	1760	2420	2350	2860	3390	3370	3950
		14 800	3660				
			25 000				
Na	Mg	Al	Si	P	S	Cl	Ar
494	736	577	786	1060	1000	1260	1520
4560	1450	1820					
	7740	2740					
		11 600					
K	Ca	Ga	Ge	As	Se	Br	Kr
418	590	577	762	966	941	1140	1350
3070	1150						
	4940						
Rb	Sr	In	Sn	Sb	Te	I	Xe
402	548	556	707	833	870	1010	1170
2650	1060						
	4120						
Cs	Ba	Tl	Pb	Bi	Po	At	Rn
376	502	812	920	1040	812	920	1040
2420	966						
3300	3390						

ionization enthalpy, and so on. For magnesium, these enthalpies refer to the processes

$$Mg(g) \longrightarrow Mg^+(g) + e^-(g) \qquad \Delta H = +738 \, kJ$$
$$Mg^+(g) \longrightarrow Mg^{2+}(g) + e^-(g) \qquad \Delta H = +1451 \, kJ$$

Note that the second ionization enthalpy is larger than the first: it takes more energy to separate an electron from a positively charged ion than from the neutral atom. Note also that enthalpies of ionization refer to the ionization of the *gas phase atom or ion*, not to the ionization of an atom or ion in a solid.

Example *Combining enthalpy changes*
The enthalpy of sublimation of magnesium at $25\,°C$ is $+148 \, kJ \, mol^{-1}$. What quantity of heat must be supplied to $1.00 \, g$ of solid magnesium metal to produce a gas composed of Mg^{2+} ions and electrons?

Strategy
The strategy is to calculate the enthalpy change for the overall process as a sum of the steps into which it can be divided (sublimation followed by the two stages of ionization). Then the heat required for the specified process is the product of the overall molar enthalpy change and the amount of atoms; the latter is calculated from the given mass and the molar mass of the substance.

Solution
The overall process is

$$Mg(s) \longrightarrow Mg^{2+}(g) + 2\,e^-(g)$$

The thermochemical equation for this process is the sum of the following thermochemical equations:

		$\Delta H/kJ$
Sublimation:	$Mg(s) \longrightarrow Mg(g)$	$+148$
First ionization:	$Mg(g) \longrightarrow Mg^+(g) + e^-(g)$	$+738$
Second ionization:	$Mg^+(g) \longrightarrow Mg^{2+}(g) + e^-(g)$	$+1451$
Overall (sum):	$Mg(s) \longrightarrow Mg^{2+}(g) + 2\,e^-(g)$	$+2337$

These processes are illustrated diagrammatically in Fig. 2.16. It follows that the overall enthalpy change per mole of Mg is $+2337 \, kJ \, mol^{-1}$. Because the molar mass of magnesium is $24.31 \, g \, mol^{-1}$, $1.0 \, g$ of magnesium corresponds to

$$n(Mg) = \frac{1.00 \, g}{24.31 \, g \, mol^{-1}} = 0.0411 \, mol$$

Therefore, the heat that must be supplied to $1.00 \, g$ of magnesium is

$$q = (0.0411 \, mol) \times (2337 \, kJ \, mol^{-1}) = +96.1 \, kJ$$

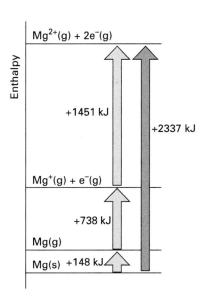

Fig. 2.16 The contributions to the enthalpy change treated in the example.

57

This quantity of heat is approximately the same as that needed to vaporize about 43 g of boiling water.

Exercise E2.8

The enthalpy of sublimation of aluminium is $+326\,\text{kJ mol}^{-1}$. Use this information and the ionization enthalpies in Table 2.3 to calculate the heat that must be supplied to convert 1.00 g of solid aluminium metal to a gas of Al^{3+} ions and electrons at 25 °C.

[*Answer*: $+203\,\text{kJ}$]

The reverse of ionization is **electron gain**, and the corresponding molar enthalpy change is called the **electron gain enthalpy**, $\Delta_{eg}H$. For example, because experiments show that

$$Cl(g) + e^-(g) \longrightarrow Cl^-(g) \qquad \Delta H = -349\,\text{kJ}$$

it follows that the electron gain enthalpy of Cl atoms is $-349\,\text{kJ mol}^{-1}$. Notice that electron gain by Cl is an *exo*thermic process, so heat is released when a Cl atom captures an electron and forms a gas phase Cl^- ion. It can be seen from Table 2.4, which lists a number of electron gain enthalpies, that

Table 2.4 Electron gain enthalpies of the main-group elements, $\Delta_{eg}H/\text{kJ mol}^{-1}$*

H							He
−72							+21
Li	Be	B	C	N	O	F	Ne
−60	+18	−28	−122	+7	−141 +844	−328	+29
Na	Mg	Al	Si	P	S	Cl	Ar
−53	+232	−44	−120	−72	−200 +532	−349	+35
K	Ca	Ga	Ge	As	Se	Br	Kr
−48	+156	−29	−117	−77	−195	−325	+39
Rb	Sr	In	Sn	Sb	Te	I	Xe
−47	+52	−29	−121	−101	−190	−295	+41

* Where two values are given, the first refers to the formation of the ion X from the neutral atom X; the second, to the formation of X^{2-} from X^-.

some electron gains are exothermic, others are endothermic. For example, electron gain by an O^- ion is strongly endothermic because it takes energy to push an electron on to an already negatively charged species:

$$O^-(g) + e^-(g) \longrightarrow O^{2-}(g) \qquad \Delta H = +844 \, kJ$$

The final atomic and molecular process that we need consider is the **dissociation**, or breaking, of a chemical bond, as in the process

$$HCl(g) \longrightarrow H(g) + Cl(g) \qquad \Delta H = +431 \, kJ$$

The corresponding molar enthalpy change is called the **bond enthalpy**, so we would report the bond enthalpy of H—Cl as $431 \, kJ \, mol^{-1}$ (because all bond enthalpies are positive, they are normally reported without the $+$ sign). Some values are given in Table 2.5. Note that the nitrogen–nitrogen bond in molecular nitrogen, N_2, is very strong, at $945 \, kJ \, mol^{-1}$, which helps to account for the chemical inertness of nitrogen and the fact that it dilutes the oxygen in the atmosphere without reacting with it. In contrast, the fluorine–fluorine bond in molecular fluorine, F_2, is relatively weak, at $155 \, kJ \, mol^{-1}$; the weakness of this bond contributes to the high reactivity of elemental fluorine. Bond enthalpies are not the full reason, because although the bond in molecular iodine is even weaker, I_2 is less reactive than F_2: the strengths of the bonds that the elements can make to *other* elements in the products of a reaction is an additional factor.

A complication when dealing with bond enthalpies is that their value depends on the molecule in which the two linked atoms occur. For instance, the total enthalpy change for the process

$$H_2O(g) \longrightarrow 2 \, H(g) + O(g) \qquad \Delta H = +927 \, kJ$$

Table 2.5 Selected bond enthalpies, $\Delta H(AB)/kJ \, mol^{-1}$

Diatomic molecules							
H—H	436	O=O	497	F—F	155	H—F	565
		N≡N	945	Cl—Cl	242	H—Cl	431
		O—H	428	Br—Br	193	H—Br	366
		C=O	1074	I—I	151	H—I	299
H—CH₃	435	H—NH₂	431			H—OH	492
H—C₆H₆	469	O₂N—NO₂	57			HO—OH	213
H₃C—CH₃	368	O=CO	531			HO—CH₃	377
H₂C=CH₂	699					Cl—CH₃	452
HC≡CH	962					Br—CH₃	293
						I—CH₃	234

is not twice an OH bond enthalpy even though two O—H bonds are dissociated. There are in fact two different dissociation steps. In the first step, an O—H bond is broken in an H_2O molecule:

$$H_2O(g) \longrightarrow HO(g) + H(g) \qquad \Delta H = +499\,\text{kJ}$$

In the second step, the O—H bond is broken in an OH radical:

$$HO(g) \longrightarrow H(g) + O(g) \qquad \Delta H = +428\,\text{kJ}$$

The sum of the two steps is the complete atomization of the molecule. As can be seen from this example, the O—H bonds in H_2O and HO have similar but not identical bond enthalpies. Although accurate calculations must use bond enthalpies for the molecule in question and its successive fragments, when such data are not available there is no choice but to make estimates by using **mean bond enthalpies**, B, which are the averages of bond enthalpies over a related series of compounds (Table 2.6). For example, the mean HO bond enthalpy, $B(\text{H—O}) = 463\,\text{kJ mol}^{-1}$, is the mean of the HO bond enthalpies in H_2O and several other similar compounds, including methanol, CH_3OH.

Example *Using mean bond enthalpies*
Estimate the enthalpy change that accompanies the reaction

$$C(s,\ \text{graphite}) + 2\,H_2(g) + \tfrac{1}{2}O_2(g) \longrightarrow CH_3OH(l)$$

Table 2.6 Mean bond enthalpies, $B\,/\text{kJ mol}^{-1}$

	H	C	N	O	F	Cl	Br	I	S	P	Si
H	436										
C	412	348 (1)									
		612 (2)									
		518 (a)									
N	388	305 (1)	163 (1)								
		613 (2)	409 (2)								
		890 (3)	945 (3)								
O	463	360 (1)	157	146 (1)							
		743 (2)		497 (2)							
F	565	484	270	185	155						
Cl	431	338	200	203	254	242					
Br	366	276				219	193				
I	299	238				210	178	151			
S	338	259			496	250	212		264		
P	322									200	
Si	318			466							226

Values are for single bonds except where otherwise stated (in parentheses).
(a) Denotes aromatic.

in which liquid methanol is formed from its elements at $25\,°C$. Use information from Appendix 1 and bond enthalpy data from Tables 2.5 and 2.6.

Strategy

In calculations of this kind, the procedure is to break the overall process down into a sequence of steps such that their sum is the chemical equation required. Always ensure, when using bond enthalpies, that all the species are in the gas phase. This means including the appropriate enthalpies of vaporization or sublimation. One approach is to atomize all the reactants and then to build the products from the atoms so produced. When explicit bond enthalpies are available (that is, data are given in the tables available), use them; otherwise, estimate atomization enthalpies from mean bond enthalpies.

Solution

The following steps are required:

		$\Delta H/\text{kJ}$
Atomization of graphite:	$C(s, \text{graphite}) \longrightarrow C(g)$	$+716.68$
Dissociation of $2\,\text{mol}\,H_2(g)$:	$2\,H_2(g) \longrightarrow 4\,H(g)$	$+871.88$
Dissociation of $\frac{1}{2}\,\text{mol}\,O_2(g)$:	$\frac{1}{2}\,O_2(g) \longrightarrow O(g)$	$+249.17$
	Overall:	$+1837.73$

At this stage, we have atoms of carbon, hydrogen, and oxygen. In the second step, three CH bonds, one CO bond, and one OH bond are formed, and we can estimate their enthalpies from the mean values of such bonds. The enthalpy change for bond formation (the reverse of dissociation) is the negative of the mean bond enthalpy B (obtained from Table 2.6):

	$\Delta H/\text{kJ}$
Formation of 3 C—H bonds:	-1236
Formation of 1 C—O bond:	-360
Formation of 1 O—H bond:	-463
Overall:	-2059

The final stage of the reaction is the condensation of methanol vapour:

$$CH_3OH(g) \longrightarrow CH_3OH(l) \qquad \Delta H = -38.00\,\text{kJ}$$

The sum of the enthalpy changes is

$$\Delta H = (+1837.73\,\text{kJ}) + (-2059\,\text{kJ}) + (-38.00\,\text{kJ}) = -259\,\text{kJ}$$

The experimental value is $-239.00\,\text{kJ}$.

Exercise E2.9

Estimate the enthalpy change for the combustion of liquid ethanol to carbon dioxide and liquid water by using the enthalpies of atomization of $CO_2(g)$ and $H_2O(g)$, which are $1609\,kJ\,mol^{-1}$ and $920\,kJ\,mol^{-1}$, respectively, and mean bond enthalpies for atomization of the alcohol.

[*Answer*: $-1348\,kJ$; the experimental value is $-1368\,kJ$]

2.7 The enthalpy of chemical change

In the remainder of this chapter we shall be considering **thermochemistry**, the study of the heat required or absorbed by chemical reactions, such as the heat output of a combustion or respiration reaction. This information is essential if we want to keep track of energy changes in reactions, such as in organisms or factories, and will also prove to be essential information for assessing the equilibrium composition of chemical reactions. The central property in all the following discussions will be the enthalpy change accompanying a chemical reaction, as in the hydrogenation reaction in which hydrogen is added to ethene,

$$CH_2\!\!=\!\!CH_2(g) + H_2(g) \longrightarrow CH_3CH_3(g) \qquad \Delta H = -137\,kJ \text{ (at } 25\,^{\circ}C)$$

The value of ΔH given here signifies that the enthalpy of the system decreases by $137\,kJ$ (and, if the reaction takes place at constant pressure, that $137\,kJ$ of heat is released into the surroundings) when $1\,mol\,CH_2CH_2$ combines with $1\,mol\,H_2$ at $25\,^{\circ}C$.

The numerical value of the reaction enthalpy depends on the conditions—the state of purity of the reactants and products, the pressure, and the temperature—under which the reaction takes place. Chemists have therefore found it convenient to report their data for a set of standard conditions at the temperature of their choice:

The **standard state** of a substance is the pure substance at exactly 1 bar.

(Remember that $1\,bar = 10^5\,Pa$ exactly. Solutions are a special case, and are dealt with in Section 4.3.) For example, the standard state of hydrogen is the pure gas at 1 bar and the standard state of calcium carbonate is the pure solid at 1 bar. The physical state needs to be specified because we can speak of the standard states of the solid, liquid, or vapour phases of methanol, which are the pure solid, the pure liquid, or the pure vapour, respectively, at 1 bar in each case. The temperature is not a part of the definition of a standard state, and it is possible to speak of the standard state of hydrogen gas at $100\,K$, $273.15\,K$, or any other temperature. It is conventional, though, for data to be reported at $298.15\,K$ ($25.00\,^{\circ}C$), and from now on, unless specified otherwise, all data will be for that temperature.

When we write $\Delta H^{\ominus}$ in a thermochemical equation, we shall always mean the change in enthalpy that occurs when the reactants in their standard states change into products in their standard states. For example, from the thermochemical equation

$$2\,H_2(g) + O_2(g) \longrightarrow 2\,H_2O(l) \qquad \Delta H^{\ominus} = -572\,kJ$$

we know that when $2\,mol\,H_2$ as pure hydrogen gas at 1 bar combines with $1\,mol\,O_2$ as pure oxygen gas at 1 bar to form $2\,mol\,H_2O$ as pure liquid water at 1 bar, the initial and final states being at $25\,°C$, then the enthalpy of the system decreases by 572 kJ, and (at constant pressure) 572 kJ of heat is released into the surroundings.

One commonly encountered reaction is **combustion**, the reaction of a compound (most commonly an organic compound) with oxygen, as in the combustion of methane in a natural gas flame:

$$CH_4(g) + 2\,O_2(g) \longrightarrow CO_2(g) + 2\,H_2O(l) \qquad \Delta H^{\ominus} = -890\,kJ$$

The **standard enthalpy of combustion**, $\Delta_{com}H^{\ominus}$, is the standard change in enthalpy per mole of combustible substance. In this example, we would write $\Delta_{com}H^{\ominus}(CH_4, g) = -890\,kJ\,mol^{-1}$. Some typical values are given in Table 2.7. Note that $\Delta_{com}H^{\ominus}$ is a *molar* quantity, and is obtained from the value of $\Delta H^{\ominus}$ by dividing by the amount of reactant consumed (in this case, by $1\,mol\,CH_4$).

One application of enthalpies of combustion is to judge the suitability of a fuel. For example, from the value of the standard enthalpy of combustion for methane, we know that for each mole of CH_4 supplied to a furnace, 890 kJ of heat can be released, whereas for each mole of iso-octane (C_8H_{18}, 2,2,4-trimethylpentane (**1**), a typical component of gasoline) supplied to an internal combustion engine, 5471 kJ of heat is released (see the data in Table 2.7). The much large value for iso-octane is a consequence of each molecule having eight C atoms to contribute to the formation of carbon dioxide whereas methane has only one.

The heating power of fuels is often expressed in terms of the **specific enthalpy**, the heat released divided by the mass of compound (typically in kilojoules per gram, $kJ\,g^{-1}$, Table 2.8). The easiest way to compute specific enthalpies is to divide the standard enthalpy of combustion (a molar

Table 2.7 Standard enthalpies of combustion

Substance*	Formula	$\Delta_{com}H^{\ominus}/$ $kJ\,mol^{-1}$
Benzene	$C_6H_6(l)$	−3268
Carbon	$C(s, graphite)$	−394
Carbon monoxide	$CO(g)$	−394
Ethanol	$C_2H_5OH(l)$	−1368
Ethyne	$C_2H_2(g)$	−1300
Glucose	$C_6H_{12}O_6(s)$	−2808
Hydrogen	$H_2(g)$	−286
Methane	$CH_4(g)$	−890
Methanol	$CH_3OH(l)$	−726
Octane	$C_8H_{18}(l)$	−5471
Propane	$C_3H_8(g)$	−2220
Sucrose	$C_{12}H_{22}O_{11}(s)$	−5645
Toluene	$C_6H_5CH_3(l)$	−3910
Urea	$CO(NH_2)_2(s)$	−632

* C is converted to $CO_2(g)$, H to $H_2O(l)$, and N to $N_2(g)$.

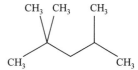

1 2,2,4−trimethylpentane

Table 2.8 Thermochemical properties of some fuels

Fuel	Combustion equation	$\Delta_{com}H^{\ominus}/$ $kJ\,mol^{-1}$	Specific enthalpy/ $kJ\,g^{-1}$	Enthalpy density*/ $kJ\,L^{-1}$
Hydrogen	$2\,H_2(g) + O_2(g)$ $\rightarrow 2\,H_2O(l)$	−286	142	13
Methane	$CH_4(g) + 2\,O_2(g)$ $\rightarrow CO_2(g) + 2\,H_2O(l)$	−890	55	40
Octane	$2\,C_8H_{18}(l) + 25\,O_2(g)$ $\rightarrow 16\,CO_2(g) + 18\,H_2O(l)$	−5471	48	3.8×10^4
Methanol	$2\,CH_3OH(l) + 3\,O_2(g)$ $\rightarrow 2\,CO_2(g) + 4\,H_2O(l)$	−726	23	1.8×10^4

* At atmospheric pressures and room temperature.

quantity) by the molar mass (another molar quantity) of the compound. For methane, for instance, the specific enthalpy is

$$\frac{890\,\text{kJ mol}^{-1}}{16.04\,\text{g mol}^{-1}} = 55.5\,\text{kJ g}^{-1}$$

(All specific enthalpies are positive, and are typically written without the $+$ sign.) Likewise, from the enthalpy of combustion of iso-octane and its molar mass of $114.23\,\text{g mol}^{-1}$ we find a specific enthalpy of $47.8\,\text{kJ g}^{-1}$. Therefore, on a mass basis, it is more economical to carry fuel as methane than as iso-octane. The specific enthalpy of hydrogen ($142\,\text{kJ g}^{-1}$) is much larger than for methane. This high value is one of the reasons why liquid hydrogen is adopted as a rocket fuel, where mass is of critical importance.

The combination of reaction enthalpies

It is often the case that a reaction enthalpy is needed but is not available in tables of data. Now the fact that the enthalpy is a state function comes in handy, because it implies that we can construct the required reaction enthalpy from the reaction enthalpies of known reactions. We have already seen a primitive example when we calculated the enthalpy of sublimation from the sum of the enthalpies of fusion and vaporization. The only difference is that we now apply the technique to a series of chemical reactions. The procedure is summarized as follows:

> **Hess's law:** The standard enthalpy of a reaction is the sum of the standard enthalpies of the reactions into which the overall reaction may be divided.

Although the procedure is given the status of a law, it hardly deserves the title because it is nothing other than a direct consequence of enthalpy being a state function, so an overall enthalpy change can be expressed as a sum of enthalpy changes for each step in an indirect path. The individual steps need not be actual reactions that can be carried out in the laboratory—they may be entirely hypothetical reactions, the only requirement being that their equations should balance.

Example *Using Hess's law*
Given the thermochemical equations

$$C_3H_6(g) + H_2(g) \longrightarrow C_3H_8(g) \qquad \Delta H^{\ominus} = -124\,\text{kJ}$$
$$C_3H_8(g) + 5\,O_2(g) \longrightarrow 3\,CO_2(g) + 4\,H_2O(l) \qquad \Delta H^{\ominus} = -2220\,\text{kJ}$$

where C_3H_6 is propene and C_3H_8 is propane, calculate the standard enthalpy of combustion of propene.

Strategy

We need to add or subtract the thermochemical equations (including the values of the enthalpy change), together with any others that are needed (from Appendix 1), so as to reproduce the thermochemical equation for the reaction required. In calculations of this type, it is quite common to need to use the synthesis of water to balance the hydrogen or oxygen atoms in the overall equation.

Solution

The overall reaction is

$$C_3H_6(g) + \tfrac{9}{2}O_2(g) \longrightarrow 3\,CO_2(g) + 3\,H_2O(l) \qquad \Delta H^{\ominus}$$

This thermochemical equation can be recreated from the following sum:

	$\Delta H^{\ominus}/kJ$
$C_3H_6(g) + H_2(g) \longrightarrow C_3H_8(g)$	-124
$C_3H_8(g) + 5\,O_2(g) \longrightarrow 3\,CO_2(g) + 4\,H_2O(l)$	-2220
$H_2O(l) \longrightarrow H_2(g) + \tfrac{1}{2}O_2(g)$	$+286$
Overall:	
$C_3H_6(g) + \tfrac{9}{2}O_2(g) \longrightarrow 3\,CO_2(g) + 3\,H_2O(l)$	-2058

It follows that the standard enthalpy of combustion of propene is $-2058\ \text{kJ mol}^{-1}$.

Exercise E2.10

Calculate the standard enthalpy of $C_6H_6(l) + 3\,H_2(g) \longrightarrow C_6H_{12}(l)$ from the standard enthalpies of combustion of benzene and cyclohexane.

[*Answer*: $-205\,\text{kJ}$]

Standard enthalpies of formation

The fact that standard reaction enthalpies can be constructed from other reaction enthalpies opens up the route to a very economical way of tabulating standard enthalpies and of calculating the values for unknown reactions. Any reaction can be considered to take place in two possibly hypothetical steps: breaking the reactants down into their elements, then building up the products from those elements (Fig. 2.17). The first of these steps is the reverse of building up the reactants from their elements. So it follows that any reaction can be expressed in terms of the reaction enthalpies for building

Fig. 2.17 An enthalpy of reaction may be expressed as the difference between the enthalpies of atomization of the products and the reactants. These enthalpies of atomization may be available in tables; if not, they may be estimated by using mean bond enthalpies. If the actual reactants or products are in a condensed phase, the enthalpies of sublimation or vaporization must be included to obtain the overall reaction enthalpy.

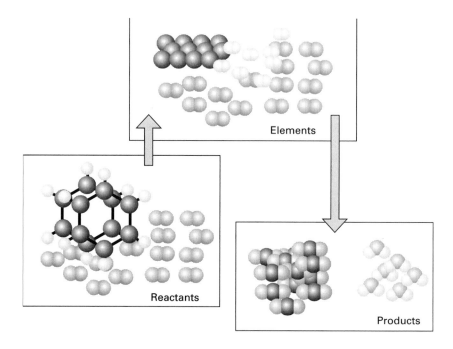

Elements

Reactants

Products

up substances from their elements. These reaction enthalpies are key quantities in thermochemistry. Specifically:

> The **standard enthalpy of formation**, $\Delta_f H^\ominus$, of a substance is the standard enthalpy (per mole of the substance) for its formation from its elements in their reference states.

By the **reference state** of an element is meant the most stable form of the element under the prevailing conditions. The reference states of some common elements at 25 °C are listed in Table 2.9. For example, the standard enthalpy of formation of liquid water (at 25 °C, as always in this text) is obtained from the thermochemical equation

$$H_2(g) + \tfrac{1}{2} O_2(g) \longrightarrow H_2O(l) \qquad \Delta H^\ominus = -286 \, \text{kJ}$$

and is $\Delta_f H^\ominus(H_2O, l) = -286 \, \text{kJ mol}^{-1}$. Note that enthalpies of formation are molar quantities, so to go from $\Delta H^\ominus$ in a thermochemical equation to $\Delta_f H^\ominus$ for that substance, we have to divide by the amount of substance formed (in this instance, by 1 mol H_2O). Similarly, the standard enthalpy of formation of liquid carbon disulfide is obtained from the experimentally determined thermochemical equation

$$C(s, \text{graphite}) + 2\, S(s, \text{rhombic}) \longrightarrow CS_2(l) \qquad \Delta H^\ominus = +90 \, \text{kJ}$$

and so $\Delta_f H^\ominus(CS_2, l) = +90 \, \text{kJ mol}^{-1}$. The values of some standard enthalpies of formation at 25 °C are given in Table 2.10, and a longer list is given in Appendix 1 at the end of the book. The standard enthalpies of formation of

Table 2.9 Reference states of some elements

Element	Reference state
Arsenic	grey arsenic
Bromine	liquid
Carbon	graphite
Hydrogen	gas
Iodine	solid
Mercury	liquid
Nitrogen	gas
Oxygen	gas
Phosphorus	white phosphorus
Sulfur	rhombic sulfur
Tin	white tin

Table 2.10 Standard enthalpies of formation at 25°C

Substance*	Formula	$\Delta_f H^{\ominus}/\text{kJ mol}^{-1}$
Inorganic compounds		
Ammonia	$NH_3(g)$	−46.11
Ammonium nitrate	$NH_4NO_3(s)$	−365.56
Carbon monoxide	$CO(g)$	−110.53
Carbon disulfide	$CS_2(l)$	+89.70
Carbon dioxide	$CO_2(g)$	−393.51
Dinitrogen tetroxide	$N_2O_4(g)$	+9.16
Dinitrogen oxide	$N_2O(g)$	+82.05
Hydrogen chloride	$HCl(g)$	−92.31
Hydrogen fluoride	$HF(g)$	−271.1
Hydrogen sulfide	$H_2S(g)$	−20.63
Nitric acid	$HNO_3(l)$	−174.10
Nitric oxide	$NO(g)$	+90.25
Nitrogen dioxide	$NO_2(g)$	+33.18
Sodium chloride	$NaCl(s)$	−411.15
Sulfur dioxide	$SO_2(g)$	−296.83
Sulfur trioxide	$SO_3(g)$	−395.72
Sulfuric acid	$H_2SO_4(l)$	−813.99
Water	$H_2O(l)$	−285.83
	$H_2O(g)$	−241.82
Organic compounds		
Benzene	$C_6H_6(l)$	+49.0
Ethane	$C_2H_6(g)$	−84.68
Ethanol	$C_2H_5OH(l)$	−277.69
Ethene	$C_2H_4(g)$	+52.26
Ethyne	$C_2H_2(g)$	+226.73
Glucose	$C_6H_{12}O_6(s)$	−1268
Methane	$CH_4(g)$	−74.81
Methanol	$CH_3OH(l)$	−238.86
Sucrose	$C_{12}H_{22}O_{11}(s)$	−2222

* A longer list is given in Appendix 1 at the end of the book.

elements in their reference states are zero by definition (because their formation is the null reaction: element $\longrightarrow$ element).

The reference states of the elements define a thermochemical 'sea level', and the enthalpies of formation can be regarded as thermochemical 'altitudes' above or below sea level. Compounds that have negative standard enthalpies of formation (such as water) are classified as **exothermic compounds**, for they lie at a lower enthalpy than their component elements (they lie below sea level). Compounds that have positive standard enthalpies of formation (such as carbon disulfide) are classified as **endothermic compounds**, and possess a higher enthalpy than their component elements (they lie above sea level).

As we have remarked, the great usefulness of standard enthalpies of formation is that they can be combined together to determine the standard enthalpy change of any reaction. We shall discuss reactions in terms of a special quantity called the standard reaction enthalpy, which is defined as follows:

The **standard reaction enthalpy**, $\Delta_r H^\ominus$, is the difference between the standard enthalpies of formation of the reactants and the products, with each term weighted by the stoichiometric number in the chemical equation.

For example, the reaction between hydrazoic acid, HN_3, and nitrogen monoxide, NO (nitric oxide), is summarized by the following equation:

$$2\,HN_3(l) + 2\,NO(g) \longrightarrow H_2O_2(l) + 4\,N_2(g)$$

To formulate its standard reaction enthalpy we write

$$\begin{aligned}
\Delta_r H^\ominus &= \Delta_f H^\ominus(H_2O_2, l) + 4\Delta_f H^\ominus(N_2, g) \\
&\quad - \{2\Delta_f H^\ominus(HN_3, l) + 2\Delta_f H^\ominus(NO, g)\} \\
&= \{(-187.78\,\text{kJ mol}^{-1}) + 4 \times 0\} - \{2 \times (264.0\,\text{kJ mol}^{-1}) \\
&\quad + 2 \times (90.25\,\text{kJ mol}^{-1})\} \\
&= -896.3\,\text{kJ mol}^{-1}
\end{aligned}$$

Note that because the units of standard enthalpy of formation are kilojoules per mole, the same is true of the standard reaction enthalpy. The calculation we have just illustrated has the form

$$\Delta_r H^\ominus = \sum_{\text{Products}} n\Delta_f H^\ominus(\text{products}) - \sum_{\text{Reactants}} n\Delta_f H^\ominus(\text{reactants}) \quad (16)$$

where the n are the stoichiometric numbers in the chemical equation for the reaction.

Example *Using standard enthalpies of formation*
Calculate the standard enthalpy of combustion of liquid benzene from the standard enthalpies of formation of the reactants and products.

Strategy
We write the chemical equation, identify the stoichiometric numbers of the reactants and products, and then use eqn (16). Note that the expression has the form 'products − reactants'. The numerical values of the standard enthalpies of formation are available in Appendix 1. The standard enthalpy of combustion is the enthalpy change per mole of substance, so we need to interpret the enthalpy change accordingly.

Solution
The chemical equation is

$$C_6H_6(l) + \tfrac{15}{2}O_2(g) \longrightarrow 6\,CO_2(g) + 3\,H_2O(l)$$

It follows that

$$\begin{aligned}
\Delta_r H^\ominus &= \{6\Delta_f H^\ominus(CO_2,g) + 3\Delta_f H^\ominus(H_2O,l)\} \\
&\quad - \{\Delta_f H^\ominus(C_6H_6,l) + \tfrac{15}{2}\Delta_f H^\ominus(O_2,g)\} \\
&= \{6 \times (-393.51\,\text{kJ mol}^{-1}) + 3 \times (-285.83\,\text{kJ mol}^{-1})\} \\
&\quad - \{(49.0\,\text{kJ mol}^{-1}) + 0\} \\
&= -3268\,\text{kJ mol}^{-1}
\end{aligned}$$

Inspection of the chemical equation shows that, in this instance, the 'per mole' is per mole of C_6H_6, which is exactly what we need for an enthalpy of combustion. It follows that the standard enthalpy of combustion of liquid benzene is $-3268\,\text{kJ mol}^{-1}$.

Exercise E2.11

Use standard enthalpies of formation to calculate the enthalpy of combustion of propane gas to carbon dioxide and water vapour.

[*Answer*: $-2220\,\text{kJ mol}^{-1}$]

2.8 The variation of enthalpy with temperature

It often happens that we have data at one temperature but need it at another temperature. For example, we might want to know the enthalpy of a particular reaction at body temperature, 37 °C, but may have data available for 25 °C. Another type of question that might arise might be whether the oxidation of glucose is more exothermic when it takes place inside an Arctic fish that inhabits water at 0 °C than when it takes place at mammalian body temperatures. Similarly, is it possible to suggest whether the synthesis of ammonia is more exothermic at a typical industrial temperature of 450 °C when we have data available only at 25 °C. In precise work, every attempt would be made to measure the reaction enthalpy at the temperature of interest, but it is useful to have a 'back-of-the-envelope' way of estimating the direction of change and even a moderately reliable numerical value. As we shall now see, we have enough information available to consider problems like these.

As a simple example of this type of calculation, consider the reaction

$$2\,H_2(g) + O_2(g) \longrightarrow 2\,H_2O(l)$$

where the standard enthalpy of reaction is known at one temperature (for example, at $25\,°C$ from the tables in this book). The standard enthalpy of reaction can be expressed as the difference in the molar enthalpies $H_m^\ominus$ (not, at this stage, the standard enthalpies of formation) of each substance:

$$\Delta_r H^\ominus = 2H_m^\ominus(H_2O,l) - \left\{2H_m^\ominus(H_2,g) + H_m^\ominus(O_2,g)\right\}$$

If the reaction takes place at a higher temperature, the molar enthalpy of each substance is increased because it stores more energy and the standard reaction enthalpy becomes

$$\Delta_r H^{\ominus\prime} = 2H_m^{\ominus\prime}(H_2O,l) - \left\{2H_m^{\ominus\prime}(H_2,g) + H_m^{\ominus\prime}(O_2,g)\right\}$$

where the prime signifies the value at the new temperature. The increase in molar enthalpy of a substance is given by eqn (13) as $C_{p,m}\Delta T$, where $C_{p,m}$ is the molar constant-pressure heat capacity of the substance and ΔT is the difference in temperature. For example, the molar enthalpy of water changes to

$$H_m^{\ominus\prime}(H_2O,l) = H_m^\ominus(H_2O,l) + C_{p,m}(H_2O,l) \times \Delta T$$

When terms like this are substituted into the expression above, we find

$$\Delta_r H^{\ominus\prime} = \Delta_r H^\ominus + \Delta_r C_p \times \Delta T \tag{17}$$

where

$$\Delta_r C_p = 2C_{p,m}(H_2O,l) - \left\{2C_{p,m}(H_2,g) + C_{p,m}(O_2,g)\right\}$$

Note that this combination has the same pattern as the reaction enthalpy, and the stoichiometric numbers occur in the same way. In general, $\Delta_r C_p$ is the difference between the weighted sums of the molar heat capacities of the products and the reactants:

$$\Delta_r C_p = \sum_{\text{Products}} nC_{p,m}(\text{products}) - \sum_{\text{Reactants}} nC_{p,m}(\text{reactants}) \tag{18}$$

The relation between reaction enthalpies that we require is given by eqn (17), which is known as **Kirchhoff's law**. We see that the standard reaction enthalpy at one temperature can be calculated from the standard reaction enthalpy at another temperature provided we know the molar constant-pressure heat capacities of all the substances: these values are also given in Appendix 1 for a range of substances. The derivation of Kirchhoff's law supposes that the heat capacities are constant over the range of temperature of interest, so the law is best restricted to small temperature differences (of no more than $100\,K$ or so).

Example *Using Kirchhoff's law*

The standard enthalpy of formation of gaseous water at 25 °C is $-241.82 \, \text{kJ mol}^{-1}$. Estimate its value at 100 °C.

Strategy

First, we write the chemical equation and identify the stoichiometric numbers. Then we calculate the value of $\Delta_r C_p$ from the data in Appendix 1 by using eqn (18), and use the result in eqn (17).

Solution

The chemical equation is

$$H_2(g) + \tfrac{1}{2}O_2(g) \longrightarrow H_2O(g)$$

and the molar constant-pressure heat capacities of $H_2O(g)$, $H_2(g)$, and $O_2(g)$ are as $33.58 \, \text{J K}^{-1} \text{mol}^{-1}$, $28.84 \, \text{J K}^{-1} \text{mol}^{-1}$, and $29.37 \, \text{J K}^{-1} \text{mol}^{-1}$, respectively. It follows that

$$
\begin{aligned}
\Delta_r C_p &= C_{p,\text{m}}(H_2O, g) - \left\{ C_{p,\text{m}}(H_2, g) + \tfrac{1}{2} C_{p,\text{m}}(O_2, g) \right\} \\
&= (33.58 \, \text{J K}^{-1} \text{mol}^{-1}) \\
&\quad - \left\{ (28.84 \, \text{J K}^{-1} \text{mol}^{-1}) + \tfrac{1}{2} \times (29.37 \, \text{J K}^{-1} \text{mol}^{-1}) \right\} \\
&= -9.95 \, \text{J K}^{-1} \text{mol}^{-1}
\end{aligned}
$$

Then, because $\Delta T = +75 \, \text{K}$, it follows from eqn (17) that

$$
\begin{aligned}
\Delta_r H^{\ominus\prime} &= (-241.82 \, \text{kJ mol}^{-1}) + (-9.95 \, \text{J K}^{-1} \text{mol}^{-1}) \times (75 \, \text{K}) \\
&= (-241.82 \, \text{kJ mol}^{-1}) - (0.75 \, \text{kJ mol}^{-1}) = -242.57 \, \text{kJ mol}^{-1}
\end{aligned}
$$

We see that the reaction is slightly more exothermic at the higher temperature.

Exercise E2.12

Estimate the standard enthalpy of formation of $NH_3(g)$ at 400 K from the data in Appendix 1.

[*Answer*: $-48.4 \, \text{kJ mol}^{-1}$]

The calculation in the example shows that the standard reaction enthalpy at 100 °C is only slightly different from the value at 25 °C. The reason is that the change in enthalpy of reaction is proportional to the *difference* between the molar heat capacities of the products and the reactants, which is usually not very large. It is generally the case that enthalpies of reactions vary only slightly with temperature (over small temperature ranges) because the heat capacities of the products differ only a little from those of the reactants. A reasonable first approximation in thermochemistry (as we shall see in later chapters), is that the standard reaction enthalpy may be assumed to be independent of temperature.

Exercises

Assume all gases are perfect unless stated otherwise. All thermochemical data are for 298 K.

2.1 Calculate the work that a person must do to raise a mass of 1.0 kg through 10 m on the surface of (a) the Earth ($g = 9.81\,\mathrm{m\,s^{-2}}$) and (b) the Moon ($g = 1.60\,\mathrm{m\,s^{-2}}$).

2.2 When we are interested in biological energy resources and metabolism, we need to know, among other things, the work that an organism has to do to carry out the normal activities of being alive. Calculate the work needed for a person of mass 65 kg to climb through 4.0 m on the surface of the Earth.

2.3 The centre of mass of a cylindrical column of liquid lies half-way along its length. Calculate the work required to raise a column of mercury (density $13.6\,\mathrm{g\,cm^{-3}}$) of diameter 1.00 cm through 760 mm on the surface of the Earth ($g = 9.81\,\mathrm{m\,s^{-2}}$).

2.4 We are all familiar with the general principles of operation of an internal combustion reaction: the combustion of fuel drives out the piston. It is possible to imagine engines that use reactions other than combustions, and we need to assess the work they can do. A chemical reaction takes place in a container of cross-sectional area $100\,\mathrm{cm^2}$; the container has a piston at one end. As a result of the reaction, the piston is pushed out through 10.0 cm against an external pressure of 100 kPa. Calculate the work done by the system.

2.5 The work done by an engine may depend on its orientation in a gravitational field, because the mass of the piston is relevant when the expansion is vertical. A chemical reaction takes place in a container of cross-sectional area $55.0\,\mathrm{cm^2}$; the container has a piston of mass 250 g at one end. As a result of the reaction, the piston is pushed out (a) horizontally, (b) vertically through 155 cm against an external pressure of 105 kPa. Calculate the work done by the system in each case.

2.6 A sample of methane of mass 4.50 g occupies 12.7 L at 310 K. (a) Calculate the work done when the gas expands isothermally against a constant external pressure of 200 Torr until its volume has increased by 3.3 L. (b) Calculate the work that

would be done if the same expansion occurred isothermally and reversibly.

2.7 In the isothermal reversible compression of 52.0 mmol of a perfect gas at 260 K, the volume of the gas is reduced from 300 mL to 100 mL. Calculate w for this process.

2.8 A sample of blood plasma occupies 0.550 L at 0 °C and 1.03 bar, and is compressed isothermally by 0.57 per cent by being subjected to a constant external pressure of 95.2 bar. Calculate w.

2.9 A strip of magnesium metal of mass 12.5 g is dropped into a beaker of dilute hydrochloric acid. Given that the magnesium is the limiting reactant, calculate the work done by the system as a result of the reaction. The atmospheric pressure is 1.00 atm and the temperature 20.2 °C.

2.10 Liquid mixtures of sodium and potassium are used in some nuclear reactors as coolants that can survive the intense radiation inside reactor cores. Calculate the heat required to melt 224 kg of sodium metal at 371 K.

2.11 A primitive air-conditioning unit for use in places where electrical power is unavailable can be made by hanging up strips of linen soaked in water: the evaporation of the water cools the air. Calculate the heat required to evaporate 1.00 kg of water at (a) 25 °C, (b) 100 °C.

2.12 What is the heat capacity of a liquid that rose in temperature by 5.23 °C when supplied with 124 J of heat?

2.13 The high heat capacity of water is ecologically benign because it stabilizes the temperatures of lakes and oceans: a large quantity of energy must be lost or gained before there is a significant change in temperature. Conversely, it means that a lot of heat must be supplied to achieve a large rise in temperature. The molar heat capacity of water is $75.3\,\mathrm{J\,K^{-1}\,mol^{-1}}$. What energy is needed to heat 250 g of water (a cup of coffee, for instance) through 40 °C?

2.14 Use the information in Tables 2.1 and 2.2 to calculate the total heat required to melt 100 g of ice at 0 °C, heat it to 100 °C, and then vaporize it at that temperature. Sketch a graph of temperature against time on the assumption that heat is supplied to the sample at a constant rate.

2.15 When 229 J of energy is supplied as heat to 3.00 mol Ar(g), the temperature of the sample increases by 2.55 K. Calculate the molar heat capacities at constant volume and constant pressure of the gas.

2.16 The heat capacity of air is much smaller than that of water, and relatively modest amounts of heat are needed to change its temperature. This is one of the reasons why desert regions, though very hot during the day, are bitterly cold at night. The heat capacity of air at room temperature and pressure is approximately 21 J K^{-1} mol^{-1}. How much energy is required to raise the temperature of a room of dimensions 5.5 m × 6.5 m × 3.0 m by 10 °C? If losses are neglected, how long will it take a heater rated at 1.5 kW to achieve that increase given that 1 W = 1 J s^{-1}?

2.17 A sample of a serum of mass 25 g is cooled from 290 K to 275 K at constant pressure by the extraction of 1.2 kJ of energy as heat. Calculate q and ΔH and estimate the heat capacity of the sample.

2.18 When 3.0 mol O$_2$(g) is heated at a constant pressure of 3.25 atm, its temperature increases from 260 K to 285 K. Given that the molar heat capacity of O$_2$ at constant pressure is 29.4 J K^{-1} mol^{-1}, calculate q, ΔH, and ΔU.

2.19 Refrigerators make use of the heat absorption required to vaporize a volatile liquid. A fluorocarbon liquid being investigated to replace a chlorofluorocarbon has $\Delta_{vap}H^{\ominus} = +26.0$ kJ mol^{-1}. Calculate q, w, ΔH, and ΔU when 1.50 mol is vaporized at 250 K and 750 Torr.

2.20 The efficient design of chemical plants depends on the designer's ability to assess and use the heat output in one process to supply another process. The standard enthalpy of reaction for N$_2$(g) + 3 H$_2$(g) $\longrightarrow$ 2 NH$_3$(g) is -92.22 kJ mol^{-1}. What is the change in enthalpy when (a) 1.00 mol N$_2$ is consumed, (b) 1.00 mol NH$_3$(g) is formed?

2.21 Ethane is flamed off in abundance from oil wells, because it is unreactive and difficult to use commercially. But would it make a good fuel? The standard enthalpy of reaction for 2 C$_2$H$_6$(g) + 7 O$_2$(g) $\longrightarrow$ 4 CO$_2$(g) + 6 H$_2$O(l) is -3120 kJ mol^{-1}. (a) What is the standard enthalpy of combustion of ethane? (b)

What is the change in enthalpy when 3.00 mol CO$_2$ is formed in the reaction?

2.22 Standard enthalpies of formation are widely available, but we might need a standard enthalpy of combustion instead. The standard enthalpy of formation of ethylbenzene is -12.5 kJ mol^{-1}. Calculate its standard enthalpy of combustion.

2.23 Combustion reactions are relatively easy to carry out and study, and their data can be combined to give enthalpies of other types of reaction. As an illustration, calculate the standard enthalpy of hydrogenation of cyclohexene to cyclohexane given that the standard enthalpies of combustion of the two compounds are -3752 kJ mol^{-1} (cyclohexene) and -3953 kJ mol^{-1} (cyclohexane).

2.24 Estimate the standard internal energy of formation of liquid methyl acetate (methyl ethanoate, CH$_3$COOCH$_3$) at 298 K from its standard enthalpy of formation, which is -442 kJ mol^{-1}.

2.25 The standard enthalpy of combustion of naphthalene is -5157 kJ mol^{-1}. Calculate its standard enthalpy of formation.

2.26 The temperature of a bomb calorimeter rose by 1.617 K when a current of 3.20 A was passed for 27.0 s from a 12.0 V source. Calculate the heat capacity of the calorimeter.

2.27 When 320 mg of naphthalene, C$_{10}$H$_8$(s), was burned in a bomb calorimeter, the temperature rose by 3.05 K. Calculate the heat capacity of the calorimeter. By how much will the temperature rise when 100 mg of phenol, C$_6$H$_5$OH(s), is burned in the calorimeter under the same conditions?

2.28 The energy resources of glucose are of major concern for the assessment of metabolic processes. When 0.3212 g of glucose was burned in a bomb calorimeter of heat capacity 641 J K^{-1} the temperature rose by 7.793 K. Calculate (a) the standard molar enthalpy of combustion, (b) the standard internal energy of combustion, and (c) the standard enthalpy of formation of glucose.

2.29 Calculate the standard enthalpy of solution of AgBr(s) in water from the standard enthalpies of formation of the solid and the aqueous ions.

2.30 The standard enthalpy of decomposition of the yellow complex NH$_3$SO$_2$ into NH$_3$ and SO$_2$ is $+40$ kJ mol^{-1}. Calculate the standard enthalpy of formation of NH$_3$SO$_2$.

2.31 Given that the enthalpy of combustion of graphite is $-393.5\,\mathrm{kJ\,mol^{-1}}$ and that of diamond is $-395.41\,\mathrm{kJ\,mol^{-1}}$, calculate the enthalpy of the $C(\mathrm{s,graphite}) \longrightarrow C(\mathrm{s,diamond})$ transition.

2.32 The pressures deep within the Earth are much greater than those on the surface, and to make use of thermochemical data in geochemical assessments we need to take the differences into account. Use the information in Exercise 2.31 together with the densities of graphite ($2.250\,\mathrm{g\,cm^{-3}}$) and diamond ($3.510\,\mathrm{g\,cm^{-3}}$) to calculate the internal energy of the transition when the sample is under a pressure of $150\,\mathrm{kbar}$.

2.33 The mass of a typical sugar (sucrose) cube is $1.5\,\mathrm{g}$. Calculate the energy released as heat when a cube is burned in air. To what height could a person of mass $68\,\mathrm{kg}$ climb on the energy a cube provides assuming 20 per cent of the energy is available for work?

2.34 Camping gas is typically propane. The standard enthalpy of combustion of propane gas is $-2220\,\mathrm{kJ\,mol^{-1}}$ and the standard enthalpy of vaporization of the liquid is $+15\,\mathrm{kJ\,mol^{-1}}$. Calculate (a) the standard enthalpy and (b) the standard internal energy of combustion of the liquid.

2.35 Classify as endothermic or exothermic (a) a combustion reaction for which $\Delta_r H^{\ominus} = -2020\,\mathrm{kJ\,mol^{-1}}$, (b) a dissolution for which $\Delta H^{\ominus} = +4.0\,\mathrm{kJ\,mol^{-1}}$, (c) vaporization, (d) fusion, (e) sublimation.

2.36 Standard enthalpies of formation are of great usefulness, for they can be used to calculate the standard enthalpies of a very wide range of reactions of interest in chemistry, biology, geology, and industry. Use the data in Appendix 1 to calculate the standard enthalpies of the following reactions:

    (a) $2\,NO_2(\mathrm{g}) \longrightarrow N_2O_4(\mathrm{g})$
    (b) $NO_2(\mathrm{g}) \longrightarrow \frac{1}{2}\,N_2O_4(\mathrm{g})$
    (c) $3\,NO_2(\mathrm{g}) + H_2O(\mathrm{l}) \longrightarrow 2\,HNO_3(\mathrm{aq}) + NO(\mathrm{g})$

    (c) $Cyclopropane(\mathrm{g}) \longrightarrow propene(\mathrm{g})$
    (d) $HCl(\mathrm{aq}) + NaOH(\mathrm{aq}) \longrightarrow NaCl(\mathrm{aq}) + H_2O(\mathrm{l})$

2.37 Calculate the standard enthalpy of formation of N_2O_5 from the following data:

$$2\,NO(\mathrm{g}) + O_2(\mathrm{g}) \longrightarrow 2\,NO_2(\mathrm{g})$$
$$\Delta_r H^{\ominus} = -114.1\,\mathrm{kJ\,mol^{-1}}$$
$$4\,NO_2(\mathrm{g}) + O_2(\mathrm{g}) \longrightarrow 2\,N_2O_5(\mathrm{g})$$
$$\Delta_r H^{\ominus} = -110.2\,\mathrm{kJ\,mol^{-1}}$$
$$N_2(\mathrm{g}) + O_2(\mathrm{g}) \longrightarrow 2\,NO(\mathrm{g})$$
$$\Delta_r H^{\ominus} = +180.5\,\mathrm{kJ\,mol^{-1}}$$

2.38 Heat capacity data can be used to estimate the reaction enthalpy at one temperature from its value at another. Use the information in Appendix 1 to predict the standard reaction enthalpy of $2\,NO_2(\mathrm{g}) \longrightarrow N_2O_4(\mathrm{g})$ at $100\,^{\circ}\mathrm{C}$ from its value at $25\,^{\circ}\mathrm{C}$.

2.39 It is often useful to be able to anticipate, without doing a detailed calculation, whether an increase in temperature will result in a raising or a lowering of a reaction enthalpy. The constant-pressure molar heat capacity of a gas of linear molecules is approximately $\frac{7}{2}R$ whereas that of a gas of nonlinear molecules is approximately $4R$. Decide whether the standard enthalpies of the following reactions will increase or decrease with increasing temperature:

    (a) $2\,H_2(\mathrm{g}) + O_2(\mathrm{g}) \longrightarrow 2\,H_2O(\mathrm{g})$
    (b) $N_2(\mathrm{g}) + 3\,H_2(\mathrm{g}) \longrightarrow 2\,NH_3(\mathrm{g})$
    (c) $CH_4(\mathrm{g}) + 2\,O_2(\mathrm{g}) \longrightarrow CO_2(\mathrm{g}) + 2\,H_2O(\mathrm{g})$

2.40 The molar heat capacity of liquid water is approximately $9R$. Decide whether the standard enthalpy of the reactions (a) and (c) in Exercise 2.39 will increase or decrease with a rise in temperature if the water is produced as a liquid.

2.41 Is the standard enthalpy of combustion of glucose likely to be higher or lower at blood temperature than at $25\,^{\circ}\mathrm{C}$?

Contents

Thermodynamics: the Second Law

Some things happen; some things don't. A gas expands to fill the vessel it occupies; a gas that already fills a vessel does not suddenly contract into a smaller volume. A hot object cools to the temperature of its surroundings; a cool object does not suddenly become hotter than its surroundings. Hydrogen and oxygen combine explosively (once their ability to do so has been liberated by a spark) and form water; water left standing in oceans and lakes does not gradually decompose into hydrogen and oxygen. These everyday observations suggest that changes can be divided into two classes:

1. **Spontaneous changes** are changes that have a natural tendency to occur.
2. **Non-spontaneous changes** are changes that have no natural tendency to occur.

Non-spontaneous changes can be *made* to occur: gas can be compressed into a smaller volume by pushing in a piston, the temperature of a cool object can be raised by forcing an electric current through a heater attached to it, and water can be decomposed by the passage of an electric current. However, in each case we need to act in some way on the system to bring the non-spontaneous change about.

There must be some feature of the world that accounts for the distinction between the two types of change. Once we have identified it, we shall be able to apply it to chemistry. Then we shall see why some reactions are spontaneous but others are not. We shall also see—and this is the principal aim of the chapter—how to predict the composition of a reaction mixture that has reached **chemical equilibrium**. At equilibrium a reaction has no further tendency to form more products, and no tendency to run in reverse and re-form reactants. A reaction at equilibrium is spontaneous in neither direction.

Throughout the chapter we shall use the terms 'spontaneous' and 'non-spontaneous' in their thermodynamic sense. That is, we use them to signify that a change does or does not have a natural tendency to occur. It is very important to note that in thermodynamics the term spontaneous has nothing to do with speed. Some spontaneous changes are very fast, such as the precipitation reaction that occurs when solutions of sodium chloride and silver nitrate are mixed. However, some spontaneous changes are so slow that there may be no observable change even after millions of years. For example, although the decomposition of benzene into carbon and hydrogen is spontaneous, it does not occur at a measurable rate under normal conditions, and benzene is a common laboratory commodity with a shelf life of (in principle) millions of years. You must never forget that spontaneity is concerned only with the *tendency* to change: it is silent on the rate at which that tendency is realized.

Entropy

A few moments thought is all that is needed to identify the reason why some changes are spontaneous and others are not. That reason is *not* the tendency of the system to move towards lower energy. This point is easily established by identifying an example of a spontaneous change in which there is no change in energy. The isothermal expansion of a perfect gas into a vacuum is spontaneous, but the total energy of the gas does not change because the molecules continue to travel at the same average speed and so keep their same total kinetic energy. Even in a process in which the energy of a system does decrease (as in the spontaneous cooling of a block of hot metal), the First Law requires the total energy to be constant. Therefore, in this case the energy of another part of the world must increase if the energy decreases in the part that interests us. For instance, a hot block of metal in contact with a cool block cools and loses energy; however, the second block becomes warmer, and increases in energy. It is equally valid to say that the second block moves spontaneously to higher energy as it is to say that the first block has a tendency to go to lower energy!

3.1 The direction of spontaneous change

We shall now show that *the apparent driving force of spontaneous change is the tendency of energy and matter to become disordered*. For example, the molecules of a gas may all be in one region of a container initially, but their ceaseless disorderly motion ensures that they spread rapidly throughout the entire volume of the container (Fig. 3.1). Because their motion is so disorderly, there is a negligibly small probability that all the molecules will find their way back simultaneously into the initial region of the container. In this instance, the natural direction of change corresponds to the dispersal of matter.

A similar explanation accounts for spontaneous cooling, but now we need to consider the dispersal of energy. In a block of hot metal, the atoms are oscillating vigorously, and the hotter the block the more vigorous their motion. The cooler surroundings also consist of oscillating atoms, but their motion is less vigorous. The vigorously oscillating atoms of the hot block jostle their neighbours in the surroundings, and the energy of the atoms in the block is handed on to the atoms in the surroundings (Fig. 3.2). The

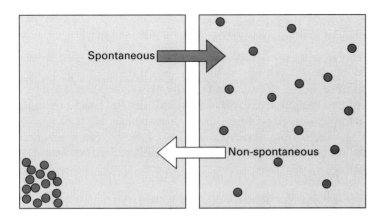

Fig. 3.1 One fundamental type of spontaneous process is the chaotic dispersal of matter. This tendency accounts for the spontaneous tendency of a gas to spread into and fill the container it occupies. It is extremely unlikely that all the particles will collect into one small region of the container. (In practice, the number of particles is of the order of 10^{23}.)

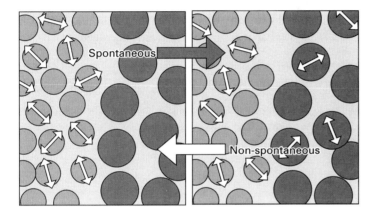

Fig. 3.2 Another fundamental type of spontaneous process is the chaotic dispersal of energy (represented by the small arrows). In these diagrams, the small spheres represent the system and the large spheres represent the surroundings. The double headed arrows represent the thermal motion of the atoms.

process continues, until the vigour with which the atoms in the system are oscillating has fallen to that of the surroundings. The opposite flow of energy is very unlikely. It is highly improbable that there will be a net flow of energy into the system as a result of jostling from less vigorously oscillating molecules in the surroundings. In this case, the natural direction of change corresponds to the dispersal of energy.

In summary, we have identified two basic types of spontaneous physical process:

1. Matter tends to become disordered.
2. Energy tends to become disordered.

We must now see how these two primitive types of physical change result in some chemical reactions being spontaneous and others not.

3.2 Entropy and the Second Law

To make progress, we need to make the discussion quantitative and make measurements, for that will help us to keep track of the degree of disorder even though it might be difficult to identify qualitatively. Disorder is often hard to identify, for instance, when one substance changes into another in the course of a chemical reaction.

The measure of the disorder of matter and energy used in thermodynamics is called the **entropy**, S. Initially, we can take entropy to be a synonym for the extent of disorder, but shortly we shall see that it can be defined precisely and quantitatively, measured, and then applied to chemical reactions. At this point, all we need know is that when matter and energy become disordered, the entropy increases. That being so, we can combine the two remarks above into a single statement known as the **Second Law of thermodynamics**:

The entropy of the universe tends to increase.

The remarkable feature of this law is that it accounts for change in all its forms: for precipitation reactions, acid–base reactions, and redox reactions, as well as the physical changes we have already considered.

The definition of entropy change

To make progress and turn the Second Law into a quantitatively useful statement, we need to define entropy precisely. We shall use the following simple definition of a change in entropy:

$$\Delta S = \frac{q_{rev}}{T} \tag{1}$$

That is, the change in entropy of a substance is equal to the energy transferred as heat to it *reversibly* (we shall explain that qualification in a moment) divided by the temperature at which the transfer takes place. The formal derivation of this expression is obtained by considering a special kind of process called a 'Carnot cycle' which was originally devised to assess the efficiency of steam engines. However, we shall not go into its formal justification. Instead, we shall show that eqn (1) is a plausible formula for the change in entropy, and then show how to use it to obtain numerical values for a range of processes.

There are three points we need to understand about the definition in eqn (1): the significance of the term 'reversible', why heat (not work) appears in the definition, and why the entropy change depends on the temperature at which the transfer takes place.

We met the concept of mechanical reversibility in Section 2.2, where we saw that it involved matching the external pressure to the pressure of the system itself. In that case, energy is transferred reversibly as expansion work between the system and its surroundings. To achieve a reversible transfer of energy as heat, we match the temperature of the surroundings to that of the system. Reversible transfer is smooth, careful, restrained transfer. By making the transfer reversible we ensure that there are no hot spots generated in the object that later disperse spontaneously and add to the entropy.

Now consider why heat and not work appears in eqn (1). Recall from Section 2.1 that to transfer energy as heat we make use of the disorderly motion of molecules whereas work involves the orderly motion of atoms in the surroundings. It should be plausible that the change in entropy—the change in the degree of disorder—is proportional to the energy transfer that takes place by making use of disorderly motion rather than orderly motion.

Finally, the presence of the temperature in the denominator in eqn (1) takes into account the disorder that is already present. If a given quantity of energy is transferred as heat to a hot object (one in which there is already a lot of disorderly thermal motion), then the additional disorder generated is less significant than if the same quantity of energy is transferred as heat to a cold object (in which the atoms have less thermal motion). The difference is like sneezing in a busy street (heat transferred to a body that already has a lot of disorderly molecular motion) and sneezing in a quiet library (heat transferred to a cold body in which there is only little disorderly motion initially). For example, transferring 100 kJ of heat to a large mass of water[†] at 0 °C results in a change in entropy of

$$\Delta S = \frac{100 \times 10^3 \, J}{273 \, K} = +366 \, J \, K^{-1}$$

whereas the same transfer at 100 °C results in

$$\Delta S = \frac{100 \times 10^3 \, J}{373 \, K} = +268 \, J \, K^{-1}$$

Notice the units of entropy: they are joules per kelvin, $J \, K^{-1}$; later, when we deal with molar entropies, their units will be joules per kelvin per mole, $J \, K^{-1} \, mol^{-1}$.

It should also be plausible that *entropy is a state function* (recall Section 2.3: a state function has a value that depends only on the present state of the system). The entropy is a measure of the current state of disorder of the system, and how that disorder was achieved is not relevant to its current value. A sample of liquid water of mass 100 g at 60 °C and 98 kPa has exactly the same degree of molecular disorder—the same entropy—regardless of what has happened to it in the past.

Entropy changes for typical processes

You can normally rely on intuition to judge whether the entropy increases or decreases when a substance undergoes a physical change. For instance, the entropy of a sample of gas increases as it expands because the molecules get to move in a greater volume and so have a greater degree of disorder. However, the advantage of eqn (1) is that it enables us to express the increase *quantitatively* and to start making numerical calculations. The equation implies that the change in entropy when a perfect gas expands isothermally from a volume V_i to a volume V_f is

$$\Delta S = nR \ln \frac{V_f}{V_i} \tag{2}$$

[†]We use a large mass of water to ensure that the temperature of the sample does not change as heat is transferred.

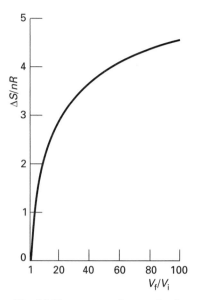

Fig. 3.3 The entropy of a sample of perfect gas increases logarithmically as its volume is increased.

It is important to learn how to read equations for their physical content. In this case we see that if $V_f > V_i$, as in an expansion, then $V_f/V_i > 1$ and the logarithm is positive. Consequently, eqn (2) predicts a positive value for ΔS, corresponding to an increase in entropy, just as we anticipated (Fig. 3.3). Perhaps surprisingly, the equation shows that the change in entropy is independent of the temperature at which expansion occurs.

Exercise E3.1

Calculate the change in molar entropy when a sample of hydrogen gas expands isothermally to twice its initial volume.

[*Answer*: $+5.8\,\mathrm{J\,K^{-1}}$]

Justification

The proof of eqn (2) for the isothermal expansion of a perfect gas makes use of the First Law expressed in the form

$$\Delta U = w + q$$

When a perfect gas expands isothermally, its internal energy remains unchanged because the molecules continue to travel at the same average speed, and therefore with the same total kinetic energy. The potential energy also remains unchanged, because it is zero at all times. Therefore, $\Delta U = 0$ for the isothermal expansion of a perfect gas. If we use this result in the equation above, we can conclude that during the isothermal expansion of a perfect gas, $q = -w$: this equation tells us that to maintain constant temperature, heat must be supplied as the gas does expansion work. To calculate the entropy change, we need to know q_{rev}, the energy transferred as heat in the course of a *reversible* change; so, according to this equation, we need to know w_{rev} (because the general result $q = -w$ implies the special case $q_{rev} = -w_{rev}$). However, we have already seen (eqn (2.6)) that

$$w_{rev} = -nRT \ln \frac{V_f}{V_i}$$

It follows that

$$\Delta S = \frac{q_{rev}}{T} = -\frac{w_{rev}}{T} = nR \ln \frac{V_f}{V_i}$$

as in eqn (2). The reason why the entropy change is independent of temperature is that more work is done if the temperature is high, so more heat must be supplied to maintain the temperature. The temperature in the denominator of eqn (1) is higher, but the 'sneeze' is greater, and the two effects cancel.

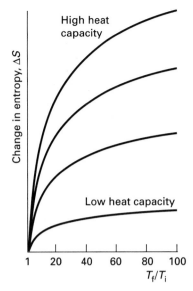

Fig. 3.4 The entropy of a sample with a heat capacity that is independent of temperature (such as a monatomic perfect gas) increases logarithmically as the temperature is increased. The increase is proportional to the heat capacity of the sample.

The second type of change we consider is raising the temperature. We should expect the entropy of a substance to increase as the temperature is raised, because the thermal disorder of the system is greater at the higher temperature, when the molecules move more vigorously. If we assume that the heat capacity of a substance is independent of temperature in the range of interest,[†] then the change in entropy when the temperature is raised from T_i to T_f at constant volume is

$$\Delta S = C_V \ln \frac{T_f}{T_i} \qquad (3)$$

This equation is in line with what we expect: when $T_f > T_i$, the value of T_f/T_i is greater than 1, which implies that ΔS is positive, and that the entropy has increased (Fig. 3.4).

Exercise E3.2

Calculate the change in molar entropy when hydrogen gas is heated from 20 °C to 30 °C at constant volume. ($C_{V,m} = 22.44\,\mathrm{J\,K^{-1}\,mol^{-1}}$.)

[*Answer*: $+0.75\,\mathrm{J\,K^{-1}\,mol^{-1}}$]

When the range of temperature is so wide that the heat capacity cannot be considered to be independent of temperature, the difference in entropy is calculated graphically. First, the heat capacity is measured over a range of temperatures, and from the table of observed values we plot a graph of C_V/T against T. The following *Justification* shows that the area under the curve between the two temperatures of interest is equal to the difference in entropy (Fig. 3.5).

Justification

To confirm this relation, we have to use calculus, and specifically to make use of the fact that an integral of a function can be identified with the area under its graph. It follows from the definition of heat capacity given in Section

[†]This assumption is strictly valid only for monatomic gases, but it is reliable for many systems over small temperature ranges.

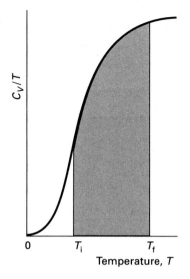

Fig. 3.5 The experimental determination of the change in entropy of a sample that has a heat capacity that varies with temperature involves measuring the heat capacity over the range of temperatures of interest, then plotting C_V/T against T and determining the area under the curve (the tinted area shown here). The heat capacity of all solids decreases toward zero as the temperature is reduced.

2.4 that when an infinitesimal energy dq is transferred as heat to a system of heat capacity C_V, the temperature increases by dT, where $dq = C_V dT$. This relation also applies when the transfer is carried out reversibly, so we can use eqn (1) written for an infinitesimal change,

$$dS = \frac{dq_{rev}}{T}$$

to write

$$dS = \frac{C_V dT}{T}$$

The total change in entropy for a change in temperature from T_i to T_f is the sum (integral) of all such infinitesimal terms:

$$\Delta S = \int_{T_i}^{T_f} \frac{C_V}{T} \, dT$$

The value of the integral, and hence the value of ΔS, is the area under the curve of C_V/T against T that lies between T_i and T_f.

If C_V is constant in the range of interest, then it may be taken outside the integral, and the latter then evaluated by using

$$\int \frac{dx}{x} = \ln x$$

to give

$$\Delta S = C_V \int_{T_i}^{T_f} \frac{dT}{T} = C_V \ln T \Big|_{T_i}^{T_f} = C_V \ln \frac{T_f}{T_i}$$

as in eqn (3).

The third common process to consider is a change of state, as in melting or boiling. We can suspect that the entropy of a substance increases when it melts and when it boils because its molecules become more disordered as it changes from solid to liquid and from liquid to vapour. Suppose a solid is at its melting temperature, then any transfer of energy as heat to the sample causes melting (fusion). Because the transition occurs at constant pressure, we can identify the heat transferred per mole of substance with the transition enthalpy (the enthalpy of fusion, in this case). Therefore, the entropy of fusion, $\Delta_{fus}S$, the change of entropy per mole of substance, at the melting temperature, T_f, is

$$\Delta_{fus}S = \frac{\Delta_{fus}H}{T_f} \tag{4}$$

All enthalpies of fusion are positive (melting is endothermic: it requires heat), so all entropies of fusion are positive too: disorder increases on melting. The entropy of water, for example, increases when it melts because the orderly structure of ice collapses as the liquid forms (Fig. 3.6).

Exercise E3.3

Calculate the entropy of fusion of ice at 0°C from the information in Table 2.2.

[*Answer*: $+22\,\mathrm{J\,K^{-1}\,mol^{-1}}$]

The entropy of other types of transition may be discussed similarly. Thus, the entropy of vaporization, $\Delta_{vap}S$, at the boiling temperature, T_b, of a liquid is related to its enthalpy of vaporization at that temperature by

$$\Delta_{vap}S = \frac{\Delta_{vap}H}{T_b} \tag{5}$$

Because vaporization is endothermic for all substances, all entropies of vaporization are positive. The increase in entropy on vaporization is in line with what we should expect when a compact liquid turns into a gas.

Exercise E3.4

Calculate the entropy of vaporization of water at 100°C.

[*Answer*: $+109\,\mathrm{J\,K^{-1}\,mol^{-1}}$]

The increase in entropy when water vaporizes at 100°C is greater than the increase when ice melts at 0°C despite the fact that the heat transfer occurs at a higher temperature. There is a much greater change in disorder on going from a liquid to a gas than from a solid to a liquid.

Entropies of vaporization shed light on an empirical relation known as **Trouton's rule**. Trouton noticed that $\Delta_{vap}H/T_b$ is approximately the same (and equal to about $+85\,\mathrm{J\,K^{-1}\,mol^{-1}}$) for all liquids except when hydrogen bonding or some other kind of specific bonding is present (see Table 3.1). We know that the quantity $\Delta_{vap}H/T_b$, however, is the entropy of vaporization of the liquid at its boiling point, and so Trouton's rule is explained if all liquids have approximately the same entropy of vaporization. This near equality is to be expected, because when a liquid vaporizes, the compact condensed phase changes into a widely dispersed gas that occupies approximately the same volume whatever its identity. To a good approximation, therefore, we expect the increase in disorder, and

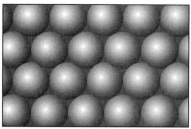

(a)

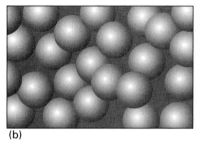

(b)

Fig. 3.6 When a solid (depicted by the orderly array of spheres, (a)) melts, the molecules form a more chaotic liquid (the disorderly array of spheres, (b)); as a result, the entropy of the sample increases.

Table 3.1 Entropies of vaporization at 1 atm and the normal boiling point

	$\Delta_{vap}S/$ $\mathrm{J\,K^{-1}\,mol^{-1}}$
Bromine, Br_2	$+88.6$
Benzene, C_6H_6	$+87.2$
Carbon tetrachloride, CCl_4	$+85.9$
Cyclohexane, C_6H_{12}	$+85.1$
Hydrogen sulfide, H_2S	$+87.9$
Ammonia, NH_3	$+97.4$
Water, H_2O	$+109.1$

therefore the entropy of vaporization, to be almost the same for all liquids at their boiling temperatures. The exceptions to Trouton's rule are liquids in which the interactions between molecules result in the liquid being less disordered than a random jumble of molecules. For example, the high value for water implies that the H_2O molecules are linked together in some kind of loose structure (by hydrogen bonding, Section 10.2), with the result that the entropy change is greater when this relatively ordered liquid forms a disordered gas. The high value for mercury has a similar explanation but stems from the presence of metallic bonding in the liquid.

Example *Using Trouton's rule*
Estimate the enthalpy of vaporization of liquid bromine from its boiling temperature, 59.2 °C.

Strategy
First, we consider whether Trouton's rule is likely to be valid by judging whether hydrogen bonding or some other specific type of interaction are not likely to play a role. If the rule is expected to be valid, we rearrange it into

$$\Delta_{vap}H \approx T_b \times (85\,\mathrm{J\,K^{-1}\,mol^{-1}})$$

and substitute the data.

Solution
No hydrogen bonding or other kind of special interaction is present, so we use the rule after converting the boiling point to 332.4 K:

$$\Delta_{vap}H \approx (332.4\,\mathrm{K}) \times (85\,\mathrm{J\,K^{-1}\,mol^{-1}}) = +28\,\mathrm{kJ\,mol^{-1}}$$

The experimental value is $+29\,\mathrm{kJ\,mol^{-1}}$.

Exercise E3.5

Estimate the enthalpy of vaporization of ethane from its boiling point, which is −88.6 °C.

[*Answer*: $+16\,\mathrm{kJ\,mol^{-1}}$]

Absolute entropies and the Third Law of thermodynamics

The graphical procedure summarized by Fig. 3.5 and eqn (3) for the determination of the difference in entropy of a substance at two temperatures has a very important application. If $T_i = 0$ (that is, the absolute zero of temperature), then the area under the graph between $T = 0$ and some temperature T gives us the value of $S(T) - S(0)$ (Fig. 3.7).[†] However, at $T = 0$, all the motion of the atoms has been eliminated, and there is no thermal disorder. Moreover, if the substance is perfectly crystalline, with every atom in a well-defined location, then there is no spatial disorder either. That is, at $T = 0$, the entropy is zero. For example, the entropy of a crystal of sucrose at $T = 0$ is zero, as all the atoms lie in a perfectly orderly array and there is no thermal motion; the same is true of a perfect crystal of sodium chloride and of solid carbon dioxide. This conclusion is generalized into the 'Third Law' of thermodynamics:

> **The Third Law of thermodynamics**: The entropy of a perfectly crystalline substance is zero at $T = 0$.

According to the Third Law, $S(0) = 0$ for all perfectly ordered crystalline materials. It therefore follows that the **Third-Law entropy** at any temperature, $S(T)$ is *equal* to the area of the graph between $T = 0$ and the temperature T.

The value of the Third-Law entropy (which is commonly called simply 'the entropy') of a substance depends on the pressure (a high pressure, for instance, would confine a gas to a smaller volume and so reduce its entropy). It is therefore common to select a standard pressure and to report the **standard molar entropy**, $S_m^\ominus$, the molar entropy of a substance in its standard state (as specified in Section 2.7: pure, and at 1 bar) at the temperature of interest. Some values at 25 °C (the conventional temperature) are given in Table 3.2.

It is worth spending a moment to look at the values in Table 3.2, to see that they are consistent with our understanding of entropy. All standard molar entropies are positive, because raising the temperature of a sample above $T = 0$ invariably increases its entropy above the value $S(0) = 0$. Another feature is that the standard molar entropy of diamond $(2.4\,\mathrm{J\,K^{-1}\,mol^{-1}})$ is lower than that of graphite $(5.7\,\mathrm{J\,K^{-1}\,mol^{-1}})$. This difference is consistent with the atoms being linked less rigidly in graphite than in diamond, and its thermal disorder being greater. The standard molar entropy of ice, water, and water vapour at 25 °C are, respectively, 45, 70, and $189\,\mathrm{J\,K^{-1}\,mol^{-1}}$, and the increase in values corresponds to the increasing disorder on going from a solid to a liquid to a gas.

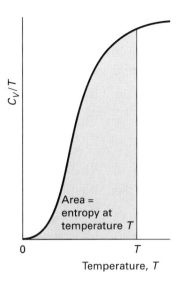

Fig. 3.7 The absolute entropy (or Third-Law entropy) of a substance is calculated by extending the measurement of heat capacities down to $T = 0$ (or as close to that value as possible), and then determining the area of the graph of C_V/T against T up to the temperature of interest. The area is equal to the absolute entropy at the temperature T.

[†]We are supposing that there are no phase transitions below the temperature T. If there are any phase transitions (for example, melting) in the temperature range of interest, then the entropy of each transition, calculated from eqns (4) or (5), must be included.

Table 3.2 Standard molar entropies of some substances at 25°C

Substance	$S_m^\ominus/$ $\mathrm{J\,K^{-1}\,mol^{-1}}$
Gases	
Ammonia, NH_3	192.5
Carbon dioxide, CO_2	213.7
Helium, He	126.2
Hydrogen, H_2	130.7
Neon, Ne	146.3
Nitrogen, N_2	191.6
Oxygen, O_2	205.1
Water vapour, H_2O	188.8
Liquids	
Benzene, C_6H_6	173.3
Ethanol, CH_3CH_2OH	160.7
Water, H_2O	69.9
Solids	
Calcium oxide, CaO	39.8
Calcium carbonate,	
$\quad CaCO_3$	92.9
Copper, Cu	33.2
Diamond, C	2.4
Graphite, C	5.7
Lead, Pb	64.8
Magnesium carbonate,	
$\quad MgCO_3$	65.7
Magnesium oxide, MgO	26.9
Sodium chloride, NaCl	72.1
Sucrose, $C_{12}H_{22}O_{11}$	360.2
Tin, Sn (white)	51.6
$\quad$ Sn (grey)	44.1

See Appendix 1 for more values.

The reaction entropy

Now we move into the arena of chemistry, where reactants are transformed into products. Although it is difficult in many cases to anticipate whether the disorder associated with the products is greater or smaller than that of the reactants, we can use the measured values of the entropies of the substances to calculate the difference without difficulty. The difference in entropy between the products and the reactants in their standard states is called the **standard reaction entropy**, $\Delta_r S^\ominus$. It can be expressed in terms of the molar entropies of the substances in much the same way as we have already used for the standard reaction enthalpy:

$$\Delta_r S^\ominus = \sum_{\text{Products}} n S_m^\ominus(\text{products}) - \sum_{\text{Reactants}} n S_m^\ominus(\text{reactants}) \qquad (6)$$

where the *n*s are the stoichiometric numbers that occur in the chemical equation. For example, for the reaction

$$2\,H_2(g) + O_2(g) \longrightarrow 2\,H_2O(l)$$

we can use the values in Appendix 1 to write

$$\begin{aligned}
\Delta_r S^\ominus &= 2 S_m^\ominus(H_2O, l) - \left\{ 2 S_m^\ominus(H_2, g) + S_m^\ominus(O_2, g) \right\} \\
&= 2 \times (70\,\mathrm{J\,K^{-1}\,mol^{-1}}) \\
&\quad - \left\{ 2 \times (131\,\mathrm{J\,K^{-1}\,mol^{-1}}) + (205\,\mathrm{J\,K^{-1}\,mol^{-1}}) \right\} \\
&= -327\,\mathrm{J\,K^{-1}\,mol^{-1}}
\end{aligned}$$

Do not make the mistake of setting the standard molar entropies of elements equal to zero: they have nonzero values (provided $T > 0$), as we have already discussed. We see from this illustration that the formation of a liquid from two gases has resulted in a decrease in entropy, as would be expected for the formation of a compact phase.

Exercise E3.6

Calculate the standard reaction entropy for

$$N_2(g) + 3\,H_2(g) \longrightarrow 2\,NH_3(g)$$

at 25 °C. What is the change in entropy when $2\,\mathrm{mol}\,H_2$ react?

[*Answer*: $-198.76\,\mathrm{J\,K^{-1}\,mol^{-1}}$, $-132.51\,\mathrm{J\,K^{-1}}$]

The spontaneity of chemical reactions

The result of the H_2O calculation should be rather surprising at first sight. We know that the reaction between hydrogen and oxygen is spontaneous and, once initiated, that it proceeds with explosive violence. Nevertheless, the entropy change that accompanies it is negative: the reaction results in less disorder, yet it is spontaneous!

The resolution of this apparent paradox underscores a feature of entropy that recurs throughout chemistry: it is always essential to consider the entropy of both the system *and its surroundings* when deciding whether a reaction is spontaneous or not. The reduction in entropy by $327 \, J \, K^{-1} \, mol^{-1}$ relates only to the system, the reaction mixture. To apply the Second Law correctly, we need to calculate the total entropy, the sum of the changes in the system and the surroundings. It may well be the case that the entropy of the system decreases when a change takes place, but there may be a more than compensating increase in entropy of the surroundings so that overall the entropy change is positive. The opposite may also be true: a large decrease in entropy of the surroundings may occur when the entropy of the system increases. In that case we would be wrong to infer from the increase of the system alone that the change is spontaneous. Whenever considering the implications of entropy, you must *always* consider the total change of the system and its surroundings.

To calculate the entropy change in the surroundings when a reaction takes place, we note that heat may enter or leave the system. For example, for the water formation reaction written above, the standard reaction enthalpy is $\Delta_r H^{\ominus} = -572 \, kJ \, mol^{-1}$, so at constant pressure 572 kJ of energy enters the surroundings as heat when $2 \, mol \, H_2O$ is produced and the reaction takes place at constant pressure. The change in entropy of the surroundings (which are maintained at $25\,°C$, the same temperature as the reaction mixture) at constant pressure is therefore

$$\Delta_r S_{surr} = \frac{572 \, kJ \, mol^{-1}}{298 \, K} = +1.92 \times 10^3 \, J \, K^{-1} \, mol^{-1}$$

Now we can see that the total entropy change is positive:

$$\Delta_r S_{total} = (-327 \, J \, K^{-1} \, mol^{-1}) + (1.92 \times 10^3 \, J \, K^{-1} \, mol^{-1})$$
$$= +1.59 \times 10^3 \, J \, K^{-1} \, mol^{-1}$$

This calculation confirms, as we know from experience, that the reaction is strongly spontaneous. In this case, the spontaneity is a result of the considerable disorder that the reaction generates in the surroundings: water is dragged into existence, even though $H_2O(l)$ has a lower entropy than the reactants, by the tendency of energy to disperse into the surroundings.

In general, when the standard reaction enthalpy is $\Delta_r H^{\ominus}$, the change in enthalpy of the surroundings is $-\Delta_r H^{\ominus}$. The entropy change of the surroundings when a change occurs at constant pressure and temperature is therefore

$$\Delta_r S_{surr}^{\ominus} = -\frac{\Delta_r H^{\ominus}}{T} \tag{7}$$

Example *Estimating the entropy change of the surroundings*
A typical resting person generates about 100 W of heat. Estimate the entropy they generate in the surroundings in the course of a day at 20 °C.

Strategy
We can estimate the approximate change in entropy from eqn (7), once we have calculated the energy transferred as heat. To find this quantity, we use the facts that $1\,W = 1\,J\,s^{-1}$ and there are $86\,400\,s$ in a day.

Solution
The heat transferred to the surroundings in the course of a day is

$$q = (86\ 400\,\text{s}) \times (100\,\text{J}\,\text{s}^{-1}) = 8.64 \times 10^6\,\text{J}$$

The increase in entropy of the surroundings is therefore approximately

$$\Delta S_{\text{surr}} = \frac{8.64 \times 10^6\,\text{J}}{293\,\text{K}} = +2.95 \times 10^4\,\text{J}\,\text{K}^{-1}$$

That is, the entropy production is about $30\,\text{kJ}\,\text{K}^{-1}$. Just to stay alive, each person on the planet contributes about $30\,\text{kJ}\,\text{K}^{-1}$ each day to the ever-increasing entropy of their surroundings. The use of transport, machinery, and communications generates far more in addition.

Exercise E3.7

Suppose a small reptile operates at 0.50 W. What entropy does it generate in the course of a day in the water in the lake that it inhabits, where the temperature is 15 °C?

[*Answer*: $+150\,\text{J}\,\text{K}^{-1}$]

The Gibbs energy

One of the problems with entropy calculations is already apparent: it is necessary to work out two entropy changes, the change in the system and the change in the surroundings, and then consider the sign of their sum. The great American theoretician J.W. Gibbs, who laid the foundations of chemical thermodynamics towards the end of the nineteenth century, found that the two calculations can be combined into one. The combination of the two procedures in fact turns out to be of much greater relevance than just saving a little labour, and throughout this text we shall see consequences of the procedure he developed.

3.3 Focusing on the system

The total entropy change that accompanies a process is

$$\Delta S_{total} = \Delta S + \Delta S_{surr} \tag{8}$$

where ΔS is the entropy change for the system; for a spontaneous change, $\Delta S_{total} > 0$. If the process occurs at constant pressure and temperature, then we have already seen that the change in entropy of the surroundings can be expressed in terms of the enthalpy change of the system, ΔH, by using eqn (7). When that expression is inserted into this one, we obtain

$$\Delta S_{total} = \Delta S - \frac{\Delta H}{T}$$

The great advantage of this formula is that it expresses the total entropy change of the system and its surroundings in terms of properties of the system alone. The only restriction is to changes at constant pressure and temperature.

Two further steps can now be taken. First, we multiply through by $-T$ to obtain

$$-T\Delta S_{total} = -T\Delta S + \Delta H$$

Then we define the change in **Gibbs energy**,[†] ΔG:

$$\Delta G = \Delta H - T\Delta S \tag{9}$$

It follows that

$$\Delta G = -T\Delta S_{total} \tag{10}$$

Thus, we see that at constant temperature and pressure the change in Gibbs energy of a system is proportional to the overall change in entropy of the system plus its surroundings.

Properties of the Gibbs energy

The difference in sign between ΔG and ΔS_{total} implies that the condition for a process being spontaneous changes from $\Delta S_{total} > 0$ in terms of the total entropy (which is universally true) to $\Delta G < 0$ in terms of the Gibbs energy

---

[†]The Gibbs energy is commonly referred to as the 'free energy'.

(for processes occurring at constant temperature and pressure). That is, *in a spontaneous change at constant temperature and pressure, the Gibbs energy decreases.* It may seem more natural to think of a system as falling to a lower value of some property. However, it must never be forgotten that to say that a system tends to fall toward lower Gibbs energy is only a modified way of saying that a system and its surroundings jointly tend towards a greater total entropy. The only criterion of spontaneous change is the total entropy of the system and its surroundings: the Gibbs energy merely contrives a way of expressing that total change in terms of the properties of the system alone, and is valid only for processes that occur at constant temperature and pressure. Every chemical reaction that is spontaneous under conditions of constant temperature and pressure, including those that drive the processes of growth, learning, and reproduction, are reactions that change in the direction of lower Gibbs energy, or—another way of expressing the same thing—result in the overall entropy of the system and its surroundings becoming greater.

A second feature of the Gibbs energy is that *its value gives the maximum amount of non-expansion work that can be extracted from a system that is undergoing a change at constant temperature and pressure.* By **non-expansion work** is meant any work other than that arising from the expansion of the system, so it may include electrical work (if the process takes place inside an electrochemical or biological cell) or other kinds of mechanical work, such as the winding of a spring or the contraction of a muscle. If we denote non-expansion work by w', then the total work, w, of a system that can do both types of work is

$$w = w' - p_{ex}\Delta V \tag{11}$$

So, an important property of the Gibbs energy is that

$$w'(\text{maximum}) = \Delta G \tag{12}$$

For example, we shall shortly see that for the formation of $1\,mol\,H_2O(l)$ at $25\,°C$ and 1 bar, $\Delta G = -237\,kJ$, so up to $237\,kJ$ of non-expansion work can be extracted from the reaction between hydrogen and oxygen to produce $1\,mol\,H_2O(l)$ at $25\,°C$. If the reaction takes place in a fuel cell (a device for using a chemical reaction to produce an electric current) like those used on the space shuttle, then up to $237\,kJ$ of electrical energy can be generated for each mole of H_2O produced.

Justification

We need to consider infinitesimal changes because dealing with reversible processes is then much easier. The aim is to derive the relation between dG and dw'. To do so, we begin by substituting the definitions of dH and dU into the expression for dG:

$dG = dH - TdS$ (from eqn (9), at constant temperature and pressure)

$\quad = dU + pdV - TdS$ (because $dH = dU + pdV$ at constant p)

$\quad = dq + dw + pdV - TdS$ (because $dU = dq + dw$)

$\quad = dq + dw' - p_{ex}dV + pdV - TdS$ (from eqn (11))

This derivation is valid for any process taking place at constant temperature and pressure. Now we specialize to a reversible change. For expansion work to be reversible, we need to match p and p_{ex}, in which case the third and fourth terms on the right of the last line cancel. Moreover, because the heat transfer is also reversible, we can replace dq by TdS, in which case the first and fifth terms also cancel. We are left with

$dG = dw'$ (at constant temperature and pressure, a reversible change)

The non-expansion work done during a reversible change is the greatest it can be, so we can write

$dG = dw'(\text{maximum})$ (at constant temperature and pressure)

Because this relation holds for each infinitesimal step between the specified initial and final states, it applies to the overall change too. Therefore, we can write

$\Delta G = w'(\text{maximum})$ (at constant temperature and pressure)

We have already seen in Section 2.7 that the enthalpy change accompanying the formation of $1\,\text{mol}\,H_2O(l)$ is $+286\,\text{kJ}$. This value shows that, for each mole of H_2O produced, $286\,\text{kJ}$ of heat can be produced. We have just seen that a different energy ($237\,\text{kJ}$) can be extracted as work. In summary, for a reaction run at constant temperature and pressure:

1. If no attempt is made to extract any energy as work, then $286\,\text{kJ}$ (in general, ΔH) of heat will be produced.

2. Alternatively, if some of the energy released is used to do work, then up to $237\,\text{kJ}$ (in general, ΔG) of non-expansion work can be obtained.

Example *Estimating a change in Gibbs energy*
Suppose a certain small bird has a mass of $30\,\text{g}$. What is the minimum mass of glucose that it must consume to fly to a branch $10\,\text{m}$ above the ground? The change in Gibbs energy that accompanies the oxidation of $1.0\,\text{mol}\,C_6H_{12}O_6$ to carbon dioxide and water vapour at $25\,°C$ is $-2828\,\text{kJ}$.

Strategy
First, we need to calculate the work needed to raise a mass m through a height h on the surface of the Earth: as we saw in eqn (2.1), this work is

equal to mgh, where g is the acceleration of free fall. This work, which is non-expansion work, can be identified with ΔG. We need to determine the amount of substance that corresponds to the required change in Gibbs energy, and then convert that amount to a mass by using the molar mass of glucose.

Solution

The work to be done is

$$w = (30 \times 10^{-3}\,\text{kg}) \times (9.81\,\text{m s}^{-2}) \times (10\,\text{m}) = 2.9\,\text{J}$$

(because $1\,\text{kg m}^2\,\text{s}^{-2} = 1\,\text{J}$). The amount of glucose molecules required for oxidation to give a change in Gibbs energy of this value given that 1 mol provides 2828 kJ is

$$n(\text{C}_6\text{H}_{12}\text{O}_6) = \frac{2.9\,\text{J}}{2828\,\text{kJ mol}^{-1}} = 1.0 \times 10^{-6}\,\text{mol}$$

Therefore, because the molar mass of glucose is $180\,\text{g mol}^{-1}$, the mass of glucose that must be oxidized is

$$m = (1.0 \times 10^{-6}\,\text{mol}) \times (180\,\text{g mol}^{-1}) = 1.8 \times 10^{-4}\,\text{g}$$

That is, the bird must consume at least 0.18 mg of glucose for the mechanical effort (and more if it thinks about it).

Exercise E3.8

A hard-working human brain, perhaps one that is grappling with physical chemistry, operates at about 25 W. What mass of glucose must be consumed to sustain that power output for an hour?

[*Answer*: 5.7 g]

The great importance of the Gibbs energy in chemistry should be beginning to become apparent. At this stage, we see that it is a measure of the non-expansion work resources of chemical reactions: if we know ΔG, then we know the maximum non-expansion work that we can do by harnessing the reaction in some way. In some cases, the non-expansion work is extracted as electrical energy: this is the case when the reaction takes place in an electrochemical cell (of which a fuel cell is a special case), as we shall see in Chapter 6. In other cases, the reaction may be used to build other molecules. This is the case in biological cells, where the Gibbs energy available from the hydrolysis of ATP (adenosine triphosphate) to ADP is used to build proteins from amino acids, to power muscular contraction, or to drive the neuronal circuits in our senses and our brains.

The reaction Gibbs energy

The change in Gibbs energy when a reaction takes place is called the **reaction Gibbs energy**. However, we need to distinguish between the reaction Gibbs energy under standard conditions and its value under other conditions. We deal with the standard value first.

> The **standard reaction Gibbs energy**, $\Delta_r G^\ominus$, is the difference in standard molar Gibbs energies of the products and the reactants weighted by the stoichiometric numbers in the chemical equation:

$$\Delta_r G^\ominus = \sum_{\text{Products}} n G_m^\ominus (\text{products}) - \sum_{\text{Reactants}} n G_m^\ominus (\text{reactants}) \qquad (13)$$

For example, the standard reaction Gibbs energy for

$$2\,H_2(g) + O_2(g) \longrightarrow 2\,H_2O(l)$$

is the change in Gibbs energy when pure hydrogen gas at 1 bar reacts with pure oxygen gas at 1 bar to produce pure liquid water at 1 bar.

We cannot calculate the standard reaction Gibbs energy from the standard molar Gibbs energies themselves, because these are not known. One practical approach is to calculate it instead from the standard reaction enthalpy and entropy by writing

$$\Delta_r G^\ominus = \Delta_r H^\ominus - T \Delta_r S^\ominus \qquad (14)$$

The standard reaction enthalpy is calculated from tables of standard enthalpies of formation, as explained in Section 2.7, and the standard reaction entropy is calculated from tables of Third-Law entropies, as described in Section 3.2. The units of the standard reaction Gibbs energy are the same as those of standard reaction enthalpy (kilojoules per mole).

Example *Determining the standard reaction Gibbs energy*
Evaluate the standard reaction Gibbs energy at 25 °C for the water formation reaction specified above.

Strategy
To use eqn (14), we need two pieces of information. One is the standard reaction enthalpy, which is obtained from tables of enthalpies of formation; the second is the standard reaction entropy, which is obtained from the standard entropies of the reactants and products.

Solution

The standard reaction enthalpy is identical to the standard enthalpy of formation of liquid water:

$$\Delta_r H^\ominus = \Delta_f H^\ominus(H_2O, l) = -285.83 \, kJ \, mol^{-1}$$

We have already calculated the standard reaction entropy:

$$\Delta_r S^\ominus = -163.34 \, J \, K^{-1} \, mol^{-1} = -0.163 \, 34 \, kJ \, K^{-1} \, mol^{-1}$$

Therefore, from eqn (14),

$$\Delta_r G^\ominus = (-285.83 \, kJ \, mol^{-1}) - (298.15 \, K) \times (-0.163 \, 34 \, kJ \, K^{-1} \, mol^{-1})$$
$$= -237.13 \, kJ \, mol^{-1}$$

Exercise E3.9

Use the information in Appendix 1 to determine the standard reaction Gibbs energy for $3\,O_2(g) \longrightarrow 2\,O_3(g)$ from standard enthalpies of formation and the Third-Law entropies.

[*Answer*: $+326.4 \, kJ \, mol^{-1}$]

Table 3.3 summarizes the conditions under which $\Delta_r G^\ominus$ may be negative and hence correspond to a spontaneous reaction. The standard reaction Gibbs energy is certainly negative if $\Delta_r H^\ominus$ is negative (an exothermic reaction) and $\Delta_r S$ is positive (a reaction system that becomes more disorderly, such as by forming a gas). It may also be negative if the reaction is endothermic ($\Delta_r H^\ominus$ is positive) and $T\Delta_r S^\ominus$ is sufficiently large and positive. This is so if the increase in disorder of the system is so great that it dominates the reduction in disorder of the surroundings (as measured by $\Delta_r H^\ominus$). Thus, in certain cases, and so long as the temperature is not too low (so that $T\Delta_r S^\ominus$ is large enough to overcome the positive value of $\Delta_r H^\ominus$), endothermic

Table 3.3 The contributions to spontaneous change

Enthalpy	Entropy	Spontaneous?
$\Delta H <$ (exothermic)	$\Delta S > 0$	Yes ($\Delta G < 0$)
$\Delta H <$ (exothermic)	$\Delta S < 0$	Yes, if $|T\Delta S| < |\Delta H|$
$\Delta H > 0$ (endothermic)	$\Delta S > 0$	Yes, if $T\Delta S > \Delta H$
$\Delta H > 0$ (endothermic)	$\Delta S < 0$	No ($\Delta G > 0$)

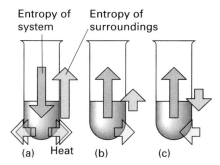

Entropy of system
Entropy of surroundings

(a) Heat (b) (c)

Fig. 3.8 The horizontal arrows represent the flow of energy as heat in exothermic (a and b) and endothermic (c) reactions and the vertical arrows represent the change in entropy of the system and the surroundings. (a) A strongly exothermic reaction results in a large increase in entropy of the surroundings, and even though the system might undergo a reduction in entropy, the reaction is spontaneous. (b) In a less exothermic reaction, the entropy of the surroundings increases a small amount but the reaction may still be spontaneous, particularly (but not necessarily) if the entropy of the system increases too. (c) If the reaction is endothermic, then the entropy of the surroundings falls, and the reaction can be spontaneous only if the entropy of the system increases (and by a more than compensating amount).

reactions may be spontaneous (Fig. 3.8). For an endothermic reaction to have a negative standard reaction Gibbs energy, its standard entropy of reaction *must* be positive. However, that alone is not sufficient: the temperature must also be high enough for $T\Delta_r S^{\ominus}$ to overcome the negative value of $\Delta_r H^{\ominus}$.

The switch of $\Delta_r G^{\ominus}$ from positive to negative, from non-spontaneous to spontaneous, occurs at a temperature given by equating $\Delta_r H^{\ominus} - T\Delta_r S^{\ominus}$ to 0, which gives:

$$T = \frac{\Delta_r H^{\ominus}}{\Delta_r S^{\ominus}} \tag{15}$$

As an example, consider the thermal decomposition of a solid, such as

$$CaCO_3(s) \xrightarrow{\Delta} CaO(s) + CO_2(g)$$

for which $\Delta_r H^{\ominus} = +178 \text{ kJ mol}^{-1}$ and $\Delta_r S^{\ominus} = +161 \text{ J K}^{-1} \text{ mol}^{-1}$. This reaction (and others like it) is endothermic because bonds are broken as the solid is decomposed; it also has a positive standard reaction entropy because a compact solid decomposes into a dispersed gas. The decomposition temperature, the temperature at which the reaction becomes spontaneous, is

$$T = \frac{178 \times 10^3 \text{ J mol}^{-1}}{161 \text{ J K}^{-1} \text{ mol}^{-1}} = 1.11 \times 10^3 \text{ K}$$

or about $837\,^{\circ}\text{C}$. Because the entropy of decomposition is similar for all such reactions (they all involve the decomposition of a solid into a gas), we can conclude that the decomposition temperature of solids increase as their enthalpy of decomposition increases, so substances with high lattice enthalpies can be expected to have high decomposition temperatures.

It will be recalled that we were able to simplify the reporting of standard reaction enthalpies by listing standard enthalpies of formation. We can

Table 3.4 Standard Gibbs energies of formation at 25°C

Substance	$\Delta_f G^{\ominus}/$ kJ mol^{-1}
Gases	
Ammonia, NH_3	−16.5
Carbon dioxide, CO_2	−394.4
Dinitrogen tetroxide, N_2O_4	+97.9
Hydrogen iodide, HI	+1.7
Nitrogen dioxide, NO_2	+51.3
Sulfur dioxide, SO_2	−300.2
Water, H_2O	−228.6
Liquids	
Benzene, C_6H_6	+124.3
Ethanol, CH_3CH_2OH	−174.8
Water, H_2O	−237.1
Solids	
Calcium carbonate, $CaCO_3$	−1128.8
Iron(III) oxide, Fe_2O_3	−742.2
Silver bromide, AgBr	−96.9
Silver chloride, AgCl	−109.8

Additional values are given in Appendix 1.

do the same for standard reaction Gibbs energies by listing the 'standard Gibbs energy of formation' of each species taking part in the reaction:

The **standard Gibbs energy of formation**, $\Delta_f G^{\ominus}$, of a species is the standard reaction Gibbs energy (per mole of the species) for its formation from the elements in their reference states.

The concept of reference state was introduced in Section 2.7; the temperature is arbitrary, but we shall almost always take it as 25 °C (298 K). For example, the standard Gibbs energy of formation of liquid water, $\Delta_f G^{\ominus}(H_2O, l)$, is the standard reaction Gibbs energy for

$$H_2(g) + \tfrac{1}{2}O_2(g) \longrightarrow H_2O(l)$$

and is -237 kJ mol^{-1}. Some standard Gibbs energies of formation are listed in Table 3.4 (this table is a brief version of a longer one in Appendix 1). Note that it follows from the definition that the values for the formation of elements in their reference states are zero because reactions such as

$$C(s, graphite) \longrightarrow C(s, graphite) \qquad \Delta_f G^{\ominus}(C, graphite) = 0$$

are null (that is, nothing happens). You should be careful, though, to note that an element may change into an allotrope, and there will then be a nonzero standard Gibbs energy of formation of the allotrope:

$$C(s, graphite) \longrightarrow C(s, diamond) \qquad \Delta_f G^{\ominus}(C, diamond) = +2.90 \text{ kJ mol}^{-1}$$

Many of the values in the tables have been compiled by combining the standard enthalpy of formation of the species with the standard entropies of the compound and the elements, as illustrated above, but there are other sources of data, and we shall encounter some of them later.

Once standard Gibbs energies of formation are available, it is easy to combine them to obtain the standard reaction Gibbs energy of almost any reaction. We use the now familiar expression

$$\Delta_r G^{\ominus} = \sum_{\text{Products}} n\Delta_f G^{\ominus}(\text{products}) - \sum_{\text{Reactants}} n\Delta_f G^{\ominus}(\text{reactants}) \qquad (16)$$

For example, to determine the standard reaction Gibbs energy for

$$2\,CO(g) + O_2(g) \longrightarrow 2\,CO_2(g)$$

we carry out the following calculation:

$$\begin{aligned}
\Delta_r G^{\ominus} &= 2\Delta_f G^{\ominus}(CO_2, g) - \{2\Delta_f G^{\ominus}(CO, g) + \Delta_f G^{\ominus}(O_2, g)\} \\
&= 2 \times (-394 \text{ kJ mol}^{-1}) - \{2 \times (-137 \text{ kJ mol}^{-1}) + 0\} \\
&= -514 \text{ kJ mol}^{-1}
\end{aligned}$$

Calculations of this kind are of considerable usefulness, for we shall now go on to see that from them we can predict the equilibrium constant of the corresponding reaction.

Exercise E3.10

Calculate the standard reaction Gibbs energy of the oxidation of ammonia to nitric oxide according to the reaction

$$4\,NH_3(g) + 5\,O_2(g) \longrightarrow 4\,NO(g) + 6\,H_2O(g)$$

[*Answer*: $-959.42\,kJ\,mol^{-1}$]

The standard Gibbs energies of formation of compounds have their own significance as well as being useful in calculations like the one just illustrated. They are like a measure of the thermodynamic altitude of a compound above or below a 'sea level' of stability represented by the elements in their reference states (Fig. 3.9). If the standard Gibbs energy of formation is positive, so that the compound lies above 'sea level' (like ozone), then the compound has a spontaneous tendency to sink towards thermodynamic sea level and decompose into the elements. We say that the compound is **thermodynamically unstable** with respect to its elements. Thus, ozone, for which $\Delta_f G^{\ominus} = +163\,kJ\,mol^{-1}$, has a spontaneous tendency to decompose into oxygen under standard conditions at 25 °C. Although ozone is thermodynamically unstable, it can survive if the reactions that convert it into oxygen are slow: that is the case in the upper atmosphere, and the O_3 molecules in the ozone layer (which help to protect us from the ultraviolet radiation of the sun) survive for long periods. Benzene ($\Delta_f G^{\ominus} = +124\,kJ\,mol^{-1}$) is also thermodynamically unstable with respect to its elements. However, the fact that bottles of benzene are everyday laboratory commodities also reminds us that *spontaneity is a thermodynamic tendency that might not be realized at a significant rate in practice.*

Another useful point that can be made about standard Gibbs energies of formation is that there is no point in searching for *direct* syntheses of thermodynamically unstable compounds from their elements (under standard conditions, at the temperature to which the determination applies), because the reaction does not occur in the required direction: the *reverse* reaction, decomposition, is spontaneous. Such compounds must be synthesized by alternative routes or under conditions (such as those of temperature or pressure) for which their Gibbs energy of formation is negative and they lie beneath thermodynamic sea level.

Compounds with negative Gibbs energies of formation are said to be **thermodynamically stable** with respect to their elements: such compounds lie below the thermodynamic sea level of the elements (under standard

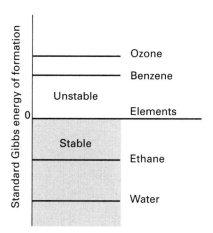

Fig. 3.9 The standard Gibbs energy of formation of compounds is like a measure of a compound's altitude above (or below) sea level: compounds that lie above sea level have a spontaneous tendency to decompose into the elements (and to revert to sea level). Compounds that lie below sea level are stable with respect to decomposition into the elements.

conditions). An example is ethane gas, with $\Delta_f G^{\ominus} = -33\,\text{kJ mol}^{-1}$: the sign of this quantity shows that the formation of ethane is spontaneous and its decomposition (under standard conditions at $25\,°C$) is non-spontaneous.

Reaction Gibbs energies at arbitrary concentrations

Because the standard reaction Gibbs energy is defined in terms of the pure reactants and products, it relates to a very specific type of change: from the pure, unmixed reactants in their standard states to the pure, unmixed products in their standard states. It follows that if $\Delta_r G^{\ominus}$ is negative, then the change from pure reactants to pure products is spontaneous. However, in chemistry we may be more interested in knowing whether an arbitrary *mixture* of reactants and products has a tendency to form more products, and at first sight $\Delta_r G^{\ominus}$ would not appear to be relevant to that question. For example, consider the synthesis of hydrogen iodide,

$$H_2(g) + I_2(s) \longrightarrow 2\,HI(g)$$

for which $\Delta_r G^{\ominus} = +3.4\,\text{kJ mol}^{-1}$. We would conclude that the formation of hydrogen iodide gas is not spontaneous at $25\,°C$. However, in practice, hydrogen and iodine do react, and produce a mixture in which hydrogen, iodine, and hydrogen iodide are all present at equilibrium. Apparently, although the Gibbs energy of pure HI is higher than that of pure $H_2(g)$ and $I_2(s)$, there is an intermediate composition, when all three substances are present, that has a lower Gibbs energy than either the pure reactants or the pure products, and that mixture forms spontaneously from the pure starting materials.

To identify the composition that corresponds to the lowest Gibbs energy, we introduce the **reaction Gibbs energy**, $\Delta_r G$. This quantity is defined like the standard reaction Gibbs energy, but refers to the conversion of reactants to products *in a mixture of fixed composition*. To imagine what this means, you could think of a reaction taking place in such a huge volume of material that the composition of the mixture does not change significantly when an additional 1 mol of reactants has been consumed. For example, consider the hydrogen iodide reaction after it has reached a stage with a certain arbitrary composition. Suppose, for example, that all three substances happen to be present in equal abundance, with $10^{10}\,\text{mol}\,H_2$, $10^{10}\,\text{mol}\,I_2$, and $10^{10}\,\text{mol}\,HI$, or some similarly vast but equal amounts. Then, when $1\,\text{mol}\,H_2$ reacts with $1\,\text{mol}\,I_2$, the reaction Gibbs energy is $\Delta_r G$, but the composition of the mixture remains virtually the same. We could choose a system with a different composition, such as one in which there are $0.5 \times 10^{10}\,\text{mol}\,H_2$, $0.5 \times 10^{10}\,\text{mol}\,I_2$, and $2 \times 10^{10}\,\text{mol}\,HI$ (or some other vast amounts in the ratio 1:1:4), and once again consider the reaction in which $1\,\text{mol}\,H_2$ reacts with $1\,\text{mol}\,I_2$. In this case, a different value of $\Delta_r G$ will be found.

It is very important to keep the distinction clear between $\Delta_r G$ and $\Delta_r G^\ominus$:

1. $\Delta_r G^\ominus$ is the change in Gibbs energy when the pure reactants change into the pure products, with each species in its standard state.
2. $\Delta_r G$ is the change in Gibbs energy when the reaction occurs under conditions of constant composition.

We have seen that (at constant temperature and pressure) a process tends to move in the direction corresponding to a decrease in Gibbs energy. Therefore, if $\Delta_r G$ is negative at a certain composition, the reactants will have a spontaneous tendency to form more products. Conversely, if $\Delta_r G$ is positive, the *reverse* reaction is spontaneous, and the composition will tend to adjust by the products that are already present tending to decompose into reactants. If, however, $\Delta_r G$ happens to be zero at a particular composition, then the reaction system will have no tendency to form either products or reactants. In other words, the system will be at equilibrium. We can conclude that the criterion for chemical equilibrium is that, at constant temperature and pressure,

$$\Delta_r G = 0 \tag{17}$$

All that remains to be done is to determine how $\Delta_r G$ varies with the composition of the system, for then we can identify the composition corresponding to this criterion. This important problem is solved in the following section.

3.4 The relation between $\Delta_r G$ and $\Delta_r G^\ominus$

First, we shall review some empirical results relating to chemical equilibria. In elementary chemistry we have been introduced to the equilibrium constant, K, which is characteristic of the equilibrium composition of a reaction. If $K \gg 1$ (typically, if K is larger than about 10^3), then at equilibrium the products dominate the reaction mixture. If $K \ll 1$ (typically, less than about 10^{-3}), then at equilibrium the reactants dominate the mixture and only a small proportion of products are formed. If $K \approx 1$, then both reactants and products are present in similar abundance. The following paragraphs put these points on a thermodynamic footing.

Definition and properties of the equilibrium constant

There are two steps in writing down the expression for the equilibrium constant of a particular reaction. First, we formulate the **reaction quotient**,

Q, of a gas phase reaction in terms of the partial pressures as illustrated in the following example:

$$N_2(g) + 3\,H_2(g) \longrightarrow 2\,NH_3(g) \qquad Q = \frac{(p_{NH_3}/p^{\ominus})^2}{(p_{N_2}/p^{\ominus})(p_{H_2}/p^{\ominus})^3}$$

In this expression, p_J is the partial pressure of the species J and $p^{\ominus}$ is the standard pressure ($p^{\ominus} = 1$ bar). The products appear in the numerator and the reactants appear in the denominator, with each partial pressure raised to a power equal to the corresponding stoichiometric number. Note that Q is dimensionless because the units of the partial pressures are cancelled by the units of the standard pressure. For a reaction in solution, the molar concentrations divided by $1\,mol\,L^{-1}$ are used in place of the partial pressures. (We refine this definition shortly.)

In the second step we allow the partial pressures (or molar concentrations) to reach their equilibrium values. The reaction quotient then has its equilibrium value, which is called the **equilibrium constant**, K, for the reaction:

$$N_2(g) + 3\,H_2(g) \rightleftharpoons 2\,NH_3(g) \qquad K = Q_{eq} = \left(\frac{(p_{NH_3}/p^{\ominus})^2}{(p_{N_2}/p^{\ominus})(p_{H_2}/p^{\ominus})^3}\right)_{eq}$$

where Q_{eq} is evaluated using the partial pressures of the gases in the equilibrium mixture. The sign $\rightleftharpoons$ signifies a state of **dynamic equilibrium** in which the forward and reverse reactions are continuing but as their rates are equal there is no net change. The appearance of the expressions for Q and K can be simplified by writing $a_J = p_J/p^{\ominus}$ for each species, for then

$$Q = \frac{a_{NH_3}^2}{a_{N_2} a_{H_2}^3}$$

There are several advantages to this change of notation: one is the simplicity of the expressions that result. The second is the ease with which this expression is generalized to equilibria that involve other phases, and the third is the ease with which it is possible to make the transition to discussing real gases (and other non-ideal systems). All we need note at this stage, though, is that a_J is a dimensionless quantity, which for gases is equal to $p_J/p^{\ominus}$.

In general, a chemical reaction and its reaction quotient have the form

$$a\,A + b\,B \longrightarrow c\,C + d\,D \qquad Q = \frac{a_C^c a_D^d}{a_A^a a_B^b} \qquad (18)$$

and at equilibrium,

$$a\,A + b\,B \rightleftharpoons c\,C + d\,D \qquad K = Q_{eq} = \left(\frac{a_C^c a_D^d}{a_A^a a_B^b}\right)_{eq} \qquad (19)$$

The a_J in these expressions is called the **activity** of species J. For our purposes, it is sufficient to use the following values of a_J:

for perfect gases: $a_J = p_J/p^{\ominus}$
for pure liquids and solids: $a_J = 1$
for solutes at low concentrations: $a_J = [J]/(1\,\mathrm{mol\,L^{-1}})$

where $[J]$ is the molar concentration of the species J. An example of the general formula is the following expression for the hydrogen iodide reaction

$$H_2(g) + I_2(s) \longrightarrow 2\,HI(g)$$

(note the appearance of *solid* iodine in this version of the reaction) for which

$$Q = \frac{a_{HI}^2}{a_{H_2}a_{I_2}} = \frac{(p_{HI}/p^{\ominus})^2}{(p_{H_2}/p^{\ominus})} = \frac{p_{HI}^2}{p_{H_2}p^{\ominus}}$$

At equilibrium,

$$H_2(g) + I_2(s) \rightleftharpoons 2\,HI(g) \qquad K = \left(\frac{p_{HI}^2}{p_{H_2}p^{\ominus}}\right)_{eq}$$

We have made use of the fact that as iodine is present as a solid, $a_{I_2} = 1$. In more advanced work, the activities take into account the effects of interactions between species (such as intermolecular interactions in reactions of real gases and interionic forces in electrolyte solutions).

Example *Writing a reaction quotient and an equilibrium constant*
An amino acid in solution is in equilibrium with its zwitterionic form, in which a proton is lost from the carboxyl group and gained by the amino group. Express the reaction quotient and equilibrium constant for the equilibrium in the case of glycine, NH_2CH_2COOH.

Strategy
Write the chemical equation for the reaction, and then use eqn (18) to express the reaction quotient in terms of the activities. For a dilute solution, the activities may be replaced by the numerical values of molar concentrations, as specified in the text (that is, the molar concentration divided by $1\,\mathrm{mol\,L^{-1}}$ to cancel the units). The equilibrium constant is the value of the reaction quotient when the reaction is at equilibrium.

Solution
The chemical equation for the equilibrium is

$$NH_2CH_2COOH(aq) \rightleftharpoons {}^+NH_3CH_2CO_2^-(aq)$$

The reaction quotient at any stage is

$$Q = \frac{a(^{+}NH_3CH_2CO_2^{-})}{a(NH_2CH_2COOH)} \approx \frac{[^{+}NH_3CH_2CO_2^{-}]}{[NH_2CH_2COOH]}$$

The second equality follows from the cancellation of the $1 \, mol \, L^{-1}$ that appears in the definition of each activity. The second equality is also valid only when the solution is so dilute that ion–ion interactions can be ignored. At equilibrium, Q is equal to a constant K, the equilibrium constant for zwitterion formation.

Exercise E3.11

Write the reaction quotient and equilibrium constant for an esterification reaction of the form $CH_3COOH + C_2H_5OH \rightleftharpoons CH_3COOC_2H_5 + H_2O$. (All four components are present in the reaction mixture: it should not be regarded as an aqueous solution.)

[*Answer*: $Q \approx [CH_3COOC_2H_5][H_2O]/[CH_3COOH][C_2H_5OH]$]

Physical transitions

The concept of equilibrium constant applies to physical transformations as well as to chemical reactions, and quantities familiar from other contexts are seen to be equilibrium constants in disguise. For example, the equilibrium constant for the vaporization of water turns out to be nothing other than its **vapour pressure**, the pressure of the vapour in equilibrium with the liquid at the temperature of the experiment. This conclusion follows from the chemical equation for the vaporization equilibrium:[†]

$$H_2O(l) \rightleftharpoons H_2O(g) \qquad K = \frac{a(H_2O, g)}{a(H_2O, l)} = \frac{p(H_2O)}{p^{\ominus}}$$

because the activity of pure liquid water is 1. It follows that

$$p(H_2O) = Kp^{\ominus} \tag{20}$$

at equilibrium. For example, the vapour pressure of water at $25\,^{\circ}C$ is $0.032 \, bar$, so we can conclude that $K = 0.032$ at this temperature. Alternatively, if we knew how K varied with temperature, then we would know the temperature variation of the vapour pressure of water.

[†]When expressions for K are given, we shall not normally mark them with an 'eq' subscript: it should be understood that all partial pressures and concentrations are the equilibrium values.

The variation of $\Delta_r G$ with composition

The justification for the use of a reaction quotient for expressing the composition of a general reaction mixture and of the equilibrium constant for reactions at equilibrium stems from the relation between $\Delta_r G$ and $\Delta_r G^{\ominus}$, which we shall now establish.

First, we need to know that the molar Gibbs energy of a perfect gas is related to its partial pressure in a mixture by

$$G_m = G_m^{\ominus} + RT \ln \frac{p}{p^{\ominus}} \qquad (21)$$

A simpler version of this expression is

$$G_m = G_m^{\ominus} + RT \ln a \qquad (22)$$

where, as usual, $a = p/p^{\ominus}$. Thus, if the gas is at a pressure of 1 bar, then $a = 1$, $\ln 1 = 0$, and so $G_m = G_m^{\ominus}$. The logarithmic term takes into account the possibility that the gas is present at a partial pressure of other than 1 bar.

Exercise E3.12

Calculate the difference between G_m and $G_m^{\ominus}$ for a gas when the pressure falls from 1.00 bar to 0.50 bar as a result of a reaction at $25\,^{\circ}\text{C}$.

[*Answer*: $G_m - G_m^{\ominus} = -1.7\,\text{kJ}\,\text{mol}^{-1}$]

Justification

The proof of eqn (21) starts at the definition of dG for a change at constant temperature, the 'infinitesimal' version of eqn (9):

$$dG = dH - TdS$$

We know that $H = U + pV$, so a change in H can be expressed in terms of changes in each of U, p, and V:

$$
\begin{aligned}
dH &= dU + pdV + Vdp \\
&= dq + dw + pdV + Vdp \qquad (\text{because } dU = dq + dw) \\
&= TdS - pdV + pdV + Vdp \qquad (\text{because } dq = TdS \\
&\qquad\qquad\qquad\text{and } dw = -pdV \text{ for reversible changes}) \\
&= TdS + Vdp
\end{aligned}
$$

103

When this expression is substituted into the one for dG given above, we obtain

$$dG = V\,dp$$

If the sample is incompressible, so that V is independent of p, this expression integrates to

$$\Delta G = V\Delta p$$

Therefore, to find the change in G of an incompressible solid or liquid when the pressure is increased isothermally, we multiply the volume of the sample by the change in pressure.

If the substance is a gas, the overall change in Gibbs energy when the pressure of the gas changes from p_i to p_f must allow for the fact that V changes with pressure. If we treat the gas as perfect, we can relate the volume to the pressure by using the perfect gas equation of state, $V = nRT/p$, and obtain

$$\Delta G = \int_{p_i}^{p_f} V\,dp = nRT \int_{p_i}^{p_f} \frac{dp}{p} = nRT\,\ln\frac{p_f}{p_i}$$

To obtain eqn (21), we set $p_i = p^{\ominus}$, at which pressure $G = G^{\ominus}$, and $p_f = p$, when the Gibbs energy has the value G. Then $\Delta G = G - G^{\ominus}$, and all we need then do is divide through by the amount n to obtain molar quantities.

In the next step we consider a reaction of the form

$$A(g) + 2\,B(g) \longrightarrow 3\,C(g)$$

(We could choose any other set of stoichiometric numbers, but these are sufficiently general.) The reaction Gibbs energy at an intermediate stage of the reaction is

$$\begin{aligned}
\Delta_r G &= 3G_m(C) - \{G_m(A) + 2G_m(B)\}\\
&= 3\big(G_m^{\ominus}(C) + RT\,\ln a_C\big) - \big\{\big(G_m^{\ominus}(A) + RT\,\ln a_A\big)\\
&\qquad\qquad + 2\big(G_m^{\ominus}(B) + RT\,\ln a_B\big)\big\}\\
&= 3G_m^{\ominus}(C) - \{G_m^{\ominus}(A) + 2G_m^{\ominus}(B)\}\\
&\qquad + RT\{3\ln a_C - \ln a_A - 2\ln a_B\}
\end{aligned}$$

The first term on the right in the last line is the standard reaction Gibbs energy:

$$\Delta_r G^{\ominus} = 3G_m^{\ominus}(C) - \{G_m^{\ominus}(A) + 2G_m^{\ominus}(B)\}$$

The second term can be rearranged as follows:[†]

$$RT\{3\ln a_C - \ln a_A - 2\ln a_B\} = RT \ln \frac{a_C^3}{a_A a_B^2}$$
$$= RT \ln Q$$

and the reaction quotient has emerged! The overall expression becomes

$$\Delta_r G = \Delta_r G^\ominus + RT \ln Q \tag{23}$$

This equation applies to all reactions, the only difference being that Q is the relevant reaction quotient. At this stage, we see that the reaction quotient appears naturally in the expression for the difference between the reaction Gibbs energy at the specified composition and the standard reaction Gibbs energy.

Exercise E3.13

Calculate the reaction Gibbs energy for $N_2(g) + 3H_2(g) \longrightarrow 2NH_3(g)$ at $25\,°C$ when the partial pressures of nitrogen, hydrogen, and ammonia are 0.20 bar, 0.42 bar, and 0.61 bar, respectively. In which direction is the reaction spontaneous under these conditions?

[*Answer*: $-25\,\text{kJ mol}^{-1}$, forward]

Reactions at equilibrium

We have arrived at the point where we are poised to derive one of the most important equations in chemical thermodynamics, the relation between the standard reaction Gibbs energy and the equilibrium constant.

3.5 The condition of equilibrium

Suppose the reaction has reached equilibrium: it has no further tendency to change and $\Delta_r G = 0$. At equilibrium, the reaction quotient has the value K. It then follows, by inserting these two values into eqn (23), that

$$0 = \Delta_r G^\ominus + RT \ln K$$

and hence that

$$\Delta_r G^\ominus = -RT \ln K \tag{24}$$

[†]For this step we have used $x \ln y = \ln y^x$, $\ln x + \ln y = \ln xy$, and $\ln x - \ln y = \ln (x/y)$.

This is one of the most important equations in the whole of chemical thermodynamics. Its principal use is to predict the value of the equilibrium constant of any reaction from tables of thermodynamic data, like those in Appendix 1. Alternatively, it can be used to measure $\Delta_r G^{\ominus}$ of a reaction following an experimental determination of the equilibrium constant of the reaction.

Example *Predicting an equilibrium constant*

Use the information in Table 3.4 to calculate the equilibrium constant of the reaction $H_2(g) + I_2(s) \rightleftharpoons 2\,HI(g)$ at 25 °C.

Strategy

To use eqn (24) we need the standard reaction Gibbs energy for the reaction. It can be calculated from the standard Gibbs energy of formation for the species (in this case, for HI). Then rearrange eqn (24) to

$$\ln K = -\frac{\Delta_r G^{\ominus}}{RT}$$

and substitute the data. Use $R = 8.3145\,\mathrm{J\,K^{-1}\,mol^{-1}}$.

Solution

The standard reaction Gibbs energy is

$$\Delta_r G^{\ominus} = 2\Delta_f G^{\ominus}(HI, g) = 2 \times (1.70\,\mathrm{kJ\,mol^{-1}}) = +3.40\,\mathrm{kJ\,mol^{-1}}$$

It follows that the equilibrium constant is

$$\ln K = -\frac{3.40 \times 10^3\,\mathrm{J\,mol^{-1}}}{(8.3145\,\mathrm{J\,K^{-1}\,mol^{-1}}) \times (298\,\mathrm{K})} = -1.37$$

Therefore,

$$K = e^{-1.37} = 0.25$$

Therefore, at equilibrium, the partial pressures of hydrogen and hydrogen iodide satisfy the relation

$$\frac{p_{HI}^2}{p_{H_2} p^{\ominus}} = 0.25$$

That is, because $p^{\ominus} = 1$ bar, at equilibrium

$$p_{HI}^2 = p_{H_2} \times 0.25\,\mathrm{bar}$$

Exercise E3.14

Calculate the equilibrium constant of the reaction $N_2(g) + 3 H_2(g) \rightleftharpoons 2 NH_3(g)$ at $25\,°C$.

[*Answer*: 6.0×10^5]

The relation between K and $\Delta_r G^\ominus$ given in eqn (24) enables us to judge the feasibility of a chemical reaction from tables of standard Gibbs energies of formation. First, we note that the equation implies that $K > 1$ if $\Delta_r G^\ominus$ is negative. Broadly speaking, an equilibrium constant greater than 1 implies that the products will be dominant at equilibrium, so we can conclude that *a reaction is thermodynamically feasible if $\Delta_r G^\ominus$ is negative*. Conversely, because eqn (24) implies that $K < 1$ when $\Delta_r G^\ominus$ is positive, then we know that the reactants will be dominant in a reaction mixture at equilibrium if for that reaction $\Delta_r G^\ominus$ is positive. Some care must be exercised with these rules, however, because the products will be significantly more abundant than reactants only if K is much larger than 1 (more than about 10^3, as remarked earlier).

3.6 The response of equilibria to the conditions

The presence of a catalyst does not change the equilibrium constant of a reaction. The justification of this statement relies on the recognition of the fact that the value of K is determined by $\Delta_r G^\ominus$, which is the difference between the standard Gibbs energies of the products and reactants, and has the same value however the reaction is brought about in practice. Therefore, even though a catalyst may speed a reaction by providing a different reaction pathway between the reactants and the products, it has no effect on $\Delta_r G^\ominus$ and therefore no effect on K: *the equilibrium composition of a reaction mixture is independent of the presence of a catalyst*.

The effect of temperature

The equilibrium constant of a reaction changes when the temperature is changed. To assess the magnitude of the effect of temperature we make use of the relation

$$\Delta_r G^\ominus = \Delta_r H^\ominus - T \Delta_r S^\ominus$$

and consider it term by term. To do so, we use the approximation that the standard reaction enthalpy and entropy are independent of temperature over the range of interest (this approximation was justified for $\Delta_r H^\ominus$ in

Section 2.8), so the entire temperature dependence of $\Delta_r G^\ominus$ stems from the T in the expression above.

At a temperature T,

$$\ln K = -\frac{\Delta_r G^\ominus}{RT} = -\frac{\Delta_r H^\ominus}{RT} + \frac{\Delta_r S^\ominus}{R}$$

At another temperature T', when

$$\Delta_r G^{\ominus\prime} = \Delta_r H^\ominus - T' \Delta_r S^\ominus$$

a similar expression holds:

$$\ln K' = -\frac{\Delta_r H^\ominus}{RT'} + \frac{\Delta_r S^\ominus}{R}$$

The difference between the two is

$$\ln K - \ln K' = -\frac{\Delta_r H^\ominus}{R}\left(\frac{1}{T} - \frac{1}{T'}\right)$$

which can be rearranged to

$$\ln K' = \ln K + \frac{\Delta_r H^\ominus}{R}\left(\frac{1}{T} - \frac{1}{T'}\right) \tag{25}$$

This formula is a version of the **van't Hoff equation**.

Consider the case when T' is higher than T. Then the term in parentheses in eqn (25) is positive. If $\Delta_r H^\ominus$ is positive, corresponding to an endothermic reaction, then the second term *adds* to $\ln K$, and so $\ln K'$ is larger than $\ln K$. That being so, we conclude that K' is larger than K for an endothermic reaction. In general, *the equilibrium constant of an endothermic reaction increases with temperature*. The opposite will be true when $\Delta_r H^\ominus$ is negative, so we can conclude that *the equilibrium constant of an exothermic reaction decreases with an increase in temperature*. These observations are consistent with an empirical observation known as **Le Chatelier's principle**:

> When a system at equilibrium is subjected to a disturbance, the composition of the system adjusts so as to tend to minimize the effect of the disturbance.

A reduction in temperature favours an exothermic reaction, for the heat released tends to oppose the lowering of temperature. An increase in temperature favours an endothermic reaction, for the heat absorbed tends to oppose the increase in temperature.

The conclusions we have outlined are of considerable commercial and environmental significance. For example, the synthesis of ammonia is exothermic, so its equilibrium constant decreases as the temperature is increased; in fact, it falls below 1 at about 200 °C. Unfortunately, the reaction is slow at low temperatures, and is commercially feasible only if the temperature exceeds about 750 °C even in the presence of a catalyst; but then K is very small. We shall see shortly how Fritz Haber, the inventor of the Haber process for the commercial synthesis of ammonia, was able to overcome this difficulty. Another example is the oxidation of nitrogen:

$$N_2(g) + O_2(g) \longrightarrow 2\,NO(g)$$

This reaction is endothermic ($\Delta_r H^{\ominus} = +180\,\text{kJ}\,\text{mol}^{-1}$) largely as a consequence of the very high bond enthalpy of N_2, and its equilibrium constant increases with temperature. It is for this reason that nitrogen monoxide (nitric oxide) is formed in significant quantities in the hot exhausts of jet engines and in the hot exhaust manifolds of internal combustion engines, and then goes on to contribute to the problems of acid rain.

The effect of pressure

First, we note that $\Delta_r G^{\ominus}$ is defined as the difference between the Gibbs energies of substances in their standard states (and therefore at 1 bar). Therefore, $\Delta_r G^{\ominus}$ has the same value whatever the actual pressure used for the reaction. Hence, because $\ln K$ is proportional to $\Delta_r G^{\ominus}$, K is *independent of pressure*. Thus, if the pressure of a reaction vessel in which ammonia is being synthesized is increased, the equilibrium constant remains unchanged.

This rather startling conclusion should not be misinterpreted. The value of K is independent of the pressure to which the system is subjected, but because partial pressures occur in the expression for K in a rather complicated way in general, that does not mean that the *individual* partial pressures or concentrations are unchanged. Suppose, for example, the volume of the reaction vessel in which the hydrogen iodide reaction has reached equilibrium is reduced by a factor of 2 and the system is allowed to reach equilibrium again. If the partial pressures were simply to double (that is, there is no adjustment of composition by further reaction), the reaction quotient would change from

$$Q_1 = \frac{p_{\text{HI}}^2}{p_{\text{H}_2} p^{\ominus}}$$

to

$$Q_2 = \frac{(2p_{\text{HI}})^2}{(2p_{\text{H}_2})p^{\ominus}} = 2Q_1$$

However, the two expressions clearly cannot both be equal to the equilibrium constant K. Therefore, the two partial pressures must adjust by different amounts. In this instance, the two reaction quotients remain equal if the partial pressure of HI changes by a smaller amount than a factor of 2 and the partial pressure of H_2 increases by more than a factor of 2; for then the numerator in Q_2 would be reduced and the denominator increased, with the result that Q_2 would be smaller than we have just calculated (and would in fact have to be equal to Q_1). In other words, the equilibrium composition must shift in the direction of the reactants in order to preserve the equilibrium constant.

The general rule of thumb for predicting the effect of increased pressure on a gas-phase reaction at equilibrium is as follows:

> When a system at equilibrium is compressed, the composition of a gas-phase equilibrium adjusts so as to reduce the number of molecules in the gas phase.

This statement can be regarded as another special case of Le Chatelier's principle.

In the hydrogen iodide reaction, one H_2 molecule forms two HI molecules, so an increase in pressure favours the decomposition of HI into hydrogen (and solid iodine). Now consider the synthesis of ammonia:

$$N_2(g) + 3 H_2(g) \rightleftharpoons 2 NH_3(g)$$

From each four reactant molecules, only two product molecules are formed, so an increase in pressure favours the formation of ammonia. Indeed, this is the key to resolving Haber's dilemma, for by working at high pressure he was able to increase the yield of ammonia even though the equilibrium constant for the reaction remained low.

Exercise E3.15

Is the formation of products in the reaction $4 NH_3(g) + 5 O_2(g) \rightleftharpoons 4 NO(g) + 6 H_2O(g)$ favoured by an increase or a decrease in pressure?

[*Answer*: decrease]

Care should be taken to note that under certain circumstances compression has *no* effect on the equilibrium composition. A simple example is the effect of compression on a reaction in which the number of gas phase molecules is the same in the reactants as in the products. An example is the oxidation of nitrogen:

$$N_2(g) + O_2(g) \rightleftharpoons 2 NO(g)$$

Another is the synthesis of hydrogen iodide in which all three substances are present in the gas phase:

$$H_2(g) + I_2(g) \rightleftharpoons 2 HI(g)$$

A more subtle example is the effect of the addition of an inert gas to a reaction mixture that is contained inside a vessel of constant volume. The overall pressure increases as the gas (such as argon) is added, but the addition of a foreign gas does not affect the partial pressures of the other gases present. (The partial pressure, recall, is the pressure a gas would exert if it alone occupied the vessel, so it is independent of the presence or absence of any other gases.) Therefore, under these circumstances, not only does the equilibrium constant remain unchanged, but the partial pressures of the reactants and products remain the same whatever the stoichiometry of the reaction.

EXERCISES

Treat all gases as perfect.

3.1 A goldfish swims in a bowl of water at $20\,^{\circ}$C. Over a period of time, the fish transfers $120\,$J to the water as a result of its metabolism. What is the change in entropy of the water?

3.2 Suppose you put a cube of ice of mass $100\,$g into a glass of water at just above $0\,^{\circ}$C. When the ice melts, about $33\,$kJ of energy is absorbed as heat. What is the change in entropy of (a) the sample (the ice), (b) the surroundings (the glass of water)?

3.3 A sample of aluminium of mass $1.25\,$kg is cooled at constant pressure from $300\,$K to $260\,$K. Calculate the energy that must be removed as heat and the change in entropy of the sample. The molar heat capacity of aluminium is $24.35\,$J$\,$K$^{-1}\,$mol^{-1}.

3.4 Calculate the change in entropy of $100\,$g of ice at $0\,^{\circ}$C as it is melted, heated to $100\,^{\circ}$C, and then vaporized at that temperature. Suppose that the changes are brought about by a heater that supplies heat at a constant rate, and sketch a graph showing (a) the change in temperature of the system, (b) the enthalpy of the system, (c) the entropy of the system as a function of time.

3.5 Whenever a gas expands—when we exhale, when a flask is opened, and so on—the gas undergoes an increase in entropy. A sample of methane gas of mass $25\,$g at $250\,$K and $185\,$kPa expands isothermally until its pressure is $2.5\,$kPa. Calculate the change in entropy of the gas.

3.6 A sample of carbon dioxide that initially occupies $15.0\,$L at $250\,$K and $1.00\,$atm is compressed isothermally. Into what volume must the gas be compressed to reduce its entropy by $10.0\,$J$\,$K^{-1}?

3.7 Calculate the change in molar entropy when a sample of argon is compressed from $2.0\,$L to $500\,$mL and simultaneously heated from $300\,$K to $400\,$K. Take $C_{V,\mathrm{m}} = \frac{3}{2}R$.

3.8 A monatomic perfect gas at a temperature T is expanded isothermally to twice its initial volume. To what temperature should it be cooled to restore its entropy to its initial value? Take $C_{V,\mathrm{m}} = \frac{3}{2}R$.

3.9 In a certain cyclic engine (technically, a Carnot cycle), a perfect gas expands isothermally and reversibly, then adiabatically ($q = 0$) and reversibly. In the adiabatic expansion step the temperature falls. At the end of the expansion stage, the sample is compressed reversibly first isothermally and then adiabatically in such a way as to end up at the starting volume and temperature. Draw a graph of entropy against temperature for the entire cycle.

3.10 Calculate the change in entropy when $100\,$g of water at $80\,^{\circ}$C is poured into $100\,$g of water at $10\,^{\circ}$C in an insulated vessel given that $C_{p,\mathrm{m}} = 75.5\,J\,K^{-1}\,$mol^{-1}.

3.11 The enthalpy of vaporization of chloroform, $CHCl_3$, is $29.4\,$kJ$\,$mol^{-1} at its normal boiling point of $334.88\,$K. (a) Calculate the entropy of vaporization of chloroform at this temperature. (b) What is the entropy change in the surroundings?

3.12 Calculate the standard reaction entropy at $298\,$K of

(a) $2\,CH_3CHO(g) + O_2(g) \longrightarrow 2\,CH_3COOH(l)$
(b) $2\,AgCl(s) + Br_2(l) \longrightarrow 2\,AgBr(s) + Cl_2(g)$
(c) $Hg(l) + Cl_2(g) \longrightarrow HgCl_2(s)$
(d) $Zn(s) + Cu^{2+}(aq) \longrightarrow Zn^{2+}(aq) + Cu(s)$
(e) $C_{12}H_{22}O_{11}(s) + 12\,O_2(g) \longrightarrow$
$$12\,CO_2(g) + 11\,H_2O(l)$$

3.13 The constant-pressure molar heat capacities of linear gaseous molecules are approximately $\frac{7}{2}R$ and those of nonlinear gaseous molecules are approximately $4R$. Estimate the change in standard reaction entropy of the following two reactions when the temperature is increased by 10 K at constant pressure:

(a) $2\,H_2(g) + O_2(g) \longrightarrow 2\,H_2O(g)$

(b) $CH_4(g) + 2\,O_2(g) \longrightarrow CO_2(g) + 2\,H_2O(g)$

3.14 Suppose that when you exercise, you consume 100 g of glucose and that all the energy released as heat remains in your body at 37 °C. What is the change in entropy of your body?

3.15 Combine the reaction entropies calculated in Exercise 3.12 with the reaction enthalpies and calculate the standard Gibbs energies of the reactions at 298 K.

3.16 Use standard Gibbs energies of formation to calculate the standard reaction Gibbs energy at 298 K of the reactions in Exercise 2.36.

3.17 Thermodynamics helps us to assess the maximum heat and work that we can expect to extract by running a particular reaction. What is the maximum energy that can be extracted as (a) heat, (b) non-expansion work when 1.0 kg of natural gas (taken to be pure methane) is burned under standard conditions at 25 °C? Take the reaction to be $CH_4(g) + 2\,O_2(g) \longrightarrow CO_2(g) + 2\,H_2O(l)$.

3.18 In assessing metabolic processes we are usually more interested in the work that may be performed for the consumption of a given mass of compound than the heat it can produce (which merely keeps the body warm). What is the maximum energy that can be extracted as (a) heat, (b) non-expansion work when 1.0 kg of glucose is burned under standard conditions at 25 °C with the production of water vapour? The reaction is $C_6H_{12}O_6(g) + 6\,O_2(g) \longrightarrow 6\,CO_2(g) + 6\,H_2O(g)$.

3.19 Is it more energy effective to ingest sucrose or glucose? Calculate the non-expansion work, the expansion work, and the total work that can be obtained from the combustion of 1.0 kg of sucrose under standard conditions at 25 °C when the product includes (a) water vapour, (b) liquid water.

3.20 The standard enthalpy of combustion of solid phenol, C_6H_5OH, is $-3054\,kJ\,mol^{-1}$ at 298 K and its standard molar entropy is $144.0\,J\,K^{-1}\,mol^{-1}$. Calculate the standard Gibbs energy of formation of phenol at 298 K.

3.21 Calculate the maximum non-expansion work per mole that may be obtained from a fuel cell in which the chemical reaction is the combustion of methane at 298 K.

3.22 The enthalpy of the graphite $\longrightarrow$ diamond phase transition, which under 100 kbar occurs at 2000 K, is $+1.9\,kJ\,mol^{-1}$. Calculate the entropy change of the transition.

3.23 The equilibrium constant for the isomerization of *cis*-2-butene to *trans*-2-butene is $K = 2.07$ at 400 K. Calculate the standard reaction Gibbs energy for the isomerization.

3.24 The standard reaction Gibbs energy of the isomerization of *cis*-2-pentene to *trans*-2-pentene at 400 K is $-3.67\,kJ\,mol^{-1}$. Calculate the equilibrium constant of the isomerization.

3.25 One biochemical reaction has a standard Gibbs energy of $-200\,kJ\,mol^{-1}$ and a second biochemical reaction has a standard Gibbs energy of $-100\,kJ\,mol^{-1}$. What is the ratio of their equilibrium constants at 310 K?

3.26 One enzyme-catalysed reaction in a biochemical cycle has an equilibrium constant that is 10 times the equilibrium constant of a second reaction. If the standard Gibbs energy of the former reaction is $-300\,kJ\,mol^{-1}$, what is the standard reaction Gibbs energy of the second reaction?

3.27 What is the value of the equilibrium constant of a reaction for which $\Delta_r G^{\ominus} = 0$?

3.28 The standard reaction enthalpy of $Zn(s) + H_2O(g) \longrightarrow ZnO(s) + H_2(g)$ is approximately constant at $+224\,kJ\,mol^{-1}$ from 920 K up to 1280 K. The standard reaction Gibbs energy is $+33\,kJ\,mol^{-1}$ at 1280 K. Assuming that both quantities remain constant, estimate the temperature at which the equilibrium constant becomes greater than 1.

3.29 The oxidation of glucose in the mitochondria in energy-hungry brain cells leads to the formation of pyruvate ions, $CH_3COCO_2^-$, which are then decarboxylated to ethanal (acetaldehyde, CH_3CHO) in the course of the ultimate formation of carbon dioxide. The standard Gibbs energies of formation of pyruvate ions in aqueous solution and gaseous ethanal are $-474\,kJ\,mol^{-1}$ and $-133\,kJ\,mol^{-1}$, respectively. Calculate the Gibbs energy of the reaction in which pyruvate ions are converted to ethanal by the action of pyruvate decarboxylase with the release of carbon dioxide.

3.30 Acetaldehyde is soluble in water. Would you expect the standard Gibbs energy of the enzyme-catalysed reaction in which pyruvate ions are decarboxylated to ethanal in solution to be larger or smaller than the value for the production of gaseous ethanal? (See the preceding exercise.)

3.31 Pyruvic acid is a weak acid with $pK_a = 2.49$. Would you expect the standard Gibbs energy of the enzyme-catalysed reaction in which pyruvate ions are decarboxylated to ethanal in solution to be larger or smaller than the value for the decarboxylation of pyruvate ions?

3.32 The equilibrium constant of the reaction $2\,C_3H_6(g) \rightleftharpoons C_2H_4(g) + C_4H_8(g)$ is found to fit the expression

$$\ln K = -1.04 - \frac{1088\,\mathrm{K}}{T} + \frac{1.51 \times 10^5\,\mathrm{K}^2}{T^2}$$

between 300 K and 600 K. Calculate the standard reaction enthalpy and standard reaction entropy at 400 K. *Hint*. Begin by calculating $\ln K$ at 390 K and 410 K; then use eqn (25).

3.33 Borneol is a pungent compound obtained from the camphorwood tree of Borneo and Sumatra. The standard reaction Gibbs energy of the isomerization of borneol (**1**) to isoborneol (**2**) in the gas phase at 503 K is $+9.4\,\mathrm{kJ\,mol^{-1}}$. Calculate the reaction Gibbs energy in a mixture consisting of 0.15 mol of borneol and 0.30 mol of isoborneol when the total pressure is 600 Torr.

1 borneol

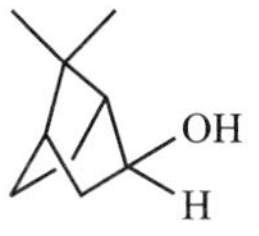

2 isoborneol

3.34 The equilibrium constant for the gas-phase isomerization of borneol, $C_{10}H_{17}OH$, to isoborneol at 503 K is 0.106. A mixture consisting of 7.50 g of borneol and 14.0 g of isoborneol in a container of volume 5.0 L is heated to 503 K and allowed to come to equilibrium. Calculate the mole fractions of the two substances at equilibrium.

3.35 The equilibrium pressure of H_2 over a mixture of solid uranium and solid uranium hydride at 500 K is 1.04 Torr. Calculate the standard Gibbs energy of formation of $UH_3(s)$ at 500 K.

3.36 Use the data in Appendix 1 to decide which of the following reactions have $K > 1$ at 298 K.

(a) $HCl(g) + NH_3(g) \longrightarrow NH_4Cl(s)$
(b) $2\,Al_2O_3(s) + 3\,Si(s) \longrightarrow 3\,SiO_2(s) + 4\,Al(s)$
(c) $Fe(s) + H_2S(g) \longrightarrow FeS(s) + H_2(g)$
(d) $FeS_2(s) + 2\,H_2(g) \longrightarrow Fe(s) + 2\,H_2S(g)$
(e) $2\,H_2O_2(l) + H_2S(g) \longrightarrow H_2SO_4(l) + 2\,H_2(g)$

3.37 Which of the products in Exercise 3.36 are favoured by a rise in temperature at constant pressure (in the sense that K increases)?

3.38 What is the standard enthalpy of a reaction for which the equilibrium constant is (a) doubled, (b) halved when the temperature is increased by 10 K at 298 K?

3.39 One of the most extensively studied reactions of industrial chemistry is the synthesis of ammonia, for its successful operation helps to govern the efficiency of the entire economy. The standard Gibbs energy of formation of $NH_3(g)$ is $-16.5\,\mathrm{kJ\,mol^{-1}}$ at 298 K. What is the reaction Gibbs energy when the partial pressure of the N_2, H_2, and NH_3 (treated as perfect gases) are 3.0 bar, 1.0 bar, and 4.0 bar, respectively? What is the spontaneous direction of the reaction in this case?

3.40 Could the synthesis of ammonia be used as the basis of a fuel cell? What is the maximum electrical energy output for the consumption of 100 g of nitrogen?

3.41 The dissociation vapour pressure (the pressure of gaseous products in equilibrium with the solid reactant) of NH_4Cl at 427 °C is 608 kPa but at 459 °C it has risen to 1115 kPa. Calculate (a) the equilibrium constant, (b) the standard reaction Gibbs energy, (c) the standard enthalpy, (d) the standard entropy of dissociation, all at 427 °C. Assume that the vapour behaves as a perfect gas and that $\Delta H^{\ominus}$ and $\Delta S^{\ominus}$ are independent of temperature in the range given.

3.42 Use the information in Appendix 1 to estimate the temperature at which (a) $CaCO_3$ decomposes spontaneously and (b) $CuSO_4 \cdot 5H_2O$ undergoes dehydration.

Phase equilibria

Boiling, freezing, and the conversion of graphite to diamond are all examples of **phase transitions**, or changes of phase without change of chemical composition. Many phase changes are common everyday phenomena, and their description is an important part of physical chemistry. They occur whenever a solid changes into a liquid (as in the melting of ice) or a liquid changes into a vapour (as in the vaporization of water in our lungs). They also occur when one solid phase changes into another, as in the conversion of graphite into diamond under high pressure, or the conversion of one phase of iron into another as it is heated in the process of steelmaking. Phase changes are important geologically too, for calcium carbonate is typically deposited as aragonite, but then gradually changes into another crystal form, calcite.

In this chapter we describe the properties of phases of pure substances and simple mixtures. Throughout, we use as our guiding principle the tendency of systems at constant temperature and pressure to adjust their Gibbs energy to the lowest possible value. Thus, if at a certain temperature and pressure the solid phase of a substance has a lower Gibbs energy than its liquid phase, then the solid is the thermodynamically stable phase and the liquid will freeze; if the opposite is true, then the liquid is the thermodynamically stable phase and the solid will melt. For example, ice is the thermodynamically stable phase of water at 1 atm and below $0\,°C$, and under these conditions water vapour and liquid water will both convert spontaneously to ice.

Exercise E4.1

The standard Gibbs energy of formation of metallic white tin (α-Sn) is 0 at $25\,°C$ and that of the nonmetallic grey tin (β-Sn) is $+0.13\,kJ\,mol^{-1}$ at the same temperature. Which is the thermodynamically stable phase at $25\,°C$?

[*Answer*: white tin]

At a given pressure, the Gibbs energy of one phase (the liquid phase, for instance) decreases as the temperature is raised, and there comes a

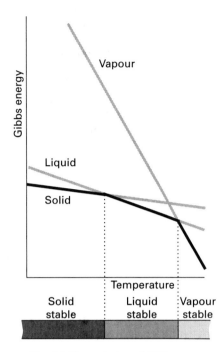

Fig. 4.1 The variation of Gibbs energy with temperature. The Gibbs energies of phases decrease with increasing temperature, but vapours decrease more rapidly than liquids, and liquids more rapidly than solids. Consequently, there are regions of temperature over which the solid, liquid, and vapour forms of a substance have the lowest Gibbs energy and which are therefore the most stable.

temperature at which it may fall below the Gibbs energy of a second phase (the solid phase, for instance), as shown in Fig. 4.1. Above this characteristic temperature the liquid phase is thermodynamically more stable; below it the solid phase is more stable, so the temperature is the **transition temperature** between the two phases. For instance, at 1 atm, the transition temperature for ice and liquid water is $0\,°C$ and the transition temperature for grey and white tin is $13\,°C$. At the transition temperature itself, the Gibbs energies of the two phases are identical, $\Delta_{trs}G = 0$, and there is no tendency for either phase to change into the other. At this temperature, therefore, the two phases are in equilibrium. Ice and liquid water are in equilibrium at 1 atm and $0\,°C$; the two allotropes of tin are in equilibrium at 1 atm and $13\,°C$.

As always when using thermodynamic arguments, it is important to keep in mind the distinction between the *spontaneity* of a phase transition and its *rate*: a phase transition that is predicted from thermodynamics to be spontaneous may occur so slowly as to be unimportant in practice. For instance, at normal temperatures and pressures the Gibbs energy of graphite is $3\,kJ\,mol^{-1}$ lower than that of diamond, and so there is a thermodynamic tendency for diamond to convert into graphite. However, for this transition to take place, the C atoms of diamond must change their locations, and because the bonds between the atoms are so strong and large numbers of bonds must change simultaneously, this process is unmeasurably slow except at high temperatures. In gases and liquids the mobilities of the molecules allow phase transitions to occur rapidly, but in solids thermodynamic instability may be frozen in and the thermodynamically unstable phase may persist for thousands of years.

Phase diagrams of single substances

The **phase diagram** of a substance is a map that shows the conditions of temperature and pressure at which its various phases are thermodynamically stable (Fig. 4.2). For example, at point A in the illustration, the gas phase of the substance is thermodynamically the most stable, but at B the liquid phase is the most stable. The boundaries between regions, which are called **phase boundaries**, show the values of p and T at which the two phases are in equilibrium. For example, if the system is arranged to have a pressure and temperature represented by point C, then the liquid and its vapour are in equilibrium (like liquid water and water vapour at 1 atm and $100\,°C$). If the temperature is reduced at constant pressure, the system moves to point B where the liquid is the thermodynamically stable phase (like water at 1 atm and at temperatures between $0\,°C$ and $100\,°C$). If the temperature is reduced still further to D, then the solid and the liquid phases are in equilibrium (like ice and water at 1 atm and $0\,°C$). A further reduction in temperature takes the system into the region where the solid is the thermodynamically stable phase.

4.1 Phase boundaries

As we saw in Section 3.4, the pressure of a vapour in equilibrium with its liquid phase is called the **vapour pressure** of the substance. It follows that the liquid–gas boundary in a phase diagram is just a plot of the vapour pressure against temperature. To determine it, we can introduce a liquid into the near-vacuum at the top of a barometer, and measure how much the mercury column is depressed: the difference in height is proportional to the pressure of the vapour (Fig. 4.3). To ensure that the pressure is the vapour pressure, we have to add enough liquid for some to remain after the vapour forms, for only then are the liquid and vapour phases in equilibrium. The temperature can then be changed, and another point on the curve determined, and so on (Fig. 4.4). The vapour pressure of a substance invariably increases with increasing temperature because as the temperature is raised a higher fraction of molecules in the liquid acquire enough energy to escape from the attractive forces that cause them to cohere and form a condensed phase (a liquid or solid).

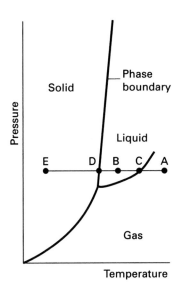

Fig. 4.2 A typical phase diagram, showing the regions of pressure and temperature at which each phase is the most stable. The phase boundaries (three are shown here) show the values of pressure and temperature at which the two phases separated by the line are in equilibrium. The significance of the letters A, B, C, D, and E is explained in the text and are referred to in Fig. 4.5.

Exercise E4.2

What would be observed when a pressure of 50 Torr is applied to a sample of water in equilibrium with its vapour at 25°C, when its vapour pressure is 23.8 Torr?

[*Answer*: condense to liquid]

The same approach can be used to plot the solid–vapour boundary, which is a graph of the vapour pressure of the solid against temperature. However, the sublimation vapour pressures of solids (the pressure of the vapour in equilibrium with a solid at a particular temperature) is usually much lower than the vapour pressure of a liquid. Special techniques have been developed for solids with very low vapour pressures: they make use of the relation between the vapour pressure and the rate of effusion of the vapour.

A more sophisticated procedure is needed to determine the locations of solid–solid phase boundaries because the transition between two solid phases is more difficult to detect. One approach is to use **thermal analysis**, which takes advantage of the heat released during a transition. In a typical thermal analysis experiment, a sample is allowed to cool and its temperature is monitored. When the transition occurs, heat is evolved and the cooling stops until the transition is complete (Fig. 4.5). The transition temperature is obvious from the shape of the graph and is used to mark a point on the phase diagram; the pressure can then be changed, and the corresponding transition temperature determined. This technique is particularly useful when simple visual inspection of the sample may be inadequate.

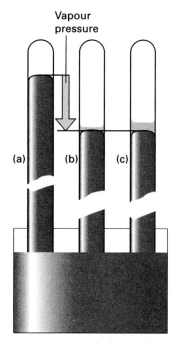

Fig. 4.3 When a small volume of water is introduced into the vacuum above the mercury in a barometer (a), the mercury is depressed (b) by an amount that is proportional to the vapour pressure of the liquid. (c) The same pressure is observed however much liquid is present (so long as some is present).

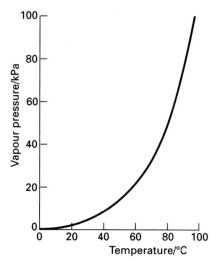

Fig. 4.4 The variation of the vapour pressure of water with temperature.

Any point lying on a phase boundary represents a pressure and temperature at which there is a **dynamic equilibrium** between the two phases. A state of dynamic equilibrium is one in which a forward process is taking place at the same rate as the reverse process; although there may be a great deal of activity at a molecular level, there is no *net* change. For example, any point on the liquid–vapour boundary represents a state of dynamic equilibrium in which vaporization and condensation continue at matching rates. Molecules are leaving the surface of the liquid at a certain rate, and molecules already in the gas phase are returning to the liquid at the same rate; as a result, there in no net change in the number of molecules in the vapour and hence no net change in its pressure. Similarly, a point on the solid–liquid curve represents conditions of pressure and temperature at which molecules are ceaselessly breaking away from the surface of the solid and contributing to the liquid. However, they are doing so at a rate that exactly matches the rate at which molecules already in the liquid are settling on to the surface of the solid and contributing to the solid phase.

Characteristic points

When a liquid is heated, its vapour pressure increases. Consider what we would observe if we heated a liquid in an open vessel. At a certain temperature, the vapour pressure (measured as we have described above) becomes equal to the external pressure. At this temperature, the vapour can drive back the surrounding atmosphere and expand indefinitely. Moreover, because there is no constraint on expansion, bubbles of vapour can form throughout the body of a liquid. This is the condition known as **boiling**. The temperature at which the vapour pressure of a liquid is equal to the external pressure is called the **boiling temperature**. When the external pressure is 1 atm, the boiling temperature is called the **normal boiling point**, T_b. It follows that we can identify the normal boiling point of a liquid by noting the temperature on the phase diagram at which its vapour pressure is 1 atm.

If the heating takes place in a closed vessel, then as the vapour pressure increases the density of the vapour increases too because more molecules are present in the gas phase. As heating continues, there comes a point at which the density of the vapour is equal to that of the remaining liquid. At this temperature, the surface between the two phases disappears, as was depicted in Fig. 1.19. The temperature at which the surface disappears is the critical temperature, T_c, which we first encountered in Section 1.10. The vapour pressure at the critical temperature is called the **critical pressure**, p_c, and the critical temperature and critical pressure together identify the **critical point** of the substance (see Table 4.1). If we exert pressure on a sample that is above its critical temperature, we produce a denser fluid, but no surface occurs separating the two parts of the sample and a single uniform phase continues to fill the container however high the pressure. That is, we have to conclude that *a liquid cannot be produced by the application of pressure to a substance if it is above its critical temperature*. That is why the liquid–vapour boundary in a phase diagram terminates at the critical

Table 4.1 Critical constants*

	p_c/atm	$V_{c,m}$/cm^3 mol^{-1}	T_c/K
Ammonia, NH_3	111	73	406
Argon, Ar	48	75	151
Benzene, C_6H_6	49	260	563
Bromine, Br_2	102	135	584
Carbon dioxide, CO_2	73	94	304
Chlorine, Cl_2	76	124	417
Ethane, C_2H_6	48	148	305
Ethene, C_2H_4	51	124	283
Hydrogen, H_2	13	65	33
Methane, CH_4	46	99	191
Oxygen, O_2	50	78	155
Water, H_2O	218	55	647

* The critical molar volume $V_{c,m}$ is the molar volume at the critical pressure and critical volume.

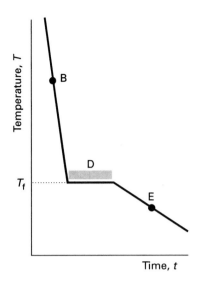

Fig. 4.5 The cooling curve for the B–E section of the horizontal line in Fig. 4.2. The halt at D corresponds to the pause in the fall in temperature while the liquid freezes and releases its enthalpy of transition. The halt enables T_f to be located even if the transition cannot be observed visually.

point. The single gas-like phase can be as dense as a normal liquid, and it is termed a **supercritical fluid** (Section 1.10)

The temperature at which the liquid and solid phases of a substance coexist in equilibrium at a specified pressure is called the **melting temperature** of the substance. Because a substance melts at the same temperature as it freezes, the melting temperature is the same as the **freezing temperature**. The solid–liquid boundary therefore shows how the melting temperature of a solid varies with pressure (Fig. 4.6). The melting temperature when the pressure on the sample is 1 atm is called the **normal melting point** or the **normal freezing point**, T_f. A liquid freezes when the energy of the molecules in the liquid is so low that they cannot escape from the attractive forces of their neighbours, and their mobility is lost.

There is a set of conditions under which three different phases (typically solid, liquid, and vapour) all simultaneously coexist in equilibrium. It is represented by the **triple point**, where the three phase boundaries meet. The triple point of a pure substance is a characteristic physical property of the substance. For water the triple point lies at 273.16 K and 611 Pa, and ice, liquid water, and water vapour coexist in equilibrium at no other combination of pressure and temperature. At the triple point, the rates of the forward and reverse processes

$$\text{solid} \rightleftarrows \text{vapour} \qquad \text{solid} \rightleftarrows \text{liquid} \qquad \text{liquid} \rightleftarrows \text{vapour}$$

are equal (but the three individual rates are not necessarily the same).

The triple point and the critical point are important features of a substance because they act as frontier posts for the existence of the liquid

119

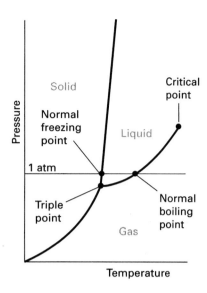

Fig. 4.6 The significant points of a phase diagram. The liquid–vapour phase boundary terminates at the critical point. At the triple point, solid, liquid, and vapour are in dynamic equilibrium. The normal freezing point is the temperature at which the liquid freezes when the pressure is 1 atm; the normal boiling point is the temperature at which the vapour pressure of the liquid is 1 atm.

phase. As we see from Fig. 4.7, if the slope of the solid–liquid phase boundary is as shown in the diagram:

1. The triple point marks the lowest temperature at which the liquid can exist.
2. The critical point marks the highest temperature at which the liquid can exist.

We shall see in the following section that for a few materials (most notably water) the solid–liquid phase boundary slopes in the opposite direction, and then only the second of these conclusions is applicable (see Fig. 4.7(b)).

The phase rule

You might wonder whether *four* phases could ever be in equilibrium (such as the two solid forms of tin, liquid tin, and tin vapour). However, it can be shown that no more than three phases of a single pure substance can coexist. This result—which can be proved from thermodynamics—is a consequence of the **phase rule**. The phase rule is celebrated as one of the most elegant results of chemical thermodynamics. It was derived by Josiah Gibbs, and is applicable to any system at equilibrium. It states that, for a system at equilibrium,

$$F = C - P + 2$$

where F is the number of degrees of freedom, C is the number of components, and P is the number of phases.

A gas, or a gaseous mixture, is a single phase, a crystal is a single phase, and two totally miscible liquids form a single phase. The **number of phases**, P, is the number of such regions in the system. Ice is a single phase ($P = 1$) even though it might be chipped into small fragments. A slurry of ice and water is a two-phase system ($P = 2$) even though it is difficult to map the boundaries between the phases. The **number of components**, C, in a system is the minimum number of independent species necessary to define the composition of all the phases present in the system. The definition is easy to apply when the species present in a system do not react, for then we simply count their number. For instance, pure water is a one-component system ($C = 1$) and a mixture of ethanol and water is a two-component system ($C = 2$). The **number of degrees of freedom**, F, of a system is the number of intensive variables (such as the pressure, temperature, or mole fractions) that can be changed independently without disturbing the number of phases in equilibrium.

For a one-component system, such as pure water,

$$F = 3 - P$$

When only one phase is present, $F = 2$ and both p and T can be varied independently. In other words, a single phase is represented by an *area* on a phase diagram. When two phases are in equilibrium $F = 1$, which implies that pressure is not freely variable if we have set the temperature. That is, the equilibrium of two phases is represented by a *line* in the phase diagram. Instead of selecting the temperature, we can select the pressure, but having done so the two phases come into equilibrium at a single definite temperature. Therefore, freezing (or any other phase transition) occurs at a definite temperature at a given pressure. When three phases are in equilibrium $F = 0$. This special invariant condition can therefore be established only at a definite temperature and pressure. The equilibrium of three phases is therefore represented by a *point*, the triple point, on the phase diagram. Four phases cannot be in equilibrium in a one-component system because F cannot be negative.

4.2 Phase diagrams of typical materials

We shall now show how these general features appear in the phase diagrams of a selection of pure substances.

Water

The phase diagram for water is shown in Fig. 4.8. The liquid–vapour phase boundary shows how the vapour pressure of liquid water varies with temperature. We can use this curve (which is shown in more detail in Fig. 4.4) to decide how the boiling temperature varies with changing external pressure. For example, when the external pressure is 149 Torr (at an altitude of 12 km), water boils at 60 °C because that is the temperature at which the vapour pressure is 149 Torr (19.9 kPa).

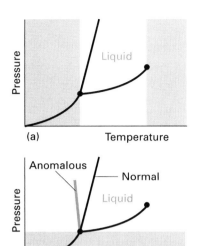

(a)   Temperature

(b)   Temperature

Fig. 4.7 (a) For substances that have phase diagrams resembling the one shown here (which is common for most substances, with the important exception of water), the triple point and the critical point mark the range of temperatures over which the substance may exist as a liquid. The shaded areas show the regions of temperature in which a liquid cannot exist as a stable phase. (b) A liquid cannot exist as a stable phase if the pressure is below that of the triple point for normal or anomalous liquids.

Exercise E4.3

What is the minimum pressure at which liquid is the thermodynamically stable phase of water at 25 °C?

[*Answer*: 23.8 Torr, 3.22 kPa]

The solid–liquid boundary line shows how the melting temperature of water depends on the pressure. For example, although ice melts at 0 °C at 1 atm, it melts at −1 °C when the pressure is 130 atm. The very steep slope of the boundary indicates that enormous pressures are needed to bring about significant changes. Notice that the line slopes down from left to right, which means that the melting temperature of ice falls as the pressure is raised. The reason for this behaviour (which is uncommon) can be traced to the decrease in volume that occurs when ice melts into water: it is favourable for the solid to transform into the denser liquid as the pressure is raised. The decrease in

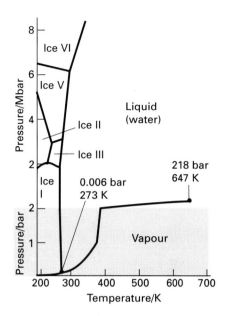

Fig. 4.8 The phase diagram for water showing the different solid phases. Note the change in the vertical scale at 2 bar.

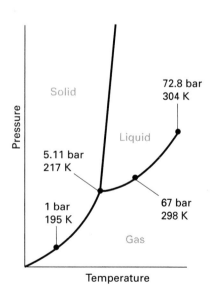

Fig. 4.9 The phase diagram for carbon dioxide. Note that, as the triple point lies well above atmospheric pressure, liquid carbon dioxide does not exist under normal conditions (a pressure of at least 5.11 bar must be applied).

volume is a result of the very open structure of the ice crystal structure: the water molecules are held apart (as well as together) by the hydrogen bonds between them, but the structure partially collapses on melting and the liquid is denser than the solid.

The phase diagram shows that water has many different solid phases other than the ordinary ice ('ice I') with which we are familiar. The solid phases differ in the arrangement of the water molecules: under the influence of very high pressures, hydrogen bonds buckle and the H_2O molecules adopt different arrangements. Some of these ices have bizarre properties: ice VII (which is not shown in the illustration) melts at $100\,°C$, but it exists only if the pressure is greater than 25 kbar. These different ices may be responsible for the advance of glaciers, for ice at the bottom of glaciers experiences very high pressures where it rests on jagged rocks. The sudden apparent explosion of Halley's comet in 1991 may have been due to the conversion of one form of ice into another in its interior.

Carbon dioxide

The phase diagram for carbon dioxide is shown in Fig. 4.9. The features to notice include the slope of the solid–liquid boundary (this positive slope is typical of almost all substances), which indicates that the melting temperature of solid carbon dioxide rises as the pressure is increased. As the triple point (217 K, 5.11 bar) lies well above ordinary atmospheric pressure, the liquid does not exist at normal atmospheric pressures whatever the temperature, and the solid sublimes when left in the open (hence the name 'dry ice'). To obtain liquid carbon dioxide, it is necessary to exert a pressure of at least 5.11 bar.

Cylinders of carbon dioxide generally contain the liquid or compressed gas; if both gas and liquid are present inside the cylinder, then at $20\,°C$ the pressure must be about 65 atm. When the gas squirts through the throttle it cools by the Joule–Thomson effect, so when it emerges into a region where the pressure is only 1 atm, it condenses into a finely divided snowlike solid.

Carbon

A simplified version of the phase diagram for carbon is shown in Fig. 4.10. It is ill-defined and incomplete because the various phases come into stability at extremes of temperature and pressure, and gathering the data is very difficult. For instance, at atmospheric pressure, carbon gas is the stable phase only at temperatures well over 4000 K. To obtain liquid carbon at 1 bar it is necessary to work at about 4000 K.

Small diamonds are synthesized and are widely used in industry, but the phase diagram does not reveal the full problem. The rate of conversion is an important factor, and pure graphite changes into diamond at a useful rate only when the temperature is about 4000 K and the pressure exceeds 200 kbar; but then the apparatus tends to disappear first! Catalysts are added in commercial syntheses, and then the conversion proceeds at 70 kbar and 2300 K, which are attainable conditions. The contamination by the metal catalysts, such as molten nickel (which also acts as a solvent for the carbon), enables commercial and natural diamonds to be distinguished at low temperatures.

Although diamond and graphite were long believed to be the only two solid phases of carbon attainable under normal conditions, a third phase, fullerite, which consists of close-packed buckminsterfullerene (C_{60}) molecules, was characterized in 1985, but its location on the phase diagram has not yet been established. It is possible to trap electrons and ions inside the C_{60} cages, and the resulting materials, such as K_3C_{60}, show unusual properties, among them superconductivity.

Helium

The phase diagram of helium is shown in Fig. 4.11. Helium behaves unusually at low temperatures. For instance, the solid and gas phases of helium are never in equilibrium however low the temperature: the atoms are so light that they vibrate with a large-amplitude motion even at very low temperatures, and the solid simply shakes itself apart. Solid helium can be obtained, but only by holding the atoms together by applying pressure. A second unique feature of helium is that pure helium-4 has two liquid phases. The phase marked He-I in the diagram behaves like a normal liquid; the other phase, He-II, is a **superfluid**; it is so called because it flows without viscosity. Helium is the only known substance with a liquid–liquid boundary in its phase diagram.

The properties of nonelectrolyte solutions

We now leave pure materials and the limited but important changes they can undergo, and examine solutions. We shall consider mainly **nonelectrolyte solutions**, where the solute is not present as ions. Examples are sucrose dissolved in water and sulfur dissolved in carbon disulfide. We shall delay until Chapter 6 the special problems of **electrolyte solutions**, in which the solute is ionized and where the ions generally interact strongly with one another.

4.3 The thermodynamic description of mixtures

In this and the following chapters we need a set of concepts that enable us to apply thermodynamics to mixtures of variable composition. We have already seen that the partial pressure, the contribution to the total pressure of one component in a gaseous mixture, is used to discuss the properties of

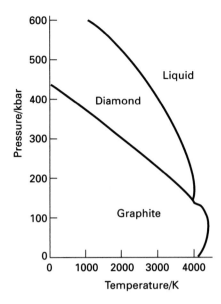

Fig. 4.10 A simplified version of the phase diagram for carbon. There are very large uncertainties about the precise form of this phase diagram because the data are so difficult to obtain.

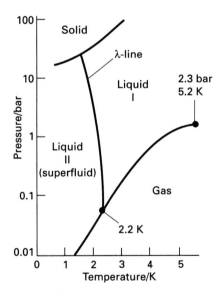

Fig. 4.11 That phase diagram for helium-4. The λ-line marks the conditions under which the two liquid phases are in equilibrium: helium-I is a conventional liquid, but helium-II is a superfluid. Note that a pressure of at least 20 bar must be exerted before solid helium can be obtained.

mixtures of gases. For a more general description of the thermodynamics of mixtures we need to introduce other 'partial' properties, each one being the contribution that a particular component makes to the overall value of the property.

Measures of concentration

We need to establish at the outset a way of reporting the composition of mixtures. There are essentially three measures of concentration that we shall employ. One, the molar concentration, is used when we need to know the amount of molecules in a sample of known volume. The other two, the molality and the mole fraction, are used when we need to know the relative numbers of solute and solvent molecules in a sample.

The **molar concentration**, [J], of a solute J in a solution is the chemical amount of J divided by the volume it occupies (for instance, the volume of solution it occupies):

$$[J] = \frac{n_J}{V} \tag{1}$$

Molar concentration is typically reported in moles per litre ($mol\,L^{-1}$, more formally, as $mol\,dm^{-3}$). The molar concentration $1\,mol\,L^{-1}$ is widely denoted by the symbol $1\,M$. In practice, a solution of given molar concentration is prepared by measuring out the appropriate mass of solute into a volumetric flask, dissolving the solute in a little solvent, and then adding enough solvent to produce the desired volume. For instance, to prepare a glucose solution of molar concentration $1.00\,M\,C_6H_{12}O_6(aq)$, $180\,g$ of glucose is dissolved in enough water to produce $1.00\,L$ of solution (note that the solute is *not* dissolved in $1.00\,L$ of water).

Exercise E4.4

What mass of glycine, NH_2CH_2COOH, should be used to make $250\,mL$ of a solution of molar concentration $0.015\,M\,NH_2CH_2COOH(aq)$?

[*Answer*: $0.282\,g$]

Once the molar concentration of a solute is known, it is easy to calculate the amount of that substance in a given volume of solution by multiplying the volume of solution by the molar concentration:

$$n_J = [J]V \tag{2}$$

The **molality**, b_J, of a solute J in a solution is the amount of substance divided by the mass of solvent used to prepare the solution:

$$b_J = \frac{n_J}{m_{solvent}} \tag{3}$$

Molality is typically reported in moles of solute per kilogram of solvent, $mol\,kg^{-1}$. There is an important distinction between molar concentration and molality: whereas the former is defined in terms of the volume of the solution, the latter is defined in terms of the mass of solvent used to prepare the solution. Thus, a solution of glucose in water of molality $1.0\,mol\,kg^{-1}$ would be prepared by measuring out $180\,g$ of glucose and dissolving it in $1.0\,kg$ of water.

As we have indicated, molality is used when it is necessary to emphasize the relative amounts of solute and solvent molecules. To see why this is so, we need to note that the mass of solvent is proportional to the chemical amount of solvent molecules that are present, so from eqn (3) we see that the molality is proportional to the ratio of the amounts of solute and solvent molecules. For example, a $1.0\,mol\,kg^{-1}\,C_6H_{12}O_6(aq)$ solution contains $1.0\,mol\,C_6H_{12}O_6$ molecules and $55.5\,mol\,H_2O$ molecules. Indeed, any $1.0\,mol\,kg^{-1}$ aqueous solution contains $1.0\,mol$ solute particles and $55.5\,mol\,H_2O$ molecules, so in each case there is 1 solute molecule per 55.5 solvent molecules.

Closely related to the molality of a solute is the mole fraction x, which was introduced in Chapter 1 in connection with mixtures of gases:

$$x_J = \frac{n_J}{n} \tag{4}$$

where n_J is the chemical amount (in moles) of a species J and n is the total amount of species (the total number of moles) in the sample. As remarked in Chapter 1, the mole fraction of a species J is the fraction of the total number of particles present that are the species J.

Example *Relating mole fraction and molality*
What is the mole fraction of glycine in an aqueous solution of molality $0.140\,mol\,kg^{-1}$?

Strategy
We consider a sample that contains (exactly) 1 kg of solvent, and hence an amount

$$n_J = b_J \times (1\,kg)$$

of solute molecules. The amount of solvent molecules in 1 kg of solvent is

$$n_{solvent} = \frac{1\,kg}{M}$$

where M is the molar mass of the solvent. Once these two amounts are available, we can calculate the mole fraction by using eqn (4) with $n = n_J + n_{solvent}$.

Solution

It follows from the discussion in the Strategy that the amount of glycine (gly) molecules in 1 kg of solvent is

$$n_{gly} = (0.140 \, \text{mol kg}^{-1}) \times (1 \, \text{kg}) = 0.140 \, \text{mol}$$

The amount of water molecules in 1 kg of water is

$$n_{water} = \frac{10^3 \, \text{g}}{18.02 \, \text{g mol}^{-1}} = 55.49 \, \text{mol}$$

The total amount of molecules present is

$$n = 0.140 \, \text{mol} + 55.49 \, \text{mol} = 55.63 \, \text{mol}$$

The mole fraction of glycine is therefore

$$x_{gly} = \frac{0.140 \, \text{mol}}{55.63 \, \text{mol}} = 2.52 \times 10^{-3}$$

Exercise E4.5

Calculate the mole fraction of sucrose in an aqueous sample of molality $1.22 \, \text{mol kg}^{-1}$.

[*Answer*: 2.15×10^{-2}]

Partial molar properties

A **partial molar property** is the contribution (per mole) that a substance makes to an overall property of a mixture. The easiest partial molar property to visualize is the **partial molar volume**, the contribution to the total volume of a mixture made by each component. We have to be alert to the fact that although 1 mol of a substance has a characteristic volume when it is pure, 1 mol of a substance can make different contributions to the total volume of a mixture because molecules pack together in different ways in the pure substances and in the mixture.

To grasp the meaning of the concept of partial molar volume, we first imagine a huge volume of pure water. When a further 1 mol H_2O is added, the volume increases by $18 \, \text{cm}^3$. However, when we add 1 mol H_2O to a huge volume of pure ethanol, the volume increases by only $14 \, \text{cm}^3$. The quantity $18 \, \text{cm}^3 \, \text{mol}^{-1}$ is the volume occupied per mole of water molecules in pure water; $14 \, \text{cm}^3 \, \text{mol}^{-1}$ is the volume occupied per mole of water molecules in virtually pure ethanol. In other words, the partial molar volume of water in pure water is $18 \, \text{cm}^3 \, \text{mol}^{-1}$ whereas the partial molar volume of water in pure ethanol is only $14 \, \text{cm}^3 \, \text{mol}^{-1}$. In the latter case there is so much ethanol present that each H_2O molecule is surrounded by ethanol molecules and the packing of the molecules results in the water molecules occupying only $14 \, \text{cm}^3$.

The partial molar volume at an intermediate composition of the water/ethanol mixture reflects in a similar way the volume the H_2O molecules occupy when they are surrounded by a mixture of molecules representative of the overall composition (half water, half ethanol, for instance, when the mole fractions are both 0.5). The partial molar volume of ethanol also varies as the composition of the mixture is changed, because the environment of an ethanol molecule changes from pure ethanol to pure water as the proportion of water increases, and the volume occupied by the ethanol molecules varies accordingly. The variation of the two partial molar volumes across the full composition range at $25\,°C$ is shown in Fig. 4.12.

Once we know the partial molar volumes V_A and V_B of the two components A and B of a mixture at the composition (and temperature) of interest we can state the total volume V of the mixture by using

$$V = n_A V_A + n_B V_B \tag{5}$$

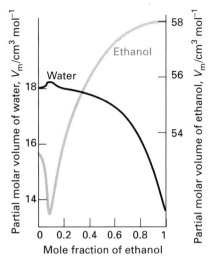

Fig. 4.12 The partial molar volumes of water and ethanol at $25\,°C$. Note the different scales (water on the left, ethanol on the right).

Justification

To derive eqn (5), consider a very large sample of the mixture of the specified composition. Then, when an amount n_A of A is added, the composition remains virtually unchanged but the volume of the sample changes by $n_A V_A$. Similarly, when an amount n_B of B is added, the volume changes by $n_B V_B$. The total change of volume is therefore $n_A V_A + n_B V_B$. The mixture now occupies a larger volume, but the proportions of the components are still the same. Next, scoop out of this enlarged volume a sample containing n_A of A and n_B of B. Its volume is $n_A V_A + n_B V_B$. Because volume is a state function, the same sample could have been prepared simply by mixing the appropriate amounts of A and B.

Example *Using partial molar volumes*
What is the total volume of a mixture of 50.0 g of ethanol and 50.0 g of water at $25\,°C$.

Strategy
To use eqn (5), we need the mole fractions of each substance and the corresponding partial molar volumes. The mole fractions are calculated in the same way as indicated in Exercise E1.7 (by using the molar masses of the components to calculate the amounts). The partial molar volumes corresponding to these mole fractions can then be found by referring to Fig. 4.12.

Solution

We find first that $n_{\text{ethanol}} = 1.09\,\text{mol}$ and $n_{\text{water}} = 2.77\,\text{mol}$, and hence that $x_{\text{ethanol}} = 0.282$ and $x_{\text{water}} = 0.718$. According to Fig. 4.12, the partial molar volumes of the two substances in a mixture of this composition are $55\,\text{cm}^3\,\text{mol}^{-1}$ and $18\,\text{cm}^3\,\text{mol}^{-1}$, respectively, so from eqn (5) the total volume of the mixture is

$$V = (1.09\,\text{mol}) \times (55\,\text{cm}^3\,\text{mol}^{-1}) + (2.77\,\text{mol}) \times (18\,\text{cm}^3\,\text{mol}^{-1})$$
$$= 60\,\text{cm}^3 + 50\,\text{cm}^3 = 110\,\text{cm}^3$$

Exercise E4.6

Use Fig. 4.12 to calculate the density of a mixture of 20 g of water and 100 g of ethanol.

[*Answer*: $0.84\,\text{g}\,\text{cm}^{-3}$]

The chemical potentials of gases

The concept of a partial molar quantity can be extended to other state properties, but they are less easy to visualize than the volume. Among the most important for our purposes is the **partial molar Gibbs energy**, G_J, of a substance J, which is the contribution of J to the total Gibbs energy of a mixture. That is, if we know the partial molar Gibbs energies of two substances A and B in a mixture of a given composition, then we can calculate the total Gibbs energy of the mixture by using an expression like eqn (5):

$$G = n_A G_A + n_B G_B \tag{6}$$

The partial molar Gibbs energy has exactly the same significance as the partial molar volume. For instance, ethanol has a particular partial molar Gibbs energy when it is pure (and every molecule is surrounded by other ethanol molecules), and it has a different partial molar Gibbs energy when it is in an aqueous solution of a certain composition (because then each ethanol molecule is surrounded by a mixture of ethanol and water molecules).

Partly because the name 'partial molar Gibbs energy' is so cumbersome, it is usually called the **chemical potential** and denoted μ_J for a species J: we shall use that name and notation from now on, and write the total Gibbs energy of a mixture of A and B as

$$G = n_A \mu_A + n_B \mu_B \tag{7}$$

In the course of this chapter and the next we shall see that 'chemical potential' is a very appropriate name, for we shall see that μ_J is a measure of the ability of J to bring about physical and chemical change: a substance with a high chemical potential has a high ability (in a sense we shall explore) to drive a reaction or some other physical process forward.

The real power and usefulness of thermodynamics becomes apparent when we do numerical calculations. To do so, we need an explicit formula for the variation of the chemical potential of a substance with the composition of the mixture. For a mixture of perfect gases, for each component J

$$\mu_J = \mu_J^\ominus + RT \ln a_J \qquad a_J = \frac{p_J}{p^\ominus} \qquad (8)$$

In this expression, p_J is the partial pressure of the gas J, $p^\ominus$ is the standard pressure (1 bar), and $\mu_J^\ominus$ is the **standard chemical potential** of the gas, its chemical potential when its pressure is 1 bar. As always, we can become familiar with an equation by listening to what it tells us. In this case, we note that as p_J increases, so does its logarithm. Therefore, eqn (8) tells us that *the higher the partial pressure of a gas, the higher its chemical potential*. This conclusion is consistent with the interpretation of the chemical potential as an indication of the potential of a substance to be active chemically: the higher the partial pressure, the more active chemically the species. In this instance the chemical potential represents the tendency of the substance to react when it is in its standard state (the significance of the term $\mu^\ominus$) plus an additional tendency that reflects whether it is at a pressure greater than 1 bar or not. Any additional pressure gives a substance more chemical 'punch', just like winding a spring gives a spring more physical punch (that is, enables it to do more work).

Exercise E4.7

Suppose that the partial pressure of a perfect gas falls from 1.00 bar to 0.50 bar as it is consumed in a reaction at 25 °C. What is the change in chemical potential of the substance?

[*Answer*: $-1.7\,\text{kJ}\,\text{mol}^{-1}$]

Justification

To derive eqn (8) we have to use calculus. First, we work out how the chemical potential changes when the pressure changes by an infinitesimal amount dP. Then we allow the pressure to increase from a low pressure to a high pressure, and add together (that is, evaluate the integral of) all these infinitesimal changes. For the first step, we use the equation derived in the *Justification* in Section 3.3, namely $dG = V dp$. However, because we are dealing with partial molar quantities, we attach a subscript J to G (and identify G_J with μ_J), and use the molar volume V_m on the right:

$$d\mu_J = V_m dp$$

The molar volume of a perfect gas at a pressure p is $V_m = RT/p$. Therefore, the change in chemical potential when the pressure changes by dp is

$$d\mu_J = RT\frac{dp}{p}$$

When the pressure changes from $p^{\ominus}$ to p_J, the chemical potential changes from $\mu_J^{\ominus}$ to μ_J. The change is equal to the sum (integral) of all such infinitesimal terms as the pressure is increased between these two values:

$$\mu_J - \mu_J^{\ominus} = RT \int_{p^{\ominus}}^{p_J} \frac{dp}{p} = RT \ln \frac{p_J}{p^{\ominus}}$$

which rearranges into eqn (8).

Ideal solutions

In chemistry we are concerned with liquids as well as gases, so we need an expression for the chemical potential of a substance in a solution. We can anticipate that the chemical potential of a species ought to increase with concentration, because the higher its concentration the greater its chemical 'punch'.

The key to setting up an expression for the chemical potential of a solute is the work done by the French chemist François Raoult, who spent most of his life measuring the vapour pressures of solutions. He measured the **partial vapour pressure** of each component in the mixture, the partial pressure of the vapour of each component in dynamic equilibrium with the solution, and established the following general result:

> **Raoult's law**: The partial vapour pressure of a substance in a mixture is proportional to its mole fraction in the solution and its vapour pressure when pure:

$$p_J = x_J p_J^* \tag{9}$$

In this expression, p_J^* is the vapour pressure of the pure substance. For example, when the mole fraction of water in an aqueous solution is 0.90, then, provided Raoult's law is obeyed, the partial vapour pressure of the water in the solution is 90 per cent that of pure water. This conclusion is approximately true whatever the identity of the solute and the solvent (Fig. 4.13). The molecular interpretation of Raoult's law is that the molecules of one substance partially block the escape of the molecules of the other species, thereby reducing the latter's vapour pressure (Fig. 4.14).

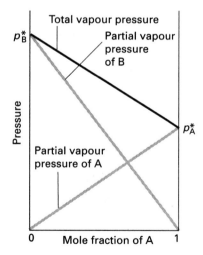

Fig. 4.13 The partial vapour pressures of the two components of an ideal binary mixture are proportional to the mole fractions of the components in the liquid. The total pressure of the vapour is the sum of the two partial vapour pressures.

130

Exercise E4.8

A solution is prepared by dissolving $1.5\,mol\,C_{10}H_8$ (naphthalene) in $1.00\,kg$ of benzene. The vapour pressure of pure benzene is 94.6 Torr at $25\,°C$. What is the partial vapour pressure of benzene in the solution?

[*Answer*: 85 Torr]

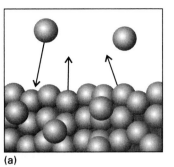

(a)

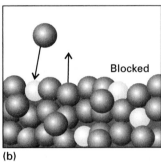

(b)

Fig. 4.14 (a) In a pure liquid, molecules may leave and return to any part of the surface. (b) When a nonvolatile solute is present, molecules may still return to any point, but their escape from the surface is partly blocked.

Raoult's law is only approximately valid. However, some mixtures obey it over a wide range of concentrations, especially when the components are structurally similar. That is, Raoult's law works best when the components of a mixture have similar molecular shapes and are held together in the liquid by similar intermolecular forces. An example is a mixture of two similar hydrocarbons (Fig. 4.15). Hypothetical solutions that obey the law throughout the composition range from pure A to pure B are called **ideal solutions**. A mixture of benzene and methylbenzene (toluene) is a good approximation to an ideal solution, for the partial vapour pressure of each component satisfies Raoult's law reasonably well throughout the composition range from pure benzene to pure toluene.

No mixture is perfectly ideal, and so all mixtures show deviations from Raoult's law. However, the deviations are small for the component of the mixture that is in large excess (the solvent) and become smaller as the concentration of solute decreases (Fig. 4.16). We can usually be confident that Raoult's law is reliable for the solvent when the solution is very dilute. More formally, Raoult's law is a limiting law, and is only strictly valid in the limit of zero concentration. Then, because the law shows that the vapour pressure of the solvent is proportional to its mole fraction, it should not be too surprising that the expression for the chemical potential of the solvent bears a close resemblance to eqn (8) for the chemical potential of a gas. Indeed, it is quite easy to show that for dilute solutions the chemical potential of the solvent A is given by the expression

$$\mu_A = \mu_A^* + RT\ln a_A \qquad a_A = x_A \qquad (10)$$

where x_A is the mole fraction of the solvent (and is close to 1) and μ^* is the chemical potential of the pure liquid. The essential feature of eqn (10) is that the chemical potential of a solvent is lower in a solution than when it is pure: x_A is then less than 1, and $\ln x_A$ is negative. A solvent in which a solute is present has less chemical 'punch' (including a lower ability to generate a vapour pressure) than when it is pure.

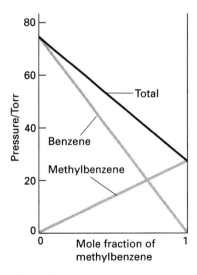

Fig. 4.15 Two similar substances, in this case benzene and toluene (methylbenzene), behave almost ideally and have vapour pressures that closely resemble those for the ideal case depicted in Fig. 4.13.

Justification

When a liquid A in a mixture is in equilibrium with its vapour at a partial pressure p_A, the chemical potentials of the two phases are equal:

$$\mu_A(l) = \mu_A(g)$$

131

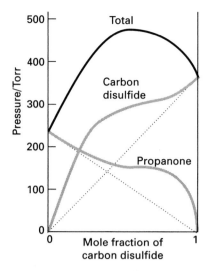

Fig. 4.16 Strong deviations from ideality are shown by dissimilar substances, in this case carbon disulfide and propanone. Note, however, that Raoult's law is obeyed by propanone when only a small amount of carbon disulfide is present (on the left) and by carbon disulfide when only a small amount of propanone is present (on the right).

However, we already have an expression for the chemical potential of a vapour, eqn (8), so at equilibrium,

$$\mu_A(l) = \mu_A^{\ominus} + RT \ln \frac{p_A}{p^{\ominus}}$$

According to Raoult's law, $p_A = x_A p_A^*$, so we can write

$$\mu_A(l) = \mu_A^{\ominus} + RT \ln \frac{x_A p_A^*}{p^{\ominus}}$$

$$= \mu_A^{\ominus} + RT \ln \frac{p_A^*}{p^{\ominus}} + RT \ln x_A$$

The first two terms on the right are constants for a given substance at a given temperature, so they may be combined into a constant μ^*. When $x_A = 1$, it follows that $\mu_A(l) = \mu_A^*$, so μ_A^* can be recognized as the chemical potential of the pure liquid. It then follows that

$$\mu_A(l) = \mu_A^* + RT \ln x_A$$

This is eqn (10). We have replaced x by a to give a more uniform appearance to all the expressions for chemical potentials.

Exercise E4.9

By how much is the chemical potential of benzene reduced at 25 °C by a solute that is present at a mole fraction of 0.10?

[*Answer*: $-0.26\,\text{kJ}\,\text{mol}^{-1}$]

Ideal-dilute solutions

Raoult's law is a good description of the vapour pressure of the solvent in a very dilute solution. However, we cannot expect it to be a good description of the partial vapour pressure of the solute (such as the partial vapour pressure of ethanol in a dilute solution of ethanol in water), for a solute is far from being in its pure form in a dilute solution. In a dilute solution, each solute molecule is surrounded by nearly pure solvent, so its environment is quite unlike that in the pure solute, and it is very unlikely that its vapour pressure will be related to that of the pure solute. However, it is found experimentally that the partial vapour pressure of the solute is in fact proportional to its mole fraction, just as for the solvent. Unlike the solvent, though, the constant of proportionality is not equal to the vapour pressure of the pure solute. This linear but different dependence was discovered by the English chemist William Henry, and is summarized as follows:

Henry's law: The vapour pressure of a volatile solute B is proportional to its mole fraction in a solution:

$$p_B = x_B K_B \qquad (11)$$

where K_B is a constant characteristic of the solute. The constant K_B is chosen so that the plot of the vapour pressure of B against its mole fraction is tangent to the experimental curve at $x_B = 0$ (Fig. 4.17).

Henry's law is usually obeyed only at low concentrations of the solute (close to $x_B = 0$), and solutions that are dilute enough for the solute to obey Henry's law are called **ideal-dilute solutions**. The essential *molecular* content of Henry's law is that the rate at which solute molecules can leave the solution is proportional to their concentration, but that the rate bears little relation to the rate at which they escape from molecules of their own kind in the pure solute.

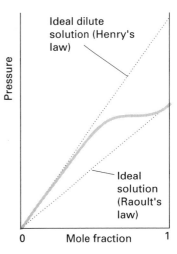

Fig. 4.17 When a component (the solvent) is almost pure, it behaves in accord with Raoult's law and has a vapour pressure that is proportional to the mole fraction in the liquid mixture, and a slope p^*, the vapour pressure of the pure substance. When the same substance is the minor component (the solute), its vapour pressure is still proportional to its mole fraction, but the constant of proportionality is now K.

Example *Verifying Raoult's and Henry's laws*

The partial vapour pressures of each component in a mixture of propanone (acetone, A) and trichloromethane (chloroform, C) were measured at 35 °C with the following results:

x_C	0	0.20	0.40	0.60	0.80	1
p_C/Torr	0	35	82	142	219	293
p_A/Torr	347	270	185	102	37	0

Confirm that the mixture conforms to Raoult's law for the component in large excess and to Henry's law for the minor component. Find the Henry's law constants.

Strategy

We need to plot the partial vapour pressures against mole fraction. Raoult's law is tested by comparing the data with the straight line $p_J = x_J p_J^*$ for each component in the region in which it is in excess (and acting as the solvent). Henry's law is tested by finding a straight line $p_J = x_J K_J$ that is tangent to each partial vapour pressure at low x_J where the component can be treated as the solute. In practice, the data are fitted to a polynomial function (using a computer) and then the tangent calculated by evaluating the first derivative of the polynomial at $x_J = 0$.

Solution

The data are plotted in Fig. 4.18 together with the Raoult's law lines. Henry's law requires $K_A = 175$ Torr and $K_C = 165$ Torr. Notice how the data deviate from both Raoult's and Henry's laws even for quite small departures from $x = 1$ and $x = 0$, respectively.

133

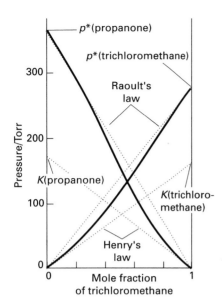

Fig. 4.18 The experimental partial vapour pressures of a mixture of trichloromethane, $CHCl_3$, and propanone, CH_3COCH_3, based on the data in the example. The values of K, the Henry's law constants for the two components, are obtained by extrapolating the dilute solution vapour pressures as explained in the example.

Exercise E4.10

The vapour pressure of chloromethane at various mole fractions in a mixture at $25\,°C$ was found to be as follows:

x	0.005	0.009	0.0019	0.0024
p/Torr	205	363	756	946

Estimate Henry's law constant.

[*Answer:* 4×10^5 Torr]

The Henry's law constants of some gases are listed in Table 4.2. They are often used in calculations relating to gas solubilities, as in the estimation of the concentration of O_2 in river water or the concentration of carbon dioxide in blood plasma. To apply Henry's law to this kind of problem, we treat the gas as the solute, and use its partial pressure above the solvent to calculate the mole fraction in the solution by rearranging eqn (11) into

$$x_B = \frac{p_B}{K_B} \qquad (12)$$

For instance, when the partial pressure of oxygen is 190 Torr and the solvent is water, the mole fraction in solution in equilibrium with the gas at $25\,°C$ is

$$x(O_2) = \frac{190\,\mathrm{Torr}}{3.30 \times 10^7\,\mathrm{Torr}} = 5.76 \times 10^{-6}$$

That is, one molecule in about 170 000 (that is, $1/5.76 \times 10^{-6}$) is an O_2 molecule, the others being water. A knowledge of Henry's law constants for gases in fats and lipids is important for the discussion of respiration, especially when the partial pressure of oxygen is abnormal, as in diving and mountaineering.

Table 4.2 Henry's law constants for gases at $25\,°C$, K/Torr

	Solvent	
	Water	Benzene
Methane, CH_4	3.14×10^5	4.27×10^5
Carbon dioxide, CO_2	1.25×10^6	8.57×10^4
Hydrogen, H_2	5.34×10^7	2.75×10^6
Nitrogen, N_2	6.51×10^7	1.79×10^6
Oxygen, O_2	3.30×10^7	

Example *Determining whether a natural water can support aquatic life*

The concentration of O_2 in water required to support aquatic life is about $4\,mg\,L^{-1}$. What is the minimum partial pressure of oxygen in the atmosphere that can achieve this concentration?

Strategy

The strategy of the calculation is to determine the partial pressure of oxygen that, according to Henry's law, corresponds to the concentration specified. To use eqn (11) we need to convert the stated mass concentration to a mole fraction of the solute. For that, we consider (exactly) 1 L of solution, calculate the mass of solute present, and then use its molar mass to convert to a chemical amount. Then we convert the mass of solvent present to a chemical amount of solvent molecules. To do so, we make the approximation that the solution is so dilute that the solvent is almost pure water, and use the fact that the density of water is approximately $1\,kg\,L^{-1}$. At that stage we can find the mole fractions, and use Henry's law to calculate the required partial pressure.

Solution

Because 1 L of solution contains $4\,mg$ of oxygen, the amount of O_2 present in 1 L of solution is

$$n(O_2) = \frac{4 \times 10^{-3}\,g}{32.00\,g\,mol^{-1}} = 1 \times 10^{-4}\,mol$$

The amount of H_2O present in 1.0 L (1.0 kg) of solution is similarly

$$n(H_2O) = \frac{1.0 \times 10^{3}\,g}{18.02\,g\,mol^{-1}} = 55\,mol$$

Therefore, the mole fraction of O_2 is

$$x(O_2) = \frac{1 \times 10^{-4}\,mol}{55\,mol + 1 \times 10^{-4}\,mol} = 2 \times 10^{-6}$$

Because the Henry's law constant for oxygen in water at $20\,°C$ is $3.3 \times 10^{7}\,Torr$, the partial pressure required to achieve the mole fraction is

$$p(O_2) = (2 \times 10^{-6}) \times (3.3 \times 10^{7}\,Torr) = 7 \times 10^{1}\,Torr$$

The partial pressure of oxygen in air at sea level is $0.21 \times 760\,Torr = 1.6 \times 10^{2}\,Torr$, so the required concentration can be maintained under normal conditions.

4.4 Colligative properties

Because a solute affects the entropy of a solution by introducing a degree of disorder that is not present in the pure solvent, we can expect it to modify its physical properties. Apart from lowering the vapour pressure of the solvent, which we have already considered, a nonvolatile solute has three main effects: it raises the boiling point of a solution, it lowers the freezing point, and it gives rise to an osmotic pressure. (The meaning of the last will be explained shortly.) All these effects depend only on the number of solute particles present, not their chemical identity, and for this reason they are called **colligative properties** (where *colligative* denotes 'depending on the collection'). Thus a $0.01 \, \text{mol kg}^{-1}$ aqueous solution of any nonelectrolyte should have the same boiling point, freezing point, and osmotic pressure.

The effect of a solute on boiling and freezing points

As indicated above, the effect of a solute is to raise the boiling point of a solvent and to lower its freezing point. It is found empirically (and can be justified thermodynamically) that the **elevation of boiling point**, ΔT_b, and the **depression of freezing point**, ΔT_f, are both proportional to the molality, b_B, of the solute:

$$\Delta T_\text{b} = K_\text{b} b_\text{B} \qquad \Delta T_\text{f} = K_\text{f} b_\text{B} \tag{13}$$

In this expression K_b is the **ebullioscopic constant** and K_f is the **cryoscopic constant** of the solvent. The two constants can be estimated from other properties of the solvent, but both are best treated as empirical constants, and some experimental values are given in Table 4.3.

Table 4.3 Cryoscopic and ebullioscopic constants

Solvent	$K_f/\text{K kg mol}^{-1}$	$K_b/\text{K kg mol}^{-1}$
Acetic acid	3.90	3.07
Benzene	5.12	2.53
Camphor	40	
Carbon disulfide	3.8	2.37
Carbon tetrachloride	30	4.95
Naphthalene	6.94	5.8
Phenol	7.27	3.04
Water	1.86	0.51

To understand the origin of these effects we shall make two simplifying assumptions:

1. The solute is not volatile, and so does not contribute to the vapour above a solution.
2. The solute does not dissolve in the solid solvent, and so any solid present is pure solvent.

For example, a solution of sucrose in water consists of a solute (sucrose, $C_{12}H_{22}O_{11}$) that is not volatile, and so never appears in the vapour (which is therefore pure water vapour) and is also left behind in the liquid solvent when ice begins to form (so the ice remains pure). The origin of all the colligative effects is the lowering of chemical potential of the solvent by the presence of a solute, as expressed by eqn (10). The chemical potentials of the solvent vapour and solid solvent are unchanged by the presence of the solute because it is nonvolatile and insoluble in the solid.

Now consider the consequences of this lowering of chemical potential for the boiling point of the solvent. At the normal boiling point of the pure solvent, the chemical potentials of the liquid and the vapour are the same (and the pressure of the latter is 1 atm). When a solute is present, it decreases the chemical potential of the solvent, so at the same temperature the liquid is the stable phase. Therefore, to recover the condition of boiling, the temperature must be increased to bring the chemical potential of the solvent back up to that of the vapour (Fig. 4.19). Consequently, the boiling point of the solution is higher than that of the pure solvent. In terms of the entropy, the disorder introduced by the solute has made the solvent relatively chemically lazy (it has a reduced tendency to vaporize, for instance), so a higher temperature must be achieved before boiling occurs.

Next, consider the freezing point of the solvent. At the normal freezing point of the pure solvent, the chemical potentials of the pure solid and liquid are equal (and the pressure is 1 atm). When a solute is present, it lowers the chemical potential of the liquid solvent but leaves the chemical potential of the solid solvent unchanged (Fig. 4.20). To reduce the chemical potential of the solid below that of the liquid, so that the solid becomes the stable phase, now requires a greater lowering of temperature. Consequently, we observe

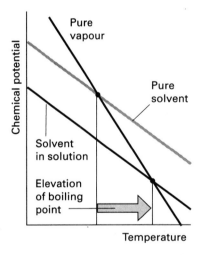

Fig. 4.19 The chemical potentials of pure solvent vapour and pure liquid solvent decrease with temperature, and the point of intersection, where the chemical potential of the vapour falls below that of the liquid, marks the boiling point of the pure solvent. A solute lowers the chemical potential of the solvent but leaves that of the vapour unchanged. As a result, the intersection point lies further to the right, and the boiling point is therefore raised.

137

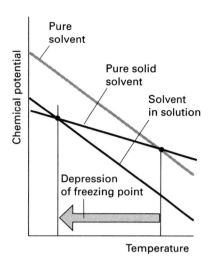

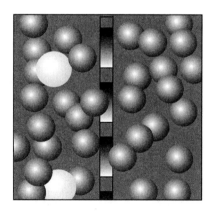

Fig. 4.20 The chemical potentials of pure solid solvent and pure liquid solvent also decrease with temperature, and the point of intersection, where the chemical potential of the liquid rises above that of the solid, marks the freezing point of the pure solvent. A solute lowers the chemical potential of the solvent but leaves that of the solid unchanged. As a result, the intersection point lies further to the left, and the freezing point is therefore lowered.

Fig. 4.21 A semipermeable membrane allows the passage of one type of molecule (such as the solvent, depicted here by small spheres) but not of another substance (the solute, depicted by large spheres). Osmosis is the net flow of solvent into the solution.

that the solvent freezes at a lower temperature when a solute is present. In terms of the entropy, the disorder introduced by the solute into the solution must be overcome by lowering the temperature more than for the pure solvent, so freezing occurs at a lower temperature. One practical consequence of the lowering of freezing point (and hence the lowering of the melting point of the pure solid) is its employment in organic chemistry to judge the purity of a sample, for any impurity lowers the melting point of a substance from its accepted value.

Osmosis

The phenomenon of **osmosis** (a name derived from the Greek word for 'push') is the passage of a pure solvent into a solution separated from it by a semipermeable membrane. A **semipermeable membrane** is a membrane that is permeable to the solvent but not to the solute (Fig. 4.21). The membrane might have microscopic holes that are large enough to allow water molecules to pass through, but not ions or carbohydrate molecules with their bulky coating of hydrating water molecules. The **osmotic pressure**, Π (pi), is the pressure that must be applied to the solution to stop the inward flow of solvent. One of the most important examples of osmosis is transport of fluids through cell membranes, but osmosis is also the basis of the technique called **osmometry**, the determination of molar mass by measurement of osmotic pressure, especially of macromolecules.

In the simple arrangement shown in Fig. 4.22, the pressure opposing the passage of solvent into the solution arises from the column of solution that the osmosis itself produces. This column is formed when the pure solvent flows through the membrane into the solution and pushes the column of solution higher up the tube. Equilibrium is reached when the downward pressure exerted by the column of solution is equal to the upward osmotic pressure. A complication of this arrangement is that the entry of solvent into the solution results in dilution of the latter, and so it is more difficult to treat than an arrangement in which an externally applied pressure opposes any flow of solvent into the solution.

The osmotic pressure of a solution is proportional to the amount n_B of solute B present in it. In fact, the expression for the osmotic pressure of an ideal solution bears an uncanny resemblance to the expression for the pressure of a perfect gas:

$$\Pi V \approx n_B RT \tag{14}$$

This equation is called the **van't Hoff equation** for the osmotic pressure. Because $n_B/V = [B]$, the molar concentration of the solute, a simpler form is

$$\Pi \approx [B]RT \tag{15}$$

This equation applies only to solutions that are sufficiently dilute to behave ideally.

Justification

The thermodynamic treatment of osmosis makes use of the fact that, at equilibrium, the chemical potential of the solvent A is the same on each side of the membrane: if it were not, then there would be a tendency for A molecules to flow from the region of high potential into the one of low potential. The starting relation is therefore

$$\mu_A(\text{solvent in the solution at pressure } p + \Pi)$$
$$= \mu_A(\text{pure solvent at pressure } p)$$

The pure solvent is at atmospheric pressure (p), and the solution is at a pressure $p + \Pi$ on account of the additional pressure Π that has to be exerted to establish equilibrium. We shall write the chemical potential of the pure solvent as $\mu_A^*(p)$. The chemical potential of the solvent in the solution is lowered by the solute, but it is raised on account of the greater pressure of the solution. We shall denote this chemical potential by $\mu_A(x_A, p + \Pi)$. The condition for equilibrium is therefore

$$\mu_A^*(p) = \mu_A(x_A, p + \Pi)$$

where x_A is the mole fraction of solvent molecules, and is related to x_B by $x_A = 1 - x_B$. The effect of the solute is taken into account by using eqn (10):

$$\mu_A(x_A, p + \Pi) = \mu_A^*(p + \Pi) + RT \ln x_A$$

The effect of pressure on the chemical potential of an (assumed incompressible) liquid is given by the expression first derived in Section 3.4 for the Gibbs energy but now expressed in terms of the chemical potential (by replacing V by the partial molar volume, V_A, of the solvent):

$$\mu_A^*(p + \Pi) = \mu_A^*(p) + V_A \Delta p$$

In this problem, the difference in pressure Δp can be identified with Π. When the last three equations are combined we get

$$-RT \ln x_A = \Pi V_A$$

As we have remarked, the mole fraction of the solvent is equal to $1 - x_B$. In a dilute solution, $\ln(1 - x_B)$ is approximately equal to $-x_B$ (for example, $\ln(1 - 0.01) = \ln 0.99 = -0.010050$), so this equation becomes

$$RT x_B \approx \Pi V_A$$

When the solution is dilute, $x_B = n_B/n \approx n_B/n_A$. Moreover, because $n_A V_A \approx V$, the total volume of the solution, this equation becomes

$$n_B RT \approx \Pi V$$

which is eqn (14).

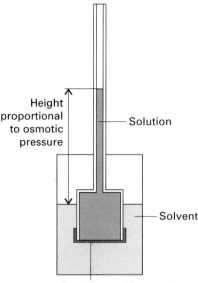

Height proportional to osmotic pressure

Solution

Solvent

Semipermeable membrane

Fig. 4.22 In a simple osmosis experiment, a solution is separated from the pure solvent by a semipermeable membrane. Pure solvent passes through the membrane and the solution rises in the inner tube. The net flow ceases when the pressure exerted by the column of liquid is equal to the osmotic pressure of the solution

One of the most common applications of osmosis is **osmometry**, the measurement of molar masses of proteins and synthetic polymers from the osmotic pressure of their solutions. As these huge molecules dissolve to produce solutions that are far from ideal, it is assumed that the van't Hoff equation is only the first term of an expansion:

$$\Pi = [B]RT\{1 + B[B] + \cdots\} \qquad (16)$$

Exactly the same expansion was used in Section 1.11 to extend the perfect gas equation to real gases. The empirical parameter B is called the **osmotic virial coefficient**. To use this expression, the osmotic pressure is measured at a series of concentrations, and the molar mass of B is found from a plot of $\Pi/[B]$ against $[B]$. The procedure is illustrated in the following example.

Example *Using osmometry to determine molar mass*

The osmotic pressures of solutions of poly(vinyl chloride), PVC, in cyclohexanone at 298 K are given below. The pressures are expressed in terms of the heights of solution (of density $\rho = 0.980\,\text{g cm}^{-3}$) in balance with the osmotic pressure. Find the molar mass of the polymer.

$c/\text{g dm}^{-3}$	1.00	2.00	4.00	7.00	9.00
h/cm	0.28	0.71	2.01	5.10	8.00

Strategy

First, we need to express eqn (16) in terms of the mass concentration, c, and the height of solution, h, so that we can use the data. The osmotic pressure is related to the height h of the solution above the level of the solvent through

$$\Pi = \rho g h$$

where ρ is the density of the solution and g is the acceleration of free fall ($9.81\,\text{m s}^{-2}$). The molar concentration $[B]$ of the solute is related to the mass concentration c (in grams per litre) by using

$$c = M[B]$$

where M is the molar mass of the solute. Then

$$\frac{h}{c} = \frac{RT}{\rho g M}\left(1 + \frac{Bc}{M} + \cdots\right)$$
$$= \frac{RT}{\rho g M} + \left(\frac{RTB}{\rho g M^2}\right)c + \cdots$$

It follows that, by plotting h/c against c, the results should fall on a straight line with intercept $RT/\rho g M$ on the vertical axis at $c = 0$. Then, from a determination of this intercept by extrapolation of the data to $c = 0$, the molar mass of the solute can be obtained. Be careful to keep units at each stage, even if they seem complicated, for that will help to ensure that the answer is correct at the end. It is best to convert all units to SI base units.

Solution

The following values of h/c can be calculated from the data:

$c/\text{g dm}^{-3}$	1.00	2.00	4.00	7.00	9.00
$(h/c)/\text{cm g}^{-1}\,\text{dm}^3$	0.28	0.36	0.503	0.739	0.889

The points are plotted in Fig. 4.23. The intercept with the vertical axis at $c = 0$ is 0.21, corresponding to $h/c = 0.21\,\text{cm g}^{-1}\,\text{dm}^3$. Because $1\,\text{cm} = 10^{-2}\,\text{m}$, $1\,\text{dm}^3 = 10^{-3}\,\text{m}^3$, and $1\,\text{g} = 10^{-3}\,\text{kg}$, this intercept corresponds to $0.21 \times 10^{-2}\,\text{m}^4\,\text{kg}^{-1}$. Therefore,

$$M = \frac{RT/\rho g}{0.21 \times 10^{-2}\,\text{m}^4\,\text{kg}^{-1}}$$

It follows that

$$M = \frac{(8.3145\,\text{J K}^{-1}\,\text{mol}^{-1}) \times (298\,\text{K})}{(980\,\text{kg m}^{-3}) \times (9.81\,\text{m s}^{-2})} \times \frac{1}{0.21 \times 10^{-2}\,\text{m}^4\,\text{kg}^{-1}}$$
$$= 1.2 \times 10^2\,\text{kg mol}^{-1}$$

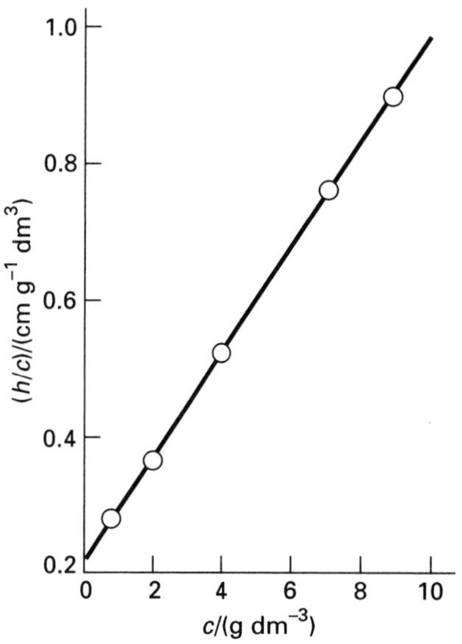

Fig. 4.23 The plot used in the determination of molar mass by osmometry. The molar mass is calculated from the intercept at $c = 0$.

Exercise E4.13

The heights of the solution in an osmometry experiment on a solution of an enzyme in water at $25\,°\text{C}$ were as follows:

$c/\text{g dm}^{-3}$	0.50	1.00	1.50	2.00	2.50
h/cm	0.18	0.35	0.53	0.71	0.90

The density of the solution is $0.9998\,\text{g cm}^{-3}$. What is the molar mass of the enzyme?

[*Answer*: $72\,\text{kg mol}^{-1}$]

Phase diagrams of mixtures

Now we consider the appearance and interpretation of the phase diagrams for a variety of mixtures. In each case it should be remembered that the phase diagram has been constructed empirically, often by making observations of transition temperatures for a series of mixtures of different compositions. To interpret the diagrams, all we have to do is to reverse the procedure in our minds and think about how they were constructed in the first place.

In the following, it will be useful to keep in mind the implications of the phase rule (Section 4.1). When two components are present in a system, $C = 2$ and

$$F = 4 - P$$

For simplicity we shall keep the pressure constant (at 1 atm, for instance), which uses up one of the degrees of freedom, and write $F' = 3 - P$ for the number of remaining degrees of freedom. One of these remaining degrees of freedom is the temperature, the other is the composition (as expressed by the mole fraction of one component). Hence we should be able to depict the phase equilibria of the system on a graph in which one axis is the temperature and the other axis is the composition. As for the phase diagrams we have already considered, a region of the graph shows the most stable state of the system for the given conditions. In a region where there is only one phase, $F' = 2$ and both the temperature and the composition can be varied. If two phases are present at equilibrium, $F' = 1$, and only one of the two variables may be changed at will. For example, if the composition is changed, then to maintain equilibrium between the two phases, the temperature must be adjusted. Such two-phase equilibria therefore define a line in the phase diagram. If three phases are present, $F' = 0$ and there is no degree of freedom for the system. That is, to achieve a three-phase equilibrium, we must adopt a specific temperature and composition. Such a condition is therefore represented by a point on the phase diagram.

4.5 Mixtures of volatile liquids

First, we consider the phase diagrams for binary (two-component) mixtures in which both components are volatile, and in particular consider the relation of the boiling point of a liquid mixture to its composition. This kind of information is important when considering the separation of liquids by fractional distillation, which is a widely used technique in industry and the laboratory. Intuitively, we might expect the boiling point of a mixture of two volatile liquids to vary smoothly from the boiling point of one pure component when only that liquid is present to the boiling point of the other pure component when only that liquid is present. This expectation is often borne out in practice, and a typical plot of boiling point against composition is shown in Fig. 4.24 (the lower curve).

The vapour in equilibrium with the boiling mixture will be a mixture of the two components, and we might also expect that it will be richer in the more volatile of the two liquids. This is also often found in practice, and the upper curve in the illustration shows the composition of the vapour in equilibrium with the boiling liquid. To identify the composition of the vapour, we note the boiling point of the liquid mixture (point a, for instance, if the mole fraction of A is 0.2), and draw a horizontal **tie line** across to the upper curve. Its point of intersection (a') gives the composition of the vapour. In

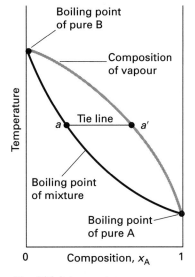

Fig. 4.24 A temperature–composition diagram for a binary mixture of volatile liquids. The tie line connects the points that represent the compositions of liquid and vapour that are in equilibrium at each temperature. The lower curve is a plot of the boiling point of the mixture against composition.

this example, we see that the mole fraction of A in the vapour is about 0.6. As expected, the vapour is richer than the liquid in the more volatile component. Graphs like these are determined empirically, by measuring the boiling points of a series of mixtures (to plot the lower curve of boiling point against composition), and measuring the composition of the vapour in equilibrium with each boiling mixture (to plot the corresponding points of the vapour-composition curve).

We can follow the changes that occur during the fractional distillation of a mixture of volatile liquids by seeing what happens when a mixture of composition a_1 is heated (Fig. 4.25). It boils at a temperature corresponding to point a_2 and its vapour has composition a_2'. This vapour condenses to a liquid of the same composition when it has risen to a cooler part of the fractionating column (a vertical column packed with glass rings or beads to give a large surface area). This condensate boils at the temperature corresponding to the point a_3 and yields a vapour of composition a_3'. This vapour is even richer in the more volatile component. That vapour condenses to a liquid which boils at the temperature corresponding to the point a_4. The cycle is repeated until almost pure A emerges from the top of the fractionating column.

Whereas many binary liquid mixtures do have temperature–composition diagrams resembling that shown in Fig. 4.24, in a number of important cases there are marked differences. For example, a maximum in the boiling point curve is sometimes found (Fig. 4.26). This behaviour is a sign that favourable interactions between the molecules of the two components reduce the vapour pressure of the mixture below the ideal value: in effect, favourable (A,B) interactions stabilize the liquid and cause it to have a higher boiling point than would be expected for a simple mingling of the two types of molecule. Examples of this behaviour include trichloromethane/propanone and nitric acid/water mixtures. Temperature–composition curves are also found that pass through a minimum (Fig. 4.27). This behaviour indicates that the (A,B) interactions are unfavourable and hence that the mixture is more volatile than expected on the basis of simple mingling of the two species. Examples include dioxane/water and ethanol/water.

There are important consequences for distillation when the temperature–composition diagram has a maximum or a minimum. Consider a liquid of composition a_1 on the right of the maximum in Fig. 4.26. It boils at a temperature corresponding to a_2 and its vapour (of composition a_2') is richer in the more volatile component A. If that vapour is removed (and condensed elsewhere), the composition of the remaining liquid moves towards a_3. The vapour in equilibrium with this boiling liquid has composition a_3': note that the two compositions are more similar than the original pair (a_3 and a_3' are closer together than a_2 and a_2'). If that vapour is removed, the composition of the boiling liquid shifts to a_4 and the vapour of that boiling mixture has an identical composition to the liquid. At this stage, evaporation occurs without change of composition. The mixture is said to form an **azeotrope** (which comes from the Greek words for 'boiling without changing').

When the azeotropic composition has been reached, distillation cannot separate the two liquids because the condensate retains the composition of

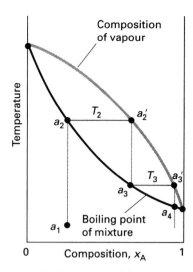

Fig. 4.25 The process of fractional distillation can be represented by a series of steps on a temperature–composition diagram like that in Fig. 4.24. The initial liquid mixture may be at a temperature and have a composition like that represented by point a_1. It boils at the temperature T_2, and the vapour in equilibrium with the boiling liquid has composition a_2'. If that vapour is condensed (to a_3 or below), the resulting condensate boils at T_3 and gives rise to a vapour of composition represented by a_3'. As the succession of vaporizations and condensations is continued, the composition of the distillate moves towards pure A (the more volatile component).

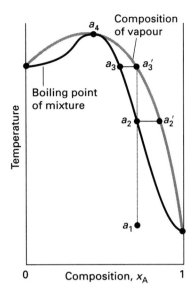

Fig. 4.26 The temperature–composition diagram for a high-boiling azeotrope. As fractional distillation proceeds, the composition of the remaining liquid moves towards a_4; however, once there, the vapour in equilibrium with that liquid has the same composition, so the mixture evaporates with an unchanged composition and no further separation can be achieved.

the liquid. One example of azeotrope formation is hydrochloric acid/water, which is azeotropic at 80 per cent water (by mass) and boils unchanged at 108.6 °C.

The system shown in Fig. 4.27 is also azeotropic, but shows its azeotropic character in a different way. Suppose we start with a mixture of composition a_1 and follow the changes in the composition of the vapour that rises through a fractionating column. The mixture boils at a_2 to give a vapour of composition a_2'. This vapour condenses in the column to a liquid of the same composition (now marked a_3). That liquid reaches equilibrium with its vapour at a_3', which condenses higher up the tube to give a liquid of the same composition. The fractionation therefore shifts the vapour towards the azeotropic composition at a_4, but the composition cannot move beyond a_4 because now the vapour and the liquid have the same composition. Consequently, the azeotropic vapour emerges from the top of the column. An example is ethanol/water, which boils unchanged when the water content is 4 per cent and the temperature is 78 °C.

4.6 Liquid–liquid phase diagrams

Diagrams like those we have been considering can also be used to discuss the composition of **partially miscible liquids**, which are liquids that do not mix together in all proportions. An example is a mixture of hexane and nitrobenzene: when the two liquids are shaken together, the liquid consists of two phases, one is a saturated solution of hexane in nitrobenzene and the other is a saturated solution of nitrobenzene in hexane. Because the two solubilities vary with temperature, the compositions and proportions of the two phases change as the temperature is changed, and we use a temperature–composition diagram to display the composition of the system at each temperature.

We shall introduce liquid–liquid phase diagrams by considering the special case of the two partially miscible liquids hexane and nitrobenzene. Suppose we add a small amount of nitrobenzene to a sample of hexane at some temperature T'. It dissolves completely; however, as more nitrobenzene is added, a stage comes when no more dissolves. The sample now consists of two phases in equilibrium with each other, the more abundant one consisting of hexane saturated with nitrobenzene, the less abundant one a trace of nitrobenzene saturated with hexane. In the temperature–composition diagram drawn in Fig. 4.28, the composition of the former is represented by the point a' and that of the latter by the point a''. The horizontal line joining a' and a'' is another example of a tie line, a line joining two phases that are in equilibrium with each other. The relative abundances of the two phases joined by a tie line are given by the **lever rule** (Fig. 4.29):

$$\frac{\text{amount of phase of composition } a''}{\text{amount of phase of composition } a'} = \frac{l'}{l''} \quad (17)$$

Justification

To prove the lever rule, we write $n = n' + n''$, where n' is the total amount of molecules in the one phase, n'' is the total amount in the other phase, and n is the total amount of molecules in the sample. The total amount of A in the sample is nx_A, where x_A is the overall mole fraction of A in the sample (this is the quantity plotted along the horizontal axis). The overall amount of A is also the sum of its amounts in the two phases, where it has the mole fractions x'_A and x''_A, respectively:

$$nx_A = n'x'_A + n''x''_A$$

Because $n = n' + n''$, we can also write

$$nx_A = n'x_A + n''x_A$$

Then, by equating these two expressions and rearranging them slightly, it follows that

$$n'(x'_A - x_A) = n''(x_A - x''_A)$$

or (as can be seen by referring to Fig. 4.29)

$$n'l' = n''l''$$

as was to be proved.

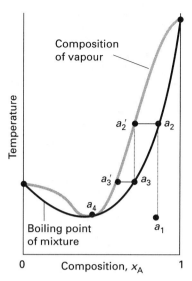

Fig. 4.27 The temperature–composition diagram for a low-boiling azeotrope. As fractional distillation proceeds, the composition of the vapour moves towards a_4; however, once there, the vapour in equilibrium with that liquid has the same composition, so no further separation of the distillate can be achieved.

When more nitrobenzene is added, hexane dissolves in it slightly. The overall composition moves to the right in the phase diagram, but the compositions of the two phases in equilibrium remain a' and a''. The difference is that the amount of the second phase increases at the expense of the first. A stage is reached when so much nitrobenzene is present that it can dissolve all the hexane, and the system reverts to a single phase. Now the point representing the overall composition and temperature lies to the right of the phase boundary in the illustration and the system is a single phase. The further addition of more nitrobenzene simply dilutes the solution, and from then on it remains a single phase.

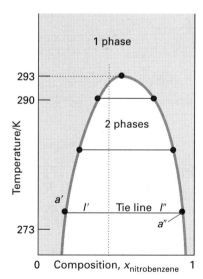

Fig. 4.28 The temperature–composition diagram for hexane and nitrobenzene at 1 atm. The upper critical solution temperature, T_{uc}, is the temperature above which no phase separation occurs. For this system it lies at 293 K (when the pressure is 1 atm).

Example *Interpreting a liquid–liquid phase diagram*
A mixture of 50 g (0.59 mol) of hexane and 50 g (0.41 mol) of nitrobenzene was prepared at 290 K. What are the compositions of the phases, and in what proportions do they occur? To what temperature must the sample be heated in order to obtain a single phase?

Strategy
The answer is based on Fig. 4.28. First, we need to identify the tie-line corresponding to the temperature specified: the points at its two ends give the compositions of the two phases in equilibrium. Next, we

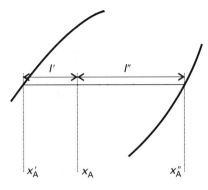

Fig. 4.29 The coordinates and compositions referred to by the lever rule.

identify the location on the horizontal axis corresponding to the overall composition of the system, and draw a vertical line. Where it cuts the tie line it divides it into the two lengths needed to use the lever rule, eqn (17). For the final part, we note the temperature at which the same vertical line cuts through the phase boundary: at that temperature and above, the system consists of a single phase.

Solution

We denote hexane by H and nitrobenzene by N. The horizontal tie-line at 290 K cuts the phase boundary at $x_N = 0.37$ and at $x_N = 0.83$, and so those mole fractions are the compositions of the two phases. The ratio of amounts of each phase is given by the lever rule:

$$\frac{l'}{l''} = \frac{0.41 - 0.37}{0.83 - 0.41} = \frac{0.04}{0.42} = 0.1$$

so the hexane-rich phase is ten times more abundant than the nitrobenzene-rich phase at this temperature. Heating the sample to 292 K takes it into the single phase region.

Exercise E4.14

Repeat the problem for 50 g hexane and 100 g nitrobenzene at 273 K.

[*Answer*: $x_N = 0.09$ and 0.95 in the ratio 1:1.3; 290 K]

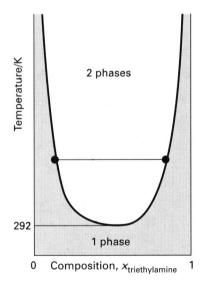

Fig. 4.30 The temperature–composition diagram for water and triethylamine. The lower critical solution temperature, T_{lc}, is the temperature below which no phase separation occurs. For this system it lies at 292 K (when the pressure is 1 atm).

As the phase diagram shows, the composition at which phase separation occurs and the compositions of the phases in equilibrium with each other depend on the temperature. The **upper critical solution temperature**, T_{uc}, is the upper limit of temperatures at which phase separation occurs. (It is sometimes called the 'upper consolute temperature'.) Above the upper critical solution temperature the two components are fully miscible. In molecular terms, this temperature exists because the greater thermal motion of the molecules leads to greater miscibility of the two components. In thermodynamic terms, the Gibbs energy of mixing becomes negative above a certain temperature, regardless of the composition.

Some systems show a **lower critical solution temperature**, T_{lc} (a 'lower consolute temperature'), below which they mix in all proportions and above which they form two phases. An example is water and triethylamine (Fig. 4.30). In this case, at low temperatures the two components are more miscible because they form a weak complex; at higher temperatures the complexes break up and the two components are less miscible.

A few systems have both upper and lower critical temperatures. The reason can be traced to the fact that after the weak complexes have been disrupted, leading to partial miscibility, the thermal motion at higher temperatures homogenizes the mixture again, just as in the case of ordinary partially miscible liquids. One example is nicotine and water, which are partially miscible between 61 °C and 210 °C (Fig. 4.31).

4.7 Liquid–solid phase diagrams

Phase diagrams are also used to show the regions of temperature and composition at which solids and liquids exist in binary systems. As we shall see, such diagrams are useful for discussing the techniques that are used to prepare the high-purity materials used in the electronics industry, and are also of great importance in metallurgy.

Eutectics

The phase diagram for a system composed of two metals that are almost completely immiscible right up to their melting points (such as antimony and bismuth) is shown in Fig. 4.32. Consider the molten liquid of composition a_1. When it is cooled to a_2 it enters the two-phase region labelled 'Liquid + A'. Almost pure solid A begins to come out of solution and the remaining liquid becomes richer in B. On cooling to a_3, more of the solid forms, and the relative amounts of the solid and liquid (which are in equilibrium) are given by the lever rule: at this stage there are roughly equal amounts of each. The liquid phase is richer in B than before (its composition is given by b_3) because A has been deposited. At a_4 there is less liquid than at a_3 and its composition is given by e. This liquid now freezes to give a two-phase system of almost pure A and almost pure B, as at point a_5.

The vertical line through e in Fig. 4.32 corresponds to the **eutectic composition**, the name coming from the Greek words for 'easily melted'. A liquid with the eutectic composition freezes at a single temperature, without previously depositing solid A or B. A solid with the eutectic composition melts, without change of composition, at the lowest temperature of any mixture. Solutions of composition to the right of e deposit A as they cool, and solutions to the left deposit B: only the eutectic mixture (apart from pure A or pure B) solidifies at a single definite temperature without gradually unloading one or other of the components from the liquid.

One technologically important eutectic is solder, which consists of 67 per cent tin and 33 per cent lead by mass and melts at 183 °C. Eutectic formation occurs in the great majority of binary alloy systems, and is of great importance for the microstructure of solid materials, for although a eutectic solid is a two-phase system, it crystallizes out in a nearly homogeneous mixture of microcrystals. The two microcrystalline phases can be distinguished by microscopy and structural techniques such as X-ray diffraction.

Thermal analysis is a very useful practical way of detecting eutectics. We can see how it is used by considering the rate of cooling down the vertical line at a_1 in Fig. 4.32. The liquid cools steadily until it reaches a_2 (Fig. 4.33),

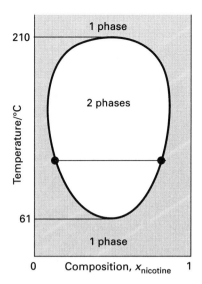

Fig. 4.31 The temperature–composition diagram for water and nicotine, which has both upper and lower critical solution temperatures. Note the high temperatures on the graph: the diagram corresponds to a sample under pressure.

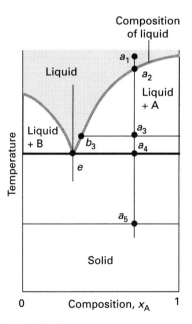

Fig. 4.32 The temperature–composition diagram for two almost immiscible solids and their completely immiscible liquids. The vertical line through e corresponds to the eutectic composition, the mixture with lowest melting point.

147

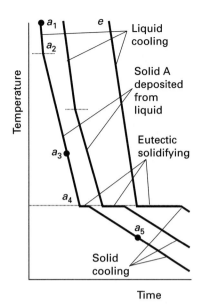

Fig. 4.33 The cooling curves for the system shown in Fig. 4.32. For a sample of composition represented by the vertical line through a_1 to a_5, the rate of cooling decreases at a_2 because solid A comes out of solution. The second cooling curve is for a sample of intermediate composition (between the vertical lines through a and e). If the experiment is repeated using a sample of composition represented by the vertical line through e, then there is a complete halt at e when the eutectic solidifies without change of composition. The halt is longest for the mixture of eutectic composition. The cooling curves can be used to construct the phase diagram.

when A begins to be deposited. Cooling is now slower because the solidification of A is exothermic and retards the cooling. When the remaining liquid reaches the eutectic composition, the temperature remains constant until the whole sample has solidified: this pause in the decrease in temperature is known as the **eutectic halt**. If the liquid has the eutectic composition e initially, then the liquid cools steadily down to the freezing temperature of the eutectic, when there is a long eutectic halt as the entire sample solidifies (like the freezing of a pure liquid).

Monitoring the cooling curves at different overall compositions gives a clear indication of the structure of the phase diagram. The solid–liquid boundary is given by the points at which the rate of cooling changes. The longest eutectic halt gives the location of the eutectic composition and its melting temperature.

Ultrapurity and controlled impurity

Advances in technology have called for materials of extreme purity. For example, semiconductor devices consist of almost perfectly pure silicon or germanium doped to a precisely controlled extent. For these materials to operate successfully, the impurity level must be kept down to less than 1 in 10^9. The technique of **zone refining** makes use of the nonequilibrium properties of mixtures. It relies on the impurities being more soluble in the molten sample than in the solid, and sweeps them up by passing a molten zone repeatedly from one end to the other along a sample (Fig. 4.34). In practice, a train of hot and cold zones are swept repeatedly from one end to the other. The zone at the end of the sample is the impurity dump: when the heater has gone by, it cools to a dirty solid that can be discarded.

We can use a phase diagram to discuss zone refining, but we have to take care to allow for the fact that the molten zone moves along the sample and the sample is uniform in neither temperature nor composition. Consider a liquid (which represents the molten zone) on the vertical line at a_1 (Fig. 4.35) and let it cool without the entire sample coming to overall equilibrium. If the temperature falls to a_2, a solid of composition b_2 is deposited and the remaining liquid (the zone where the heater has moved on) is at a_2'. Cooling that liquid down a vertical line passing through a_2' deposits solid of composition b_3 and leaves liquid at a_3'. The process continues until the last drop of liquid to solidify is heavily contaminated with A. There is plenty of everyday evidence that impure liquids freeze in this way. For example, an ice cube is clear near the surface but misty in the core. This is because the water used to make ice normally contains dissolved air; freezing proceeds from the outside, and air is accumulated in the retreating liquid phase. It cannot escape from the interior of the cube, and so when that freezes it occludes the air in a mist of tiny bubbles.

A modification of zone refining is **zone levelling**. It is used to introduce controlled amounts of impurity (for example, of indium into germanium). A sample rich in the required dopant is put at the head of the main sample, and made molten. The zone is then dragged repeatedly in alternate directions through the sample, where it deposits a uniform distribution of the impurity.

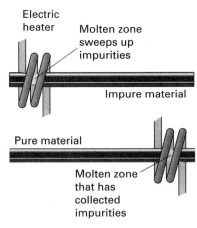

Fig. 4.34 In the zone refining procedure, a heater is used to melt a small region of a long cylindrical sample of the impure solid, and that zone is swept to the other end of the rod. As it moves, it collects impurities. If a series of passes are made, the impurities accumulate at one end of the rod and can be discarded.

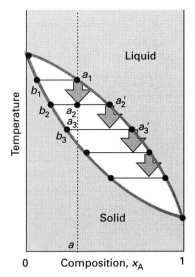

Fig. 4.35 A binary temperature–composition diagram can be used to discuss zone-refining, as explained in the text.

EXERCISES

4.1 The material in this chapter lets us estimate the contamination of air in closed regions, such as laboratories. An open vessel containing (a) water, (b) benzene, (c) mercury stands in a laboratory measuring $6.0\,m \times 5.3\,m \times 3.2\,m$ at $25\,°C$. What mass of each substance will be found in the air if there is no ventilation? (The vapour pressures are (a) 24 Torr, (b) 98 Torr, (c) 1.7 mTorr.)

4.2 On a cold, dry morning after a frost, the temperature was $-5\,°C$ and the partial pressure of water in the atmosphere fell to 2 Torr. Will the frost sublime? What partial pressure of water would ensure that the frost remained?

4.3 Refer to Fig. 4.8 and describe the changes that would be observed when water vapour at 1.0 atm and 400 K is cooled at constant pressure to 260 K. Suggest the appearance of a plot of temperature against time if energy is removed at a constant rate. To judge the relative slopes of the cooling curves, you need to know that the constant-pressure molar heat capaci-

ties of water vapour, liquid, and solid are approximately $4R$, $9R$, and $4.5R$; the enthalpies of transition are given in Table 2.2.

4.4 Refer to Fig. 4.8 and describe the changes that would be observed when cooling takes place at the pressure of the triple point.

4.5 Use the phase diagram in Fig. 4.9 to state what would be observed when a sample of carbon dioxide, initially at 1.0 atm and 298 K is subjected to the following cycle: (a) constant-pressure heating to 320 K, (b) isothermal compression to 100 atm, (c) constant-pressure cooling to 210 K, (d) isothermal decompression to 1.0 atm, constant-pressure heating to 298 K.

4.6 The partial molar volumes of propanone and trichloromethane in a mixture in which the mole fraction of $CHCl_3$ is 0.4693 are $74.166\,cm^3\,mol^{-1}$ and $80.235\,cm^3\,mol^{-1}$, respectively. What is the volume of a solution of total mass 1.000 kg?

4.7 At 300 K, the vapour pressure of dilute solutions of HCl in liquid $GeCl_4$ are as follows:

$x(HCl)$	0.005	0.012	0.019
p/kPa	32.0	76.9	121.8

Show that the solution obeys Henry's law in this range of mole fractions and calculate Henry's law constant at 300 K.

4.8 At 90 °C the vapour pressure of toluene (methylbenzene) is 400 Torr and that of o-xylene (1,2-dimethylbenzene) is 150 Torr. What is the composition of the liquid mixture that boils at 25 °C when the pressure is 0.50 atm? What is the composition of the vapour produced?

4.9 The vapour pressure of a sample of benzene is 400 Torr at 60.6 °C, but it fell to 386 Torr when 0.125 g of an organic compound was dissolved in 5.00 g of the solvent. Calculate the molar mass of the compound.

4.10 The addition of 28.0 g of a compound to 750 g of tetrachloromethane, CCl_4, lowered the freezing point of the solvent by 5.40 K. Calculate the molar mass of the compound.

4.11 A compound A existed in equilibrium with its dimer, A_2, in propanone solution. Derive an expression for the equilibrium constant in terms of the depression in vapour pressure caused by a given concentration of compound. *Hint.* Suppose that a fraction f of the A molecules are present as the dimer. The depression of vapour pressure is proportional to the total concentration of A and A_2 molecules regardless of their chemical identities.

4.12 The osmotic pressure of an aqueous solution of urea at 300 K is 120 kPa. Calculate the freezing point of the same solution.

4.13 The rise in atmospheric carbon dioxide results in higher concentrations of dissolved carbon dioxide in natural waters. Use Henry's law and the data in Table 4.2 to calculate the solubility of CO_2 in water at 25 °C when its partial pressure is (a) 4.0 kPa, (b) 100 kPa.

4.14 The mole fractions of N_2 and O_2 in air at sea level are approximately 0.78 and 0.21. Calculate the molalities of the solution formed in an open flask of water at 25 °C.

4.15 A water-carbonating plant is available for use in the home and operates by providing carbon dioxide

at 3.0 atm. Estimate the molar concentration of the CO_2 in the soda water it produces.

4.16 Estimate the freezing point of 150 cm^3 of water sweetened with 7.5 g of sucrose.

4.17 The osmotic pressure of a solution of polystyrene in toluene was measured at 25 °C and the pressure was expressed in terms of the height of the solvent of density 1.004 $g\,cm^{-3}$:

$c/(g\,L^{-1})$	2.042	6.613	9.521	12.602
h/cm	0.592	1.910	2.750	3.600

Calculate the molar mass of the polymer.

4.18 The molar mass of an enzyme was determined by dissolving it in water, measuring the osmotic pressure at 20 °C and extrapolating the data to zero concentration. The following data were used:

$c/(mg\,cm^{-3})$	3.221	4.618	5.112	6.722
h/cm	5.746	8.238	9.119	11.990

Calculate the molar mass of the enzyme.

4.19 The following temperature/composition data were obtained for a mixture of octane (O) and toluene (T) at 760 Torr, where x is the mole fraction in the liquid and y the mole fraction in the vapour at equilibrium.

$\theta/°C$	110.9	112.0	114.0	115.8	117.3	119.0	120.0	123.0
x_T	0.908	0.795	0.615	0.527	0.408	0.300	0.203	0.097
y_T	0.923	0.836	0.698	0.624	0.527	0.410	0.297	0.164

The boiling points are 110.6 °C for toluene and 125.6 °C for octane. Plot the temperature–composition diagram of the mixture. What is the composition of the vapour in equilibrium with the liquid of composition (a) $x_T = 0.250$ and (b) $x_O = 0.250$?

4.20 Sketch the phase diagram of the system NH_3/N_2H_4 given that the two substances do not form a compound with each other, that NH_3 freezes at −78 °C and N_2H_4 freezes at +2 °C, and that a eutectic is formed when the mole fraction of N_2H_4 is 0.07 and that the eutectic melts at −80 °C.

4.21 Figure 4.36 shows the phase diagram for two partially miscible liquids, which can be taken to be that for water (A) and 2-methyl-1-propanol (B). Describe what will be observed when a mixture of composition b_3 is heated, at each stage giving the number, composition, and relative amounts of the phases present.

4.22 Figure 4.37 is the phase diagram for silver/tin. Label the regions, and describe what will be observed when liquids of compositions a and b are cooled to 200 °C.

4.23 Sketch the cooling curves for the compositions a and b in Fig. 4.37.

4.24 Use the phase diagram in Fig. 4.37 to determine (a) the solubility of silver in tin at 800 °C, (b) the solubility of Ag_3Sn in silver at 460 °C, and (c) the solubility of Ag_3Sn in silver at 300 °C.

4.25 Hexane and perfluorohexane show partial miscibility below 22.70 °C. The critical concentration at the upper critical temperature is $x = 0.355$, where x is the mole fraction of C_6F_{14}. At 22.0 °C the two solutions in equilibrium have $x = 0.24$ and $x = 0.48$ respectively, and at 21.5 °C the mole fractions are 0.22 and 0.51. Sketch the phase diagram. Describe the phase changes that occur when perfluorohexane is added to a fixed amount of hexane at (a) 23 °C, (b) 25 °C.

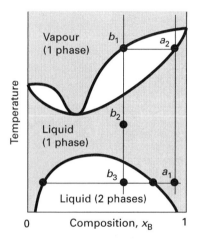

Fig. 4.36 The phase diagram for Exercise 4.21

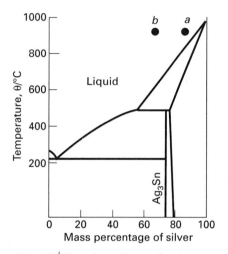

Fig. 4.37 That phase diagram for the binary silver/tin system referred to in Exercise 4.22

Chemical equilibrium

In this chapter we examine some of the consequences of dynamic chemical equilibria. Among them are the equilibria that exist in solutions of acids, bases, and their salts in water, where rapid proton transfer between species ensures that equilibrium is maintained at all times. Proton transfer equilibrium is maintained in living cells, and helps to keep proteins viable. Even small drifts in the equilibrium concentration of hydrogen ions can result in disease, cell damage, and death, and we shall see the basis of our bodies' mechanism for maintaining correct concentrations. In many cases in industry chemical systems are not allowed to reach equilibrium, but knowing whether equilibrium lies in favour of reactants or products under certain conditions is a good indication of the feasibility of a process, and indicates whether it would be worthwhile building a plant to achieve the reaction. We shall see, for instance, that even such large-scale processes as ironmaking can be discussed in terms of dynamic equilibrium, even though equilibrium is not fully achieved.

The ideas developed in this chapter are all based on the equilibrium constant K or, what is equivalent, the standard reaction Gibbs energy $\Delta_r G^{\ominus}$. As we have seen (Section 3.5), the two quantities are related by

$$\Delta_r G^{\ominus} = -RT \ln K \tag{1}$$

As explained in Section 3.5, the equilibrium constant of a reaction is defined as follows:

$$a\,A + b\,B \rightleftharpoons c\,C + d\,D \qquad K = \frac{a_C^c a_D^d}{a_A^a a_B^b} \tag{2}$$

where a_J is the activity of species J. For pure liquids and solids, $a = 1$, and so such species do not appear in the expression for the equilibrium constant. In elementary work it is common to confine attention to ideal systems in which $a_J = p_J/p^{\ominus}$ if J is a gas at a partial pressure p_J, and $a_J = [J]/(\mathrm{mol\,L^{-1}})$ if J is a solute at molar concentration [J]. Both approximations are reliable for gases at low pressures, and are reasonably reliable for dilute solutions of nonelectrolytes. However, chemists are commonly interested in electrolyte solutions. Ion–ion interactions are so strong, and extend over such a long range that electrolyte solutions can be regarded as ideal only at very low concentrations (at well below $10^{-3}\,\mathrm{mol\,L^{-1}}$), and extreme caution bordering on scepticism must be exercised when carrying out calculations on anything more concentrated.

These somewhat pessimistic remarks do not mean that equilibrium calculations cannot be applied to electrolyte solutions. The discussion of equilibria in terms of activities is always valid, and it is often the case that a measurable property is in fact related to the activity and not to a concentration. That is the case with the pH of a solution, which is defined in terms of activities, as we shall see. The material we cover in the next chapter, on electrochemistry, is also expressed in terms of activities. It is only when we want to take the step from an exact calculation in terms of activities into a statement about concentrations that we have to be cautious. In that case it is common to express the activity as

$$a_J = \gamma[J]/(mol\,L^{-1}) \tag{3}$$

where γ is an empirical parameter called the **activity coefficient**, and to hope to find some way of estimating or measuring the activity coefficient so that the calculated activity can be converted into a molar concentration.

The interpretation of equilibrium constants

The magnitude of an equilibrium constant is a good *qualitative* indication of the feasibility of a reaction regardless of whether the system is ideal or not. Broadly speaking, if K is much larger than 1 (typically $K > 10^3$), then a reaction has a strong tendency to form products. If K is much smaller than 1 (that is, for $K < 10^{-3}$), then the equilibrium composition will consist of largely unchanged reactants. If K is comparable to 1 (in the range 10^{-3} to 10^3), then significant amounts of both reactants and products will be present at equilibrium.

An equilibrium constant expresses the composition of an equilibrium mixture as a ratio of products of activities. Even if we confine our attention to ideal systems, it is still necessary to do some work to extract the actual equilibrium concentrations or partial pressures of the reactants and products. This work can be organized into a systematic procedure resembling a spreadsheet by constructing a table that shows, in successive rows:

The reactant and product species

---

1. The initial molar concentrations or partial pressures of the species.
2. The changes in these quantities that must take place for the system to reach equilibrium.
3. The resulting equilibrium values.

In most cases, we do not know the explicit change that must occur for the system to reach equilibrium, and so it is written as x: the reaction stoichiometry is then used to write the changes in each species given a particular change x in one of them. When the values at equilibrium are substituted into the expression for the equilibrium constant, we obtain an equation for K in terms of x. This equation can be rearranged into an equation for x in terms of K, and then solved for x, and hence the concentrations of all the species at equilibrium may be found. In simple cases the equation can be solved exactly. In other cases the equation has to be solved either by using a computer with mathematical software or by making an approximation. Making approximations should not be scorned: the resulting expressions are often very reliable and, when thought about, are very helpful for understanding what is happening in the system as it approaches equilibrium. We shall use the approximation technique very frequently when considering acids and bases.

As an illustration of the procedure, suppose that in an industrial process N_2 at a partial pressure of 1.00 bar is mixed with H_2 at a partial pressure of 3.00 bar and the two gases are allowed to come to equilibrium with the product ammonia (in the presence of a catalyst) in a reactor of constant volume. The reaction and the expression for the equilibrium constant are

$$N_2(g) + 3\,H_2(g) \rightleftharpoons 2\,NH_3(g) \qquad K = \frac{a(NH_3)^2}{a(N_2)a(H_2)^3}$$

At the temperature of the reaction, it has been determined experimentally that $K = 977$. The equilibrium table has the form

Species:	N_2	H_2	NH_3
Initial partial pressure/bar	1.00	3.00	0
Change to reach equilibrium/bar	$-x$	$-3x$	$+2x$
Equilibrium partial pressure/bar	$1.00 - x$	$3.00 - 3x$	$2x$

This procedure is valid when the volume of the reaction vessel is constant, because then each partial pressure is proportional to the amount of its molecules present $(p_J = n_J RT/V)$, and so the stoichiometric relations, which in general apply only to amounts, apply to partial pressures too. The equilibrium constant for the reaction is expressed in terms of the following activities:

$$a(N_2) = 1.00 - x \qquad a(H_2) = 3.00 - 3x \qquad a(NH_3) = 2x$$

and is

$$K = \frac{(2x)^2}{(1.00 - x)(3.00 - 3x)^3}$$

Then, with $K = 977$, this equation rearranges first to

$$977 = \frac{4}{27}\left(\frac{x}{(1.00 - x)^2}\right)^2$$

and, after taking the square root of both sides, to

$$\sqrt{977 \times \frac{27}{4}} = \frac{x}{(1.00 - x)^2}$$

This expression rearranges into the quadratic equation

$$gx^2 - (2.00g + 1)x + 1.00g = 0 \qquad g = \sqrt{977 \times \frac{27}{4}} = 81.208\cdots$$

At this stage we can use the quadratic formula:

$$ax^2 + bx + c = 0 \qquad x = \frac{-b \pm \sqrt{b^2 - 4ac}}{2a} \tag{4}$$

For this example, $x = 1.12$ or $x = 0.895$. Because $a(N_2)$ cannot be negative (it is proportional to the partial pressure of nitrogen), and $a(N_2) = 1.00 - x$ (from the equilibrium table), we know that x cannot be greater than 1.00. Therefore, we select $x = 0.895$ as the acceptable solution. It then follows from the last line of the equilibrium table that

$$p(N_2) = 0.10\,\text{bar} \qquad p(H_2) = 0.32\,\text{bar} \qquad p(NH_3) = 1.8\,\text{bar}$$

as the composition of the reaction mixture at equilibrium. Note that, because K is large (of the order of 10^3), the products dominate at equilibrium. It is always wise to confirm the accuracy of the calculation by substituting the calculated equilibrium partial pressures into the expression for the equilibrium constant to verify that the value so calculated is equal to the experimental value used in the calculation:

$$\frac{a(NH_3)^2}{a(N_2)a(H_2)^3} = \frac{(1.8)^2}{(0.10)(0.32)^3} = 9.9 \times 10^2$$

which is close to the experimental value (the discrepancy stems from rounding errors).

All equilibrium calculations can be treated similarly, and it is often the case that simple equations result if the reactants are initially present in stoichiometric proportions (the nitrogen and hydrogen partial pressures in the example above were in the ratio 1:3, like their stoichiometric numbers). When the reactants are not present in stoichiometric proportions initially, or when the expression for the rate constant is complicated, we get much more difficult expressions to solve for x. The procedure to adopt in such cases is illustrated in the following example.

Example *Solving an equilibrium calculation by approximation*
Suppose that in an industrial process, nitrogen and hydrogen were introduced into a vessel at partial pressures of 2.00 bar in each case, and the two gases were allowed to come to equilibrium with their

product, ammonia, in a vessel of fixed volume and at a temperature at which $K = 1.0 \times 10^{-6}$. What is the equilibrium composition of the reaction mixture?

Strategy

The equilibrium table is drawn up in the same way as before, and the expression for K written in terms of the unknown quantity x. This procedure leads to an equation that is best solved on a computer. For quick estimates, though, we may be able to use an approximation based on the observation that K is small, so only a small conversion to ammonia takes place. If we ignore x in expressions like $2.00 - 3x$ (or whatever the calculation brings up), the expression for K in terms of x is greatly simplified. At the end of the calculation we need to check the validity of the approximation by verifying that x is indeed very small compared to other terms.

Solution

The equilibrium table has the form

Species:	N_2	H_2	NH_3
Initial partial pressure/bar	2.00	2.00	0
Change to reach equilibrium/bar	$-x$	$-3x$	$+2x$
Equilibrium partial pressure/bar	$2.00 - x$	$2.00 - 3x$	$2x$

The equilibrium constant for the reaction is expressed in terms of the quantities in the last line of the table:

$$K = \frac{(2x)^2}{(2.00 - x)(2.00 - 3x)^3}$$

with $K = 1.0 \times 10^{-6}$. This expression is very awkward to solve for x. However, if $x \ll 2.00$ and $3x \ll 2.00$ (if the second is true, then the first certainly is), we can use

$$a(N_2) \approx 2.00 \qquad a(H_2) \approx 2.00 \qquad a(NH_3) = 2x$$

Then,

$$K \approx \frac{4x^2}{(2.00)(2.00)^3} = \frac{4x^2}{(2.00)^4}$$

This expression is easily rearranged into

$$x = \left(\frac{K \times (2.00)^4}{4}\right)^{1/2} = 2.0 \times 10^{-3}$$

This value of x is consistent with the approximations made. Therefore, at equilibrium, the partial pressures of N_2, H_2, and NH_3 are 2.00 bar, 2.00 bar, and 4.0 mbar, respectively.

Exercise E5.1

Find the equilibrium partial pressures for the same reaction, but starting with partial pressures of 3.00 bar, 1.00 bar, and 0.500 bar of N_2, H_2, and NH_3, respectively, and at a temperature for which $K = 2.50 \times 10^{-4}$.

[*Answer*: 2.99 bar, 0.96 bar, 0.528 bar]

The approximation procedure is valid when the equilibrium composition differs only slightly from the initial (given) composition. This is typically the case when reactants are mixed initially and K is much smaller than 1 ($K < 10^{-3}$), for then only a small amount of product need be formed to reach equilibrium (so x is very small). It may also be the case when products are mixed initially and the equilibrium constant is much larger than 1 ($K > 10^3$), because only a small amount of reactants are formed to reach equilibrium. Whenever the approximation procedure is employed, it should be verified that the answer is consistent with the approximations that have been made. For example, if it is assumed that the change in concentration x is much smaller than the initial concentration or partial pressure of a reactant, then it should be verified that the calculated change is no more than about 5 per cent of the initial concentration. If the calculated change is more than 5 per cent of the initial concentration, then there is no choice but to use a computational approach to the solution.

Acids and bases

One of the most important examples of chemical equilibrium is the one that exists when acids and bases are present in solution. As we shall see, the equilibration of acids and bases depends on the transfer of protons between species. In the context of acids and bases, a 'proton' is a hydrogen ion, H^+, and the theory of acids and bases that we shall develop depends on the transfer of hydrogen ions between species. This transfer is normally so facile that we can be confident that any solution of acid or base is at equilibrium and that it is appropriate to calculate the composition of the solution using the equilibrium constant for the proton transfer reaction.

The principal reason for carrying out a calculation of the equilibrium composition of a solution of an acid or a base is to find the concentration of hydrogen ions in solution. This concentration is of the greatest importance in many applications of chemistry, for hydrogen ions govern the processes of life, modify landscapes by their influence on geochemical processes, and determine the success of many industrial syntheses. Much of our environment, outside us in our physical surroundings and inside us in our biochemical composition, is a manifestation of the role of hydrogen ion concentration.

The concentration of hydrogen ions in a solution is normally reported in terms of the pH of the solution, where

$$pH = -\log a(H^+) \qquad (5)$$

In elementary work, it is common to replace the hydrogen ion activity by the numerical value of its molar concentration. For example, if the molar concentration of H^+ is $2.0 \times 10^{-3}\,mol\,L^{-1}$ then

$$pH \approx -\log(2.0 \times 10^{-3}) = 2.70$$

If the molar concentration were ten times less, at $2.0 \times 10^{-4}\,mol\,L^{-1}$, then the pH would be 3.70. Notice that *the higher the pH, the lower the concentration of hydrogen ions in the solution*, and that a change in pH by 1 unit corresponds to a 10-fold change in molar concentration.

It should never be forgotten that the replacement of the activities of ions by molar concentration is invariably hazardous, and because ions interact over long distances is unreliable for all but the most dilute solutions. However, this does not mean that pH is useless: many observable properties depend on the activity, not the concentration, and so for them working with pH is exactly what is needed. It is only when conversions between concentration and activity need to be done that approximations enter. In those cases, the resulting pH value is reliable only to about one significant figure, representing an uncertainty in molar concentration of approximately a factor of 10.

Exercise E5.2

Death is likely if the pH of human blood plasma changes by more than ± 0.4 from its normal value of 7.4. What is the approximate range of molar concentrations of hydrogen ions for which life can be sustained?

[*Answer*: $16\,nmol\,L^{-1}$ to $100\,nmol\,L^{-1}$ ($1\,nmol = 10^{-9}\,mol$)]

5.1 The Brønsted–Lowry theory

Most modern work on the reactions of acids and bases is expressed in terms of a theory proposed by the Danish chemist Johannes Brønsted and the

English chemist Thomas Lowry in 1923. According to the **Brønsted–Lowry theory** of acids and bases,

A Brønsted acid is a proton donor.
A Brønsted base is a proton acceptor.

Hydrogen chloride, HCl, is an acid because it can donate a proton to another molecule. Methane, CH_4, is not a Brønsted acid because, despite its hydrogen atoms, it is not a proton donor. Ammonia, NH_3, is a base because it can accept a proton from another molecule and become NH_4^+. The definitions make no mention of the solvent (and apply even if no solvent is present); however, by far the most important medium is aqueous solution, and we shall confine our attention to that.

Exercise E5.3

Identify H_2SO_4, HSO_4^-, and SO_4^{2-} as an acid or a base.

[*Answer*: acid, acid and base, base]

An acid HA (for example, HCl or CH_3COOH) takes part in the following proton-transfer equilibrium in water:

$$HA(aq) + H_2O(l) \rightleftharpoons H_3O^+(aq) + A^-(aq)$$

In the forward reaction, the H_2O molecule accepts a proton from the acid, and therefore acts as a Brønsted base. The H_3O^+ ion, which is formed when H_2O accepts a proton from an acid, is called a **hydronium ion**. In the reverse reaction, the A^- ion accepts a proton from H_3O^+ and is converted back into the acid HA. Therefore, in the reverse reaction, H_3O^+ acts as a Brønsted acid and A^- acts as a Brønsted base.

The proton donor that results from the transfer of a proton to a species is called the **conjugate acid** of the original base. In the forward reaction above, the H_3O^+ ion that results from proton transfer to H_2O is the conjugate acid of the base H_2O. Similarly, the proton acceptor A^- that remains after a Brønsted acid has donated a proton is called the **conjugate base** of the acid HA. For example, in the equilibrium established when HF is present in water,

$$HF(aq) + H_2O(l) \rightleftharpoons H_3O^+(aq) + F^-(aq)$$

the H_3O^+ ion is the conjugate acid of the base H_2O and the F^- ion is the conjugate base of the acid HF. Similarly, in the equilibrium

$$CH_3COOH(aq) + H_2O(l) \rightleftharpoons H_3O^+(aq) + CH_3CO_2^-(aq)$$

the acetate ion, $CH_3CO_2^-$, is the conjugate base of the acid CH_3COOH. The general form of the Brønsted proton transfer equilibrium in water is therefore

$$acid_1 + base_2 \rightleftharpoons acid_2 + base_1 \qquad (6)$$

with $(acid_1, base_1)$ one conjugate acid–base pair and $(acid_2, base_2)$ another conjugate acid–base pair.

Exercise E5.4

What is the conjugate base of ammonia when it acts as an acid?

[*Answer*: NH_2^-, the amide ion]

For a base B, such as ammonia, NH_3, in water, the characteristic proton transfer equilibrium is

$$H_2O(l) + B(aq) \rightleftharpoons BH^+(aq) + OH^-(aq)$$

The proton donor formed when the base B accepts a proton and becomes BH^+ is another example of a conjugate acid of a base. Likewise, OH^-, the ion that results from the loss of a proton from the proton donor H_2O, is another example of a conjugate base. An example of this equilibrium is

$$H_2O(l) + NH_3(aq) \rightleftharpoons NH_4^+(aq) + OH^-(aq)$$

The conjugate acid of NH_3 is the ammonium ion, NH_4^+. As can be seen, the form of this equilibrium is exactly the same as in the general case, but now H_2O acts as aBrønsted acid:

$$H_2O(l) + NH_3(aq) \rightleftharpoons NH_4^+(aq) + OH^-(aq)$$
$$acid_1 + base_2 \rightleftharpoons acid_2 + base_1$$

The fact that water can act as both a Brønsted acid and a Brønsted base means that even in the absence of any solute, hydronium ions and hydroxide ions exist in water as a result of the **autoprotolysis equilibrium** in which protons are transferred between neighbouring water molecules:

$$H_2O(l) + H_2O(l) \rightleftharpoons H_3O^+(aq) + OH^-(aq)$$
$$acid_1 + base_2 \rightleftharpoons acid_2 + base_1$$

This equilibrium is present at all times in water, and you can never be sure that a hydrogen atom belonging to one H_2O molecule will still be a part of the same molecule a microsecond later. However, although there is this unceasing exchange, at any instant the *overall* concentration of hydronium ions and hydroxide ions in a sample of pure water is constant. At 25 °C the molar concentration of hydronium ions in water (and the molar concentration of hydroxide ions) arising from this autoprotolysis equilibrium is only 1.0×10^{-7} mol L^{-1}, so at any instant only 1 in 550 million H_2O molecules has donated a proton to another H_2O molecule. Nevertheless, this tiny concentration of hydronium ions is of crucial importance to the properties of aqueous solutions of acids, bases, and salts, as we shall see.

It follows that, as a result of autoprotolysis, the pH of pure water at 25 °C is

$$pH \approx -\log(1.0 \times 10^{-7}) = 7.00$$

This approximation (of replacing activities by molar concentrations) is very good in pure water because the concentrations of ions are so low. A neutral solution is an aqueous solution with this pH at 25 °C. However, because the autoprotolysis reaction is endothermic, the formation of hydronium ions is favoured by increased temperature, and at 37 °C (body temperature) the equilibrium concentration of hydronium ions has risen to 1.5×10^{-7} mol L^{-1}; the pH of a neutral solution is then 6.82.

5.2 Acidity constants

Like any equilibrium, the proton transfer equilibrium of a Brønsted acid in water can be expressed in terms of an equilibrium constant:

$$HA(aq) + H_2O(l) \rightleftharpoons H_3O^+(aq) + A^-(aq) \qquad K = \frac{a(H_3O^+)a(A^-)}{a(HA)a(H_2O)}$$

where, as usual, $a(J)$ is best regarded as an activity, but which can cautiously be set equal to $[J]/(\text{mol L}^{-1})$ for a solute at concentrations so low that deviations from ideality can be ignored; $a(J)$ is equal to 1 for a pure liquid or solid. The same expression applies to the proton transfer equilibrium of the conjugate acid of a Brønsted base—there is no fundamental distinction between an acid and a conjugate acid of a base: both act as proton donors and exchange protons with the surrounding solvent molecules. Strictly speaking, there are no pure liquids or solids in the proton transfer equilibrium, but the solutions we shall consider will always be so dilute that it is almost always the case that the water present can be regarded as being a nearly pure liquid. If, therefore, we make the approximation that $a(H_2O) = 1$ for all the solutions we consider, then the resulting equilibrium constant is called the **acidity constant**, K_a, of the acid HA:

$$K_a = \frac{a(H_3O^+)a(A^-)}{a(HA)} \qquad (7)$$

An explicit example of an acidity constant is that for hydrofluoric acid:

$$HF(aq) + H_2O(l) \rightleftharpoons H_3O^+(aq) + F^-(aq) \qquad K_a = \frac{a(H_3O^+)a(F^-)}{a(HF)}$$

It is found experimentally that $K_a = 3.5 \times 10^{-4}$ at 25 °C. The acidity constants of a number of acids (and the conjugate acids of bases) are given in Table 5.1 together with additional information that will be explained below.

Table 5.1 Acidity and basicity constants* at 25°C

Acid/Base	K_b	pK_b	K_a	pK_a
Strongest weak acids				
Trichloroacetic acid, CCl_3COOH	3.3×10^{-14}	13.48	3.0×10^{-1}	0.52
Benzenesulfonic acid, $C_6H_5SO_3H$	5.0×10^{-14}	13.30	2×10^{-1}	0.70
Iodic acid, HIO_3	5.9×10^{-14}	13.23	1.7×10^{-1}	0.77
Sulfurous acid, H_2SO_3	6.3×10^{-13}	12.19	1.6×10^{-2}	1.81
Chlorous acid, $HClO_2$	1.0×10^{-12}	12.00	1.0×10^{-2}	2.00
Phosphoric acid, H_3PO_4	1.3×10^{-12}	11.88	7.6×10^{-3}	2.12
Chloroacetic acid, $CH_2ClCOOH$	7.1×10^{-12}	11.15	1.4×10^{-3}	2.85
Lactic acid, $CH_3CH(OH)COOH$	1.2×10^{-11}	10.92	8.4×10^{-4}	3.08
Nitrous acid, HNO_2	2.3×10^{-11}	10.63	4.3×10^{-4}	3.37
Hydrofluoric acid, HF	2.9×10^{-11}	10.55	3.5×10^{-4}	3.45
Formic acid, HCOOH	5.6×10^{-11}	10.25	1.8×10^{-4}	3.75
Benzoic acid, C_6H_5COOH	1.5×10^{-10}	9.81	6.5×10^{-5}	4.19
Acetic acid, CH_3COOH	5.6×10^{-10}	9.25	5.6×10^{-5}	4.75
Carbonic acid, H_2CO_3	2.3×10^{-8}	7.63	4.3×10^{-7}	6.37
Hypochlorous acid, HClO	3.3×10^{-7}	6.47	3.0×10^{-8}	7.53
Hypobromous acid, HBrO	5.0×10^{-6}	5.31	2.0×10^{-9}	8.69
Boric acid, $B(OH)_3$[†]	1.4×10^{-5}	4.86	7.2×10^{-10}	9.14
Hydrocyanic acid, HCN	2.0×10^{-5}	4.69	4.9×10^{-10}	9.31
Phenol, C_6H_5OH	7.7×10^{-5}	4.11	1.3×10^{-10}	9.89
Hypoiodous acid, HIO	4.3×10^{-4}	3.36	2.3×10^{-11}	10.64
Weakest weak acids				
Weakest weak bases				
Urea, $CO(NH_2)_2$	1.3×10^{-14}	13.90	7.7×10^{-1}	0.10
Aniline, $C_6H_5NH_2$	4.3×10^{-10}	9.37	2.3×10^{-5}	4.63
Pyridine, C_5H_5N	1.8×10^{-9}	8.75	5.6×10^{-6}	5.35
Hydroxylamine, NH_2OH	1.1×10^{-8}	7.97	9.1×10^{-7}	6.03
Nicotine, $C_{10}H_{11}N_2$	1.0×10^{-6}	5.98	1.0×10^{-8}	8.02
Morphine, $C_{17}H_{19}O_3N$	1.6×10^{-6}	5.79	6.3×10^{-9}	8.21
Hydrazine, NH_2NH_2	1.7×10^{-6}	5.77	5.9×10^{-9}	8.23
Ammonia, NH_3	1.8×10^{-5}	4.75	5.6×10^{-10}	9.25
Trimethylamine, $(CH_3)_3N$	6.5×10^{-5}	4.19	1.5×10^{-10}	9.81
Methylamine, CH_3NH_2	3.6×10^{-4}	3.44	2.8×10^{-11}	10.56
Dimethylamine, $(CH_3)_2NH$	5.4×10^{-4}	3.27	1.9×10^{-11}	10.73
Ethylamine, $C_2H_5NH_2$	6.5×10^{-4}	3.19	1.5×10^{-11}	10.81
Triethylamine, $(C_2H_5)_3N$	1.0×10^{-3}	2.99	1.0×10^{-11}	11.01
Strongest weak bases				

*Values for polyprotic acids—those capable of donating more than one proton—refer to the first ionization. Base equilibria and K_b are introduced in Section 5.3.
[†]The proton transfer equilibrium is $B(OH)_3(aq) + 2 H_2O(l) \rightleftharpoons H_3O^+(aq) + B(OH)_4^-(aq)$.

Exercise E5.5

Write the expression for the acidity constant of $H_2PO_4^-$.

[*Answer*: $K_a = a(H_3O^+)a(HPO_4^{2-})/a(H_2PO_4^-)$]

The value of the acidity constant indicates the extent to which proton transfer has occurred: the smaller the value of K_a, the lower the concentration of deprotonated molecules, and hence the weaker the proton donating power of the acid. In the context of the Brønsted–Lowry theory, the term 'ionization' is used to denote proton transfer, so the small value of K_a for HF indicates that hydrogen fluoride is only slightly ionized—deprotonated—in solution. The value of K_a for hydrocyanic acid, HCN, is only 4.9×10^{-10}, which indicates that at a similar concentration in water it is ionized to a far smaller extent even than hydrofluoric acid.

Because acidity constants are typically very small, it proves convenient to report them as their logarithms, by analogy with the definition of pH. Therefore, we define

$$pK_a = -\log K_a \tag{8}$$

For hydrofluoric acid

$$pK_a = -\log(3.5 \times 10^{-4}) = 3.46$$

Note that the lower the value of the acidity constant, the higher the value of pK_a. For example, hydrocyanic acid, HCN, has $K_a = 4.9 \times 10^{-10}$, which is less than the value for hydrofluoric acid, and its pK_a is 9.31, which is larger than for hydrofluoric acid. In general, the higher the value of pK_a, the weaker the proton donating power of the acid. The HCN molecule is a far weaker proton donor (to water) than an HF molecule; a CH_4 molecule has a completely negligible proton donating power to water.

The autoprotolysis constant of water

The equilibrium constant for the water autoprotolysis equilibrium is

$$K = \frac{a(H_3O^+)a(OH^-)}{a(H_2O)^2}$$

However, in the dilute solutions that we shall always consider, the water is almost pure, and to a very good approximation we can replace $a(H_2O)$ by 1. The resulting expression is called the **autoprotolysis constant** of water, and is denoted K_w:

$$K_w = a(H_3O^+)a(OH^-) \tag{9}$$

We have already seen that the molar concentrations of H_3O^+ and OH^- ions in pure water at $25\,^\circ C$ are $1.0 \times 10^{-7}\,mol\,L^{-1}$, so it follows that at $25\,^\circ C$

$$K_w = (1.0 \times 10^{-7}) \times (1.0 \times 10^{-7}) = 1.0 \times 10^{-14}$$

The corresponding logarithmic expression is

$$pK_w = -\log K_w = 14.00 \text{ at } 25\,^\circ C$$

The importance of the autoprotolysis equilibrium is that, although the individual hydronium ion and hydroxide ion concentrations may change as acid or base is added to a solution, the product of the concentrations must remain equal to K_w (for otherwise the proton transfer between water molecules would not be at equilibrium).

Exercise E5.6

The molar concentration of OH^- ions in a certain solution is $1.0 \times 10^{-4}\,mol\,L^{-1}$. What is the pH of the solution?

[*Answer*: 10.00]

The relation between the pH and the concentration of OH^- ions in an aqueous solution is most easily expressed by introducing the **pOH** of the solution, which is defined as

$$pOH = -\log a(OH^-) \tag{10}$$

Then, by taking negative logarithms of both sides of the expression for the autoprotolysis constant, eqn (9), we obtain

$$-\log K_w = -\log a(H_3O^+) - \log a(OH^-)$$

because $\log xy = \log x + \log y$. It follows that

$$pK_w = pH + pOH \tag{11}$$

This expression is very useful for calculating the pH of an alkaline solution provided it is very dilute and we can write $a(OH^-) \approx [OH^-]/(mol\,L^{-1})$. For example, in a solution in which the molar concentration of OH^- ions is $1.0 \times 10^{-4}\,mol\,L^{-1}$, so $pOH = 4.00$, the pH is $pH = 14.00 - 4.00 = 10.00$, as stated in the exercise above.

Weak and strong acids

A species is classified as a **strong acid** if it is almost completely deprotonated in solution (usually in water). A species is classified as a **weak acid** if it is only partially deprotonated in solution. Hydrogen chloride is a strong acid in water. When hydrogen chloride dissolves in water, every HCl molecule that dissolves donates a proton to a water molecule and the solution consists almost entirely of H_3O^+ and Cl^- ions. Therefore, if the concentration of HCl is reported as 1.0×10^{-3} mol L^{-1}, then we know that in fact the solution consists of virtually no HCl molecules but 1.0×10^{-3} mol L^{-1} of H_3O^+ and the same concentration of Cl^- ions. The pH of the solution will be close to 3.00 because the actual concentration of H_3O^+ ions will be the same as the nominal concentration of HCl (that is, the stated concentration of the solution).[†] There are very few strong acids in water: the ones to remember are HCl, HBr, HI, HNO_3, H_2SO_4 (with respect to the donation of one proton), and $HClO_4$.

Most acids are weak (see Table 5.1), and in solution exist largely as the non-ionized acid molecules with only a small proportion ionized. The extent of ionization depends on the acidity constant and the concentration of the solution, and it may be estimated by using the equilibrium-table technique presented earlier in the chapter. As an example, consider acetic acid,

$$CH_3COOH(aq) + H_2O(l) \rightleftharpoons H_3O^+(aq) + CH_3CO_2^-(aq)$$

for which $pK_a = 4.74$ (corresponding to $K_a = 1.8 \times 10^{-5}$, a value that indicates a low degree of ionization into the products H_3O^+ and $CH_3CO_2^-$). To estimate the pH and the proportion of CH_3COOH molecules that are ionized in a solution of molar concentration A mol L^{-1} we draw up the following equilibrium table:

Species:	CH_3COOH	H_3O^+	$CH_3CO_2^-$
Initial concentration/(mol L^{-1})	A	0	0
Change to reach equilibrium/(mol L^{-1})	$-x$	$+x$	$+x$
Equilibrium concentration/(mol L^{-1})	$A - x$	x	x

The value of x is found by inserting the equilibrium concentrations into the expression for the acidity constant:

$$K_a = \frac{a(H_3O^+)a(CH_3CO_2^-)}{a(CH_3COOH)} \approx \frac{x \times x}{A - x}$$

This expression rearranges to the quadratic equation

$$x^2 + K_a x - AK_a = 0$$

[†]Measurement of the pH of such a solution gives pH $= 3.02$, so the approximation we have used is reliable in this instance.

with the solution

$$x = \frac{-K_a \pm \sqrt{K_a^2 + 4AK_a}}{2}$$

Because x is equal to the molar concentration of hydronium ions (see the last line of the equilibrium table), it must be positive, and so we must take the solution with the positive square root (for only that solution can give a positive value of x). Suppose that $A = 0.010$ (corresponding to an initial CH_3COOH concentration of $0.010 \, mol \, L^{-1}$) then

$$x = \frac{-(1.8 \times 10^{-5}) + \sqrt{(1.8 \times 10^{-5})^2 + 4(0.010)(1.8 \times 10^{-5})}}{2}$$
$$= 4.2 \times 10^{-4}$$

from which it follows that

$$pH \approx -\log{(4.2 \times 10^{-4})} = 3.38$$

pH calculations of this kind are rarely accurate to more than one decimal place (and even that may be over optimistic) because the effects of ion–ion interactions have been ignored, so this answer would be reported as $pH = 3.4$. Notice that although the acid concentration is 10 times greater than the HCl concentration treated above, the hydronium ion concentration is approaching a factor of 10 lower because so few of the acid molecules have given up a proton.

Exercise E5.7

Estimate the pH of a solution of 0.010 M lactic acid using the data in Table 5.1. Before carrying out the numerical calculation, decide whether you expect the pH to be higher or lower than that calculated for the same concentration of acetic acid.

[*Answer*: 2.7]

A second application of acidity constants is to the estimation of the extent to which an acid is ionized in solution. To report this information, we calculate the **fraction ionized**, the fraction of acetic acid molecules that have donated a proton:

$$\text{fraction ionized} = \frac{\text{molar concentration of conjugate base}}{\text{initial molar concentration of acid}} \quad (12)$$

The last line of the equilibrium table gives us x as the molar concentration of conjugate base at equilibrium, so we can write

$$\text{fraction ionized} \approx \frac{x}{A}$$

$$\approx \frac{4.2 \times 10^{-4}}{0.010} = 4.2 \times 10^{-2}$$

That is, only 4.2 per cent of the acetic acid molecules have donated a proton. For lactic acid, a somewhat stronger acid, the fraction is larger, at 9.1 per cent.

It is very rare in chemistry that you will ever need to calculate a pH: in practice, the only reliable procedure is to measure it (by using techniques to be described in Section 6.3). However, it is important to understand in general terms how the pH varies with the nominal concentration of the acid and to know when to expect pH larger or smaller than 7 (neutrality). To gauge these general trends without needing numerically exact values, we can make use of the approximation procedure outlined earlier. Approximations are often reliable for weak acids with low acidity constants because the extent of deprotonation is small. Specifically, the calculation of the pH of a weak acid is greatly simplified if we assume that the fraction ionized is so small that the denominator $A - x$ in the expression for K_a can be approximated by A itself. Then

$$K_a \approx \frac{x^2}{A}$$

which solves to

$$x \approx \sqrt{AK_a} \tag{13}$$

Hence, by taking logarithms and changing the sign throughout,

$$-\log x \approx -\tfrac{1}{2}\log K_a - \tfrac{1}{2}\log A$$

Because $-\log x$ is the pH of the solution, it follows that

$$\text{pH} \approx \tfrac{1}{2}\text{p}K_a - \tfrac{1}{2}\log A \tag{14}$$

For example, the pH of $0.010\,\text{M}\,CH_3COOH(aq)$ is predicted to be

$$\text{pH} \approx \tfrac{1}{2} \times 4.75 - \tfrac{1}{2} \times \log 0.010 = 2.38 + 1.00 = 3.4$$

in good agreement with the value found without making the approximation. This approximation procedure can be used when the fraction of ionization is smaller than about 0.05 (5 per cent).

To verify that the fraction ionized is small, we can substitute the approximate expression for x in eqn (13) into eqn (12), and obtain

$$\text{fraction ionized} = \frac{x}{A} \approx \frac{\sqrt{AK_a}}{A} = \sqrt{\frac{K_a}{A}} \tag{15}$$

It is now easy to verify that the fraction of CH_3COOH molecules ionized in $0.010\,M\,CH_3COOH(aq)$ is

$$\text{fraction ionized} \approx \sqrt{\frac{1.8 \times 10^{-5}}{0.010}} = 0.042$$

which is within the limit of 5 per cent and in good agreement with the 'precise' value.

The procedure for calculating the pH of the solution of an acid can now be summarized as follows:

1. If the acid is strong, assume that it is fully ionized in solution, and calculate the pH from the stated molar concentration of the acid by using $[H_3O^+] = [HA]_{\text{added initially}}$.
2. If the acid is weak, use eqn (15) to decide whether, for the concentration specified and the numerical value of K_a of the acid, the fraction ionized is less than 0.05.
3. If the fraction ionized is greater than 0.05, use the 'exact' equilibrium table procedure to calculate x and thence the pH.
4. If the fraction ionized is not greater than 0.05, use the approximate formula, eqn (14), for the pH.

Even if step 4 is permissible, it is good practice to set up the equilibrium table and make the approximations rather than to rely on memory to recall eqn (14). This procedure is valid only for very dilute solutions, for which activity coefficients are close to 1; because ions are present in a lower concentration in solutions of weak acids than they are in solutions of strong acids of the same nominal concentration, the procedure is generally more reliable for weak acids than for strong.

Example *Calculating the pH of a solution of a weak acid*
What is the pH of $0.25\,M\,HCN(aq)$ at $25\,^\circ C$?

Strategy
We follow the 'weak-acid procedure' outlined above. The data are in Table 5.1.

Solution
The pK_a of hydrocyanic acid, HCN, is given in Table 5.1 as 9.31, which corresponds to a very small value of K_a (4.9×10^{-10}, in fact). We can expect the fraction of acid ionized to be very small, and confirm that this is so by substituting the data into eqn (15), to obtain

$$\text{fraction ionized} \approx \sqrt{\frac{4.9 \times 10^{-10}}{0.25}} = 4.4 \times 10^{-5}$$

Because this fraction is very much smaller than 1, it is valid to use eqn (14) for the pH:

$$\text{pH} \approx \tfrac{1}{2} \times 9.31 - \tfrac{1}{2}\log 0.25 = 4.96$$

or about 5.0.

Exercise E5.8

Calculate the pH of 0.20 M HClO(aq).

[*Answer*: 4.1]

Polyprotic acids

A **polyprotic acid** is a compound that can donate more than one proton. Two examples are sulfuric acid, H_2SO_4, which can donate up to two protons, and phosphoric acid, H_3PO_4, which can donate up to three. A polyprotic acid is best considered to be a molecular species that can give rise to a series of Brønsted acids as it donates its succession of protons. Thus, sulfuric acid is the parent of two Brønsted acids, H_2SO_4 itself and HSO_4^-, and phosphoric acid is the parent of three Brønsted acids, namely H_3PO_4, $H_2PO_4^-$, and HPO_4^{2-}.

For a species with two acidic (donatable) protons (such as H_2SO_4), the successive equilibria we need to consider are

$$H_2A(aq) + H_2O(l) \rightleftharpoons H_3O^+(aq) + HA^-(aq) \quad K_{a1} = \frac{a(H_3O^+)a(HA^-)}{a(H_2A)}$$

$$HA^-(aq) + H_2O(l) \rightleftharpoons H_3O^+(aq) + A^{2-}(aq) \quad K_{a2} = \frac{a(H_3O^+)a(A^{2-})}{a(HA^-)}$$

In the first of these expressions, HA^- is the conjugate base of H_2A; in the second HA^- acts as the acid and A^{2-} is its conjugate base. In all cases, K_{a2} is smaller than K_{a1}, typically by three orders of magnitude for small molecular species (Table 5.2), because the second proton is more difficult to remove, partly on account of the negative charge on HA^-. Enzymes are polyprotic acids, for they possess many protons that can be donated to a substrate molecule or to the surrounding aqueous medium of the cell. For them, successive acidity constants vary much less because the molecules are so large that the loss of a proton from one part of the molecule has little effect on the ease with which another may be lost from a location some distance away.

Table 5.2 Successive acidity constants of polyprotic acids

Acid	K_{a1}	pK_{a1}	K_{a2}	pK_{a2}	K_{a3}	pK_{a3}
Carbonic acid, H_2CO_3	4.3×10^{-7}	6.37	5.6×10^{-11}	10.25		
Hydrosulfuric acid, H_2S	1.3×10^{-7}	6.88	7.1×10^{-15}	14.15		
Oxalic acid, $(COOH)_2$	5.9×10^{-2}	1.23	6.5×10^{-5}	4.19		
Phosphoric acid, H_3PO_4	7.6×10^{-3}	2.12	6.2×10^{-8}	7.21	2.1×10^{-13}	12.67
Phosphorous acid, H_2PO_3	1.0×10^{-2}	2.00	2.6×10^{-7}	6.59		
Sulfuric acid, H_2SO_4	Strong		1.2×10^{-2}	1.92		
Sulfurous acid, H_2SO_3	1.5×10^{-2}	1.81	1.2×10^{-7}	6.91		
Tartaric acid, $C_2H_4O_2(COOH)_2$	6.0×10^{-4}	3.22	1.5×10^{-5}	4.82		

Example *Calculating the concentration of carbonate ion in carbonic acid*

Estimate the molar concentration of CO_3^{2-} ions in carbonic acid.

Strategy

When calculating the molar concentration of a species formed by the loss of successive protons (A^{2-}), start with the equilibrium that produces the ion, and write its activity in terms of the equilibrium constant for that reaction (K_{a2} for a diprotic acid). That expression will contain the term HA^-, so we express that activity in terms of the conjugate acid (H_2A) by using K_{a1}. If the ion of interest is A^{3-} derived from a triprotic acid, then K_{a3} will be the first acidity constant to appear in the chain. The first proton loss equilibrium (from H_2A in the case of a diprotic acid) dominates all the rest (for small molecules with marked differences between their acidity constants), so it may be possible to make an approximation at this stage.

Solution

The CO_3^{2-} ion, the conjugate base of the acid HCO_3^-, is produced in the equilibrium

$$HCO_3^-(aq) + H_2O(l) \rightleftharpoons H_3O^+(aq) + CO_3^{2-}(aq)$$

$$K_{a2} = \frac{a(H_3O^+)a(CO_3^{2-})}{a(HCO_3^-)}$$

Hence,

$$a(CO_3^{2-}) = \frac{a(HCO_3^-)K_{a2}}{a(H_3O^+)}$$

The HCO_3^- ions are produced in the equilibrium

$$H_2CO_3(aq) + H_2O(l) \rightleftharpoons H_3O^+(aq) + HCO_3^-(aq)$$

This equilibrium dominates the second deprotonation, and so it also dominates the molar concentrations of HCO_3^- and H_3O^+. (They are not exactly the same, because a little HCO_3^- has been lost in the second ionization and the amount of H_3O^+ has been increased by it; however, those secondary changes can safely be ignored in an approximate calculation.) Because the molar concentrations are approximately the same, we can suppose that the activities are also approximately the same, and set $a(HCO_3^-) \approx a(H_3O^+)$. When this equality is substituted into the preceding expression, we obtain

$$a(CO_3^{2-}) \approx K_{a2}$$

Because we know from Table 5.2 that $pK_{a2} = 10.25$, it follows that $a(CO_3^{2-}) = 5.6 \times 10^{-11}$, and therefore that

$$[CO_3^{2-}] \approx 5.6 \times 10^{-11} \, mol \, L^{-1}$$

Note that the concentration of this ion is independent of the concentration of the acid.

Exercise E5.9

Calculate the molar concentration of S^{2-} ions in $H_2S(aq)$.

[*Answer*: $7.1 \times 10^{-15} \, mol \, L^{-1}$]

5.3 Weak and strong bases

According to the Brønsted–Lowry theory, when a base (a proton acceptor) such as ammonia is dissolved in water it takes part in the proton transfer equilibrium

$$H_2O(l) + NH_3(aq) \rightleftharpoons NH_4^+(aq) + OH^-(aq)$$
$$acid_1 + base_2 \rightleftharpoons acid_2 + base_1$$

Proton transfer is so fast that a solution of a base is always at equilibrium with its conjugate acid, and the concentrations of the species are described by the equilibrium constant

$$K = \frac{a(NH_4^+)a(OH^-)}{a(NH_3)a(H_2O)}$$

As usual, if we confine our attention to dilute solutions, the water may be treated as pure liquid with $a = 1$, and the resulting equilibrium constant is called the **basicity constant** (or 'base ionization constant'), K_b:

$$K_b = \frac{a(NH_4^+)a(OH^-)}{a(NH_3)}$$

In general, a basicity constant is defined as follows:

$$H_2O(l) + B(aq) \rightleftharpoons HB^+(aq) + OH^-(aq) \qquad K_b = \frac{a(HB^+)a(OH^-)}{a(B)} \quad (16)$$

The basicity constants of a number of species (including the conjugate bases of some acids) are given in Table 5.1. For ammonia in water, for example, $K_b = 1.8 \times 10^{-5}$ at 25 °C. As in the case of acidity constants, it is convenient to report the values of basicity constants in terms of their negative logarithms:

$$pK_b = -\log K_b \qquad (17)$$

Thus, for ammonia at 25 °C,

$$pK_b = -\log(1.8 \times 10^{-5}) = 4.75$$

A **strong base** is a species that is fully protonated in solution. One example is the oxide ion, which cannot survive in water but is immediately converted into its conjugate acid, OH^-. A **weak base** is a species that is not fully protonated in water. Ammonia, NH_3, and its organic derivatives, the amines, are all weak bases in water, and only a small proportion of their molecules exist as the conjugate acid (NH_4^+ or RNH_3^+). The numerical value of the basicity constant indicates the extent to which a base is protonated in aqueous solution: the greater the value of K_b, the stronger the proton accepting power of the base. Conversely, the greater the value of pK_b, the weaker the proton accepting power of the base. The weak base morphine, an amine, for example, has $K_b = 1.6 \times 10^{-6}$ and $pK_b = 5.79$, and is a weaker proton acceptor than ammonia. As a result, when morphine is present in water at the same concentration as ammonia, a smaller proportion of its molecules are protonated.

Exercise E5.10

Which will be more fully protonated in aqueous solutions of the same concentration, methylamine or ethylamine?

[*Answer*: ethylamine]

The extent of protonation

The extent to which a base is protonated is reported in terms of the **fraction protonated**:

$$\text{fraction protonated} = \frac{\text{molar concentration of conjugate acid}}{\text{initial molar concentration of base}} \qquad (18)$$

For strong bases, the fraction protonated is close to 1; for weak bases it is typically much less than 1. As an example of the use of basicity constants, we can estimate the fraction of NH_3 protonated in a solution of concentration B mol L^{-1}. This type of calculation is another version of the equilibrium calculations presented earlier in the chapter, and we draw up the following equilibrium table for the reaction

$$H_2O(l) + NH_3(aq) \rightleftharpoons NH_4^+(aq) + OH^-(aq) \qquad K_b = \frac{a(NH_4^+)a(OH^-)}{a(NH_3)}$$

Species:	NH_3	NH_4^+	OH^-
Initial concentration/(mol L^{-1})	B	0	0
Change to reach equilibrium/(mol L^{-1})	$-x$	$+x$	$+x$
Equilibrium concentration/(mol L^{-1})	$B - x$	x	x

As usual, to make an estimate, we make the approximation that activities can be replaced by molar concentrations, in which case the expression for the basicity constant becomes:

$$K_b \approx \frac{x \times x}{B - x}$$

This expression rearranges into the quadratic equation

$$x^2 + K_b x - BK_b = 0$$

with the solutions

$$x = \frac{-K_b \pm \sqrt{K_b^2 + 4BK_b}}{2}$$

The positive root must be selected because x is the molar concentration of OH^- ions (see the equilibrium table), which is positive. For an 0.10 M $NH_3(aq)$ solution,

$$x = \frac{-(1.8 \times 10^{-5}) + \sqrt{(1.8 \times 10^{-5})^2 + 4(0.10)(1.8 \times 10^{-5})}}{2} = 1.3 \times 10^{-3}$$

It follows that the fraction of NH_3 protonated is

$$\text{fraction protonated} = \frac{x}{B} = \frac{1.3 \times 10^{-3}}{0.10} = 1.3 \times 10^{-2}$$

That is, 1.3 per cent of the NH_3 molecules (just over 1 in 100) is protonated.

Just as in the calculation of the pH of a weak acid, we can estimate the fraction protonated more quickly. To do so, we anticipate that, because $K_b \ll 1$, the molar concentration of protonated species is so low that the molar concentration of unprotonated base is almost the same as the molar concentration of base added to prepare the solution. If that is so, we can replace $B - x$ in the expression for the basicity constant by B. Then

$$x \approx \sqrt{K_b B} \qquad (19)$$

This approximation is valid so long as x does not exceed about 5 per cent of B. For the example we have just treated exactly, we would estimate

$$x \approx \sqrt{(1.8 \times 10^{-5}) \times 0.10} = 1.3 \times 10^{-3}$$

in excellent agreement with the exact value (and with far less work). It follows that the fraction protonated is

$$\text{fraction protonated} \approx \frac{x}{B} \approx \sqrt{\frac{K_b}{B}} \qquad (20)$$

Exercise E5.11

Estimate the fraction of morphine that is protonated in a solution of molar concentration $0.010 \, \text{mol} \, \text{L}^{-1}$.

[*Answer*: 0.013]

The pH of a basic solution

The calculation of the pH of a solution of a base involves one more step than that for the pH of a solution of an acid. The first step is to calculate the concentration of OH^- ions in the solution and to express it as the pOH of the solution. The additional step is to convert that pOH into a pH by using the water autoprotolysis equilibrium equation, eqn (11), in the form

$$\text{pH} = \text{p}K_w - \text{pOH} \qquad (21)$$

with $\text{p}K_w = 14.00$ at $25 \,°\text{C}$.

The procedure is straightforward in the case of a strong base, such as a hydroxide, for almost every formula unit (such as $NaOH$ or $Ca(OH)_2$) present in the solution is present as OH^- ions. For example, in $0.0010 \, \text{M} \, Ca(OH)_2(\text{aq})$ the concentration of OH^- ions is $0.0020 \, \text{mol} \, \text{L}^{-1}$, and so the pOH of the solution is

$$\text{pOH} \approx -\log 0.0020 = 2.70$$

The result is only approximate, because we are using molar concentration instead of activity. The pH of the solution can then be deduced from the autoprotolysis equilibrium:

$$\text{pH} = 14.00 - 2.70 = 11.30$$

Care should be taken to note that if a compound such as $Ca(OH)_2$ is present in solution, each formula unit that dissolves gives rise to two OH^- ions.

For a solution of a weak base in water, we show in the following *Justification* that the pH is given by

$$pH \approx pK_w - \tfrac{1}{2}pK_b + \tfrac{1}{2}\log B \qquad (22)$$

when the fraction of protonation is low. This expression shows that the pH of the solution increases as the concentration of base increases (because $\log B$ increases) and as the pK_b of the base decreases. The latter trend is consistent with the strength of a base increasing as pK_b becomes smaller (as from trimethylamine to methylamine).

Justification

When the fraction of protonation of a weak base is low, the molar concentration of OH^- ions is equal to $x\,mol\,L^{-1}$, with x given by eqn (19). It follows that

$$
\begin{aligned}
pOH &\approx -\log x \approx -\log \sqrt{K_b B} \qquad [\text{because } x \approx \sqrt{K_b B}] \\
&\approx -\tfrac{1}{2}\log K_b B \qquad [\text{because } \log \sqrt{y} = \tfrac{1}{2}\log y] \\
&\approx -\tfrac{1}{2}\log K_b - \tfrac{1}{2}\log B \qquad [\text{because } \log yz = \log y + \log z] \\
&\approx \tfrac{1}{2}pK_b - \tfrac{1}{2}\log B \qquad [\text{because } -\log K_b = pK_b]
\end{aligned}
$$

It then follows that

$$pH \approx pK_w - (\tfrac{1}{2}pK_b - \tfrac{1}{2}\log B) \qquad [\text{because } pH = pK_w - pOH]$$

which can be rearranged into eqn (22).

Exercise E5.12

Estimate the pH of $0.10\,M\,NH_3(aq)$ at $25\,^\circ C$.

[*Answer*: 11.1]

5.4 Conjugate acids and bases

There is an important (and exact) relation between the acidity and basicity constants of conjugate acid–base pairs:

$$pK_a + pK_b = pK_w \qquad (23)$$

The proof of the relation is very straightforward: we need to show that

$$K_a K_b = K_w \qquad (24)$$

because taking the logarithm of both sides of this equation and changing the signs throughout leads to the relation above. To confirm the latter expression, we multiply together the explicit expressions for K_a for the acid HB^+ and K_b and the base B:

$$K_a K_b = \frac{a(H_3O^+)a(B)}{a(HB^+)} \times \frac{a(HB^+)a(OH^-)}{a(B)}$$
$$= a(H_3O^+)a(OH^-) = K_w$$

As you can see, we do not have to make any approximation about the interpretation of the activities in terms of molar concentrations. The relation implies that from the basicity constant of NH_3:

$$H_2O(l) + NH_3(aq) \rightleftharpoons NH_4^+(aq) + OH^-(aq) \qquad pK_b = 4.75$$

we can conclude that the acidity constant of NH_4^+, the conjugate acid of NH_3,

$$NH_4^+(aq) + H_2O(l) \rightleftharpoons H_3O^+(aq) + NH_3(aq)$$

is given by

$$pK_a = 14.00 - 4.75 = 9.25$$

The great advantage of eqn (23) is that *all proton transfer equilibrium constants can be expressed as acidity constants, even those of bases.* For example, instead of treating ammonia as a weak base, we can think of it as the conjugate base of an acid (NH_4^+, for which $pK_a = 9.25$) which is slightly stronger than hydrocyanic acid ($pK_a = 9.31$).

Exercise E5.13

The acidity constant of HPO_4^{2-} is reported as $pK_a = 12.67$. Write the equilibrium of its conjugate base and give the value of pK_b for that base.

[*Answer*: $H_2O(l) + PO_4^{3-}(aq) \rightleftharpoons HPO_4^{2-}(aq) + OH^-(aq)$, $pK_b = 1.33$]

The relation in eqn (23) implies that the stronger an acid, the weaker its conjugate base, and vice versa. This see-saw like relation follows from the fact that the sum of pK_a and pK_b is a constant for a given conjugate acid–base pair, so a large pK_a implies a small pK_b and vice versa. That is,

- The stronger an acid, the weaker its conjugate base.
- The stronger a base, the weaker its conjugate acid.

It follows that *the conjugate base of a strong acid is a very weak base*, and *the conjugate acid of a strong base is a very weak acid*. For example, the fact that HCl is a strong acid in water implies that the Cl⁻ ion is a very weak base and has almost no tendency to acquire a proton: that is why almost all the HCl molecules are deprotonated. On the other hand, the conjugate base of a weak acid has a tendency to accept a proton, and the weaker the acid, the greater this proton-accepting ability. Similar remarks may be made about bases: the greater the tendency of a base to accept a proton, the weaker its conjugate acid.

5.5 Salts in water

The remarks in the last paragraph explain why a solution of ammonium chloride is acidic even though the salt provides both an acid (NH_4^+) and a base (Cl^-) when it dissolves in water. The NH_4^+ ion is the conjugate acid of a weak base, and although it is still a weak acid, it does have an appreciable proton-donating power (p$K_a = 9.25$). The Cl^- ion is the conjugate base of a strong acid, and so it is a very weak proton acceptor. The solution therefore consists of a weak acid (NH_4^+) and a very weak base (Cl^-), and the net effect is that the solution is acidic. Similarly, a solution of sodium acetate consists of an essentially neutral ion (the Na^+ ion) and a base ($CH_3CO_2^-$). The net effect is that the solution is basic, and its pH is greater than 7.

Exercise E5.14

Is an aqueous solution of potassium tartrate likely to be acidic or basic?

[*Answer*: basic]

When an ammonium salt is dissolved in water, the NH_4^+ ions supplied by the salts participate in the equilibrium

$$NH_4^+(aq) + H_2O(l) \rightleftharpoons H_3O^+(aq) + NH_3(aq)$$

and the equilibrium composition of the solution can be estimated from the acidity constant of the NH_4^+ ion exactly as we have illustrated for other weak acids:

$$K_a = \frac{a(H_3O^+)a(NH_3)}{a(NH_4^+)}$$

There is no distinction between acids and conjugate acids: in the Brønsted–Lowry theory, any proton donor is an acid.

Example *Estimating the pH of a solution of a salt*
Estimate the pH of 0.010 M NH₄Cl(aq) at 25 °C.

Strategy
Judge from the relative strengths of the acids and bases present (that is, the two kinds of ion) whether the solution is likely to be acidic or basic. Next, if the acid present is very weak, use the approximate formula in eqn (14) to calculate its pH.

Solution
Because NH_4^+ is a weak acid and Cl^- is effectively neutral, we can anticipate that the solution will be acidic, with pH < 7. Next, we note from Table 5.1 that the pK_a of NH_4^+ is 9.25. This value (which signifies that NH_4^+ is a very weak acid) suggests that only a small fraction of NH_4^+ will have undergone proton donation. We can therefore use eqn (14):

$$pH \approx \tfrac{1}{2} \times 9.25 - \tfrac{1}{2}\log 0.010 = 5.63$$

That is, the pH of the solution will be about 5.6, on the acidic side of neutrality.

Exercise E5.15

Estimate the pH of 0.0025 M [NH(CH₃)₃]Cl(aq) at 25 °C.

[*Answer*: 6.2]

The addition of ammonium ions to water (by dissolving an ammonium salt) is the addition of an acid, and as a result the pH of the solution is lowered below 7. The same is true of all salts that contain the conjugate acid of a base, but ammonium salts are the most common. In the earlier literature (and still sometimes today), the influence of a dissolved salt on the pH of a solution is ascribed to the **hydrolysis** of the salt, in which the ions are supposed to react with water and produce an acidic or basic solution; however, it is far simpler, and much more in the spirit of Brønsted–Lowry theory, to regard the ions of the salt as acids or bases in their own right.

The pH of a solution that contains ions that are weak bases (such as solutions of sodium acetate) can be estimated in much the same way as we have demonstrated for solutions of weak acids. The difference is that we use a table that allows for the protonation of a base, as in the reaction

$$H_2O(l) + CH_3CO_2^-(aq) \rightleftharpoons CH_3COOH(aq) + OH^-(aq)$$

For example, to estimate the pH of a solution of acetate ions of molar concentration $B\,\mathrm{mol\,L^{-1}}$, we use eqn (22):

$$pH \approx pK_w - \tfrac{1}{2}pK_b + \tfrac{1}{2}\log B$$

When rewritten in terms of the pK_a of the conjugate acid of the base (that is, in terms of the pK_a of acetic acid if the salt is an acetate) by using eqn (23), this expression becomes

$$pH \approx pK_w - \tfrac{1}{2}(pK_w - pK_a) + \tfrac{1}{2}\log B$$

which simplifies to

$$pH \approx \tfrac{1}{2}pK_w + \tfrac{1}{2}pK_a + \tfrac{1}{2}\log B \qquad (25)$$

We see that the pH of the solution increases as the concentration of the salt increases, which is as expected because the salt is providing a base (the acetate ion, for instance). Similarly, the pH increases as the the pK_a of the parent acid increases: this is also as expected, because a higher pK_a indicates a stronger conjugate base.

Exercise E5.16

Estimate the pH of $0.010\,\mathrm{M\ NaCH_3CO_2(aq)}$.

[*Answer*: 8.4]

5.6 Acid–base titrations

Acidity constants play an important role in acid–base titrations, for they can be used to decide the value of the pH that signals the **stoichiometric point**, the stage at which a stoichiometrically equivalent amount of acid has been added to a given amount of base.[†] The plot of the pH of the **analyte solution** (the solution in the flask that is being analysed) against the volume of **titrant solution** (the solution in the burette) added is called the **pH curve**. It shows a number of features that are still of interest even nowadays when many titrations are carried out in automatic titrators with the pH monitored electronically: automatic titration equipment is built to make use of the concepts we describe here.

[†]The stoichiometric point is widely called the 'equivalence point' of a titration.

First, consider the titration of a strong acid with a strong base, such as the titration of hydrochloric acid with sodium hydroxide. The reaction is

$$HCl(aq) + NaOH(aq) \longrightarrow NaCl(aq) + H_2O(l)$$

Initially, the analyte (hydrochloric acid) has a low pH. The ions present at the stoichiometric point (the Na^+ ions from the strong base and the Cl^- ions from the strong acid) barely affect the pH, so the pH is that of almost pure water, namely pH = 7. After the stoichiometric point, when base is added to a neutral solution, the pH rises sharply to a high value. The pH curve for such a titration is shown in Fig. 5.1.

At the stoichiometric point of a titration of a weak acid (such as CH_3COOH) and strong base (NaOH), the solution contains $CH_3CO_2^-$ ions and Na^+ ions together with any ions stemming from autoprotolysis. The presence of the Brønsted base $CH_3CO_2^-$ in the solution means that we can expect pH > 7. In a titration of a weak base (such as NH_3) and a strong acid (HCl), the solution contains NH_4^+ ions and Cl^- ions at the stoichiometric point. Because Cl^- is only a very weak Brønsted base and NH_4^+ is a weak Brønsted acid the solution is acidic and its pH will be less than 7.

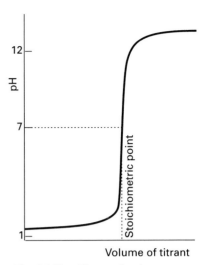

Fig. 5.1 The pH curve for the titration of a strong acid (the analyte) with a strong base (the titrant). There is an abrupt change in pH near the stoichiometric point; the stoichiometric point itself is at pH = 7. The final pH of the medium approaches that of the titrant.

The pH curve for a weak-acid–strong-base titration

Now we consider the shape of the pH curve in terms of the acidity constants of the species involved. The approximations we make are based on the fact that the acid is weak, and therefore that HA is more abundant than any A^- ions in the solution. Furthermore, when HA is present, it provides so many H_3O^+ ions (even though it is a weak acid) that they greatly outnumber any H_3O^+ ions that come from the very feeble autoprotolysis of water. Finally, when excess base is present after the stoichiometric point has been passed, the OH^- ions it provides dominate any that come from the water autoprotolysis.

To be definite, we shall suppose that we are titrating 25.00 mL of 0.10 M HClO(aq) with 0.20 M NaOH(aq) at 25 °C. The pH at the start of a titration of a weak acid with a strong base can be calculated from eqn (14). Because $pK_a = 7.53$ for hypochlorous acid, it follows that initially

$$pH \approx \tfrac{1}{2} \times 7.53 - \tfrac{1}{2}\log 0.10 \approx 4.3$$

The value is only a crude estimate of the actual pH because we are treating activities as molar concentrations, which is highly dubious in solutions of this concentration. However, all we want is a general understanding of the shape of the pH curve, not a precise graph (which could be obtained far more simply experimentally!).

The addition of titrant converts some of the acid to its conjugate base in the reaction

$$HClO(aq) + OH^-(aq) \longrightarrow H_2O(l) + ClO^-(aq)$$

181

Suppose we add enough titrant to produce a concentration [base] of the conjugate base and reduce the concentration of acid to [acid]. Then (because the solution remains at equilibrium)

$$K_a = \frac{a(H_3O^+)a(ClO^-)}{a(HClO)} \approx \frac{a(H_3O^+)[\text{base}]}{[\text{acid}]}$$

which rearranges first to

$$a(H_3O^+) \approx \frac{K_a[\text{acid}]}{[\text{base}]}$$

and then, by taking negative logarithms, to

$$\text{pH} \approx pK_a - \log\frac{[\text{acid}]}{[\text{base}]} \tag{26}$$

This approximate expression is widely called the **Henderson–Hasselbalch equation**.

Example *Estimating the pH at an intermediate stage in a titration*
Estimate the pH of the solution after the addition of 5.00 mL of the titrant to the analyte in the titration described above.

Strategy
The first step involves deciding the amount of OH^- ions added in the titrant, and then to use that amount to calculate the amount of HClO remaining. Notice that because the ratio of acid to base molar concentrations occurs in eqn (26), the volume of solution cancels, and we can equate the ratio of concentrations to the ratio of amounts.

Solution
The addition of 5.00 mL of titrant corresponds to the addition of

$$n(OH^-) = (5.00 \times 10^{-3}\,\text{L}) \times (0.200\,\text{mol L}^{-1}) = 1.00 \times 10^{-3}\,\text{mol}$$

This amount of OH^- converts 1.00×10^{-3} mol HClO to the base ClO^-. The initial amount of HClO in the analyte is

$$n(HClO) = (25.00 \times 10^{-3}\,\text{L}) \times (0.100\,\text{mol L}^{-1}) = 2.50 \times 10^{-3}\,\text{mol}$$

so the amount remaining after the addition of titrant is 1.50×10^{-3} mol. It then follows from the Henderson–Hasselbalch equation that

$$\text{pH} \approx 7.53 - \log\frac{1.50 \times 10^{-3}}{1.00 \times 10^{-3}} = 7.4$$

As expected, the addition of base has resulted in an increase in pH from 4.3.

Exercise E5.17

Estimate the pH after the addition of a further 5.00 mL of titrant.

[*Answer*: 8.1]

Half way to the stoichiometric point (when enough base has been added to neutralize half the acid), the concentrations of acid and base are equal and the Henderson–Hasselbalch equation gives

$$pH \approx pK_a \tag{27}$$

In the present titration, we see that at this stage of the titration, $pH \approx 7.5$. Equation (27) implies that the pK_a of the acid can be measured directly from the pH of the mixture. In practice this is done by recording the pH during a titration and then examining the record for the pH half way to the stoichiometric point.

At the stoichiometric point, enough base has been added to convert all the acid to its base, and so the solution consists only of ClO^- ions. These ions are Brønsted bases, so we can expect the solution to be basic with a pH of well above 7. We have already seen how to estimate the pH of a solution of a weak base in terms of its concentration B (eqn (22); here B is the concentration of ClO^- ions), so all that remains to be done is to calculate the concentration of ClO^- at the stoichiometric point. Because the analyte initially contained 2.50×10^{-3} mol HClO, the volume of titrant needed to neutralize it is the volume that contains the same amount of base:

$$V(\text{base}) = \frac{2.50 \times 10^{-3}\,\text{mol}}{0.200\,\text{mol L}^{-1}} = 1.25 \times 10^{-2}\,\text{L}$$

or 12.5 mL. The total volume of the solution at this stage is therefore 37.5 mL, so the concentration of base is

$$[ClO^-] = \frac{2.50 \times 10^{-3}\,\text{mol}}{37.5 \times 10^{-3}\,\text{L}} = 6.67 \times 10^{-2}\,\text{mol L}^{-1}$$

It then follows from eqn (25) that the pH of the solution at the stoichiometric point is

$$pH \approx \tfrac{1}{2} \times 14.00 + \tfrac{1}{2} \times 7.53 + \tfrac{1}{2}\log(6.67 \times 10^{-2}) \approx 10.2$$

183

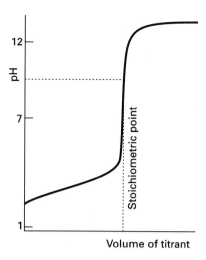

Fig. 5.2 The pH curve for the titration of a weak acid (the analyte) with a strong base (the titrant). Note that the stoichiometric point occurs at pH > 7 and that the change in pH near the stoichiometric point is less abrupt than in Fig. 5.1.

It is very important to note that *the pH at the stoichiometric point of a weak-acid–strong-base titration is on the basic side of neutrality*. At the stoichiometric point, the solution consists of a weak base (the conjugate base of the weak acid, here the ClO^- ions) and neutral cations (the Na^+ ions from the titrant).

The general form of the pH curve suggested by these estimates throughout a weak-acid–strong-base titration is illustrated in Fig. 5.2. The pH rises from the value given by eqn (14), passing through the values given by the Henderson–Hasselbalch equation (eqn (26)) when the acid and its conjugate base are both present, until the stoichiometric point is approached. It then changes rapidly to and through the value given by eqn (25), which takes into account the effect on the pH of a solution of a weak base (the conjugate base of the original acid). The pH then climbs less rapidly towards the value corresponding to a solution consisting of excess base, and finally approaches the pH of the original base solution when (a point never reached in practice) so much titrant has been added that the solution is virtually the same as the titrant itself. The stoichiometric point is detected by observing where the pH changes rapidly through the value given by eqn (25).

A similar sequence of changes occurs when the analyte is a weak base (such as ammonia) and the titrant is a strong acid (such as hydrochloric acid). In this case the pH curve is like that shown in Fig. 5.3: the pH falls as acid is added, plunges through the pH corresponding to a solution of a weak acid (the conjugate acid of the original base, such as NH_4^+), and then slowly approaches the pH of the original strong acid. The pH of the stoichiometric point is that of a solution of a weak acid, and is calculated as described in Section 5.2.

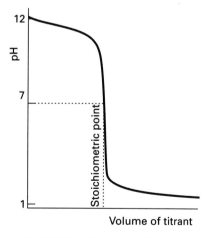

Fig. 5.3 The pH curve for the titration of a weak base (the analyte) with a strong acid (the titrant). The stoichiometric point occurs at pH < 7. The final pH of the solution approaches that of the titrant.

Buffers

The slow variation of the pH when the concentrations of the conjugate acid and base are equal, when $pH = pK_a$, is the basis of **buffer action**, the ability of a solution to oppose changes in pH when small amounts of acids and bases are added (Fig. 5.4). An **acid buffer** solution, one that stabilizes the solution at a pH below 7, is typically prepared by making a solution of a weak acid (such as acetic acid) and a salt that supplies its conjugate base (such as sodium acetate). A **base buffer**, one that stabilizes a solution at a pH above 7, is prepared by making a solution of a weak base (such as ammonia) and a salt that supplies its conjugate acid (such as ammonium chloride).

The mathematical basis of buffer action is the logarithmic dependence of the pH as given by the Henderson–Hasselbalch equation (eqn (26)). This logarithmic dependence results in a very flat curve near $pH = pK_a$ because $\log x$ changes much more slowly than x (for example, if x changes from 10 to 1000, $\log x$ changes only from 1 to 3). The physical basis of the action of an acid buffer is much more interesting: the existence of an abundant supply of A^- ions (supplied by the salt) can remove any H_3O^+ ions brought by additional acid; furthermore, the existence of an abundant supply of HA molecules can supply H_3O^+ ions to react with any base that is added. Similarly, the physical basis of the action of a base buffer is the ability of the base to accept protons when an acid is added, and the ability of the conjugate acid (NH_4^+, for instance) to supply protons if a base is added.

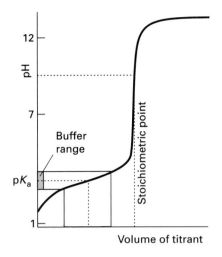

Fig. 5.4 The pH of a solution changes only slowly in the region of half way to the stoichiometric point, and in this region the solution is buffered to a pH that is close to pK_a.

Example *Estimating the pH of a buffer solution*

You may be called on to select a mixture to buffer a solution close to a particular value; to do so, you need to call on tables of pK_a values. Estimate the pH of a buffer formed from equal amounts of $KH_2PO_4(aq)$ and $K_2HPO_4(aq)$.

Strategy

The buffer region is at pH close to pK_a. Therefore, we need to identify the acid in the solution, and then identify its pK_a from Table 5.1. The acid is the species with an additional acidic hydrogen atom.

Solution

The two anions present are $H_2PO_4^-$ and HPO_4^{2-}. The former is the conjugate acid of the latter, so we need its pK_a. We recognize it as the pK_{a2} of phosphoric acid, and take it from Table 5.2. In either case, $pK_a = 7.21$. Hence, the solution should buffer close to pH = 7.

Exercise E5.18

Calculate the pH of an aqueous buffer solution which contains equal amounts of NH_3 and NH_4Cl.

[*Answer*: 8.95; more realistically: 9]

Indicators

The rapid change of pH near the stoichiometric point of an acid–base titration is the basis of indicator detection. An **acid–base indicator** is a large, water-soluble, organic molecule with acid (HIn) and conjugate base (In^-) forms that differ in colour. The two forms are in equilibrium in solution:

$$HIn(aq) + H_2O(l) \rightleftharpoons H_3O^+(aq) + In^-(aq) \qquad K_{In} = \frac{a(H_3O^+)a(In^-)}{a(HIn)}$$

and the ratio of the concentrations is

$$\frac{[In^-]}{[HIn]} \approx \frac{K_{In}}{a(H_3O^+)}$$

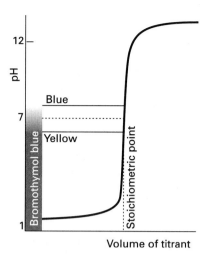

Fig. 5.5 The range of pH over which an indicator changes colour is depicted by the tinted band. For a strong acid–strong base titration, the stoichiometric point is indicated accurately by an indicator that changes colour at pH = 7 (such as bromothymol blue). However, the change in pH is so sharp that accurate results are also obtained even if the indicator changes colour in neighbouring values. Thus, phenolphthalein (which has $pK_{In} = 9.4$, see Table 5.3) is also often used.

This expression can be rearranged (after taking logarithms) to

$$\log \frac{[\text{In}^-]}{[\text{HIn}]} \approx \text{pH} - pK_{In} \tag{28}$$

The pK_{In} of some indicators are listed in Table 5.3. We see that as the pH swings from higher than pK_{In} to lower than pK_{In} as acid is added to the solution, the ratio of In^- to HIn swings from well above 1 to well below 1 (Fig. 5.5).

Exercise 5.19

What is the ratio of the yellow and blue forms of bromocresol green in solution of pH (a) 3.7, (b) 4.7, and (c) 5.7?

[*Answer*: (a) 10:1, (b) 1:1, (c) 1:10]

At the stoichiometric point, the pH changes sharply through several pH units, so the molar concentration of H_3O^+ changes through several orders of magnitude. The indicator equilibrium changes so as to accommodate the change of pH, with HIn the dominant species on the acid side of the stoichiometric point, when H_3O^+ ions are abundant, and In^- dominant on the basic side, when the base can remove protons from HIn. The accompanying colour change signals the stoichiometric point of the titration. The colour in fact changes over a range of pH (typically from pH $\approx pK_{In} - 1$ when HIn is ten times as abundant as In^-, to pH $\approx pK_{In} + 1$, when In^- is ten times as

Table 5.3 Indicator color changes

Indicator	Acid colour	pH range of colour change	pK_{In}	Base colour
Thymol blue	Red	1.2 to 2.8	1.7	Yellow
Methyl orange	Red	3.2 to 4.4	3.4	Yellow
Bromophenol blue	Yellow	3.0 to 4.6	3.9	Blue
Bromocresol green	Yellow	4.0 to 5.6	4.7	Blue
Methyl red	Red	4.8 to 6.0	5.0	Yellow
Bromothymol blue	Yellow	6.0 to 7.6	7.1	Blue
Litmus	Red	5.0 to 8.0	6.5	Blue
Phenol red	Yellow	6.6 to 8.0	7.9	Red
Thymol blue	Yellow	9.0 to 9.6	8.9	Blue
Phenolphthalein	Colourless	8.2 to 10.0	9.4	Pink
Alizarin yellow	Yellow	10.1 to 12.0	11.2	Red
Alizarin	Red	11.0 to 12.4	11.7	Purple

abundant as HIn). The pH half way through a colour change (when pH $\approx$ pK_{In}, and the two forms, HIn and In$^-$, are in equal abundance) is the **end point** of the indicator. In a well designed experiment, the end point of the indicator coincides with the stoichiometric point of the titration.

Care must be taken to use an indicator that changes colour at the pH appropriate to the type of titration. Specifically, the end point should be matched to the stoichiometric point, and so an indicator should be selected for which pK_{In} is close to the pH at the stoichiometric point. Thus, in a weak-acid–strong-base titration, the stoichiometric point lies at the pH given by eqn (25), and so an indicator that changes at that pH must be selected (Fig. 5.6). Similarly, in a strong-acid–weak-base titration, an indicator changing near the pH given by eqn (14) should be used. Qualitatively, an indicator with p$K_{In} \approx 7$ should be chosen for strong-acid–strong-base titrations, one with p$K_{In} < 7$ should be chosen for strong-acid–weak-base titrations, and one with p$K_{In} > 7$ should be chosen for weak-acid–strong-base titrations. For strong-acid–strong-base titrations, the pH curve changes so rapidly through such a wide range of pH that the choice of indicator is not very critical.

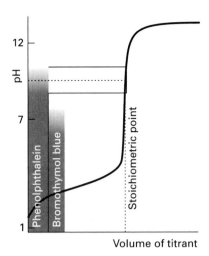

Fig. 5.6 In a weak acid–strong base titration, an indicator with p$K_{In} \approx 7$ (the lower band, like bromothymol blue) would give a false indication of the stoichiometric point; it is necessary to use an indicator that changes colour close to the pH of the stoichiometric point. If that lies at about pH = 9, then phenolphthalein would be appropriate.

Exercise E5.20

Vitamin C is a weak acid (ascorbic acid), and the amount in a sample may be determined by titration with sodium hydroxide solution. Should you use methyl red or phenolphthalein as the indicator?

[*Answer*: phenolphthalein]

Solubility equilibria

A solid dissolves in a solvent until the solution and the solid solute are in equilibrium. At this stage, the solution is said to be **saturated**, and its molar concentration is the **molar solubility**, S, of the solid. That the two phases—the solid solute and the solution—are in dynamic equilibrium implies that we can use equilibrium concepts to discuss the composition of the saturated solution. The properties of aqueous solutions of electrolytes are commonly treated in terms of equilibrium constants, and in this section we shall confine our attention to them. We shall also limit our attention to **sparingly soluble** compounds, which are compounds that dissolve only slightly in water. This restriction is applied because the effects of ion–ion interactions are a complicating feature of more concentrated solutions, and more advanced techniques are needed before the calculations are reliable. Once again, we shall concentrate on general trends and properties rather than expecting to obtain numerically precise results.

5.7 The solubility constant

The equilibrium between a sparingly soluble ionic compound, such as calcium hydroxide, $Ca(OH)_2$, and its ions in aqueous solution is

$$Ca(OH)_2(s) \rightleftharpoons Ca^{2+}(aq) + 2\,OH^-(aq)$$

The equilibrium constant for an ionic equilibrium such as this, bearing in mind that the solid does not appear in the equilibrium expression because its activity is 1, is called the **solubility constant** (sometimes the 'solubility product constant' or simply the 'solubility product'), and denoted K_s:

$$K_s = a(Ca^{2+})a(OH^-)^2$$

As usual, for very dilute solutions, the activity $a(J)$ of a species J can be approximated by the molar concentration of J divided by $1\,mol\,L^{-1}$. Experimental values for solubility constants are given in Table 5.4.

Example *Writing the expression for a solubility constant*
Write the expression for the solubility constant of aluminium sulfide, Al_2S_3.

Strategy
As always in equilibrium calculations, begin by writing the chemical equation and identify the stoichiometric numbers. Pure solids do not appear in expressions for equilibrium constants because they are at unit activity.

Solution
The solubility equilibrium is

$$Al_2S_3(s) \rightleftharpoons 2\,Al^{3+}(aq) + 3\,S^{2-}(aq)$$

The equilibrium constant is therefore

$$K_s = a(Al^{3+})^2 a(S^{2-})^3$$

where $a(Al^{3+}) \approx [Al^{3+}]/(mol\,L^{-1})$ and $a(S^{2-}) \approx [S^{2-}]/(mol\,L^{-1})$.

Exercise E5.21

Write the expression for the solubility constant of mercury(I) sulfate, Hg_2SO_4.

[*Answer*: $K_s = a(Hg_2^{2+})a(SO_4^{2-})$]

Table 5.4 Solubility constants at 25°C

Compound	Formula	K_s
Aluminium hydroxide	$Al(OH)_3$	1.0×10^{-33}
Antimony sulfide	Sb_2S_3	1.7×10^{-93}
Barium carbonate	$BaCO_3$	8.1×10^{-9}
fluoride	BaF_2	1.7×10^{-6}
sulfate	$BaSO_4$	1.1×10^{-10}
Bismuth sulfide	Bi_2S_3	1.0×10^{-97}
Calcium carbonate	$CaCO_3$	8.7×10^{-9}
fluoride	CaF_2	4.0×10^{-11}
hydroxide	$Ca(OH)_2$	5.5×10^{-6}
sulfate	$CaSO_4$	2.4×10^{-5}
Copper(I) bromide	$CuBr$	4.2×10^{-8}
chloride	$CuCl$	1.0×10^{-6}
iodide	CuI	5.1×10^{-12}
sulfide	Cu_2S	2.0×10^{-47}
Copper(II) iodate	$Cu(IO_3)_2$	1.4×10^{-7}
oxalate	CuC_2O_4	2.9×10^{-8}
sulfide	CuS	8.5×10^{-45}
Iron(II) hydroxide	$Fe(OH)_2$	1.6×10^{-14}
sulfide	FeS	6.3×10^{-18}
Iron(III) hydroxide	$Fe(OH)_3$	2.0×10^{-39}
Lead(II) bromide	$PbBr_2$	7.9×10^{-5}
chloride	$PbCl_2$	1.6×10^{-5}
fluoride	PbF_2	3.7×10^{-8}
iodate	$Pb(IO_3)_2$	2.6×10^{-13}
iodide	PbI_2	1.4×10^{-8}
sulfate	$PbSO_4$	1.6×10^{-8}
sulfide	PbS	3.4×10^{-28}
Magnesium		
ammonium phosphate	$MgNH_4PO_4$	2.5×10^{-13}
carbonate	$MgCO_3$	1.0×10^{-5}
fluoride	MgF_2	6.4×10^{-9}
hydroxide	$Mg(OH)_2$	1.1×10^{-11}
Mercury(I) chloride	Hg_2Cl_2	1.3×10^{-18}
iodide	Hg_2I_2	1.2×10^{-28}
Mercury(II) sulfide	HgS black:	1.6×10^{-52}
	red:	1.4×10^{-53}
Nickel(II) hydroxide	$Ni(OH)_2$	6.5×10^{-18}
Silver bromide	$AgBr$	7.7×10^{-13}
carbonate	Ag_2CO_3	6.2×10^{-12}
chloride	$AgCl$	1.6×10^{-10}
hydroxide	$AgOH$	1.5×10^{-8}
iodide	AgI	1.5×10^{-16}
sulfide	Ag_2S	6.3×10^{-51}
Zinc hydroxide	$Zn(OH)_2$	2.0×10^{-17}
sulfide	ZnS	1.6×10^{-24}

The numerical value of a solubility constant can be interpreted in terms of the solubility S of a sparingly soluble substance. For instance, it follows from the stoichiometry of the equilibrium equation written above that the molar concentration of Ca^{2+} ions in solution is equal to that of the $Ca(OH)_2$ dissolved in solution, so, $[Ca^{2+}] = S$. Likewise, because the concentration of OH^- ions is twice that of $Ca(OH)_2$ formula units, it follows that $[OH^-] = 2S$. Therefore, provided it is permissible to replace activities by molar concentrations,

$$K_s \approx (S/\text{mol L}^{-1}) \times (2S/\text{mol L}^{-1})^2 = 4(S/\text{mol L}^{-1})^3$$

from which it follows that

$$S \approx \left(\frac{K_s}{4}\right)^{1/3} \text{mol L}^{-1} \tag{29}$$

This expression should be regarded as very approximate because ion–ion interactions have been ignored. However, because the solid is sparingly soluble, the concentrations of the ions are low and the inaccuracy is moderately low. Thus, from Table 5.4, $K_s = 5.5 \times 10^{-6}$, so $S \approx 1 \times 10^{-2} \text{mol L}^{-1}$. Solubility constants (which are determined by electrochemical measurements of the kind described in Chapter 6) provide a more accurate way of measuring solubilities of very sparingly soluble compounds than the direct measurement of the mass that dissolves.

Exercise E5.22

Copper occurs in many minerals, one of which is chalcocite, Cu_2S. What is the approximate solubility of this compound in water at $25\,^\circ C$?

[*Answer*: $1.7 \times 10^{-16} \text{mol L}^{-1}$]

5.8 The common-ion effect

The principle that an equilibrium constant remains unchanged whereas the individual concentrations of species may change is applicable to solubility constants, and may be used to assess the effect of the addition of species to solutions. An example of particular importance is the effect on the solubility of a sparingly soluble compound of the presence of a freely soluble compound that has an ion in common with the former. For example, we may consider the effect on the solubility of adding sodium chloride to a saturated solution of silver chloride, the common ion in this case being Cl^-.

The molar solubility of silver chloride in pure water is related to its solubility constant by

$$S \approx \sqrt{K_s}\,\text{mol L}^{-1}$$

(This equation is obtained from the expression $K_s = a(Ag^+)a(Cl^-)$ and the fact that $S = [AgCl] = [Ag^+] = [Cl^-]$.) To assess the effect of the common ion, we suppose that Cl^- ions are added to a concentration $C\,\text{mol}\,L^{-1}$, which greatly exceeds the concentration of the same ion that stems from the presence of the silver chloride. Therefore, we can write

$$K_s = a(Ag^+)a(Cl^-) \approx a(Ag^+)C$$

It is very dangerous to neglect deviations from ideal behaviour in ionic solutions and the possibility of complex formation, so from now on the calculation will only be indicative of the kinds of changes that occur when a common ion is added to a solution of a sparingly soluble salt: the qualitative trends are reproduced, but the quantitative calculations are unreliable. With these cautionary remarks in mind, it follows that the solubility S' of silver chloride in the presence of added chloride ions is

$$S' \approx \frac{K_s}{C}\,\text{mol}\,L^{-1}$$

The solubility is greatly reduced by the presence of the common ion. For example, whereas the solubility of silver chloride in water is $1.3 \times 10^{-5}\,\text{mol}\,L^{-1}$, in the presence of $0.10\,M\,NaCl(aq)$ it is only

$$S' \approx \frac{1.6 \times 10^{-10}}{0.10}\,\text{mol}\,L^{-1} \approx 2 \times 10^{-9}\,\text{mol}\,L^{-1}$$

which is nearly ten thousand times less. The reduction of the solubility of a sparingly soluble salt by the presence of a common ion is called the **common-ion effect**.

Exercise E5.23

Estimate the molar solubility of calcium fluoride, CaF_2, in (a) water, (b) $0.010\,M\,NaF(aq)$.

[*Answer:* (a) $2 \times 10^{-4}\,\text{mol}\,L^{-1}$; (b) $4 \times 10^{-7}\,\text{mol}\,L^{-1}$]

Sometimes, the common-ion effect appears in a disguised form, but the same principles still apply. An example is the removal of Ca^{2+} ions from hard water by the addition of more Ca^{2+} ions. The reduction in solubility occurs as a result of the increase in concentration of anions that accompany the Ca^{2+} ions, and not directly as a result of the cations themselves. The Ca^{2+} ions are supplied to the hard water in the form of calcium hydroxide (lime). The role of the lime is to supply OH^- ions that act as a Brønsted base and remove protons from HCO_3^- ions:

$$HCO_3^-(aq) + OH^-(aq) \longrightarrow H_2O(l) + CO_3^{2-}(aq)$$

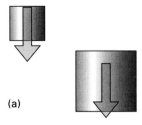

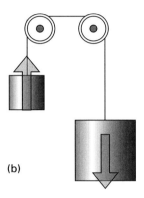

Fig. 5.7 (a) The two weights shown here both have a tendency to fall downwards in a gravitational field, and will do so if released. (b) However, if the two weights are coupled as shown here, then the heavier weight will move the lighter weight in its non-spontaneous direction: overall, the process is still spontaneous. The weights are the analogues of two chemical reactions: a reaction with a large negative ΔG can force another reaction with a smaller ΔG to run in its non-spontaneous direction.

The formation of CO_3^{2-} ions results in the precipitation of $CaCO_3$ in the reaction

$$Ca^{2+}(aq) + CO_3^{2-}(aq) \longrightarrow CaCO_3(s)$$

which removes both the Ca^{2+} ions present initially and those added as lime; overall there is a net reduction of Ca^{2+} in the water.

Coupled reactions

A special kind of approach to equilibrium occurs when a reaction that is not spontaneous is driven forward by a reaction that is more aggressively spontaneous. Although the equilibrium composition of the first reaction may lie strongly in favour of the reactants, that of the overall pair of coupled reactions lies strongly in favour of the products. A simple mechanical analogy is a pair of weights joined by a string (Fig. 5.7): the smaller of the pair of weights will be pulled up as the heavier weight falls, and although the smaller weight has a natural tendency to move downwards, its coupling to the heavier weight results in it being raised. The thermodynamic analogue is an **endergonic reaction**, a reaction with a positive Gibbs energy, $\Delta_r G$ (the analogue of the smaller weight), being forced to occur by virtue of its coupling to an **exergonic reaction**, a reaction with a negative Gibbs energy, $\Delta_r G'$ (the analogue of the heavier weight falling to the ground), because the sum $\Delta_r G + \Delta_r G'$ is negative. The whole of life's activities depend on coupling of this kind, for the oxidation reactions of food act as the heavy weights that drive other reactions forward and result in the formation of proteins from amino acids, the actions of muscles for propulsion, and even the activities of the brain for reflection, learning, and imagination. We shall consider two examples of the 'coupled weights' analogy: one is drawn from biology and the other from industrial chemistry.

5.9 Biological activity: the thermodynamics of ATP

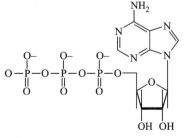

1 ATP, adenosinetriphosphate

The function of adenosine triphosphate, ATP (**1**), is to store the energy made available when food is metabolized and then to supply it on demand to a wide variety of processes, including muscular contraction, reproduction, and vision. The essence of the action of ATP is its ability to lose its terminal phosphate group by hydrolysis and to form adenosine diphosphate, ADP (**2**):

$$ATP(aq) + H_2O(l) \longrightarrow ADP(aq) + P_i^-(aq) + H^+(aq)$$

(The symbol P_i^- denotes an inorganic phosphate group, such as $H_2PO_4^-$.) This reaction is exergonic and can drive an endergonic reaction forward if suitable enzymes are available to couple the reactions.

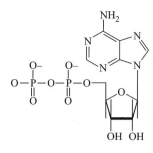

2 ADP, adenosinediphosphate

Biological standard states

The conventional standard state of hydrogen ions ($a = 1$, corresponding to pH = 0, an acidic solution) is not appropriate to normal biological conditions inside cells, where the cell medium has a pH close to 7. Therefore, in biochemistry it is common to adopt the biological standard state, in which pH = 7, a neutral solution. We shall adopt this convention in this section, and label the corresponding standard thermodynamic functions as $G^{\oplus}$, $H^{\oplus}$, and $S^{\oplus}$. (For a further discussion of the biological standard state, see Section 6.3.)

The standard values for the ATP hydrolysis at $37\,°C$ ($310\,K$, body temperature) are

$$\Delta_r G^{\oplus} = -30\,kJ\,mol^{-1} \quad \Delta_r H^{\oplus} = -20\,kJ\,mol^{-1} \quad \Delta_r S^{\oplus} = +34\,J\,K^{-1}\,mol^{-1}$$

The hydrolysis is therefore exergonic ($\Delta_r G < 0$) under these conditions, and $30\,kJ\,mol^{-1}$ is available for driving other reactions. Moreover, because the reaction entropy is large, the reaction Gibbs energy is sensitive to temperature (recall that $\Delta G = \Delta H - T\Delta S$, so ΔS acts as an amplification factor for changes in T). On account of its exergonic character, the ADP–phosphate bond has been called a **high-energy phosphate bond**. The name is intended to signify a high tendency to undergo reaction, and should not be confused with 'strong' bond in its normal chemical sense (that of a high bond enthalpy). In fact, even in the biological sense it is not of very 'high energy'. The action of ATP depends on the bond being intermediate in activity. Thus ATP acts as a phosphate donor to a number of acceptors (such as glucose), but is recharged with a new phosphate group by more powerful phosphate donors in the respiration cycle.

Anaerobic and aerobic metabolism

The efficiency of some biological processes can be gauged in terms of the value of $\Delta_r G^{\oplus}$ given above. The energy source of anaerobic cells is glycolysis, the enzymatic conversion of glucose to lactic acid:

$$C_6H_{12}O_6(aq) \longrightarrow 2\,CH_3CH(OH)COOH(aq) \qquad \Delta_r G^{\oplus} = -218\,kJ\,mol^{-1}$$
$$\text{at } 37\,°C$$

The glycolysis (the 'heavy weight') is coupled to a reaction (the 'light weight') in which two ADP molecules are converted into two ATP molecules:

$$C_6H_{12}O_6(aq) + 2\,P_i^-(aq) + 2\,ADP(aq)$$
$$\longrightarrow 2\,CH_3CH(OH)CO_2^-(aq) + 2\,ATP(aq) + 2\,H_2O(l)$$

The biological standard Gibbs energy for this reaction is $(-218\,\text{kJ mol}^{-1}) - 2(-30\,\text{kJ mol}^{-1}) = -158\,\text{kJ mol}^{-1}$. The overall reaction is exergonic, and therefore spontaneous: the metabolism of the food has been used to 'recharge' the ATP.

Metabolism by aerobic respiration is much more efficient, as is indicated by the standard Gibbs energy of combustion of glucose, which is $-2880\,\text{kJ mol}^{-1}$, and so terminating its oxidation at lactic acid is a poor use of resources. In aerobic respiration the oxidation is carried out to completion, and an extremely complex set of reactions preserves as much of the energy released as possible. In the overall reaction, 38 ATP molecules are generated for each glucose molecule consumed. Each mole of ATP extracts 30 kJ from the 2880 kJ supplied by $1\,\text{mol}\,C_6H_{12}O_6$ (180 g of glucose), and so 1140 kJ per mole of glucose molecules has been stored for later use.

Each ATP molecule can be used to drive an endergonic reaction for which $\Delta_r G^\oplus$ does not exceed $30\,\text{kJ mol}^{-1}$. For example, the biosynthesis of sucrose from glucose and fructose can be driven (if a suitable enzyme system is available) because the reaction is endergonic to the extent $\Delta_r G^\oplus = +23\,\text{kJ mol}^{-1}$. The biosynthesis of proteins is strongly endergonic, not only on account of the enthalpy change but also on account of the large decrease in entropy that occurs when many amino acids are assembled into a precisely determined sequence. For instance, the formation of a peptide link is endergonic, with $\Delta_r G^\oplus = +17\,\text{kJ mol}^{-1}$, but the biosynthesis occurs indirectly and is equivalent to the consumption of three ATP molecules for each link. In a moderately small protein like myoglobin, with about 150 peptide links, the construction alone requires 450 ATP molecules, and therefore about 12 mol of glucose molecules for 1 mol of protein molecules.

5.10 The extraction of metals from their oxides

The extraction of metals from their ores is another example where a reaction must be driven in its non-spontaneous direction. Almost all metal oxides are exergonic, in the sense that their formation is spontaneous, so the task of industry is to find economical ways of driving oxidation reactions in reverse, to release the free metal. Carbon and carbon monoxide are both very cheap reducing agents, so we need to determine the conditions—if any—under which they can successfully reduce metal oxides. Although equilibrium is rarely reached in such reductions, the equilibrium composition does at least indicate whether such a reduction is likely to be feasible and what change in conditions can enhance the yield. Therefore, we need to examine the equilibria

$$MO(s) + C(s) \rightleftharpoons M(s) + CO(g)$$
$$MO(s) + \tfrac{1}{2}C(s) \rightleftharpoons M(s) + \tfrac{1}{2}CO_2(g)$$

The products are favored if $K > 1$ or, equivalently, $\Delta_r G^\ominus < 0$.

These equilibria can be discussed in terms of the reactions

$$
\begin{aligned}
&\text{(i)} && \mathrm{M(s)} + \tfrac{1}{2}\,\mathrm{O_2(g)} \longrightarrow \mathrm{MO(s)} \\
&\text{(ii)} && \tfrac{1}{2}\,\mathrm{C(s)} + \tfrac{1}{2}\,\mathrm{O_2(g)} \longrightarrow \tfrac{1}{2}\,\mathrm{CO_2(g)} \\
&\text{(iii)} && \mathrm{C(s)} + \tfrac{1}{2}\,\mathrm{O_2(g)} \longrightarrow \mathrm{CO(g)} \\
&\text{(iv)} && \mathrm{CO(g)} + \tfrac{1}{2}\,\mathrm{O_2(g)} \longrightarrow \mathrm{CO_2(g)}
\end{aligned}
$$

All these reactions have been written for the reaction of 1 mol O (in the form of $\tfrac{1}{2}$ mol O_2), for then they can be added and subtracted without further modification.

We shall need to assess whether the reaction Gibbs energy for a metal oxide reduction reaction increases or decreases with temperature. To do so, we make use of the fact that the temperature dependences of the standard Gibbs energies for these four reactions depend on their reaction entropy through

$$
\text{Change in } \Delta_r G^{\ominus} = -(\text{change in } T) \times \Delta_r S^{\ominus} \tag{30}
$$

This equation comes from $\Delta G = \Delta H - T\Delta S$ and the assumption that both ΔH and ΔS are independent of temperature, so all the change in ΔG stems from the change in T. To decide whether the reaction Gibbs energy increases or decreases with temperature we need to decide whether there is an increase or decrease in entropy, and to do that we have to note whether there is a net formation or net consumption of gas. Because in reaction (iii) there is a net increase in the amount of gas molecules, from $\tfrac{1}{2}$ mol to 1 mol, the standard reaction entropy is large and positive; therefore, $\Delta_r G^{\ominus}$ for this reaction decreases sharply with increasing temperature. In reaction (iv), there is a similar net decrease in the amount of gas molecules, from $\tfrac{3}{2}$ mol to 1 mol, and so $\Delta_r G^{\ominus}$ for this reaction increases sharply with increasing temperature. In reaction (ii), the amount of gas is constant, and so the entropy change is small and $\Delta_r G^{\ominus}$ for this reaction changes only slightly with temperature. These remarks are summarized in Fig. 5.8, which is called an **Ellingham diagram** (note that $\Delta_r G^{\ominus}$ decreases upwards!).

The standard Gibbs energy for reaction (i) is a measure of the metal's affinity for oxygen. At room temperature the contribution of the reaction entropy to $\Delta_r G^{\ominus}$ is dominated by the reaction enthalpy, and so the order of increasing $\Delta_r G^{\ominus}$ is the same as the order of increasing $\Delta_r H^{\ominus}$ (for this conclusion we are using $\Delta G = \Delta H - T\Delta S$ again). The reaction enthalpy therefore gives the order of values on the left of the diagram (Al_2O_3 is most exothermic, Ag_2O is least). The standard reaction entropy is similar for all metals because in each case gaseous oxygen is eliminated and a compact, solid oxide is formed. This similarity implies that the temperature dependence of the standard Gibbs energy of oxidation should be similar for all metals, as is shown by the similar slopes of the lines in the diagram. The kinks at high temperatures correspond to the evaporation of the metals; less pronounced kinks occur at the melting temperatures of the metals and the oxides.

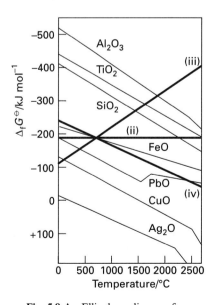

Fig. 5.8 An Ellingham diagram for a number of common metal oxides.

The standard Gibbs energies of the overall metal oxide reduction reactions can be expressed in terms of the standard Gibbs energies of the reactions above:

$$MO(s) + C(s) \longrightarrow M(s) + CO(g) \qquad \Delta_r G^{\ominus} = \Delta_r G^{\ominus}(iii) - \Delta_r G^{\ominus}(i)$$
$$MO(s) + \tfrac{1}{2} C(s) \longrightarrow M(s) + \tfrac{1}{2} CO_2(g) \qquad \Delta_r G^{\ominus} = \Delta_r G^{\ominus}(ii) - \Delta_r G^{\ominus}(i)$$
$$MO(s) + CO(g) \longrightarrow M(s) + CO_2(g) \qquad \Delta_r G^{\ominus} = \Delta_r G^{\ominus}(iv) - \Delta_r G^{\ominus}(i)$$

The equilibrium lies to the right if $\Delta_r G^{\ominus} < 0$. In terms of the Ellingham diagram, this is the case when the line for reaction (i) lies below (is more positive than) the line for one of the carbon reactions (ii) to (iv).

We can now predict the success of a reduction at any temperature simply by looking at the diagram: a metal oxide is reduced by any carbon reaction lying above it because the overall reaction then has $\Delta_r G^{\ominus} < 0$. For example, copper(II) oxide, CuO, can be reduced to copper at any temperature above room temperature. Even in the absence of carbon, Ag_2O decomposes when heated above $200\,°C$ because then the standard Gibbs energy of reaction (i) becomes positive (so that the reverse reaction is then spontaneous).

Exercise E5.24

What is the minimum temperature at which carbon could be used to reduce alumina, Al_2O_3, to aluminium?

[*Answer*: 2000 °C]

EXERCISES

5.1 Write the expressions for the equilibrium constants of the following reactions:
(a) $CO(g) + Cl_2(g) \rightleftharpoons COCl(g) + Cl(g)$
(b) $2\,SO_2(g) + O_2(g) \rightleftharpoons 2\,SO_3(g)$
(c) $H_2(g) + Br_2(g) \rightleftharpoons 2\,HBr(g)$
(d) $2\,O_3(g) \rightleftharpoons 3\,O_2(g)$

5.2 If the equilibrium constant for the reaction $A + B \rightleftharpoons C$ is reported as 0.224, what would be the equilibrium constant for the reaction written as $C \rightleftharpoons A + B$?

5.3 The equilibrium constant for the reaction $A + B \rightleftharpoons 2\,C$ is reported as 3.4×10^4. What would it be for the reaction written as (a) $2\,A + 2\,B \rightleftharpoons 4\,C$, (b) $\tfrac{1}{2}\,A + \tfrac{1}{2}\,B \rightleftharpoons C$?

5.4 The bond in molecular iodine is quite weak, and hot iodine vapour contains a proportion of atoms.

When $1.00\,g$ of I_2 is heated to $1000\,K$ in a sealed container of volume $1.00\,L$, the resulting equilibrium mixture contains $0.830\,g$ of I_2. Calculate K for the dissociation equilibrium $I_2(g) \rightleftharpoons 2\,I(g)$.

5.5 In a gas-phase equilibrium mixture of $SbCl_5$, $SbCl_3$, and Cl_2 at $500\,K$, $p(SbCl_5) = 0.15\,bar$ and $p(SbCl_3) = 0.20\,bar$. Calculate the equilibrium partial pressure of Cl_2 given that $K = 3.5 \times 10^{-4}$ for the reaction $SbCl_5(g) \rightleftharpoons SbCl_3(g) + Cl_2(g)$.

5.6 The equilibrium constant $K = 0.36$ for the reaction $PCl_5(g) \rightleftharpoons PCl_3(g) + Cl_2(g)$ at $400\,K$. (a) Given that $2.0\,g$ of PCl_5 was initially placed in a reaction vessel of volume $250\,mL$, determine the molar concentrations in the mixture at equilibrium. (b) What is the percentage of PCl_5 decomposed at $400\,K$?

5.7 In the Haber process for ammonia, $K = 0.036$ for the reaction $N_2(g) + 3H_2(g) \rightleftharpoons 2NH_3(g)$ at 500 K. If a reactor is charged with partial pressures of 0.020 bar of N_2 and 0.020 bar of H_2, what will be the equilibrium partial pressure of the components?

5.8 Express the equilibrium constant for $N_2O_4(g) \rightleftharpoons 2NO_2(g)$ in terms of the fraction α of N_2O_4 that has dissociated and the total pressure p of the reaction mixture, and show that when the extent of dissociation is small ($\alpha \ll 1$), α is inversely proportional to the square root of the total pressure ($\alpha \propto 1/\sqrt{p}$).

5.9 Write the proton transfer equilibria for the following acids in aqueous solution and identify the conjugate acid–base pairs in each one: (a) H_2SO_4, (b) HF (hydrofluoric acid), (c) $C_6H_5NH_3^+$ (anilinium ion), (d) $H_2PO_4^-$ (dihydrogenphosphate ion), (e) HCOOH (formic acid), (f) $NH_2NH_3^+$ (hydrazinium ion).

5.10 Numerous acidic species are found in living systems. Write the proton transfer equilibria for the following biochemically important acids in aqueous solution: (a) lactic acid ($CH_3CHOHCOOH$), (b) glutamic acid (**3**), (c) glycine (NH_2CH_2COOH), (d) oxalic acid (HOOCCOOH).

3 glutamic acid

5.11 For biological and medical applications we often need to consider proton transfer equilibria at body temperature (37 °C). The value of K_w for water at body temperature is 2.5×10^{-14}. (a) What is the value of $[H_3O^+]$ and the pH of neutral water at 37 °C. (b) What is the molar concentration of OH^- ions and the pOH of neutral water at 37 °C?

5.12 Suppose that something had gone wrong in the Big Bang, and instead of ordinary hydrogen there was an abundance of deuterium in the universe. There would be many subtle changes in equilibria, particularly the deuteron transfer equilibria of heavy atoms and bases. The K_w for heavy water at 25 °C is 1.35×10^{-15}. (a) Write the chemical equation for the autoprotolysis of D_2O. (b) Evaluate pK_w for D_2O at 25 °C. (c) Calculate the molar concentrations of D_3O^+ and OD^- in neutral heavy water at 25 °C. (d) Evaluate the pD and pOD of neutral heavy water at 25 °C. (e) Find the relation between pD, pOD, and $pK_w(D_2O)$.

5.13 The molar concentration of H_3O^+ ions in the following solutions was measured at 25 °C. Calculate the pH and pOH of the solution: (a) 1.5×10^{-5} mol L^{-1} (a sample of rain water), (c) 1.5 mmol L^{-1}, (d) 5.1×10^{-14} mol L^{-1}, (e) 5.01×10^{-5} mol L^{-1}.

5.14 Calculate the molar concentration of H_3O^+ ions and the pH of the following solutions: (a) 25.0 mL of 0.144 M HCl(aq) was added to 25.0 mL of 0.125 M NaOH(aq), (b) 25.0 mL of 0.15 M HCl(aq) was added to 35.0 mL of 0.15 M KOH(aq), (c) 21.2 mL of 0.22 M HNO_3(aq) was added to 10.0 mL of 0.30 M NaOH(aq).

5.15 Determine whether aqueous solutions of the following salts have a pH equal to, greater than, or less than 7; if pH > 7 or pH < 7, write a chemical equation to justify your answer. (a) NH_4Br, (b) Na_2CO_3, (c) KF, (d) KBr, (e) $AlCl_3$, (f) $Co(NO_3)_2$.

5.16 (a) A sample of potassium acetate, KCH_3CO_2, of mass 8.4 g is used to prepare 250 mL of solution. What is the pH of the solution? (b) What is the pH of a solution when 3.75 g of ammonium bromide, NH_4Br, is used to make 100 mL of solution? (c) An aqueous solution of volume 1.0 L contains 10.0 g of potassium bromide. What is the percentage of Br^- ions that are protonated?

5.17 There are many organic acids and bases in our cells, and their presence modifies the pH of the fluids inside them. It is useful to be able to assess the pH of solutions of acids and bases and to make inferences from measured values of the pH. A solution of equal concentrations of lactic acid and sodium lactate was found to have pH = 3.08. (a) What are the values of pK_a and K_a of lactic acid? (b) What would the pH be if the acid had twice the concentration of the salt?

5.18 Sketch reasonably accurately the pH curve for the titration of 25.0 mL of 0.15 M $Ba(OH)_2$(aq) with 0.22 M HCl(aq). Mark on the curve (a) the initial pH, (b) the pH at the stoichiometric point.

5.19 Determine the fraction of solute ionized in (a) 0.25 M C_6H_5COOH(aq), (b) 0.150 M NH_2NH_2(aq) (hydrazine), (c) 0.112 M $(CH_3)_3N$(aq) (trimethylamine).

5.20 Calculate the pH, pOH, and fraction of solute ionized in the following aqueous solutions: (a) 0.120 M $CH_3CH(OH)COOH$(aq) (lactic acid), (b) 1.4×10^{-4} M $CH_3CH(OH)COOH$(aq), (c) 0.10 M $C_6H_5SO_3H$(aq) (benzenesulfonic acid).

5.21 Calculate the pH of the following acid solutions at $25\,^\circ$C; ignore second ionizations only when that approximation is justified. (a) 1.0×10^{-4} M H$_3$BO$_3$(aq) (boric acid acts as a monoprotic acid), (b) 0.015 M H$_3$PO$_4$(aq), (c) 0.10 M H$_2$SO$_3$(aq).

5.22 The weak base colloquially known as Tris, and more precisely as tris(hydroxymethyl)aminomethane, has $pK_a = 8.3$ at $20\,^\circ$C and is commonly used to produce a buffer for biochemical applications. At what pH would you expect Tris to act as a buffer in a solution that has equal molar concentrations of Tris and its conjugate acid?

5.23 The amino acid tyrosine has $pK_a = 2.20$ for deprotonation of its carboxylic acid group. What are the relative concentrations of tyrosine and its conjugate base at a pH of (a) 7, (b) 2.2, (c) 1.5?

5.24 (a) Calculate the molar concentrations of (COOH)$_2$, HOOCCO$_2^-$, (CO$_2$)$_2^{2-}$, H$_3$O$^+$, and OH$^-$ in 0.15 M (COOH)$_2$(aq). (b) Calculate the molar concentrations of H$_2$S, HS$^-$, S^{2-}, H$_3$O$^+$, and OH$^-$ in 0.065 M H$_2$S(aq).

5.25 A sample of 0.10 M CH$_3$COOH(aq) of volume 25.0 mL is titrated with 0.10 M NaOH(aq). The K_a for CH$_3$COOH is 1.8×10^{-5}. (a) What is the pH of 0.10 M CH$_3$COOH(aq)? (b) What is the pH after the addition of 10.0 mL of 0.10 M NaOH(aq)? (c) What volume of 0.10 M NaOH(aq) is required to reach halfway to the stoichiometric point? (d) Calculate the pH at that half-way point. (e) What volume of 0.10 M NaOH(aq) is required to reach the stoichiometric point? (f) Calculate the pH at the stoichiometric point.

5.26 A buffer solution of volume 100 mL consists of 0.10 M CH$_3$COOH(aq) and 0.10 M NaCH$_3$CO$_2$(aq). (a) What is its pH? (b) What is the pH after the addition of 3.3 mmol NaOH to the buffer solution? (c) What is the pH after the addition of 6.0 mmol HNO$_3$ to the initial buffer solution?

5.27 Predict the pH region in which each of the following buffers will be effective, assuming equal molar concentrations of the acid and its conjugate base: (a) sodium lactate and lactic acid, (b) sodium benzoate and benzoic acid, (c) potassium hydrogenphosphate and potassium phosphate, (d) potassium hydrogenphosphate and potassium dihydrogenphosphate, (e) hydroxylamine and hydroxylammonium chloride.

5.28 At the half-way point in the titration of a weak acid with a strong base the pH was measured as 4.66. What is the acidity constant and the pK_a of the acid? What is the pH of the solution that is 0.015 M in the acid?

5.29 Calculate the pH of (a) 0.15 M NH$_4$Cl(aq), (b) 0.15 M NaCH$_3$CO$_2$(aq), (c) 0.150 M CH$_3$COOH(aq).

5.30 Calculate the pH at the stoichiometric point of the titration of 25.00 mL of 0.100 M lactic acid with 0.175 M NaOH(aq).

5.31 Sketch the pH curve of a solution containing 0.10 M NaCH$_3$CO$_2$(aq) and a variable amount of acetic acid.

5.32 From the information in Tables 5.1 and 5.2, select suitable buffers for (a) pH = 2.2 and (b) pH = 7.0.

5.33 Write the expression for the solubility constants of the following compounds: (a) AgI, (b) Hg$_2$S, (c) Fe(OH)$_3$, (d) Ag$_2$CrO$_4$.

5.34 Use the data in Table 5.4 to determine the molar solubilities of (a) BaSO$_4$, (b) Ag$_2$CO$_3$, (c) Fe(OH)$_2$, (d) Hg$_2$Cl$_2$.

5.35 Use the data in Table 5.4 to estimate the solubility of each sparingly soluble substance in its respective solution: (a) silver bromide in 1.4×10^{-3} M NaBr(aq), (b) magnesium carbonate in 1.1×10^{-5} M Na$_2$CO$_3$(aq), (c) lead(II) sulfate in a 0.10 M CaSO$_4$(aq), (d) nickel(II) hydroxide in 2.7×10^{-5} M NiSO$_4$(aq).

5.36 Calculate the equilibrium constant of the reaction that results in the formation of glutamine (G; $^-$O$_2$CCH(NH$_3^+$)CH$_2$CH$_2$COOH) from glutamate (G'; $^-$O$_2$CCH(NH$_3^+$)CH$_2$CH$_2$CONH$_2$) in living cells, which is facilitated by the enzyme glutamine synthetase, which is a means of transporting ammonia from the kidneys to other cells. Base your answer on the fact that the standard reaction Gibbs energy for G'(aq)+ NH$_4^+$(aq) $\longrightarrow$ G(aq) is $+15.7$ kJ mol^{-1} at blood temperature, $37\,^\circ$C. The reaction is in fact driven by coupling with ATP(aq) $\longrightarrow$ ADP(aq) + P$_i$(aq), for which the standard Gibbs energy is -31.0 kJ mol^{-1}. What is the equilibrium constant for the overall reaction G'(aq) + NH$_4^+$ + ATP(aq) $\longrightarrow$ G(aq) + ADP(aq) + P$_i$(aq)?

5.37 The native (active) form of an enzyme is in equilibrium with its denatured (inactive) form, and their relative abundances change with temperature. In a study of ribonuclease, the ratio of concentrations (active over inactive) was found to be 390 at $50\,^\circ$C and 6.2 at $100\,^\circ$C. Estimate the enthalpy of denaturation.

Electrochemistry

Such apparently unrelated processes as combustion, respiration, photosynthesis, and corrosion are actually all closely related, for in each of them an electron (sometimes accompanied by a group of atoms) is transferred from one species to another. Indeed, together with acid–base reactions, in which a proton is transferred, reactions in which electrons are transferred, the so-called **redox reactions**, account for many of the reactions encountered in chemistry. Redox reactions—the principal topic of this chapter—are of immense practical significance, not only because they underlie many biochemical and industrial processes, but also because they are the basis of the generation of electricity by chemical reactions and the investigation of reactions by making electrical measurements. Measurements like the ones we describe in this chapter lead to a collection of data that are very useful for discussing the characteristics of electrolyte solutions and of a wide range of different types of equilibria in solution. They are also used throughout inorganic chemistry to assess the thermodynamic feasibility of reactions and the stabilities of compounds, and in physiology to discuss the details of the propagation of signals in neurons.

Electrochemical cells

The device used to study reactions electrically is called an **electrochemical cell**. Such a cell consists of two electronic conductors dipping into an electrolyte (an ionic conductor), which may be a solution, a liquid, or a solid. The electronic conductor and its surrounding electrolyte is an **electrode**, and the physical structure containing them is called an **electrode compartment**. The two electrodes may share the same compartment (Fig. 6.1). If the electrolytes are different, then the two compartments may be joined by a **salt bridge**, which is an electrolyte solution that completes the electrical circuit by permitting ions to move between the compartments, and so enables the cell to function (Fig. 6.2). Alternatively, the two solutions may be in direct physical contact (for example, through a porous membrane), but the presence of such **liquid junctions** introduces complications into the interpretation of measurements, and we shall not consider them further.

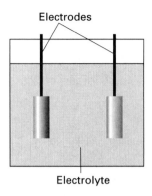

Fig. 6.1 The arrangement for an electrochemical cell in which the two electrodes share a common electrolyte.

Electrodes

Electrolyte

Salt bridge

Electrode

Electrode

Electrode compartments

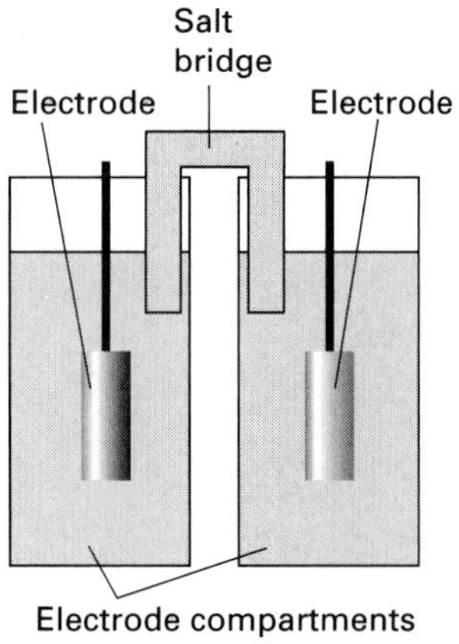

Fig. 6.2 When the electrolytes in the electrode compartments of a cell are different, they need to be joined so that ions can travel from one compartment to another. One device for joining the two compartments is a salt bridge.

An electrochemical cell that produces electricity as a result of the spontaneous reaction occurring inside it is called a **galvanic cell**. An electrochemical cell in which a non-spontaneous reaction is driven by an external source of direct current is called an **electrolytic cell**. The commercially available dry cells, mercury cells, and nickel–cadmium cells that are used to power electrical equipment are all galvanic cells, and produce electricity as a result of the spontaneous chemical reaction between the substances built into them at manufacture. A **fuel cell** is a galvanic cell in which the reagents (such as hydrogen and oxygen or methane and oxygen) are supplied from outside. Fuel cells are used on manned spacecraft, and gas supply companies hope that one day they may be used as a convenient, compact source of electricity in homes. Electric eels and electric catfish are biological versions of fuel cells in which the fuel is food and the cells are adaptations of muscle cells. Electrolytic cells include the arrangement used to electrolyse water into hydrogen and oxygen (a non-spontaneous process) and to obtain aluminium from its oxide in the Hall process. Electrolysis is the only commercially viable means for the production of fluorine.

The link between this chapter and the earlier chapters on thermodynamics lies in the fact that a galvanic cell is a device for extracting the energy of a spontaneous reaction as non-expansion work. We saw in Section 3.3 that the maximum non-expansion work is given by the value of ΔG for a process; because we shall now identify that non-expansion work with electrical work, the key equation of this chapter will have the form

$$\text{maximum electrical work available} = \text{change in Gibbs energy}$$

which is valid at constant temperature and pressure. (The precise form of this expression, and its signs, will be given later.) Most of this chapter will be based on this relation.

This chapter necessarily makes use of a number of concepts related to electricity; they are reviewed in *Further information 5*.

6.1 Half-reactions and electrodes

As we have remarked, a redox reaction is a process in which there is a transfer of electrons from one species to another. As a result of this transfer of electrons, there is a change in the oxidation numbers of the elements of the species. The concept of oxidation number is reviewed in *Further information 7*.

A redox reaction is the outcome of two contributions, the loss of electrons (and perhaps atoms) from one species, and their gain by another species. The loss of electrons is called **oxidation**, and is identified by noting whether an element has undergone an increase in oxidation number. The gain of electrons is called **reduction**, and it is identified by noting whether an element has undergone a decrease in oxidation number. As we also remarked, electron transfer may be accompanied by atom transfer (as in the conversion of PCl_3 to PCl_5 or of NO_2^- to NO_3^-); the requirement to break and form covalent bonds in some redox reactions is one of the reasons why they often achieve equilibrium quite slowly (often much more slowly than acid–base proton transfer reactions). The **reducing agent** (or 'reductant') is the species that acts as electron donor, and at least one of the elements in the species undergoes an increase in oxidation number when it acts. The **oxidizing agent** (or 'oxidant') is the species that acts as an electron acceptor, and at least one of its elements undergoes a decrease in oxidation number.

Examples of redox reactions include the combustion of magnesium in oxygen

$$2\,Mg(s) + O_2(g) \longrightarrow 2\,MgO(s)$$

in which magnesium is the reducing agent (and becomes oxidized to Mg^{2+} ions) and oxygen is the oxidizing agent (and becomes reduced to O^{2-} ions). Another example is the reaction of a metal oxide with hydrogen:

$$CuO(s) + H_2(g) \longrightarrow Cu(s) + H_2O(g)$$

In this reaction hydrogen is the reducing agent and copper(II) oxide the oxidizing agent. Both these reactions involve the transfer of atoms as well as electrons, but the net effect is a change in oxidation numbers. A reaction which is purely a transfer of electrons is the displacement of copper from solution by zinc

$$Cu^{2+}(aq) + Zn(s) \longrightarrow Cu(s) + Zn^{2+}(aq)$$

in which Cu^{2+} ions are the oxidizing agents and zinc metal is the reducing agent.

Exercise E6.1

Identify the species that have undergone oxidation and reduction in the reaction

$$CuS(s) + O_2(g) \longrightarrow Cu(s) + SO_2(g)$$

[*Answer*: Cu(II) reduced, S^{2-} oxidized to S(IV), O reduced]

The breadth of the scope of redox reactions can be judged by noting that because a combustion reaction is a redox reaction, then its reverse is also a redox reaction (with the electrons and atoms transferred in the opposite direction). The reverse of the combustion of glucose,

$$C_6H_{12}O_6(s) + 6\,O_2(g) \longrightarrow 6\,CO_2(g) + 6\,H_2O(l)$$

is nothing other than the net outcome of the photosynthesis reaction

$$6\,CO_2(g) + 6\,H_2O(l) \longrightarrow C_6H_{12}O_6(s) + 6\,O_2(g)$$

Therefore, the net reaction that we call photosynthesis is a redox reaction. The actual reaction proceeds by a very complicated mechanism, some steps of which involve proton transfer and others electron and atom transfer; but the net outcome is a redox reaction. Therefore, the generation of carbohydrates on this planet and their consumption in respiration are both aspects or redox reactions.

Half-reactions

An important step in the analysis of redox reactions draws an analogy with acid–base reactions, which, as we saw in Chapter 5, may be expressed in terms of proton loss from one species and proton gain by another. Thus, any redox reaction may be expressed as the sum of two **half-reactions**, one involving electron loss by a species and the other electron gain. Two examples are

Oxidation of Zn: $\qquad Zn(s) \longrightarrow Zn^{2+}(aq) + 2\,e^-$

Reduction of Cu^{2+} : $\qquad Cu^{2+}(aq) + 2\,e^- \longrightarrow Cu(s)$

A half-reaction in which atom transfer accompanies electron transfer is

Reduction of MnO_4^- :

$$MnO_4^-(aq) + 8\,H^+(aq) + 5\,e^- \longrightarrow Mn^{2+}(aq) + 4\,H_2O(l)$$

Half-reactions are conceptual reactions showing the loss and gain of electrons: the reactions do not actually occur (except in special cases) by one species releasing an electron and then another species accepting it; electron transfer reactions normally proceed by a much more complex mechanism in which the electron is never free. The electrons in these conceptual reactions are regarded as being 'in transit' and are not ascribed a state.

A redox reaction is the sum of an oxidation and a reduction half-reaction. For the reduction of Cu^{2+} ions by zinc metal, for example, the redox reaction is

$$Cu^{2+}(aq) + Zn(s) \longrightarrow Cu(s) + Zn^{2+}(aq)$$

It is common practice, however, to write all half-reactions as reductions; so the zinc oxidation half-reaction is reversed and expressed as a reduction:

$$Zn^{2+}(aq) + 2\,e^- \longrightarrow Zn(s)$$

Now the overall reaction is the *difference* of the two reduction half-reactions.

The oxidized and reduced substances in a half-reaction form a **redox couple**, denoted Ox/Red. Thus, the redox couples mentioned so far are

Cu^{2+}/Cu, Zn^{2+}/Zn, and $MnO_4^-, H^+/Mn^{2+}$. In general, we adopt the notation

Couple: Ox/Red   Half reaction: $Ox + \nu e^- \longrightarrow Red$

The overall reaction need not be a redox reaction for it to be expressed in terms of reduction half-reactions. For instance, the expansion of a gas

$$H_2(g, p_i) \longrightarrow H_2(g, p_f)$$

is not a redox reaction (there is no change of oxidation number) but it can be expressed as the difference of two reductions:

$$2\,H^+(aq) + 2\,e^- \longrightarrow H_2(g, p_f)$$
$$2\,H^+(aq) + 2\,e^- \longrightarrow H_2(g, p_i)$$

The two couples in this case are both H^+/H_2 but the gas is at a different pressure in each case.

Example *Expressing a reaction in terms of half-reactions*
Express the dissolution of AgCl(s), as the difference of two reduction half-reactions. This process is not a redox reaction because the oxidation numbers of the elements do not change in the overall reaction. Nevertheless, it can be expressed in terms of half-reactions.

Strategy
To decompose a reaction into reduction half-reactions, identify one reactant species that undergoes reduction, its corresponding reduction product, and write the half-reaction for this process. To find the second half-reaction, subtract the overall reaction from this half-reaction and rearrange the species so that all the stoichiometric numbers are positive. When the overall reaction is not a redox reaction, the reduction product does not actually appear on the right of the equation, but its identity can usually be inferred.

Solution
The overall chemical equation is

$$AgCl(s) \longrightarrow Ag^+(aq) + Cl^-(aq)$$

If we select AgCl as the species to be reduced, then the silver in the AgCl is reduced to metallic silver in the half-reaction

$$AgCl(s) + e^- \longrightarrow Ag(s) + Cl^-(aq)$$

Subtraction of the overall equation gives the odd-looking equation

$$e^- \longrightarrow Ag(s) - Ag^+(aq)$$

However, by adding $Ag^+(aq)$ to both sides, it becomes more recognizable as

$$Ag^+(aq) + e^- \longrightarrow Ag(s)$$

Exercise E6.2

Express the formation of H_2O from H_2 and O_2 in acidic solution as the difference of two reduction half-reactions.

[*Answer*: $4\,H^+(aq) + 4\,e^- \longrightarrow 2\,H_2(g)$,

$O_2(g) + 4\,H^+(aq) + 4\,e^- \longrightarrow 2\,H_2O(l)$]

We have already seen that, for thermodynamic considerations, a natural way to express the composition of a system is in terms of the reaction quotient Q (because Q occurs in a number of thermodynamic formulas, particularly the formula for the reaction Gibbs energy, eqn (3.23)). We shall in fact find it useful to express the reaction at an electrode in terms of the reaction quotient Q for the half-reaction that takes place there. The half-reaction quotient is defined like the quotient for the overall reaction, but with the electrons ignored. Thus, for the half-reaction of the Cu^{2+}/Cu couple we would write

$$Cu^{2+}(aq) + 2\,e^- \longrightarrow Cu(s) \qquad Q = \frac{1}{a(Cu^{2+})}$$

where, as explained in Chapter 3, $a(Cu^{2+}) \approx [Cu^{2+}]/(mol\,L^{-1})$, if the solution is very dilute, and $a = 1$ for a pure solid (the copper metal). The replacement of activities by molar concentrations is very hazardous for ionic solutions, and so wherever possible we shall delay taking that final step. The step is not always necessary: as we shall see, electrochemical measurements are widely used to determine pH and equilibrium constants, both of which are defined in terms of activities themselves.

Exercise E6.3

Express the oxidation of NADH (nicotinamide adenine dinucleotide, which participates in the chain of oxidations that constitutes respiration) to NAD^+ by oxygen, when the latter is reduced to H_2O_2, in aqueous solution as the difference of two reduction half-reactions.

[*Answer*: $O_2(g) + 2\,H^+(aq) + 2\,e^- \longrightarrow H_2O_2(l)$;

$NAD^+(aq) + H^+(aq) + 2\,e^- \longrightarrow NADH(aq)$]

Reactions at electrodes

In an electrochemical cell, the oxidation half-reaction takes place at one electrode and the reduction half-reaction takes place at the other electrode. As the reaction proceeds, the electrons released in the oxidation half-reaction

$$\text{reduced species} \longrightarrow \text{oxidized species} + \nu\,e^-$$

in one electrode compartment travel through the external circuit, and enter the cell at the other electrode, where they bring about the reduction

$$\text{oxidized species} + \nu\,e^- \longrightarrow \text{reduced species}$$

For example, if one electrode is zinc in aqueous zinc sulfate, then the oxidation reaction is the release of two electrons from each atom in the metal, with the formation of Zn^{2+} ions, which go into solution:

$$Zn(s) \longrightarrow Zn^{2+}(aq) + 2\,e^-$$

These electrons travel through the external circuit and, if the other electrode is copper in aqueous copper(II) sulfate, reduce the Cu^{2+} ions in solution:

$$Cu^{2+}(aq) + 2\,e^- \longrightarrow Cu(s)$$

The electrode where oxidation occurs is called the **anode**; the electrode where reduction occurs is called the **cathode** (Fig. 6.3):

Anode reaction (oxidation):
$$\text{reduced species} \longrightarrow \text{oxidized species} + \nu\,e^-$$

Cathode reaction (reduction):
$$\text{oxidized species} + \nu\,e^- \longrightarrow \text{reduced species}$$

In the example just described, the zinc electrode is the anode (the site of oxidation) and the copper electrode is the cathode (the site of reduction).

In a galvanic cell, the cathode has a higher potential than the anode because the species undergoing reduction withdraws electrons from its electrode (the cathode), so leaving a relative positive charge on it (corresponding to a high potential). At the anode, oxidation results in the transfer of electrons to the electrode, so giving it a relative negative charge (corresponding to a low potential). In an electrolytic cell, the anode is also the location of oxidation (by definition), but now electrons must be withdrawn from the species in that compartment because oxidation does not occur spontaneously, and at the cathode there must be a supply of electrons to drive the reduction. Therefore, in an electrolytic cell the anode must be made relatively positive to the cathode so that electrons are effectively sucked out of the anode and pushed on to the cathode (Fig. 6.4).

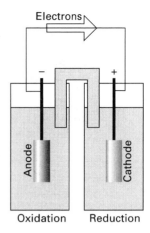

Fig. 6.3 The flow of electrons in the external circuit is from the anode of a galvanic cell, where they have been lost in the oxidation reaction, to the cathode, where they are used in the reduction reaction. Electrical neutrality is preserved in the electrolytes by the flow of cations and anions in opposite directions through the salt bridge.

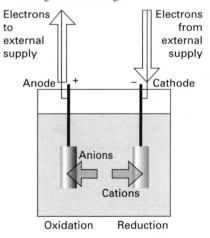

Fig. 6.4 The flow of electrons and ions in an electrolytic cell. An external supply forces electrons into the cathode, where they are used to bring about a reduction, and withdraws them from the anode, which results in an oxidation reaction at that electrode. Cations migrate towards the negatively charged cathode and anions migrate towards the positively charged anode. An electrolytic cell usually consists of a single compartment, but a number of industrial versions have two compartments.

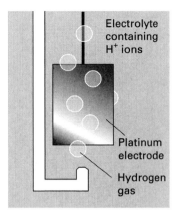

Electrolyte
containing
H^+ ions

Platinum
electrode

Hydrogen
gas

Fig. 6.5 The schematic structure of a hydrogen electrode (which is like other gas electrodes). Hydrogen is bubbled over a black platinum surface that is in contact with a solution containing hydrogen ions. The platinum, as well as acting as a source or sink for electrons speeds the electrode reaction because hydrogen attaches to (adsorbs on) the surface as atoms.

Varieties of electrodes

In a **gas electrode** (Fig. 6.5), a gas is in equilibrium with a solution of its ions in the presence of an inert metal. The inert metal, which is often platinum, acts as a source or sink of electrons, but takes no other part in the reaction (but may act as a catalyst for it). One important example is the **hydrogen electrode**, in which hydrogen is bubbled through an aqueous solution of hydrogen ions and the redox couple is H^+/H_2. This electrode is denoted

$$Pt\,|\,H_2(g)\,|\,H^+(aq)$$

where the vertical lines denote junctions between phases (in this electrode, the junction between the metal and the gas, and between the gas and the liquid containing its ions). Note that the electrode description runs in the order Red | Ox, which is opposite to the order in which the couple is denoted.

The hydrogen electrode may be either a cathode or an anode, depending on the other electrode in the cell and the direction in which the overall reaction is taking place. The half-reaction at the electrode when it is acting as a cathode (undergoing reduction) is

$$H^+(aq) + e^- \longrightarrow \tfrac{1}{2}H_2(g) \qquad Q = \frac{a(H_2)^{1/2}}{a(H^+)}$$

where $a(H_2) \approx p(H_2)/p^{\ominus}$. The replacement of the gas activity by its partial pressure is a good approximation for pressures close to atmospheric.

Example *Writing the half-reaction for a gas electrode*
Write the half-reaction and the reaction quotient for the reduction of oxygen to water in acidic solution.

Strategy
Write the chemical equation for the half-reaction, and then express the reaction quotient in terms of the activities and the corresponding stoichiometric numbers, with products in the numerator and reactants in the denominator. Pure (and nearly pure) solids and liquids do not appear in Q; nor does the electron.

Solution
The balanced equation for the reduction of O_2 in acidic solution is

$$O_2(g) + 4\,H^+(aq) + 4\,e^- \longrightarrow 2\,H_2O(l)$$

The reaction quotient for the half-reaction is therefore

$$Q = \frac{1}{a(O_2)a(H^+)^4} \approx \frac{p^{\ominus}}{p(O_2)a(H^+)^4}$$

Exercise E6.4

Write the half-reaction and the reaction quotient for a chlorine gas electrode.

[*Answer*: $Cl_2(g) + 2e^- \longrightarrow 2Cl^-(aq)$, $Q = a(Cl^-)^2/a(Cl_2)$]

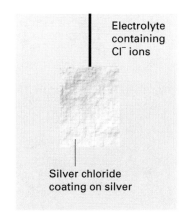

Silver chloride coating on silver

Electrolyte containing Cl^- ions

Fig. 6.6 The schematic structure of a silver chloride electrode (as an example of an insoluble-salt electrode). The electrode consists of metallic silver coated with a layer of silver chloride in contact with a solution containing Cl^- ions.

An **insoluble-salt electrode** consists of a metal M covered by a porous layer of insoluble salt MX, the whole being immersed in a solution containing X^- ions (Fig. 6.6). The electrode is denoted $M \mid MX \mid X^-$. An example is the silver–silver-chloride electrode, $Ag \mid AgCl \mid Cl^-$, for which the reduction half-reaction is

$$AgCl(s) + e^- \longrightarrow Ag(s) + Cl^-(aq) \qquad Q = a(Cl^-)$$

Note that the reaction quotient (and consequently, as we shall see later, the potential of the electrode) depends on the activity of chloride ions in the electrolyte solution.

Example *Writing the half-reaction for an insoluble-salt electrode*

Write the half-reaction and the reaction quotient for the lead–lead-sulfate electrode of the lead–acid battery, in which Pb(II), as lead(II) sulfate, is reduced to metallic lead in the presence of hydrogensulfate ions in the electrolyte.

Strategy

Begin by identifying the species that is reduced, and writing the half-reaction: balance it by using H_2O molecules if O atoms are required, hydrogen ions (because the solution is acidic) if H atoms are needed, and electrons for the charge. Then write the reaction quotient in terms of the stoichiometric numbers and activities of the species present, except for any pure solids and liquids and the electrons; products appear in the numerator, reactants in the denominator.

Solution

The electrode is

$$Pb \mid PbSO_4(s) \mid HSO_4^-(aq)$$

in which Pb(II) is reduced to metallic lead. The equation for the reduction half-reaction is therefore

$$PbSO_4(s) + H^+(aq) + 2e^- \longrightarrow Pb(s) + HSO_4^-(aq)$$

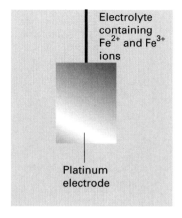

Fig. 6.7 The schematic structure of a redox electrode. The platinum metal acts as a source or sink for electrons required for the interconversion of (in this case) Fe^{2+} and Fe^{3+} ions in the surrounding solution.

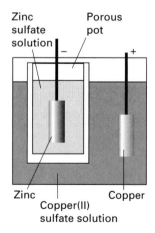

Fig. 6.8 A Daniell cell consists of copper in contact with copper(II) sulfate solution and zinc in contact with zinc sulfate solution; the two compartments are in contact through the porous pot that contains the zinc sulfate solution. The copper electrode is the cathode and the zinc electrode is the anode.

and so the reaction quotient is

$$Q = \frac{a(HSO_4^-)}{a(H^+)}$$

Exercise E6.5

Write the half-reaction and the reaction quotient for the 'calomel electrode', $Hg(l)\,|\,Hg_2Cl_2(s)\,|\,Cl^-(aq)$, in which mercury(I) chloride (calomel) is reduced to mercury metal in the presence of chloride ions. This electrode is a component of instruments used to measure pH, as explained later.

[*Answer*: $Hg_2Cl_2(s) + 2e^- \longrightarrow 2Hg(l) + 2Cl^-(aq)$, $Q = a(Cl^-)^2$]

The term **redox electrode** is normally reserved for an electrode in which the couple consists of two oxidation states of the same element (Fig. 6.7). An example is an electrode in which the couple is Fe^{3+}/Fe^{2+}. In general, the equilibrium is

$$Ox + \nu e^- \longrightarrow Red \qquad Q = \frac{a(Red)}{a(Ox)}$$

A redox electrode is denoted $M\,|\,Red, Ox$, where M is an inert metal (typically platinum) making electrical contact with the solution. The electrode corresponding to the Fe^{3+}/Fe^{2+} couple is therefore denoted $Pt\,|\,Fe^{2+}(aq), Fe^{3+}(aq)$ and the reduction half-reaction is

$$Fe^{3+}(aq) + e^- \longrightarrow Fe^{2+}(aq) \qquad Q = \frac{a(Fe^{2+})}{a(Fe^{3+})}$$

6.2 Varieties of cell

The simplest type of galvanic cell has a single electrolyte common to both electrodes (as in Fig. 6.1). In some cases it is necessary to immerse the electrodes in different electrolytes, as in the Daniell cell (Fig. 6.8), in which the redox couple at one electrode is Cu^{2+}/Cu and at the other is Zn^{2+}/Zn. In an **electrolyte concentration cell** (which would be constructed like the cell in Fig. 6.2), the electrode compartments are of identical composition except for the concentrations of the electrolytes. An electrolyte concentration cell is a model of a neuron, which consists of a cell membrane with different concentrations of Na^+ and K^+ ions on either side. In an **electrode concentration cell** the electrodes themselves have different concentrations, either because they are gas electrodes operating at different pressures or because they are amalgams (solutions in mercury) with different concentrations.

Liquid junction potentials

In a cell with two different electrolyte solutions in contact, as in the Daniell cell or in a cell in which the compartments have different concentrations of hydrochloric acid, there is an additional source of potential difference, the **liquid junction potential**, E_j, the potential difference across the interface of the two electrolytes. Electrolyte concentration cells always have a liquid junction; electrode concentration cells do not.

The contribution of the liquid junction to the potential can be decreased (to about 1 to 2 mV) by joining the electrolyte compartments through a salt bridge consisting of a saturated electrolyte solution (usually KCl) in agar jelly (as in Fig. 6.2). The reason for the success of the salt bridge is that the liquid junction potentials at either end are largely independent of the concentrations of the two more dilute solutions in the electrode compartments, and so nearly cancel.

Notation

In the notation for cells, an interface between phases is denoted by a vertical bar. For example, a cell in which the left-hand electrode is a hydrogen electrode and the right-hand electrode is a silver–silver-chloride electrode is denoted

$$Pt \mid H_2(g) \mid HCl(aq) \mid AgCl(s) \mid Ag$$

A double vertical line $\parallel$ denotes an interface for which it is assumed that the junction potential has been eliminated. Thus a cell in which the left-hand electrode, in an arrangement like that in Fig. 6.2, is zinc in contact with aqueous zinc sulfate and the right-hand electrode is copper in contact with aqueous copper(II) sulfate is denoted

$$Zn(s) \mid ZnSO_4(aq) \parallel CuSO_4(aq) \mid Cu(s)$$

Exercise E6.6

Give the notation for a cell in which the oxidation of NADH by oxygen could be studied.

[*Answer*:
$$Pt \mid NADH(aq), NAD^+(aq), H^+(aq) \parallel H_2O_2(aq), H^+(aq) \mid O_2(g) \mid Pt]$$

The cell reaction

The current produced by a galvanic cell arises from the spontaneous reaction taking place inside it. The **cell reaction** is the reaction in the cell written on the assumption that the right-hand electrode is the cathode, and hence that reduction is taking place in the right-hand compartment. Later we see how to predict if the right-hand electrode is in fact the cathode; if it is, then the cell reaction is spontaneous as written. If the left-hand electrode turns out to be the cathode, then the reverse of the cell reaction is spontaneous.

To write the cell reaction corresponding to the cell diagram, we first write the half-reactions at both electrodes as reductions, and then subtract the left-hand equation from the right-hand equation. Thus, in the cell

$$Zn(s)\,|\,ZnSO_4(aq)\,\|\,CuSO_4(aq)\,|\,Cu(s)$$

the two reduction half-reactions are

$$\text{Right (R)}: \qquad Cu^{2+}(aq) + 2\,e^- \longrightarrow Cu(s)$$
$$\text{Left (L)}: \qquad Zn^{2+}(aq) + 2\,e^- \longrightarrow Zn(s)$$

Hence, the equation for the cell reaction is the difference:

$$\text{Overall (R} - \text{L)}: \qquad Cu^{2+}(aq) + Zn(s) \longrightarrow Cu(s) + Zn^{2+}(aq) \qquad (1)$$

In other cases, it may be necessary to match the numbers of electrons in the two half-reactions by multiplying one of the equations through by a numerical factor: there should be no electrons showing in the overall equation.

Exercise E6.7

Write the chemical equation for the cell in Exercise E6.6.

[*Answer*: $NADH(aq) + O_2(g) + H^+(aq) \longrightarrow NAD^+(aq) + H_2O_2(aq)$]

The cell potential

A galvanic cell operates by the oxidation half-reaction depositing electrons in the anode which are then used by the reduction half-reaction at the cathode. So long as the overall reaction is not at equilibrium, the oxidation half-reaction effectively pushes the electrons into the external circuit, and the reduction half-reaction effectively pulls them out of it. If the cell reaction is not at equilibrium, the cell can do electrical work as the reaction drives electrons through an external circuit. The work that a given transfer of electrons can do depends on the potential difference between the two electrodes. This potential difference is called the **cell potential** and is measured in volts, V. When the cell potential is large (for instance, 2 V), a given number of electrons travelling between the electrodes can do a large amount of electrical work; when the cell potential is small (such as 2 mV), the same number of electrons can do only a small amount of work. A cell in which the reaction is at equilibrium can do *no* work, and its potential is zero.

According to the discussion in Section 3.3, we know that the maximum electrical work, w', that a system (in this context, the cell) can do is given by the value of ΔG, and in particular that

$$w'(\text{maximum}) = \Delta G \qquad \text{at constant temperature and pressure}$$

However, maximum work is achieved only when a process occurs reversibly. In the present context, reversibility means that the cell should be connected to an external source of potential difference (another cell or a stable power supply) that opposes and exactly matches the potential generated by the cell of interest. Then an infinitesimal change of the external potential will allow the reaction to proceed in its spontaneous direction and an opposite infinitesimal change will drive the reaction in its reverse direction. (As explained in Section 2.2, the reversal of a process by an infinitesimal change in the external conditions is the criterion of thermodynamic reversibility.) The potential difference measured when a cell is balanced against an external source of potential is called the **zero-current cell potential**, E (Fig. 6.9). An older name for this quantity (which is still widely used), is the 'electromotive force', or emf, of the cell.

The zero-current cell potential is related to the reaction Gibbs energy, $\Delta_r G$, at a particular composition of the cell compartments by

$$\Delta_r G = -\nu F E \qquad (2)$$

where ν is the stoichiometric number of the electrons in the half-reactions contributing to the overall cell reaction. Because the overall equation must balance, the stoichiometric numbers of the electrons in the two half-reactions must be the same when they are combined, so there is no ambiguity in the value of ν. For example, in the Daniell cell, for which the reaction is given in eqn (1), $\nu = 2$. The constant F is Faraday's constant, and is the magnitude of electric charge per mole of electrons:

$$F = eN_A = 96.485 \text{ kC mol}^{-1}$$

The symbol C denotes the coulomb, the unit of electric charge. For future reference it will be useful to note that

$$1 \text{ C} \times 1 \text{ V} = 1 \text{ J}$$

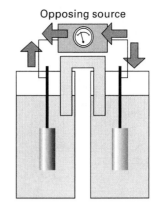

Fig. 6.9 The zero-current cell potential is measured by balancing the cell against an external potential that opposes the reaction in the cell. When there is no current flow, the external potential difference is equal to the cell potential.

Justification

When the reaction takes place, νN_A electrons are transferred from the reducing agent to the oxidizing agent per mole of reaction events, so the charge transferred between the electrodes is $\nu N_A \times (-e)$, or $-\nu F$. The electrical work w' done when this charge travels from the anode to the cathode is equal to the product of the charge and the potential difference E:

$$w' = -\nu F \times E$$

Finally, provided the work is done reversibly at constant temperature and pressure, we can equate this electrical work to the reaction Gibbs energy:

$$-\nu F E = \Delta_r G$$

Equation (2) means that, if we know the reaction Gibbs energy at a specified composition, then we can state the zero-current cell potential at that composition, and vice versa. It follows that we now have an electrical method for measuring a reaction Gibbs energy at any composition of the reaction system, because a measurement of E can be converted into a value of $\Delta_r G$. A point to note is that the negative sign in eqn (2) means that a negative reaction Gibbs energy, which corresponds to a spontaneous cell reaction, results in a positive zero-current cell potential.

It is interesting to use eqn (2) to estimate the cell potential for a typical reaction for which $\Delta_r G \approx -10^2\,\mathrm{kJ\,mol^{-1}}$ and $\nu = 1$:

$$E = -\frac{\Delta_r G}{\nu F} = -\frac{(-1 \times 10^5\,\mathrm{J\,mol^{-1}})}{1 \times (9.6485 \times 10^4\,\mathrm{C\,mol^{-1}})} \approx 1\,\mathrm{V}$$

Most electrochemical cells bought commercially are indeed rated at between 1 and 2 V.

The Nernst equation

We saw in Chapter 3 that the reaction Gibbs energy varies with the composition of the reaction mixture, and that

$$\Delta_r G = \Delta_r G^{\ominus} + RT \ln Q \tag{3}$$

In this expression (which is eqn (3.23)), $\Delta_r G^{\ominus}$ is the standard reaction Gibbs energy and Q is the reaction quotient for the cell reaction. We can express the reaction Gibbs energy in terms of the cell potential by dividing through by $-\nu F$ and using eqn (2), which gives

$$E = E^{\ominus} - \frac{RT}{\nu F} \ln Q \tag{4}$$

In this expression $E^{\ominus}$ is the **standard cell potential**:

$$E^{\ominus} = -\frac{\Delta_r G^{\ominus}}{\nu F} \tag{5}$$

Equation (4), which expresses the cell potential in terms of the composition, is called the **Nernst equation**. Because $RT/F = 25.7\,\text{mV}$ at $25\,°\text{C}$, a practical form of the Nernst equation at this temperature is

$$E \approx E^{\ominus} - \frac{25.7\,\text{mV}}{\nu}\ln Q \qquad (6)$$

Hence, for a reaction in which $\nu = 1$, if Q is increased by a factor of 10, then the cell potential becomes more negative by $59.2\,\text{mV}$:

$$E' - E = -(25.7\,\text{mV}) \times (\ln Q' - \ln Q) = -(25.7\,\text{mV}) \times \ln\frac{Q'}{Q}$$
$$= -(25.7\,\text{mV}) \times \ln 10 = -59.2\,\text{mV}$$

Cells at equilibrium

A special case of the Nernst equation has great importance in chemistry. Suppose the reaction has reached equilibrium; then $Q = K$, where K is the equilibrium constant of the cell reaction. However, we have remarked that a chemical reaction at equilibrium cannot do work, and hence a cell in which the reaction is at equilibrium generates zero potential difference between the electrodes. Therefore, setting $E = 0$ and $Q = K$ in the Nernst equation gives

$$\ln K = \frac{\nu F E^{\ominus}}{RT} \qquad (7)$$

This very important equation lets us predict equilibrium constants from standard cell potentials. For example, because the standard potential of the Daniell cell is $+1.10\,\text{V}$, the equilibrium constant for the cell reaction (eqn (1)) is

$$\ln K = \frac{2 \times (9.6485 \times 10^4\,\text{C mol}^{-1}) \times (1.10\,\text{V})}{(8.3145\,\text{J K}^{-1}\,\text{mol}^{-1}) \times (298.15\,\text{K})} = 85.6$$

and therefore $K = 1.5 \times 10^{37}$. Hence, the displacement of copper by zinc goes virtually to completion in the sense that the ratio of concentrations of Zn^{2+} ions to Cu^{2+} ions at equilibrium is about 10^{37}. Note that if $E^{\ominus}$ is positive, then K is greater than 1 and at equilibrium the cell reaction lies in favour of products. The opposite is true if $E^{\ominus}$ is negative, for then K is less than 1 and the reactants are favoured at equilibrium.

Concentration cells

We can use the same procedure to derive an expression for the potential of an electrolyte concentration cell. Consider the cell

$$M\,|\,M^+(\text{aq, L})\,\|\,M^+(\text{aq, R})\,|\,M$$

where the solutions L and R have different concentrations; we shall use this cell as a model of the potential difference across a neuron cell membrane. The cell reaction is

$$M^+(aq, R) \longrightarrow M^+(aq, L) \qquad Q = \frac{a_L}{a_R} \qquad \nu = 1$$

The standard potential of an electrolyte concentration cell is zero, because the cell cannot drive a current through a circuit when the two electrode compartments are identical, which is the case when they both have their standard concentrations. Therefore, because $E^\ominus = 0$ and $\nu = 1$, the cell potential is

$$E = -\frac{RT}{F} \ln \frac{a_L}{a_R} \tag{8}$$

If R is the more concentrated solution, then $E > 0$. Physically, a positive potential arises because positive ions tend to be reduced, so tending to withdraw electrons from the external circuit, and this process is dominant in the right-hand electrode compartment.

As we have indicated, one important example of a membrane system that resembles a concentration cell is a biological cell wall, which is more permeable to K^+ ions than to either Na^+ or Cl^- ions. The concentration of K^+ ions inside the cell is about 20 to 30 times that on the outside, and is maintained at that level by a specific pumping operation fuelled by ATP and governed by enzymes. If the system is approximately at equilibrium, the potential difference between the two sides is predicted to be

$$E \approx -(25.7\,\text{mV}) \times \ln\tfrac{1}{20} = 77\,\text{mV}$$

which is broadly correct.

The potential difference across a cell membrane plays a particularly interesting role in the transmission of nerve impulses. When a neuron is inactive, there is a high concentration of K^+ ions inside the cell and a high concentration of Na^+ ions outside. The potential difference across the cell wall is about $-70\,\text{mV}$. When the cell wall is subjected to a pulse of about $20\,\text{mV}$, the structure of the membrane adjusts and it becomes permeable to Na^+ ions. As a result, there is a decrease in membrane potential as the Na^+ ions flood into the interior of the cell. The change in potential difference triggers the adjacent part of the cell membrane, and the pulse of collapsing potential passes along the nerve. Behind the pulse the sodium and potassium pumps restore the concentration difference ready for the next pulse.

6.3 Reduction potentials

Each electrode in a galvanic cell makes a characteristic contribution to the overall cell potential. Although it is not possible to measure the contribution of a single electrode, one electrode can be assigned a value zero and the others assigned values on that basis. The specially selected electrode is the **standard hydrogen electrode** (SHE):

$$Pt \,|\, H_2(g) \,|\, H^+(aq) \qquad E^{\ominus} = 0 \text{ at all temperatures}$$

The **standard potential**, $E^{\ominus}(Ox/Red)$, of a couple Ox/Red is then measured by constructing a cell in which the couple of interest forms the right-hand electrode and the standard hydrogen electrode is used as the left-hand electrode. For example, the standard potential of the Ag^+/Ag couple is the standard potential of the following cell:

$$Pt \,|\, H_2(g) \,|\, H^+(aq) \| Ag^+(aq) \,|\, Ag(s)$$

and is $+0.80\,V$. Similarly, the standard potential of the $AgCl/Ag, Cl^-$ couple is the standard potential of the cell

$$Pt \,|\, H_2(g) \,|\, H^+(aq) \| Cl^-(aq) \,|\, AgCl(s) \,|\, Ag(s)$$

and is $+0.22\,V$. Table 6.1 lists a selection of standard potentials; a longer list will be found in Appendix 2.

The standard potential of a cell in terms of reduction potentials

We can predict the standard potential of a cell formed from any two electrodes by taking the difference of their standard potentials:

$$E^{\ominus} = E_R^{\ominus} - E_L^{\ominus} \qquad (9)$$

where $E_R^{\ominus}$ is the standard potential of the right-hand electrode and $E_L^{\ominus}$ is that of the left. This conclusion follows from the fact that a cell such as

$$Ag(s) \,|\, Ag^+(aq) \| Cl^-(aq) \,|\, AgCl(s) \,|\, Ag(s)$$

is equivalent to two cells joined back-to-back:

$$Ag(s) \,|\, Ag^+(aq) \| H^+(aq) \,|\, H_2(g) \,|\, Pt\text{---}\!\!\rceil$$
$$\lceil\!\!\text{---}Pt \,|\, H_2(g) \,|\, H^+(aq) \| Cl^-(aq) \,|\, AgCl(s) \,|\, Ag(s)$$

The overall potential of this composite cell, and therefore of the cell of interest, is

$$E^{\ominus} = E^{\ominus}(AgCl/Ag, Cl^-) - E^{\ominus}(Ag^+/Ag) = -0.58\,V$$

Because $\Delta_r G^{\ominus} = -\nu F E$, it follows that if $E^{\ominus} < 0$ (as it is in this example), then the corresponding cell reaction is spontaneous in the reverse direction (in the sense that $K < 1$).

Table 6.1 Standard potentials at 25°C

Reduction half-reaction			$E^{\ominus}/V$
Oxidizing agent		Reducing agent	
Strongly oxidizing			
F_2	$+2\,e^-$	$\rightarrow 2\,F^-$	$+2.87$
$S_2O_8^{2-}$	$+2\,e^-$	$\rightarrow 2\,SO_4^{2-}$	$+2.05$
Au^+	$+e^-$	$\rightarrow Au$	$+1.69$
Pb^{4+}	$+2\,e^-$	$\rightarrow Pb^{2+}$	$+1.67$
Ce^{4+}	$+e^-$	$\rightarrow Ce^{3+}$	$+1.61$
$MnO_4^- + 8\,H^+$	$+5\,e^-$	$\rightarrow Mn^{2+} + 4\,H_2O$	$+1.51$
Cl_2	$+2\,e^-$	$\rightarrow 2\,Cl^-$	$+1.36$
$Cr_2O_7^{2-} + 14\,H^+$	$+6\,e^-$	$\rightarrow 2\,Cr^{3+} + 7\,H_2O$	$+1.33$
$O_2 + 4\,H^+$	$+4\,e^-$	$\rightarrow 2\,H_2O$	$+1.23$, +0.81 at pH = 7
Br_2	$+2\,e^-$	$\rightarrow 2\,Br^-$	$+1.09$
Ag^+	$+e^-$	$\rightarrow Ag$	$+0.80$
Hg_2^{2+}	$+2\,e^-$	$\rightarrow 2\,Hg$	$+0.79$
Fe^{3+}	$+e^-$	$\rightarrow Fe^{2+}$	$+0.77$
I_2	$+2\,e^-$	$\rightarrow 2\,I^-$	$+0.54$
$O_2 + 2\,H_2O$	$+4\,e^-$	$\rightarrow 4\,OH^-$	$+0.40$, +0.81 at pH = 7
Cu^{2+}	$+2\,e^-$	$\rightarrow Cu$	$+0.34$
$AgCl$	$+e^-$	$\rightarrow Ag + Cl^-$	$+0.22$
$2\,H^+$	$+2\,e^-$	$\rightarrow H_2$	0, by definition
Fe^{3+}	$+3\,e^-$	$\rightarrow Fe$	-0.04
$O_2 + H_2O$	$+2\,e^-$	$\rightarrow HO_2^- + OH^-$	-0.08
Pb^{2+}	$+2\,e^-$	$\rightarrow Pb$	-0.13
Sn^{2+}	$+2\,e^-$	$\rightarrow Sn$	-0.14
Fe^{2+}	$+2\,e^-$	$\rightarrow Fe$	-0.44
Zn^{2+}	$+2\,e^-$	$\rightarrow Zn$	-0.76
$2\,H_2O$	$+2\,e^-$	$\rightarrow H_2 + 2\,OH^-$	-0.83, −0.42 at pH = 7
Al^{3+}	$+3\,e^-$	$\rightarrow Al$	-1.66
Mg^{2+}	$+2\,e^-$	$\rightarrow Mg$	-2.36
Na^+	$+e^-$	$\rightarrow Na$	-2.71
Ca^{2+}	$+2\,e^-$	$\rightarrow Ca$	-2.87
K^+	$+e^-$	$\rightarrow K$	-2.93
Li^+	$+e^-$	$\rightarrow Li$	-3.05
		Strongly reducing	

For a more extensive table, see Appendix 2.

Example *Identifying the spontaneous direction of a reaction*
One of the reactions important in corrosion in an acidic environment is

$$Fe(s) + 2\,H^+(aq) + \tfrac{1}{2}O_2(g) \longrightarrow Fe^{2+}(aq) + H_2O(l)$$

Does the equilibrium constant favour the formation of $Fe^{2+}(aq)$?

Strategy
We need to decide whether $E^{\ominus}$ for the reaction is positive or negative: the reaction has $K > 1$ if $E^{\ominus} > 0$. To do so, we break the overall reaction into two reduction half-reactions (using the technique described in Section 6.1), look up the standard potentials of the two couples, and take their difference.

Solution
The two reduction half-reactions are

$$\text{Right:} \quad 2\,H^+(aq) + \tfrac{1}{2}O_2(g) + 2\,e^- \longrightarrow H_2O(l)$$
$$E^{\ominus}(H^+, O_2, H_2O) = +1.23\,V$$
$$\text{Left:} \quad Fe^{2+}(aq) + 2\,e^- \longrightarrow Fe(s) \quad E^{\ominus}(Fe^{2+}, Fe) = -0.44\,V$$

The difference Right − Left is

$$Fe(s) + 2\,H^+(aq) + \tfrac{1}{2}O_2(g) \longrightarrow Fe^{2+}(aq) + H_2O(l) \quad E^{\ominus} = +1.67\,V$$

Therefore, because $E^{\ominus} > 0$, it follows that $K > 1$, favouring products.

Exercise E6.8

Is the equilibrium constant for the displacement of copper by zinc greater or smaller than 1?

[*Answer*: greater]

Example *Calculating an equilibrium constant 1*
Calculate the equilibrium constant for the disproportionation reaction

$$2\,Cu^+(aq) \rightleftharpoons Cu(s) + Cu^{2+}(aq)$$

at 298 K.

Strategy

The aim is to find the values of $E^{\ominus}$ and ν corresponding to the reaction, for then we can use eqn (7). To do so, we express the equation as the difference of two reduction half-reactions. The stoichiometric number of the electron in these matching half-reactions is the value of ν we require. We can then look up the standard potentials for the couples corresponding to the half-reactions, and calculate their difference to find $E^{\ominus}$. In calculations of this kind a useful value is

$$\frac{RT}{F} = 25.69 \, \text{mV}$$

Solution

The two half-reactions are

Right: $\quad Cu^+(aq) + e^- \longrightarrow Cu(aq) \quad E^{\ominus}(Cu^+, Cu) = +0.52 \, V$

Left: $\quad Cu^{2+}(aq) + e^- \longrightarrow Cu^+(aq) \quad E^{\ominus}(Cu^{2+}, Cu^+) = +0.15 \, V$

The value of ν is 1, and the standard potential for the overall reaction is

$$E^{\ominus} = (0.52 \, V) - (0.15 \, V) = +0.37 \, V$$

It then follows from eqn (7) that

$$\ln K = \frac{0.37 \, V}{25.69 \times 10^{-3} \, V} = \frac{0.37}{25.69 \times 10^{-3}}$$

Therefore, $K = 1.8 \times 10^6$. The equilibrium lies strongly towards the right of the reaction as written, and so Cu^+ disproportionates almost totally in solution.

Exercise E6.9

Calculate the equilibrium constant for the reaction

$$Sn^{2+}(aq) + Pb(s) \rightleftharpoons Sn(s) + Pb^{2+}(aq)$$

at 298 K.

[*Answer*: 0.46]

Example *Calculating an equilibrium constant 2*

The reduced and oxidized forms of riboflavin form a couple with $E^{\ominus} = -0.21 \, V$ in a solution in which pH = 7, and the acetate/acetaldehyde couple has $E^{\ominus} = -0.60 \, V$ under the same conditions. What is the equilibrium constant for the reduction of riboflavin

(Rib) by acetaldehyde in neutral solution at $25\,^{\circ}$C? The reaction can be symbolized

$$RibO(aq) + CH_3CHO(aq) \rightleftharpoons Rib(aq) + CH_3COOH(aq)$$

where RibO is the oxidized form of riboflavin and Rib is the reduced form.

Strategy

The procedure is the same as before: we express the overall reaction as the difference of two half-reactions, and identify the value of ν required to match them. The standard potential for the reaction is then the difference of the two standard potentials for the half-reactions. Note that pH $= 7$ does not correspond to a true standard state. However, because H^+ does not appear in the overall equation, the effect of the pH on the individual 'standard' potentials cancels and the $E^{\ominus}$ may be treated as true standard potentials. This point is treated in more detail in the following section on biological standard states.

Solution

The two reduction half-reactions are

Right : $\quad RibO(aq) + 2\,H^+(aq) + 2\,e^- \longrightarrow Rib(aq) + H_2O(l)$
$$E^{\ominus} = -0.21\,V$$

Left : $\quad CH_3COOH(aq) + 2\,H^+(aq) + 2\,e^- \longrightarrow CH_3CHO(aq) + H_2O(l)$
$$E^{\ominus} = -0.60\,V$$

and their difference is the redox reaction required. Note that $\nu = 2$. The corresponding standard potential is

$$E^{\ominus} = (-0.21\,V) - (-0.60\,V) = +0.39\,V$$

It follows that

$$\ln K = \frac{2 \times (0.39\,V)}{25.69 \times 10^{-3}\,V} = \frac{2 \times 0.39}{25.69 \times 10^{-3}}$$

from which we find $K = 1.5 \times 10^{13}$. We conclude that riboflavin can be reduced by acetaldehyde in neutral solution; however, there may be mechanistic reasons (the energy required to break covalent bonds) that make the reduction too slow to be feasible in practice.

Exercise E6.10

What is the equilibrium constant for the reduction of riboflavin with rubredoxin in the reaction

$$riboflavin(ox) + rubredoxin(red) \rightleftharpoons riboflavin(red) + rubredoxin(ox)$$

given that at $pH = 7$, the reduction potential for rubredoxin is -0.06 V?

[*Answer*: 8.5×10^{-6}; the reactants are favoured]

The variation of potential with pH

The half-reactions of many redox couples involve hydrogen ions. For example, the reduction of fumaric acid, HOOCCHCHCOOH, to succinic acid, $HOOCCH_2CH_2COOH$, is

$$HOOCCHCHCOOH(aq) + 2\,H^+(aq) + 2\,e^-$$
$$\longrightarrow HOOCCH_2CH_2COOH(aq)$$

Half-reactions of this kind have potentials that depend on the pH of the medium. In this example, where the hydrogen ions occur as reactants, an increase in pH (corresponding to a decrease in hydrogen ion activity) favours the formation of reactants, so the fumaric acid has a lower thermodynamic tendency to become reduced: we expect, therefore, that the potential of the fumaric/succinic acid couple should decrease as the pH is increased.

The quantitative variation of reduction potential with pH for a reaction can be established quite readily by using the Nernst equation for the half-reaction and noting that

$$\ln a(H^+) = (\ln 10) \times \log a(H^+) = -\ln 10 \times pH$$

with $\ln 10 = 2.303 \cdots$. (We have used the mathematical relation that $\ln x = \ln 10 \times \log x$, where log is the logarithm to base 10.) If we suppose that the fumaric acid and succinic acid have their standard concentrations, with only the hydrogen ion concentration variable, then the reduction potential for the fumaric/succinic redox couple is

$$E^{\ominus\,\prime} = E^{\ominus} - \frac{RT}{2F} \ln Q \qquad Q = \frac{1}{a(H^+)^2}$$

which is easily rearranged into

$$E^{\ominus\,\prime} = E^{\ominus} - \frac{RT \ln 10}{F} pH \tag{10}$$

At 25 °C,

$$E^{\ominus\,\prime} = E^{\ominus} - (59.2\,\text{mV}) \times pH \tag{11}$$

We see that each increase of 1 unit in pH decreases the reduction potential by 59.2 mV, which is in agreement with the remark above, that the reduction of fumaric acid is discouraged by an increase in pH.

Arguments of this type can be used to convert standard potentials to **biological standard potentials**, $E^\oplus$, which correspond to neutral solution ($pH = 7$). If the hydrogen ions appear as reactants in the reduction half-reaction, then the potential is decreased below its standard value (for the fumaric/succinic couple, by $7 \times 59.2\,mV = 414\,mV$, or about $0.4\,V$). If the hydrogen ions appear as products, then the biological standard potential is higher than the thermodynamic standard potential. The precise change depends on the number of electrons and protons participating in the half-reaction.

Example *Converting a standard potential to a biological standard value*

Estimate the biological standard potential of the $NAD^+/NADH$ couple at $25\,°C$, where NAD^+ is the oxidized form of nicotinamide adenine dinucleotide and NADH is the reduced form of this species that is involved in electron transfer in biological systems. The reduction half-reaction is

$$NAD^+(aq) + H^+(aq) + 2\,e^- \longrightarrow NADH(aq) \qquad E^\ominus = -0.11\,V$$

Strategy

We write the Nernst equation for the potential, and express the reaction quotient in terms of the activities of the species. All species except H^+ are in their standard states, so their activities are all equal to 1. The remaining task is to express the hydrogen ion activity in terms of the pH, exactly as was done in the text, and set $pH = 7$.

Solution

The Nernst equation for the half-reaction with $\nu = 2$ is

$$E^{\ominus\prime} = E^\ominus - \frac{RT}{2F}\ln Q \qquad Q = \frac{a(NADH)}{a(H^+)a(NAD^+)} = \frac{1}{a(H^+)}$$

This expression rearranges to

$$E^{\ominus\prime} = E^\ominus + \frac{RT}{2F}\ln a(H^+) = E^\ominus - \frac{RT\ln 10}{2F}pH$$
$$= E^\ominus - (29.59\,mV) \times pH$$

On substitution of the data, the biological standard potential (at $pH = 7$) is

$$E^\oplus = (-0.11\,V) - (29.59 \times 10^{-3}\,V) \times 7 = -0.32\,V$$

Exercise E6.11

Calculate the biological standard potential of the half-reaction $O_2(g) + 4\,H^+(aq) + 4\,e^- \longrightarrow 2\,H_2O(l)$ at $25\,^\circ C$ given its value $+1.23\,V$ under thermodynamic standard conditions.

[*Answer:* $+0.81\,V$]

The hydrogen electrode and pH

The potential of a hydrogen electrode is directly proportional to the pH of the solution. For example, for the cell

$$Hg(l)\,|\,Hg_2Cl_2(s)\,|\,Cl^-(aq)\,\|\,H^+(aq)\,|\,H_2(g)\,|\,Pt$$

in which the cell reaction is

$$Hg_2Cl_2(s) + H_2(g) \longrightarrow 2\,Hg(l) + 2\,Cl^-(aq) + 2\,H^+(aq)$$

the Nernst equation gives

$$E = E^{\ominus} - \frac{RT}{2F}\ln Q \qquad Q = \frac{a(H_2)a(Cl^-)^2}{a(H^+)^2}$$

We shall suppose that the hydrogen pressure has its standard value of $p^{\ominus}$ (small variations in pressure make very little change to the cell potential). We also note that because the activity of the Cl^- ions is constant (it depends on the composition of the calomel electrode, which is independent of that of the hydrogen electrode), we find that

$$\begin{aligned} E &= E^{\ominus} - \frac{RT}{F}\ln a(Cl^-) - \frac{RT}{2F}\ln\frac{1}{a(H^+)^2} \\ &= E' + \frac{RT}{F}\ln a(H^+) \end{aligned}$$

where E' is a constant. It follows that

$$E = E' - \frac{RT\ln 10}{F} \times pH \tag{12}$$

and the pH of a solution can be measured by determining the potential of a cell in which a hydrogen electrode is one component.

Exercise E6.12

What range should a voltmeter have (in volts) to display changes of pH from 1 to 14 at 25 °C if it is arranged to give a reading of 0 when pH = 7?

[*Answer*: from −0.41 V to +0.36 V, a range of 0.77 V]

In practice, indirect methods are much more convenient to use than one based on the standard hydrogen electrode, and the hydrogen electrode is replaced by a **glass electrode** (Fig. 6.10). This electrode is sensitive to hydrogen ion activity, and has a potential proportional to pH. It is filled with a phosphate buffer containing Cl^- ions, and conveniently has $E \approx 0$ when the external medium is at pH = 7. The glass electrode is much more convenient to handle than the gas electrode itself, and can be calibrated using solutions of known pH (for example, one of the buffer solutions described in Section 5.6).

Finally, it should be noted that we now have a method for measuring the pK_a of an acid electrically. As explained in Section 5.6, the pH of a solution containing equal amounts of the acid and its conjugate base is pH = pK_a; but we now know how to determine pH electrochemically, and hence can determine pK_a in the same way.

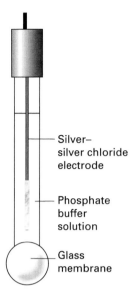

Fig. 6.10 A glass electrode has a potential that varies with the hydrogen ion concentration in the medium in which it is immersed. It consists of a thin glass membrane containing an electrolyte and a silver chloride electrode. The electrode is used in conjunction with a calomel (Hg_2Cl_2) electrode that makes contact with the test solution through a salt bridge.

Silver–silver chloride electrode

Phosphate buffer solution

Glass membrane

Applications of reduction potentials

The measurement of zero-current cell potential is a convenient source of data on the Gibbs energies, enthalpies, and entropies of reactions. In practice the standard values (and the biological standard values) of these quantities are the ones normally determined.

6.4 The electrochemical series

We have seen that a cell reaction is spontaneous as written if it has a positive cell potential, $E^\ominus > 0$. We have also seen that $E^\ominus$ may be written as the difference of the standard potentials of the redox couples in the right and left electrodes:

$$E^\ominus = E_R^\ominus - E_L^\ominus$$

The cell reaction is therefore spontaneous as written if $E_R^\ominus > E_L^\ominus$. Because the reduced species in the left hand electrode compartment (the anode, the site of oxidation) reduces the oxidized species in the right hand electrode compartment (the cathode, the site of reduction) in the cell reaction, we can conclude that

A species with a low standard potential has a thermodynamic tendency to reduce a species with a high standard potential.

More briefly: *low reduces high* and, equivalently, *high oxidizes low*. For example,

$$E^{\ominus}(Zn^{2+}, Zn) = -0.76\,V < E^{\ominus}(Cu^{2+}, Cu) = +0.34\,V$$

and Zn(s) has a thermodynamic tendency to reduce Cu^{2+}(aq) under standard conditions. Hence, the reaction

$$Zn(s) + CuSO_4(aq) \rightleftharpoons ZnSO_4(aq) + Cu(s)$$

can be expected to have $K > 1$ (in fact, as we have seen, $K = 1.5 \times 10^{37}$ at 298 K).

Exercise E6.13

Does acidified dichromate $(Cr_2O_7^{2-})$ have a thermodynamic tendency to oxidize mercury to mercury(I)?

[*Answer*: yes]

Table 6.2 shows a part of the **activity series** of the metals, the redox couples of the metals arranged in the order of their reducing power. In the activity series:

The reduced member of a couple higher in the series (as printed on the page) can reduce the oxidized member of couples lower in the series.

This is a qualitative conclusion. The quantitative value of K is then obtained by doing the calculations we have introduced previously. Appendix 2 lists a larger selection of standard potentials of a wider variety of redox couples.

Example *Using the activity series*
Can zinc displace magnesium from aqueous solutions at 298 K?

Strategy
Displacement of a metal corresponds to the reduction of its cations. So, to decide whether one metal can displace another, we have to note whether the displaced metal lies below the displacing metal in the activity series as it is printed in Table 6.2.

Table 6.2 The metal activity series

Element	Reduced form	Oxidized form
Most strongly reducing		
Potassium	K	K^+
Calcium	Ca	Ca^{2+}
Sodium	Na	Na^+
Magnesium	Mg	Mg^{2+}
Aluminium	Al	Al^{3+}
Zinc	Zn	Zn^{2+}
Chromium	Cr	Cr^{2+}
Iron	Fe	Fe^{2+}
Nickel	Ni	Ni^{2+}
Tin	Sn	Sn^{2+}
Lead	Pb	Pb^{2+}
(Hydrogen)	H_2	H^+
Copper	Cu	Cu^{2+}
Mercury	Hg	Hg^{2+}
Silver	Ag	Ag^+
Platinum	Pt	Pt^{2+}
Gold	Au	Au^+
Least strongly reducing		

Solution

We see that magnesium lies above zinc in the activity series, so zinc cannot displace magnesium.

Exercise E6.14

Can lead displace (a) iron(II) ions, (b) copper(II) ions from solution at 298 K?

[*Answer*: (a) no, (b) yes]

6.5 Thermodynamic functions from cell potential measurements

The standard cell potential is related to the standard reaction Gibbs energy by eqn (5):

$$\Delta_r G^\ominus = -\nu F E^\ominus \tag{13}$$

Therefore, by measuring the standard potential of a cell driven by the reaction of interest we can obtain the standard reaction Gibbs energy. If we were interested in the biological standard state, then we would use the same expression, but with the standard potential at pH $= 7$:

$$\Delta_r G^{\oplus} = -\nu F E^{\oplus} \tag{14}$$

where $E^{\oplus}$ is the biological standard potential.

Exercise E6.15

Estimate the standard reaction Gibbs energy of $Ag^+(aq) + \frac{1}{2} H_2(g) \longrightarrow H^+(aq) + Ag(s)$ given that the standard potential of the cell $H_2 \,|\, H^+(aq) \| Ag^+(aq) \,|\, Ag$ is $E^{\ominus} = +0.7996\,V$.

[*Answer*: $-77.10\,kJ\,mol^{-1}$]

The relation between the standard potential of a cell and the standard reaction Gibbs energy is a convenient route for the evaluation of a standard potential from two others. For example, given that the standard potentials of the Cu^{2+}/Cu and Cu^+/Cu couples are $E^{\ominus}(Cu^{2+}, Cu) = +0.340\,V$ and $E^{\ominus}(Cu^+, Cu) = +0.522\,V$, we can evaluate $E^{\ominus}(Cu^{2+}, Cu^+)$ by converting the $E^{\ominus}$ values to $\Delta_r G^{\ominus}$ values by using eqn (13), adding them appropriately, and converting the overall $\Delta_r G^{\ominus}$ so obtained to the required $E^{\ominus}$ by using eqn (13) again. Because the Fs cancel at the end of the calculation, we carry them through. The electrode reactions are as follows:

(a) $Cu^{2+}(aq) + 2\,e^- \longrightarrow Cu(s) \qquad E^{\ominus} = +0.340\,V$
$\Delta_r G^{\ominus}(a) = -2F \times (0.340\,V) = (-0.680\,V) \times F$

(b) $Cu^+(aq) + e^- \longrightarrow Cu(s) \qquad E^{\ominus} = +0.522\,V$
$\Delta_r G^{\ominus}(b) = -F \times (0.522\,V) = (-0.522\,V) \times F$

The required reaction is

(c) $Cu^{2+}(aq) + e^- \longrightarrow Cu^+(aq) \qquad \Delta_r G^{\ominus}(c) = -FE^{\ominus}$

Because (c) = (a) − (b), it follows that

$$\Delta_r G^{\ominus}(c) = \Delta_r G^{\ominus}(a) - \Delta_r G^{\ominus}(b)$$

Therefore, from eqn (13),

$$E^{\ominus}(c) = -\frac{(-0.680\,V)F - (-0.522\,V)F}{F} = +0.158\,V$$

Note that we cannot combine the $E^\ominus$ values directly because they are not extensive properties, and we must always work via the reaction Gibbs energy.

The entropy of the cell reaction can be obtained from the change in the cell potential with temperature by using the relation

$$\Delta_r S^\ominus = \nu F \times \left(\frac{E^\ominus(T') - E^\ominus(T)}{T' - T} \right) \qquad (15)$$

where $E^\ominus(T')$ is the standard cell potential at a temperature T' and $E^\ominus(T)$ is its value at a temperature T. Hence, we now have an electrochemical technique for obtaining standard reaction entropies of reactions that can be studied in a galvanic cell.

Justification

The thermodynamic relation on which eqn (15) is based is

Change in $G = -S \times$ change in T (at constant pressure)

or, more formally,

$dG = -SdT$ (at constant pressure)

This equation is derived from the definition $G = H - TS$ by changing the temperature infinitesimally. Because this equation applies to both the reactants and the products, it follows that

$$d(\Delta_r G^\ominus) = -\Delta_r S^\ominus dT$$

Substitution of $\Delta_r G^\ominus = -\nu F E^\ominus$ then gives

$$\nu F dE^\ominus = \Delta_r S^\ominus dT$$

This equation is exact, but it applies only to infinitesimal changes in the temperature. The equation may be integrated if we suppose that the reaction entropy is constant in the temperature range of interest:

$$\nu F \int_{E^\ominus(T)}^{E^\ominus(T')} dE^\ominus = \Delta_r S^\ominus \int_{T}^{T'} dT$$

Therefore,

$$\nu F \times \left\{ E^\ominus(T') - E^\ominus(T) \right\} = \Delta_r S^\ominus \times (T' - T)$$

which is easily rearranged into eqn (15).

Finally, we can combine the results obtained so far by using $G = H - TS$ in the form $H = G + TS$ to obtain the standard reaction enthalpy:

$$\Delta_r H^\ominus = \Delta_r G^\ominus + T\Delta_r S^\ominus \qquad (16)$$

with $\Delta_r G^{\ominus}$ determined from the cell potential and $\Delta_r S^{\ominus}$ from its temperature variation. Thus, we now have a noncalorimetric method of measuring a reaction enthalpy.

Example *Using the temperature coefficient of the cell potential*
The standard potential of the cell

$$\text{Pt} \,|\, H_2(g) \,|\, HCl(aq) \,|\, Hg_2Cl_2(s) \,|\, Hg(l)$$

was found to be $+0.2699\,V$ at $293\,K$ and $+0.2669\,V$ at $303\,K$. Evaluate the standard Gibbs energy, enthalpy, and entropy at $298\,K$ of the reaction

$$Hg_2Cl_2(s) + H_2(g) \longrightarrow 2\,Hg(l) + 2\,HCl(aq)$$

Strategy
We find the standard reaction Gibbs energy from the standard potential by using eqn (13) and making a linear interpolation between the two temperatures (in this case, we take the mean $E^{\ominus}$ because 298 K lies midway between 293 K and 303 K). The standard reaction entropy is obtained by substituting the data into eqn (15), and then the standard reaction enthalpy is obtained by combining these two quantities by using eqn (16).

Solution
Because the mean standard cell potential is $+0.2684\,V$ and $\nu = 2$ for the reaction,

$$\Delta_r G^{\ominus} = -\nu F E^{\ominus} = -2 \times (9.6485 \times 10^4 \, C\,mol^{-1}) \times (0.2684\,V)$$
$$= -51.79 \, kJ\,mol^{-1}$$

Then, from eqn (15), the standard reaction entropy is

$$\Delta_r S^{\ominus} = 2 \times (9.6485 \times 10^4 C\,mol^{-1}) \times \left(\frac{0.2699\,V - 0.2669\,V}{293\,K - 303\,K} \right)$$
$$= -57.9 \, J\,K^{-1}\,mol^{-1}$$

For the next stage of the calculation it is convenient to write the last value as $-5.79 \times 10^{-2} \, kJ\,K^{-1}\,mol^{-1}$. Then, from eqn (16) we find

$$\Delta_r H^{\ominus} = (-51.79 \, kJ\,mol^{-1}) + (298\,K) \times (-5.79 \times 10^{-2} \, kJ\,K^{-1}\,mol^{-1})$$
$$= -69.0 \, kJ\,mol^{-1}$$

One difficulty with this procedure lies in the accurate measurement of small temperature coefficients of cell potential. Nevertheless, it is another example of the striking ability of thermodynamics to relate the apparently unrelated, in this case to relate electrical measurements to thermal properties.

Exercise E6.16

Predict the standard potential of the Harned cell

$$Pt \,|\, H_2(g) \,|\, HCl(aq) \,|\, AgCl(s) \,|\, Ag(s)$$

at 303 K from tables of thermodynamic data for 298 K.

[Answer: +0.2191 V]

EXERCISES

6.1 Thermodynamic data can be used to predict the solubilities of compounds which would be very difficult to measure directly. Calculate the solubility of mercury(II) chloride at 25 °C from standard Gibbs energies of formation.

6.2 Consider a hydrogen electrode in aqueous HBr solution at 25 °C operating at 1.45 bar. Estimate the change in the electrode potential when the solution is changed from $5.0 \, \text{mmol L}^{-1}$ to $25.0 \, \text{mmol L}^{-1}$.

6.3 A hydrogen electrode can, in principle, be used to monitor changes in the molar concentrations of weak acids in biologically active solutions. Consider a hydrogen electrode in a solution of lactic acid as part of an overall galvanic cell at 25 °C and 1 bar. Estimate the change in the electrode potential when the solution is changed from $5.0 \, \text{mmol L}^{-1}$ to $25.0 \, \text{mmol L}^{-1}$.

6.4 Devise a cell in which the cell reaction is $Mn(s) + Cl_2(g) \longrightarrow MnCl_2(aq)$. Give the half-reactions for the electrodes and from the standard cell potential of $+2.54 \, V$ deduce the standard potential of the Mn^{2+}/Mn couple.

6.5 Write the cell reactions and electrode half-reactions for the following cells:
(a) $Ag|AgNO_3(aq, m_L)\|AgNO_3(aq, m_R)|Ag$
(b) $Pt|H_2(g, p_L)|HCl(aq)|H_2(g, p_L)|Pt$
(c) $Pt|K_3[Fe(CN)_6](aq),K_4[Fe(CN)_6](aq)\|$
$Mn^{2+}(aq),H^+(aq)|MnO_2(s)|Pt$

(d) $Pt|Cl_2(g)|HCl(aq)\|HBr(aq)|Br_2(l)|Pt$
(e) $Pt|Fe^{3+}(aq), Fe^{2+}(aq)\|Sn^{4+}(aq), Sn^{2+}(aq)|Pt$
(f) $Fe|Fe^{2+}(aq)\|Mn^{2+}(aq), H^+(aq)|MnO_2(s)|Pt$

6.6 Write the Nernst equations for the cells in the preceding exercise.

6.7 Devise cells in which the following are the reactions. In each case state the value for ν to use in the Nernst equation.
(a) $Fe(s) + PbSO_4(aq) \longrightarrow FeSO_4(aq) + Pb(s)$
(b) $Hg_2Cl_2(s) + H_2(g) \longrightarrow 2\,HCl(aq) + 2\,Hg(l)$
(c) $2\,H_2(g) + O_2(g) \longrightarrow 2\,H_2O(l)$
(d) $H_2(g) + O_2(g) \longrightarrow H_2O_2(aq)$
(e) $H_2(g) + I_2(g) \longrightarrow 2\,HI(aq)$
(f) $2\,CuCl(aq) \longrightarrow Cu(s) + CuCl_2(aq)$

6.8 Use the standard potentials of the electrodes to calculate the standard potentials of the cells in Exercise 6.5.

6.9 Use the standard potentials of the electrodes to calculate the standard potentials of the cells devised in Exercise 6.7.

6.10 State what you would expect to happen to the cell potential when the following changes are made to the corresponding cells in Exercise 6.5. Confirm your prediction by using the Nernst equation in each case.
(a) The molar concentration of silver nitrate in the left-hand compartment is increased.

(b) The pressure of hydrogen in the left-hand compartment is increased.

(c) The pH of the right-hand compartment is decreased.

(d) The concentration of HCl is increased.

(e) Some iron(III) chloride is added to both compartments.

(f) Acid is added to both compartments.

6.11 State what you would expect to happen to the cell potential when the following changes are made to the corresponding cells devised in Exercise 6.7. Confirm your prediction by using the Nernst equation in each case.

(a) The molar concentration of $FeSO_4$ is increased.

(b) Some nitric acid is added to both cell compartments.

(c) The pressure of oxygen is increased.

(d) The pressure of hydrogen is increased.

(e) Some (i) hydrochloric acid, (ii) hydroiodic acid is added to both compartments.

(f) Hydrochloric acid is added to both compartments.

6.12 (a) Calculate the standard potential of the cell $Hg(l)|HgCl_2(aq)\|TlNO_3(aq)|Tl$ at 25°C. (b) Calculate the cell potential when the molar concentration of the Hg^{2+} ion is $0.150\,mol\,L^{-1}$ and that of the Tl^+ ion is $0.93\,mol\,L^{-1}$.

6.13 Calculate the standard Gibbs energies at 25°C of the following reactions from the standard potential data in Appendix 2.

(a) $Ca(s) + 2\,H_2O(l) \longrightarrow Ca(OH)_2(aq) + H_2(g)$

(b) $2\,Ca(s) + 4\,H_2O(l) \longrightarrow 2\,Ca(OH)_2(aq) + 2\,H_2(g)$

(c) $Fe(s) + 2\,H_2O(l) \longrightarrow Fe(OH)_2(aq) + H_2(g)$

(d) $Na_2S_2O_8(aq) + 2\,NaI(aq) \longrightarrow I_2(s)$
$+2\,Na_2SO_4(aq)$

(e) $Na_2S_2O_8(aq) + 2\,KI(aq) \longrightarrow I_2(s)+$
$Na_2SO_4(aq) + K_2SO_4(aq)$

(f) $Pb(s) + Na_2CO_3(aq) \longrightarrow PbCO_3(aq) + 2\,Na(s)$

6.14 Tabulated thermodynamic data can be used to predict the standard potential of a cell even if it cannot be measured directly. The standard Gibbs energy of the reaction

$K_2CrO_4(aq) + 2\,Ag(s) + 2\,FeCl_3(aq) \longrightarrow$
$Ag_2CrO_4(s) + 2\,FeCl_2(aq) + 2\,KCl(aq)$

is $-62.5\,kJ\,mol^{-1}$ at 298 K. (a) Calculate the standard potential of the corresponding galvanic cell and (b) the standard potential of the $Ag_2CrO_4/Ag, CrO_4^{2-}$ couple.

6.15 Estimate the potential of the cell

$Ag|AgCl(s)|KCl(aq, 0.025\,mol\,kg^{-1})\|AgNO_3(aq, 0.010\,mol\,kg^{-1})|Ag$

at 25°C.

6.16 Use the information in Appendix 2 to calculate the standard potential of the cell $Ag|AgNO_3(aq)\|Cu(NO_3)_2(aq)|Cu$ and the standard Gibbs energy and enthalpy of the cell reaction at 25°C. Estimate the value of $\Delta_r G^\ominus$ at 35°C.

6.17 One ecologically important equilibrium is that between carbonate and hydrogencarbonate (bicarbonate) ions in natural water. (a) The standard Gibbs energies of formation of $CO_3^{2-}(aq)$ and $HCO_3^-(aq)$ are $-527.81\,kJ\,mol^{-1}$ and $-586.77\,kJ\,mol^{-1}$, respectively. What is the standard potential of the $HCO_3^-/CO_3^{2-}, H_2$ couple? (b) Calculate the standard potential of a cell in which the cell reaction is $Na_2CO_3(aq) + H_2O(l) \longrightarrow NaHCO_3(aq) + NaOH(aq)$. (c) Write the Nernst equation for the cell, and (d) predict and calculate the change in potential when the pH is changed to 7.0. (e) Calculate the value of pK_a for $HCO_3^-(aq)$.

6.18 Calcium phosphate is the principal inorganic component of bone. Its solubility characteristics are important for the stabilities of skeletons, and they can be assessed electrochemcically. The solubility constant of $Cu_3(PO_4)_2$ is 1.3×10^{-37}. Calculate (a) the solubility of $Cu_3(PO_4)_2$, (b) the potential of the cell

$Pt|H_2(g)|HCl(aq, pH = 0)\|Cu_3(PO_4)_2(aq, satd.)|Cu$

at 25°C.

6.19 Calculate the equilibrium constants of the following reactions at 25°C from standard potential data:

(a) $Sn(s) + Sn^{4+}(aq) \rightleftharpoons 2\,Sn^{2+}(aq)$

(b) $Sn(s) + 2\,AgBr(s) \rightleftharpoons SnBr_2(aq) + 2\,Ag(s)$

(c) $Fe(s) + Hg(NO_3)_2(aq) \rightleftharpoons Hg(l) + Fe(NO_3)_2(aq)$

(d) $Cd(s) + CuSO_4(aq) \rightleftharpoons Cu(s) + CdSO_4(aq)$

(e) $Cu^{2+}(aq) + Cu(s) \rightleftharpoons 2\,Cu^+(aq)$

6.20 The molar solubilities of AgCl and $BaSO_4$ in water are $1.34 \times 10^{-5}\,mol\,L^{-1}$ and $9.51 \times 10^{-4}\,mol\,L^{-1}$, respectively, at 25°C. Calculate their solubility constants from the appropriate standard potentials.

6.21 The dichromate ion in acidic solution is a common oxidizing agent for organic compounds. Derive an expression for the potential of an electrode for which the half-reaction is the reduction of $Cr_2O_7^{2-}$ ions to Cr^{3+} ions in acidic solution.

6.22 The zero-current potential of the cell $Pt|H_2(g)|HCl(aq)|AgCl(s)|Ag$ is $0.312\,V$ at $25\,°C$. What is the pH of the electrolyte solution?

6.23 The molar solubility of AgBr is $2.6\,\mu mol\,L^{-1}$ at $25\,°C$. What is the zero-current potential of the cell $Ag|AgBr(aq)|AgBr(s)|Ag$ at that temperature?

6.24 The standard potential of the cell $Ag|AgI(s)|AgI(aq)|Ag$ is $+0.9509\,V$ at $25\,°C$. Calculate (a) the molar solubility of AgI and (b) its solubility constant.

6.25 Devise a cell in which the overall reaction is $Pb(s) + Hg_2SO_4(s) \longrightarrow PbSO_4(s) + 2\,Hg(l)$. What is its potential when the electrolyte is saturated with both salts at $25\,°C$?

6.26 A fuel cell develops an electric potential from the chemical reaction between reagents supplied from an outside source. What is the zero-current potential of a cell fuelled by (a) hydrogen and oxygen, (b) the complete oxidation of benzene at $1.0\,bar$ and $298\,K$?

6.27 A fuel cell is constructed in which both electrodes make use of the oxidation of methane. The left-hand electrode makes use of the complete oxidation of methane to carbon dioxide and water; the right-hand electrode makes use of the partial oxidation of methane to carbon monoxide and water. (a) Which electrode is the cathode? (b) What is the zero-current cell potential at $25\,°C$ when all gases are at 1 bar?

6.28 The permanganate ion is a common oxidizing agent. What is the standard potential of the $MnO_4^-, H^+/Mn^{2+}$ couple at $pH = 6.00$?

6.29 The biological standard potential of the redox couple pyruvic acid/lactic acid is $-0.19\,V$ and that of fumaric acid/succinic acid is $+0.03\,V$ at $25\,°C$. What is the equilibrium constant for the reaction $P + S \rightleftharpoons L + F$ in $pH = 7$ (where P = pyruvic acid, S = succinic acid, L = lactic acid, and F = fumaric acid)?

6.30 The biological standard potential of the couple pyruvic acid/lactic acid is $-0.19\,V$. What is the thermodynamic standard potential of the couple? Pyruvic acid is $CH_3COCOOH$ and lactic acid is $CH_3CH(OH)COOH$.

The rates of reactions

The branch of physical chemistry called **chemical kinetics** is concerned with the rates of chemical reactions: how rapidly reactants are consumed and products formed, how the rate responds to changes in the conditions or the presence of a catalyst (including enzymes), and the identification of the steps by which the reaction takes place. One reason for studying the rates of reactions is the practical importance of being able to predict how quickly a reaction mixture approaches equilibrium. The rate might depend on variables under our control, such as the pressure, the temperature, and the presence of a catalyst, and we might be able to optimize it by the appropriate choice of conditions. Another reason is that the study of reaction rates leads to an understanding of the **mechanism** of a reaction, its analysis into a sequence of elementary steps. For example, we might discover that the reaction of hydrogen and bromine to form hydrogen bromide proceeds by the dissociation of a Br_2 molecule, the attack of a Br atom on an H_2 molecule, and several subsequent steps, and not by a single event in which an H_2 molecule encounters a Br_2 molecule and the atoms exchange partners to form two HBr molecules. The analysis of the effect of an enzyme on the rate of the reaction it catalyses is one of the principal means by which its mode of action may be determined. **Enzyme kinetics**, the study of the effect of enzymes on the rates of biochemically significant reactions, is also an important window on to the manner in which the action of enzymes is inhibited, such as by poisons or (from a human perspective) pharmaceuticals.

Empirical chemical kinetics

The first stage in the investigation of the rate and mechanism of a reaction is the determination of the overall stoichiometry of the reaction and the identification of any side reactions. Once those features have been established, the next step is to determine experimentally how the concentrations of the reactants and products change with time after the reaction has been initiated. Because the rates of chemical reactions are sensitive to the temperature, the temperature of the reaction mixture must be held constant throughout the course of the reaction, for otherwise the observed rate would be a meaningless average of rates at different temperatures. Indeed, one of the reasons why we fight infection with a fever is to upset the balance of reaction rates in the infecting organism, and hence destroy it, by the increase in temperature. There is a fine line, though, between killing an invader and killing the invaded!

7.1 Experimental techniques

The method selected to monitor the concentrations of reactants and products and their variation with time depends on the substances involved and the rapidity with which their concentrations change. Many reactions reach thermodynamic equilibrium over periods of minutes or hours, but some reactions reach equilibrium in fractions of a second. Under special conditions, modern techniques are capable of studying reactions that are complete within a few femtoseconds (1 fs $= 10^{-15}$ s).

Monitoring the progress of a reaction

A reaction in which at least one component is a gas might result in an overall change in pressure if the reaction vessel has a constant volume, and so its progress may be followed by recording the variation of pressure with time. An example is the decomposition of nitrogen(V) oxide,

$$2\,N_2O_5(g) \longrightarrow 4\,NO_2(g) + O_2(g)$$

For each mole of N_2O_5 molecules destroyed, $\frac{5}{2}$ mol of gas molecules is formed, and so the total pressure of the sample will increase as the reaction proceeds. A disadvantage of this method is that it is not specific: all the gas phase particles contribute to the pressure.

Example *Monitoring the variation in pressure*
Predict how the total pressure varies during the gas-phase decomposition of N_2O_5.

Strategy
The pressure is proportional to the chemical amount of gas-phase molecules present, regardless of their chemical identities. Therefore, we have to determine the amount of gas molecules at any stage of the reaction. The simplest procedure is to suppose that a fraction α of reactant molecules has decomposed, and then to use the reaction stoichiometry to calculate the amounts of the other species. The total pressure is proportional to the total amount of molecules.

Solution
Let the initial pressure be p_0 and the initial amount of N_2O_5 molecules present be n. When a fraction α of the N_2O_5 molecules has decomposed, the amount remaining is $(1 - \alpha)n$. It follows from the

reaction stoichiometry (specifically, $1\,mol\,N_2O_5 \equiv 2\,mol\,NO_2$ and $2\,mol\,N_2O_5 \equiv 1\,mol\,O_2$) that the amount of NO_2 produced will be twice that amount, or $2\alpha n$, and that the amount of O_2 present will be $\frac{1}{2}\alpha n$. That is, at an arbitrary stage of the reaction, the amounts of the components in the reaction mixture are:

	N_2O_5	NO_2	O_2	Total
Amount:	$(1-\alpha)n$	$2\alpha n$	$\frac{1}{2}\alpha n$	$(1+\frac{3}{2}\alpha)n$

The total pressure (at constant volume and temperature, and assuming perfect gas behaviour) is proportional to the number of gas-phase molecules. When $\alpha = 0$ the pressure is p_0, and so at any later stage the total pressure is

$$p = (1+\tfrac{3}{2}\alpha)p_0$$

For example, when the reaction is complete, $\alpha = 1$ and the pressure will have risen to $\frac{5}{2}$ times its initial value (Fig. 7.1). If the initial amount of N_2O_5 is known, then the amounts of N_2O_5, NO_2, and O_2 can be calculated from

$$\alpha = \tfrac{2}{3}\left(\frac{p}{p_0} - 1\right)$$

and the expressions in the table above.

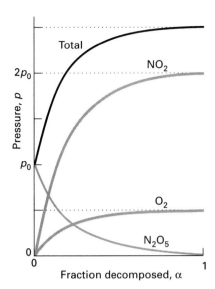

Fig. 7.1 The variation with time of the partial pressures of nitrogen dioxide, dinitrogen pentoxide, and oxygen in a reactor, and their total pressure, the sum of the partial pressures.

Exercise E7.1

Repeat the calculation for the decomposition

$$2\,NOBr(g) \longrightarrow 2\,NO(g) + Br_2(g)$$

[*Answer*: $p = (1+\tfrac{1}{2}\alpha)p_0$]

Spectrophotometry, the measurement of the intensity of absorption in a particular spectral region, is widely applicable to the monitoring of concentration, and is especially useful when one substance (and only one) in the reaction mixture has a strong characteristic absorption in a conveniently accessible region of the spectrum. For example, the reaction

$$H_2(g) + Br_2(g) \longrightarrow 2\,HBr(g)$$

can be followed by measuring the absorption of visible light by bromine.

If a reaction changes the number or type of ions present in a solution, then it may be followed by monitoring the conductivity of the solution.

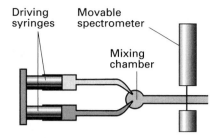

Fig. 7.2 The arrangement used in the flow technique for studying reaction rates. The reactants are squirted into the mixing chamber at a steady rate from the syringes or by using peristaltic pumps (pumps that squeeze the fluid through flexible tubes, like in our intestines). The location of the spectrometer corresponds to different times after initiation.

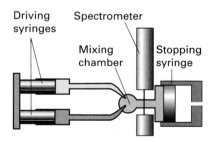

Fig. 7.3 In the stopped-flow technique the reagents are driven quickly into the mixing chamber and then the time-dependence of the concentrations is monitored.

Reactions that change the concentration of hydrogen ions may be studied by monitoring the pH of the solution with a glass electrode. Other methods of monitoring the composition include titration, mass spectrometry, gas chromatography, and magnetic resonance (Chapter 11). Polarimetry, the observation of the optical activity of a reaction mixture, is occasionally applicable.

Application of the techniques

In a **real-time analysis**, the composition of a system is analysed while the reaction is in progress, either by direct spectroscopic observation of the reaction mixture or by withdrawing a small sample and analysing it. In the **quenching method**, the reaction is stopped after it has been allowed to proceed for a certain time, and the composition is analysed at leisure. The quenching (of the entire mixture or of a sample drawn from it) can be achieved either by cooling suddenly, by adding the mixture to a large volume of solvent, or by rapid neutralization of an acid reagent. This method is suitable only for reactions that are slow enough for there to be little reaction during the time it takes to quench the mixture.

In the **flow method**, the reactants are mixed as they flow together in a chamber (Fig. 7.2). The reaction continues as the thoroughly mixed solutions flow through the outlet tube, and different points along the tube correspond to different times after the start of the reaction. Spectroscopic determination of the composition at different positions along the tube is equivalent to the determination of the composition of the reaction mixture at different times after mixing. The disadvantage of conventional flow techniques is that a large volume of reactant solution is necessary, because the mixture must flow continuously through the apparatus. This disadvantage is particularly important for reactions that take place very rapidly, because to spread the reaction over an appreciable length of tube the flow must be rapid. The **stopped-flow technique** avoids this disadvantage (Fig. 7.3). The two solutions are mixed very rapidly by injecting them into a mixing chamber designed to ensure that the flow is turbulent and that complete mixing occurs very quickly. Behind the reaction chamber there is an observation cell fitted with a plunger that moves back as the liquids flood in, but which comes up against a stop after a certain volume has been admitted. The filling of that chamber corresponds to the sudden creation of an initial sample of the reaction mixture. The reaction then continues in the thoroughly mixed solution and is monitored spectrophotometrically. Because only a small, single charge of the reaction chamber is prepared, the technique is much more economical than the flow method. The suitability of the stopped-flow technique to the study of small samples means that it is appropriate for biochemical reactions, and it has been widely used to study the kinetics of enzyme action.

In **flash photolysis**, the gaseous or liquid sample is exposed to a brief photolytic flash of light, and then the contents of the reaction chamber are monitored spectrophotometrically. Most work is now done with lasers, which can be used to generate nanosecond flashes routinely, picosecond flashes quite readily, and flashes as brief as a few femtoseconds in special arrangements. Either emission or absorption spectroscopy can be used to monitor the reaction, and the spectra are recorded electronically at a series of times following the flash.

7.2 Reaction rates

The raw data from experiments to measure reaction rates are the concentrations of reactants and products at a series of times after the reaction is initiated. Ideally, information on any intermediates should also be obtained, but they cannot always be studied because their existence is so fleeting or their concentration so low. More information about the reaction can be extracted if data are obtained at a series of different temperatures. The next few sections look at these observations in more detail.

The definition of rate

The rate of a reaction is defined in terms of the rate of change of the concentration of a designated species. However, because the rates at which reactants are consumed and products are formed change in the course of a reaction, it is necessary to consider the **instantaneous rate** of the reaction, its rate at a specific instant. The instantaneous rate of consumption of a reactant is the slope of a graph of its molar concentration plotted against the time, with the slope evaluated at the instant of interest (Fig. 7.4). The steeper the slope, the greater the rate of consumption of the reactant. Similarly, the rate of formation of a product is the slope of the graph of its concentration plotted against time. With the concentration measured in moles per litre and the time in seconds, the reaction rate is reported in moles per litre per second ($mol\,L^{-1}\,s^{-1}$).

In general, because of the reaction stoichiometry, the various reactants in a given reaction are consumed at different rates, and the various products are also formed at different rates. For example, in a reaction of the form

$$N_2(g) + 3\,H_2(g) \longrightarrow 2\,NH_3(g)$$

the rate of formation of NH_3 is twice the rate of disappearance of N_2 (because for every mole of N_2 consumed, *two* moles of NH_3 are formed), and the H_2 disappears three times as rapidly as the N_2 (because $3\,mol\,H_2 \equiv 1\,mol\,N_2$). Once the rate of formation or consumption of one substance is known, the reaction stoichiometry can be used to deduce the rates of formation or consumption of the other participants in the reaction.

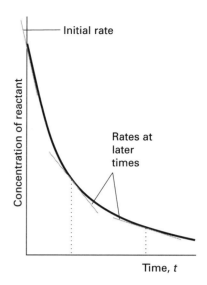

Fig. 7.4 The rate of a chemical reaction is the slope of the tangent to the curve showing the variation of concentration of a species with time. This graph is a plot of the concentration of a reactant, which is consumed as the reaction progresses. The rate of consumption decreases in the course of the reaction as the concentration of reactant decreases.

Example *Reporting rates of reaction*
The rate of formation of NO(g) in the reaction

$$2\,NOBr(g) \longrightarrow 2\,NO(g) + Br_2(g)$$

was reported as $1.6 \times 10^{-4}\,mol\,L^{-1}\,s^{-1}$. What is the rate of consumption of NOBr and the rate of formation of Br_2?

Strategy

The rates are related by the stoichiometric numbers of the species; all rates are positive.

Solution

Because $1\,mol\,NOBr \equiv 1\,mol\,NO$, the rate of consumption of NOBr is the same as the rate of formulation of NO. Because $\frac{1}{2}\,mol\,Br_2 \equiv 1\,mol\,NOBr$, the rate of formation of Br_2 is only one-half the rate of consumption of NOBr, and is therefore $8.0 \times 10^{-5}\,mol\,L^{-1}\,s^{-1}$.

Exercise E7.2

The rate of consumption of CH_3 radicals in the reaction $2\,CH_3(g) \longrightarrow CH_3CH_3(g)$ was reported as $1.2\,mol\,L^{-1}\,s^{-1}$ under a certain set of conditions. What is the rate of formation of CH_3CH_3?

[*Answer*: $0.60\,mol\,L^{-1}\,s^{-1}$]

Rate laws and rate constants

An empirical observation of the greatest importance is that *the rate of reaction is often found to be proportional to the molar concentrations of the reactants raised to a simple power*. For example, it may be found that the rate is directly proportional to the concentration of the reactant A and to the concentration of another reactant B, so that

$$\text{rate of reaction} = k[\text{A}][\text{B}] \tag{1}$$

The coefficient k, which is characteristic of the reaction being studied, is called the **rate constant**. It is independent of the concentrations but depends on the temperature. An *experimentally determined* equation of this kind is called the 'rate law', of the reaction. More formally:

A **rate law** is an equation that expresses the rate of reaction in terms of the molar concentrations of the species in the overall reaction (including, possibly, the products).

The units of k are always such as to convert the product of concentrations into a rate expressed as a change in concentration divided by time. For example, if the rate law is the one shown in eqn (1), with concentrations expressed in moles per litre ($mol\,L^{-1}$), then the units of k will be $L\,mol^{-1}\,s^{-1}$ because

$$(k, \text{ in } L\,mol^{-1}\,s^{-1}) \times ([\text{A}], \text{ in } mol\,L^{-1}) \times ([\text{B}], \text{ in } mol\,L^{-1})$$
$$= \text{rate, in } mol\,L^{-1}\,s^{-1}$$

Rate laws have three main applications. Once we know the rate law and the rate constant we can use them to predict the rate of reaction for any given composition of the reaction mixture. We shall also see that we can use a rate law to predict the concentrations of the reactants and products at any time after the start of the reaction. Moreover, a rate law is a guide to the mechanism of the reaction, for any proposed mechanism must be consistent with the observed rate law.

Reaction order

A rate law provides a basis for the classification of reactions according to their kinetics. The advantage of having such a classification is that reactions belonging to the same class will have similar kinetic behaviour—their rates will vary with composition in a similar way. The classification of reactions is based on their 'order':

The **order** of a reaction with respect to each species is the power to which the concentration of that species is raised in the rate law.

For example, a reaction with the rate law

$$\text{rate} = k[\text{A}][\text{B}] \tag{2}$$

is *first order* in A and first order in B. A reaction with the rate law

$$\text{rate} = k[\text{A}]^2 \tag{3}$$

is *second order* in A.

The **overall order** of a reaction is the sum of the orders of all the components. The two rate laws just quoted both correspond to reactions that are *second order* overall. Examples of the two different types of reaction are the reaction between persulfate and iodide ions,

$$S_2O_8^{2-}(aq) + 3\,I^-(aq) \longrightarrow 2\,SO_4^{2-}(aq) + I_3^-(aq) \qquad \text{rate} = k[S_2O_8^{2-}][I^-]$$

which is first-order in $S_2O_8^{2-}$ ions, first-order in I^- ions, and second-order overall, and the reduction of nitrogen dioxide by carbon monoxide,

$$NO_2(g) + CO(g) \longrightarrow NO(g) + CO_2(g) \qquad \text{rate} = k[NO_2]^2$$

which is second-order in NO_2 and, because no other species occurs in the rate law, second-order overall. In the latter reaction, the rate is independent of the concentration of CO in the sense that, so long as *some* CO is present, then the rate is independent of the precise concentration. This independence of concentration is expressed by saying that the reaction is *zero order* in CO, because a concentration raised to the power zero is 1 ($[CO]^0 = 1$, just as $x^0 = 1$ in algebra).

Exercise E7.3

The reaction between the amino acid tyrosine (Tyr) and iodine obeys the rate law: rate $= k[\text{Tyr}][\text{I}_2]$. Classify it by order.

[*Answer*: first order in Tyr, first order in I_2, and second order overall]

A reaction need not have an integral order, and many gas-phase reactions do not. For example, if a reaction is found to have the rate law

$$\text{rate} = k[\text{A}]^{1/2}[\text{B}]$$

then it is *half order* in A, first order in B, and three-halves order overall. If a rate law is not of the form $[\text{A}]^x[\text{B}]^y[\text{C}]^z \cdots$ then the reaction does not have an order. Thus, the experimentally determined rate law for the gas-phase reaction $\text{H}_2(\text{g}) + \text{Br}_2(\text{g}) \longrightarrow 2\,\text{HBr}(\text{g})$ is

$$\text{rate of formation of HBr} = \frac{k[\text{H}_2][\text{Br}_2]^{3/2}}{[\text{Br}_2] + k'[\text{HBr}]} \qquad (4)$$

Although the reaction is first order in H_2, it has an indefinite order with respect to both Br_2 and HBr and an indefinite order overall. Similarly, a typical rate law for the action of an enzyme E on a substrate S is

$$\text{rate of formation of product} = \frac{k[\text{E}][\text{S}]}{[\text{S}] + K_\text{M}} \qquad (5)$$

where K_M is a constant. This rate law is first order in the enzyme E, but it does not have a specific order with respect to the substrate S.

Under certain circumstances a complicated rate law without an overall order may simplify into a law with a definite order. For example, if the concentration of Br_2 is so high that $[\text{Br}_2] \gg k'[\text{HBr}]$, then the denominator in eqn (4) is equal to $[\text{Br}_2]$ to a good approximation, and the rate law simplifies to

$$\text{rate of formation of HBr} = k[\text{H}_2][\text{Br}_2]^{1/2}$$

which is first order in H_2, half order in Br_2, and three-halves order overall. Likewise, if the substrate concentration in the enzyme catalysed reaction is so low that $[\text{S}] \ll K_\text{M}$, then eqn (5) simplifies to

$$\text{rate of formation of product} = \left(\frac{k}{K_\text{M}}\right)[\text{E}][\text{S}]$$

which is first order in S, first order in E, and second order overall.

Rate laws may also be expressed in terms of quantities that are proportional to the molar concentrations of the species. Among the most common of these alternatives is the use of the partial pressure of a gas phase species. For example, a gas phase reaction might have a rate law expressed in the form

$$\text{rate of formation of J} = k p_A p_B^2$$

where the rate of formation of the species J would be interpreted as the rate of change of the partial pressure of J. Most of the time, though, we shall express rate laws in terms of molar concentrations.

It is very important to note that *a rate law is established experimentally, and cannot in general be inferred from the chemical equation for the reaction.* The reaction of hydrogen and bromine, for example, has a very simple stoichiometry, but its rate law (eqn (4)) is very complicated. Similarly, the thermal decomposition of nitrogen(V) oxide specified at the beginning of Section 7.1 has the rate law

$$\text{rate of consumption of } N_2O_5 = k[N_2O_5]$$

and the reaction is first-order in N_2O_5. In some cases, however, the rate law does happen to reflect the reaction stoichiometry. This is the case with the oxidation of nitrogen oxide, NO, which under certain conditions is found to have a third-order rate law:

$$2\,NO(g) + O_2(g) \longrightarrow 2\,NO_2(g) \quad \text{rate of formation of } NO_2 = k[NO]^2[O_2]$$

Some reactions obey a zero-order rate law, with a rate that is independent of the concentration of the reactant (so long as some is present). Thus, the catalytic decomposition of phosphine, PH_3, on hot tungsten at high pressures is found to obey the rate law

$$\text{rate of decomposition of } PH_3 = k, \text{independent of concentration}$$

The PH_3 decomposes at a constant rate until it has entirely disappeared, when the reaction stops abruptly.

The determination of the rate law

The determination of a rate law is simplified by the **isolation method**, in which all the reactants except one are present in large excess. If a reactant B is in large excess, for example, it is a good approximation to take its concentration as constant throughout the reaction. Then, although the true rate law might be

$$\text{rate} = k[A][B]^2$$

we can approximate [B] by its initial value $[B]_0$ (which barely changes in the course of the reaction) and write

$$\text{rate} = k'[A], \text{with } k' = k[B]_0^2$$

which has the form of a first-order rate law. Because the true rate law has been forced into first-order form by assuming a constant B concentration, the effective rate law is classified as **pseudofirst order**. If, instead, the concentration of A were in large excess, and hence effectively constant, the rate law would simplify to

$$\text{rate} = k''[\text{B}]^2, \text{with } k'' = k[\text{A}]_0$$

This **pseudosecond-order** rate law is also much easier to analyse and identify than the complete law. Many reactions in aqueous solution that are reported as first or second order are actually pseudofirst or pseudosecond order, because the solvent water participates in the reaction, but it is in such large excess that its concentration remains constant. The dependence of the rate on all the reactants may be found by isolating each of them in turn (by having all the other substances present in large excess), and piecing together a picture of the overall rate law.

In the method of **initial rates**, which is often used in conjunction with the isolation method, the instantaneous rate is measured at the beginning of the reaction for several different initial concentrations of reactants. For example, suppose the rate law for a reaction with A isolated is

$$r = k'[\text{A}]^a$$

where r denotes rate. Then the initial rate of the reaction, r_0, is given by the initial concentration of A:

$$r_0 = k'[\text{A}]_0^a$$

Taking logarithms gives

$$\log r_0 = \log k' + a \log [\text{A}]_0 \tag{6}$$

This equation has the form of the equation for a straight line:

$$y = \text{intercept} + \text{slope} \times x$$

with $y = \log r_0$ and $x = \log [\text{A}]_0$. It follows that, for a series of initial concentrations, a plot of the logarithms of the initial rates against the logarithms of the initial concentrations of A should be a straight line, and that the slope of the graph will be a, the order of the reaction with respect to the species A (Fig. 7.5).

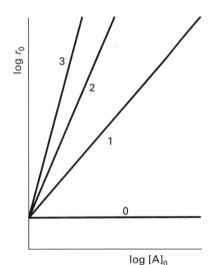

Fig. 7.5 The plot of $\log r_0$ against $\log [\text{A}]_0$ gives straight lines with slopes equal to the order of the reaction.

Example *Using the method of initial rates*
The recombination of I atoms in the gas phase in the presence of argon (which removes the energy released by the formation of an I—I bond, and so prevents the immediate dissociation of a newly formed I_2 molecule) was investigated and the order of the reaction was determined by

the method of initial rates. The initial rates of reaction of $2\,I(g) + Ar(g) \longrightarrow I_2(g) + Ar(g)$ were as follows:

$[I]_0/(10^{-5}\,mol\,L^{-1})$	1.0	2.0	4.0	6.0
$r_0/(mol\,L^{-1}\,s^{-1})$ (a)	8.70×10^{-4}	3.48×10^{-3}	1.39×10^{-2}	3.13×10^{-2}
(b)	4.35×10^{-3}	1.74×10^{-2}	6.96×10^{-2}	1.57×10^{-1}
(c)	8.69×10^{-3}	3.47×10^{-2}	1.38×10^{-1}	3.13×10^{-1}

The Ar concentrations are (a) $1.0 \times 10^{-3}\,mol\,L^{-1}$, (b) $5.0 \times 10^{-3}\,mol\,L^{-1}$, and (c) $1.0 \times 10^{-2}\,mol\,L^{-1}$. Find the orders of reaction with respect to I and Ar and the rate constant.

Strategy

We need to make a plot of $\log r_0$ against $\log [I]_0$ for a given $[Ar]_0$ and against $\log [Ar]_0$ for a given $[I]_0$. The intercepts give $\log k$ and the slopes give the orders.

Solution

The graphs are shown in Fig. 7.6. The slopes are 2 and 1 respectively, so the (initial) rate law is

$$r_0 = k[I]_0^2[Ar]_0$$

This rate law signifies that the reaction is second order in [I], first order in [Ar], and third order overall. The intercept corresponds to $\log (k/mol^{-2}\,L^2\,s^{-2}) = 9.9$, so $k = 8.7 \times 10^9\,mol^{-2}\,L^2\,s^{-1}$.

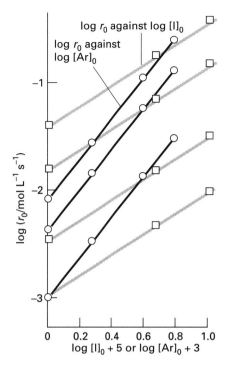

Fig. 7.6 The plots of the data in the example.

Exercise E7.4

The initial rate of a certain reaction depended on concentration of a substance J as follows:

$[J]_0/(10^{-3}\,mol\,L^{-1})$	5.0	8.2	17	30
$r_0/(10^{-7}\,mol\,L^{-1}\,s^{-1})$	3.6	9.6	41	130

Find the order of the reaction with respect to J and the rate constant.

[*Answer*: 2, $1.4 \times 10^{-2}\,mol^{-1}\,L\,s^{-1}$]

7.3 Integrated rate laws

The method of initial rates might not reveal the entire rate law, for in a complex reaction the products themselves might affect the rate. For example, products participate in the synthesis of HBr, for eqn (4) shows that the rate law depends on the concentration of HBr, none of which is present initially. To avoid this difficulty, the rate law should be fitted to the data throughout the reaction. The fitting may be done, in simple cases at least, by using a proposed rate law to predict the concentration of any component at any time, and then comparing it with the data.

Because rate laws are differential equations (equations for the rate $d[A]/dt$ in terms of $[A]$), they must be integrated to find the concentration as a function of time. An **integrated rate law** is an expression that gives the concentration of a species as a function of the time. Now that computers are so widely available, even the most complex rate laws may be integrated numerically. However, in a number of simple cases analytical solutions are easily obtained, and prove to be very useful. They have two principal uses. One is to let us predict the concentration of a species at any time after the start of the reaction. The other is to help find the rate constant and order of the reaction. Indeed, although we have introduced rate laws through a discussion of the determination of reaction rates, these rates are rarely measured directly because slopes are so difficult to determine accurately. Almost all experimental work in chemical kinetics deals with integrated rate laws; their great advantage being that they are expressed in terms of the experimental observables of concentration and time.

First-order reactions

The concentration of a reactant A at a time t for a first-order reaction in which the rate law is

$$\text{rate of consumption of A} = k[A] \tag{7}$$

and for which the initial concentration of A, at $t = 0$, is $[A]_0$, is found from

$$\ln \frac{[A]_0}{[A]} = kt \tag{8a}$$

Alternative forms of this expression are

$$\ln [A] = \ln [A]_0 - kt \tag{8b}$$

$$[A] = [A]_0 e^{-kt} \tag{8c}$$

The latter has the form of an **exponential decay** (Fig. 7.7). A common feature of all first-order reactions, therefore, is that *the concentration of the reactant decays exponentially with time.*

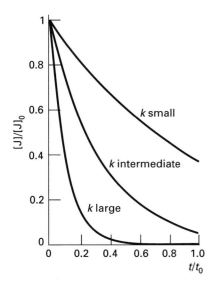

Fig. 7.7 The exponential decay of the reactant in a first-order reaction. The greater the rate constant, the more rapid the decay.

Justification

The formal definition of the instantaneous rate is expressed in terms of the slope, dx/dt, of a concentration versus time graph at a specified time after the start of the reaction:

$$\text{rate of formation of product P} = \frac{d[P]}{dt}$$

$$\text{rate of consumption of reactant R} = -\frac{d[R]}{dt}$$

The minus sign is included in the second expression because the concentration of R decreases with time: the slope of its graph is negative, and the negative sign converts that negative slope into a positive quantity.

The rate of consumption of a specific reactant A is $-d[A]/dt$, so a first-order rate equation has the form

$$\frac{d[A]}{dt} = -k[A]$$

This expression can be rearranged to

$$\frac{d[A]}{[A]} = -k \, dt$$

Integration from $t = 0$, when the concentration of A is $[A]_0$, to the time of interest, t, when the molar concentration of A is $[A]$, gives

$$\int_{[A]_0}^{[A]} \frac{d[A]}{[A]} = -\int_0^t k \, dt$$

and therefore, because the integral of $1/x$ is $\ln x$,

$$\ln [A] \Big|_{[A]_0}^{[A]} = -kt$$

Equation (8) follows by rearranging this expression.

At this point we can see how to use the integrated rate law to investigate a reaction. Equation (8) shows that if $\ln ([A]_0/[A])$ is plotted against t, then a first-order reaction will give a straight line. If in fact the experimental data do give a straight line when plotted in this way, then we know that the reaction is indeed first order; if the line is curved, then the reaction is not first order. If the line is straight, it follows from eqn (8) that its slope is k, so we can also determine that important quantity from the graph. Some rate constants determined in this way are given in Table 7.1.

Table 7.1 Kinetic data for first-order reactions

Reaction	Phase	$\theta/^\circ C$	k/s^{-1}	$t_{1/2}$
$2\,N_2O_5 \rightarrow 4\,NO_2 + \mathbf{O_2}$	g	25	3.38×10^{-5}	2.85 h
$2\,N_2O_5 \rightarrow 4\,NO_2 + \mathbf{O_2}$	$Br_2(l)$	25	4.27×10^{-5}	2.25 h
$\mathbf{C_2H_6} \rightarrow 2\,CH_3$	g	700	5.46×10^{-4}	21.2 min
$\mathbf{Cyclopropane} \rightarrow$ propene	g	500	6.17×10^{-4}	17.2 min

The rate constant is for the rate of formation or consumption of the species in bold type. The rate laws for the other species may be obtained from the reaction stoichiometry.

The rates of reactions

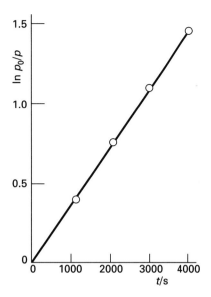

Fig. 7.8 The determination of the rate constant of a first-order reaction. A straight line is obtained when $\ln[A]$ (or $\ln p$, where p is the partial pressure of the species of interest) is plotted against t; the slope gives the rate constant as $-k$. In this case, the data have been plotted as $\ln p_0/p$, with the result that the slope is k itself (because $\ln 1/p = -\ln p$).

Example *Analysing a first-order reaction*
The variation in the partial pressure p of azomethane with time was followed at 460 K, with the results given below. Confirm that the decomposition

$$CH_3N_2CH_3(g) \longrightarrow CH_3CH_3(g) + N_2(g)$$

is first order in $CH_3N_2CH_3$, and find the rate constant at this temperature.

t/s	0	1000	2000	3000	4000
$p/(10^{-2}\,\text{Torr})$	8.20	5.72	3.99	2.78	1.94

Strategy
The easiest procedure is to plot the dimensionless quantity $\ln p_0/p$ against the dimensionless quantity t/s and expect to obtain a straight line. If the graph is straight, then the slope is $k \times s$.

Solution
The graph of the data is shown in Fig. 7.8. The plot is straight, confirming a first-order reaction. Its slope is 3.6×10^{-4}, so $k = 3.6 \times 10^{-4}\,\text{s}^{-1}$.

Exercise E7.5

The concentration of N_2O_5 in liquid bromine varied with time as follows:

t/s	0	200	400	600	1000
$[N_2O_5]/(\text{mol L}^{-1})$	0.110	0.073	0.048	0.032	0.014

Show that the reaction is first-order in N_2O_5 and determine the rate constant.

[*Answer*: $2.1 \times 10^{-3}\,\text{s}^{-1}$]

Second-order reactions

If the rate law is

$$\text{rate of consumption of A} = k[A]^2 \tag{9}$$

and the concentration of A at $t = 0$ is $[A]_0$, then at time t the concentration $[A]$ can be obtained from the expression

$$\frac{1}{[A]} = \frac{1}{[A]_0} + kt \tag{10}$$

This integrated rate law can be rearranged into

$$[A] = \frac{[A]_0}{1 + kt[A]_0} \qquad (11)$$

Justification

The differential equation for the rate law is

$$\frac{d[A]}{dt} = -k[A]^2$$

It can be rearranged into

$$-\frac{d[A]}{[A]^2} = k\,dt$$

This expression is now integrated between $t = 0$, when the concentration of A is $[A]_0$, and the time of interest t, when the concentration of A is $[A]$:

$$-\int_{[A]_0}^{[A]} \frac{d[A]}{[A]^2} = \int_0^t k\,dt$$

The integral on the left is evaluated by using

$$\int \frac{1}{x^2}\,dx = -\frac{1}{x}$$

which gives

$$\left.\frac{1}{[A]}\right|_{[A]_0}^{[A]} = kt$$

and a straightforward rearrangement leads to eqn (10).

Equation (10) shows that to test for a second-order reaction we should plot $1/[A]$ against t and expect a straight line. If it is straight, then the reaction is second-order in A and the slope of the line is equal to the rate constant. Some rate constants determined in this way are given in Table 7.2. Equation (11) lets us predict the concentration of A at any time after the start of the reaction. A point to note is that when [A] is plotted against t, the concentration of A approaches zero more slowly than in a first-order reaction with the same initial rate (Fig. 7.9). That is, reactants that decay by a second-order process, die away more slowly at low concentrations than would be expected if the decay was first order.

Table 7.3 summarizes the integrated rate laws for a variety of simple reaction types.

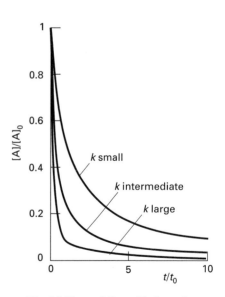

Fig. 7.9 The variation with time of the concentration of a reactant in a second-order reaction. Note that although the initial decay may be rapid, later the concentration approaches zero more slowly than in a first-order reaction with the same initial rate (compare Fig. 7.7).

Table 7.2 Kinetic data for second-order reactions

Reaction	Phase	$\theta/°C$	$k/\text{L mol}^{-1}\text{ s}^{-1}$
$2\ NOBr \rightarrow 2\ NO + \mathbf{Br_2}$	g	10	0.80
$2\ NO_2 \rightarrow 2\ NO + \mathbf{O_2}$	g	300	0.54
$\mathbf{H_2} + \mathbf{I_2} \rightarrow 2\ HI$	g	400	2.42×10^{-2}
$\mathbf{D_2} + HCl \rightarrow DH + DCl$	g	600	0.141
$2\ I \rightarrow \mathbf{I_2}$	g	23	7×10^9
	hexane	50	1.8×10^{10}
$\mathbf{CH_3Cl} + CH_3O^-$	$CH_3OH(l)$	20	2.29×10^{-6}
$\mathbf{CH_3Br} + CH_3O^-$	$CH_3OH(l)$	20	9.23×10^{-6}
$\mathbf{H^+} + OH \rightarrow H_2O$	water	25	1.5×10^{11}

The rate constant is for the rate of formation or consumption of the species in bold type. The rate laws for the other species may be obtained from the reaction stoichiometry.

Table 7.3 Integrated rate laws

Order	Reaction type	Rate law	Integrated rate law
0	$A \rightarrow P$	$r = k$	$[P] = kt$ for $kt \leq [A]_0$
1	$A \rightarrow P$	$r = k[A]$	$[P] = [A]_0\,(1 - e^{-kt})$
2	$A \rightarrow P$	$r = k[A]^2$	$[P] = \dfrac{kt[A]_0^2}{1 + kt[A]_0}$
	$A + B \rightarrow P$	$r = k[A][B]$	$[P] = \dfrac{[A]_0[B]_0(1 - e^{([B]_0-[A]_0)kt})}{[A]_0 - [B]_0 e^{([B]_0-[A]_0)kt}}$

7.4 Half-lives

A useful indication of the rate of a first-order chemical reaction is the **half-life**, $t_{1/2}$, of a reactant, the time it takes for the concentration of the species to fall to half its initial value.

The half-life of a species A that decays in a first-order reaction can be found by substituting $[A] = \frac{1}{2}[A]_0$ and $t = t_{1/2}$ into eqn (8):

$$kt_{1/2} = -\ln \frac{\frac{1}{2}[A]_0}{[A]_0} = -\ln\frac{1}{2} = \ln 2$$

from which it follows that

$$t_{1/2} = \frac{\ln 2}{k} \tag{12}$$

with $\ln 2 = 0.693 \cdots$. For example, because the rate constant for the first-order reaction

$$2\,N_2O_5(g) \longrightarrow 4\,NO_2(g) + O_2(g)$$
$$\text{rate of consumption of } N_2O_5 = k[N_2O_5]$$

is equal to $6.76 \times 10^{-5}\,s^{-1}$ at $25\,°C$, the half-life of N_2O_5 is $2.85\,h$. Hence, the concentration of N_2O_5 falls to half its initial value in $2.85\,h$, and then to half that concentration again in a further $2.85\,h$, and so on (Fig. 7.10).

The main point to note about eqn (12) is that *for a first-order reaction, the half-life of a reactant is independent of its initial concentration*. It follows that if the concentration of A at some arbitrary stage of the reaction is [A], then the concentration will fall to $\frac{1}{2}[A]$ after an interval of $0.693/k$ whatever the actual value of [A] (Fig. 7.11). Some half-lives are given in Table 7.1.

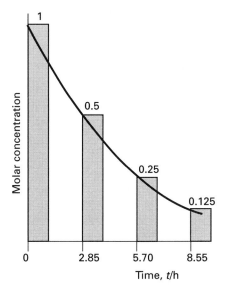

Fig. 7.10 The molar concentration of N_2O_5 after a succession of half-lives.

Example *Using the half-life of a species*

In acidic solution, the disaccharide sucrose (cane sugar) is converted to a mixture of the monosaccharides glucose and fructose in a pseudo-first-order reaction. Under certain conditions of pH, the half-life of sucrose is $28.4\,min$. How long will it take for the concentration of a sample to fall from $8.0\,mmol\,L^{-1}$ to $1.0\,mmol\,L^{-1}$?

Strategy

Each successive half-life corresponds to the halving of the molar concentration of a species, so n successive half-life intervals reduce the concentration by a factor of $\left(\frac{1}{2}\right)^n$. To tackle this problem, we need to determine the value of n corresponding to the stated reduction.

Solution

Under the conditions specified, the sequence is

$$\text{Molar concentration/mmol L}^{-1}: \quad 8.0 \xrightarrow{28.4\,min} 4.0 \xrightarrow{28.4\,min} 2.0 \xrightarrow{28.4\,min} 1.0$$

The total time required is $3 \times 28.4\,min = 85.2\,min$.

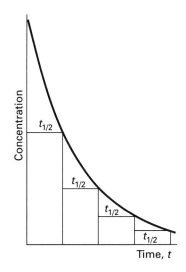

Fig. 7.11 In each successive period of duration $t_{1/2}$, the concentration of a reactant in a first-order reaction decays to half its value at the start of that period. After n such periods, the concentration is $\left(\frac{1}{2}\right)^n$ of its initial concentration.

Exercise E7.6

The half-life of a substrate in a certain enzyme-catalysed first order reaction is $138\,s$. How long is required for the initial concentration of substrate, which was $1.28\,mmol\,L^{-1}$, to fall to $0.040\,mmol\,L^{-1}$?

[*Answer*: $690\,s$]

Exercise E7.7

Derive an expression for the half-life of a second order reaction in terms of the rate constant k.

[*Answer*: $t_{1/2} = 1/k[A]_0$]

In contrast to first-order reactions, the half-life of a second-order reaction does depend on the initial concentration of the reactant (see the preceding exercise). It is therefore not characteristic of the reaction itself, and for that reason is rarely used.

One application of the concentration-independence of the half-life of a first-order reaction is to the identification of such reactions. Thus, if in a set of data of composition against time it is seen that the initial concentration falls to half its value in a certain time, and that another concentration falls to half its value in the same time, then we can infer that the reaction is first order. The first-order character can then be confirmed by plotting $\ln[A]$ against t and obtaining a straight line, as indicated earlier.

7.5 The temperature dependence of reaction rates

It is found that the rates of most chemical reactions increase as the temperature is raised. Many organic reactions in solution fall somewhere in the range spanned by the hydrolysis of methyl ethanoate (where the rate constant at $35\,°C$ is 1.8 times that at $25\,°C$) and the hydrolysis of sucrose (where the factor is 4.1). Reactions in the gas phase typically have rates that are only weakly sensitive to the temperature.

The Arrhenius parameters

As data on reaction rates were accumulated towards the end of the nineteenth century, the Swedish chemist Svante Arrhenius noted that almost all of them showed a similar dependence on the temperature. In particular, he noted that a graph of $\ln k$, where k is the rate constant for the reaction, against $1/T$, where T is the (absolute) temperature at which k is measured, gives a straight line with a slope that is characteristic of the reaction (Fig. 7.12). The mathematical expression of this conclusion is that the rate constant varies with temperature in accord with the expression

$$\ln k = \text{intercept} + \text{slope} \times \frac{1}{T}$$

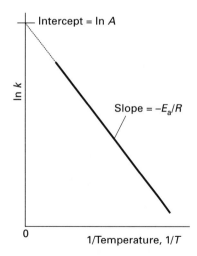

Fig. 7.12 The general form of an Arrhenius plot of $\ln k$ against $1/T$. The slope is equal to $-E_a/R$ and the intercept at $1/T = 0$ is equal to $\ln A$.

250

This expression is normally written as the **Arrhenius equation**:

$$\ln k = \ln A - \frac{E_a}{RT} \tag{13}$$

The parameter A (which has the same units as k) is called the **pre-exponential factor**, and E_a (which is a molar energy and has the units of kilojoules per mole) is called the **activation energy**. Collectively the two parameters are called the **Arrhenius parameters** of the reaction; some experimental values are given in Table 7.4. The Arrhenius equation is often written as

$$k = Ae^{-E_a/RT} \tag{14}$$

A practical point to note from Fig. 7.12 is that *a high activation energy corresponds to a reaction rate that is very sensitive to temperature* (the slope of the Arrhenius plot is steep). Conversely, a small activation energy indicates a reaction rate that varies only slightly with temperature. A reaction with zero activation energy (such as for some radical recombination reactions in the gas phase) has a rate that is largely independent of temperature.

Table 7.4 Arrhenius parameters

Reactions	A/s^{-1}	$E_a/\text{kJ mol}^{-1}$
First order		
Cyclopropene $\rightarrow$ propane	1.58×10^{15}	272
$CH_3NC \rightarrow CH_3CN$	3.98×10^{13}	160
cis-CHD=CHD $\rightarrow$ *trans*-CHD=CHD	3.16×10^{12}	256
cyclobutane $\rightarrow$ 2 C_2H_4	3.98×10^{15}	261
2 $N_2O_5 \rightarrow$ 4 NO_2 + O_2	4.94×10^{13}	103
$N_2O \rightarrow N_2$ + O	7.94×10^{11}	250

	$A/\text{L mol}^{-1}\,s^{-1}$	$E_a/\text{kJ mol}^{-1}$
Second-order, gas phase		
O + $N_2 \rightarrow$ NO + N	1×10^{11}	315
OH + $H_2 \rightarrow H_2O$ + H	8×10^{10}	42
Cl + $H_2 \rightarrow$ HCl + H	8×10^{10}	23
CH_3 + $CH_3 \rightarrow C_2H_6$	2×10^{10}	0
NO + $Cl_2 \rightarrow$ NOCl + Cl	4×10^9	85
Second order, solution		
NaC_2H_5O + CH_3I in ethanol	2.42×10^{11}	81.6
C_2H_5Br + OH^- in water	4.30×10^{11}	89.5
CH_3I + $S_2O_3^{2-}$ in water	2.19×10^{12}	78.7
Sucrose + H_2O in acidic water	1.50×10^{15}	107.9

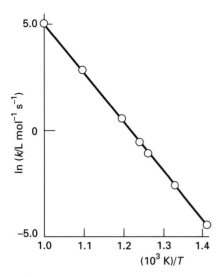

Fig. 7.13 The Arrhenius plot for the decomposition of CH_3CHO, and the best (least squares) straight line fitted to the data points.

Example *Determining the Arrhenius parameters*
The rate of the second-order decomposition of acetaldehyde (ethanal, CH_3CHO) was measured over the temperature range 700 to 1000 K, and the rate constants that were found are reported below. Find the activation energy and the pre-exponential factor.

T/K	700	730	760	790	810	840	910	1000
$k/(mol^{-1}\,L\,s^{-1})$	0.011	0.035	0.105	0.343	0.789	2.17	20.0	145

Strategy
We plot $\ln k$ against $1/T$ and expect a straight line. The slope is $-E_a/R$ and the intercept of the extrapolation to $1/T = 0$ is $\ln A$. It is best to do a least-squares fit of the data to a straight line.

Solution
The Arrhenius plot is shown in Fig. 7.13. The least-squares best fit of the line has slope -2.21×10^4 and intercept (which is well off the graph) 27.0. Therefore,

$$E_a = (2.21 \times 10^4\,K) \times (8.3145\,J\,K^{-1}\,mol^{-1}) = 184\,kJ\,mol^{-1}$$

and

$$A = e^{27.0}\,mol^{-1}\,L\,s^{-1} = 5.3 \times 10^{11}\,mol^{-1}\,L\,s^{-1}$$

Exercise E7.8

Determine A and E_a from the following data

T/K	300	350	400	450	500
$k/(mol^{-1}\,L\,s^{-1})$	7.9×10^6	3.0×10^7	7.9×10^7	1.7×10^8	3.2×10^8

[*Answer*: $8 \times 10^{10}\,mol^{-1}\,L\,s^{-1}$, $23\,kJ\,mol^{-1}$]

Once the activation energy of a reaction is known, it is a simple matter to predict the value of a rate constant k' at a temperature T' from its value k at another temperature T. To do so, we write

$$\ln k' = \ln A - \frac{E_a}{RT'}$$

and then subtract eqn (13), so obtaining

$$\ln k' - \ln k = -\frac{E_a}{RT'} + \frac{E_a}{RT}$$

This expression can be rearranged to

$$\ln\frac{k'}{k} = \frac{E_a}{R}\left(\frac{1}{T} - \frac{1}{T'}\right) \tag{15}$$

As an illustration, for a reaction with an activation energy of 50 kJ mol^{-1}, an increase in the temperature from 25 °C to 37 °C (body temperature) corresponds to

$$\ln\frac{k'}{k} = \frac{50 \times 10^3 \, \text{J mol}^{-1}}{8.3145 \, \text{J K}^{-1} \, \text{mol}^{-1}} \times \left(\frac{1}{298 \, \text{K}} - \frac{1}{310 \, \text{K}}\right) = 0.78$$

Then, by taking natural antilogarithms,

$$k' = e^{0.78} \times k = 2.18k$$

which corresponds to slightly more than a doubling of the rate constant.

Exercise E7.9

The activation energy of one of the reactions in the Krebs citric acid cycle is 87 kJ mol^{-1}. What is the change in rate constant when the temperature falls from 37 °C to 15 °C?

[*Answer*: $k' = 0.076k$]

The origin of the Arrhenius parameters: collision theory

The origin of the Arrhenius parameters can be understood most simply in terms of a certain class of gas-phase reactions in which reaction occurs when two molecules encounter one another.[†] In this **collision theory** it is supposed that such a reaction occurs only if two molecules collide with a certain minimum kinetic energy along their line of approach (Fig. 7.14). In collision theory, a reaction is supposed to resemble the collision of two billiard balls: the balls bounce apart if they collide with only a small energy, but might smash each other into fragments (products) if they collide with more than a certain minimum energy.

First, we consider the form of the **reaction profile** for the encounter of two molecules. A reaction profile in collision theory is a graph showing the

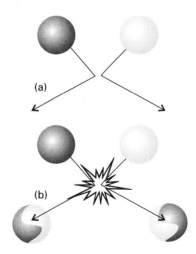

Fig. 7.14 In the collision theory of gas phase chemical reactions, reaction occurs when two molecules collide, but only if the collision is sufficiently vigorous. (a) An insufficiently vigorous collision: the reactant molecules collide but bounce apart unchanged. (b) A sufficiently vigorous collision results in a reaction.

[†]In the terminology to be introduced in Section 7.6, we are considering *bimolecular gas-phase reactions*.

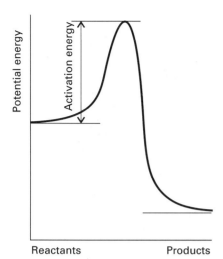

Potential energy

Activation energy

Reactants   Products

Fig. 7.15 A reaction profile. The graph depicts schematically the changing potential energy of two species that approach, collide, and then go on to form products. The activation energy is the height of the barrier above the potential energy of the reactants.

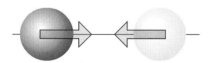

Fig. 7.16 The criterion for a successful collision is that the two reactant species should collide with a kinetic energy along their line of approach that exceeds a certain minimum value E_a that is characteristic of the reaction. The two molecules might also have components of velocity (and an associated kinetic energy) in other directions (for example, the two molecules depicted here might be moving up the page as well as towards each other); but only the energy associated with their mutual approach can be used to overcome the activation energy.

potential energy[†] of two molecules as they approach, react, and then separate as products. A typical example is shown in Fig. 7.15. On the left, the horizontal line represents the energy of the two stationary reactant molecules that are far apart from one another. The potential energy rises from this value only when the separation of the molecules is so small that they are in contact, when it rises as bonds bend and start to break. The potential energy reaches a peak when the two molecules are highly distorted. Then it starts to decrease as new bonds are formed. At separations to the right of the maximum, the potential energy rapidly falls to a low value as the product molecules separate. For the reaction to be successful, therefore, the molecules must approach with sufficient kinetic energy along their line of approach to carry them over the **activation barrier**, the peak in the reaction profile. As we shall see, we can identify the height of the activation barrier with the activation energy of the reaction.

With the reaction profile in mind, it is quite easy to establish that collision theory accounts for Arrhenius behaviour. Thus, the rate of collisions between species A and B is proportional to both of their concentrations: if the concentration of B is doubled, then the rate at which A molecules collide with B molecules is doubled, and if the concentration of A is doubled, then the rate at which B molecules collide with A molecules is also doubled. It follows that the rate of collision of A and B molecules is directly proportional to the concentrations of the two types of molecule, and we can write

$$\text{rate of collision} \propto [A][B]$$

Next, we need to multiply the collision rate by a factor f that represents the fraction of collisions that occur with at least a kinetic energy E_a along the line of approach (Fig. 7.16), for only these collisions will lead to the formation of products. Molecules that approach with less than a kinetic energy E_a will behave like a ball that rolls toward the activation barrier, fails to surmount it, and rolls back to form reactants. It follows from very general arguments (see the discussion of the Boltzmann distribution in *Further information 10*) concerning the probability that a molecule has a specified energy, that the fraction of collisions that occur with at least a kinetic energy E_a is

$$f = e^{-E_a/RT} \tag{16}$$

Exercise E7.10

What is the fraction of collisions that have sufficient energy for reaction if the activation energy is $50\,\text{kJ}\,\text{mol}^{-1}$ and the temperature is (a) $25\,°\text{C}$, (b) $500\,°\text{C}$?

[*Answer*: (a) 1.7×10^{-9}, (b) 4.2×10^{-4}]

[†]Remember that the potential energy of an object is the energy arising from its position (not speed), in this case the separation of the two reactant molecules.

At this stage we can conclude that the rate of reaction, which is proportional to the rate of collision multiplied by the fraction of successful collisions, is

$$\text{reaction rate} \propto [A][B]e^{-E_a/RT}$$

If we compare this expression with a second-order rate law,

$$\text{reaction rate} = k[A][B]$$

it follows that

$$k \propto e^{-E_a/RT} \qquad (17)$$

This expression has exactly the Arrhenius form if we identify the constant of proportionality with A. Collision theory therefore suggests the following interpretations:

- The pre-exponential factor A is the constant of proportionality between the concentrations of the reactants and the rate at which they collide.
- The activation energy E_a is the minimum kinetic energy required for a collision to result in reaction.

The value of A can be calculated from the kinetic theory of gases (Chapter 1). However, it is often found that the experimental value of A is smaller than that calculated. One possible explanation is that not only must the molecules collide with sufficient kinetic energy, but they must also come together in a specific relative orientation (Fig. 7.17). It follows that the reaction rate is proportional to the probability that the encounter occurs in the correct relative orientation. The pre-exponential factor A should therefore include a **steric factor**, P, which usually lies between 0 (no relative orientations lead to reaction) and 1 (all relative orientations lead to reaction). As an example, for the reactive collision

$$NOCl + NOCl \longrightarrow NO + NO + Cl_2$$

in which two NOCl molecules collide and break apart into two NO molecules and a Cl_2 molecule, $P \approx 0.16$. For the hydrogen addition reaction

$$H_2 + H_2C{=}CH_2 \longrightarrow H_3C{-}CH_3$$

in which a hydrogen molecule attaches directly to an ethene molecule, to form an ethane molecule, P is only 1.7×10^{-6}, which suggests that the reaction has very stringent orientational requirements.

Some reactions have $P > 1$. Such a value may seem absurd, because it appears to suggest that the reaction occurs more often than the molecules meet! An example of a reaction of this kind is

$$K + Br_2 \longrightarrow KBr + Br$$

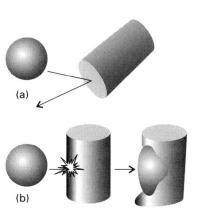

Fig. 7.17 Energy is not the only criterion of a successful reactive encounter, for relative orientation may also play a role. (a) In this schematic collision, the reactants approach in an inappropriate relative orientation, and no reaction occurs even though their energy is sufficient. (b) In this encounter, both the energy and the orientation are suitable for reaction.

255

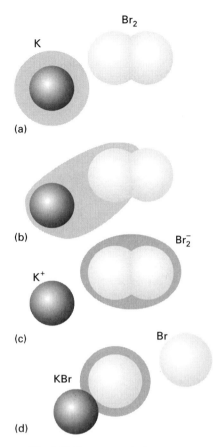

Fig. 7.18 The sequence of events that are supposed to take place in the harpoon mechanism of chemical reaction. (a) The atom and molecule approach; (b) the electron moves from the K atom on to the Br_2 molecule; (c) the two ions are drawn together by their Coulombic attraction; (d) the K^+ ion extracts a Br^- ion, and the other Br atom is released.

in which a K atom plucks a Br atom out of a Br_2 molecule; for this reaction the experimental value of P is 4.8. In this reaction, the distance of approach at which reaction can occur seems to be considerably larger than the distance needed for deflection of the path of the approaching molecules in a non-reactive collision! To explain this surprising conclusion, it has been proposed that the reaction proceeds by a **harpoon mechanism**. This brilliant name is based on a model of the reaction which pictures the K atom as approaching the Br_2 molecules, and when the two are close enough an electron (the harpoon) flips across to the Br_2 molecule. In place of two neutral particles there are now two ions, and so there is a Coulombic attraction between them: this attraction is the line on the harpoon. Under its influence the ions move together (the line is wound in), the reaction takes place, and KBr and Br emerge (Fig. 7.18). The harpoon extends the cross-section for the reactive encounter, and we greatly underestimate the reaction rate by taking for the collision cross-section the value for simple mechanical contact between K and Br_2.

Activated complex theory

A more sophisticated theory of reaction rates can be applied to reactions taking place in solution as well as in the gas phase, and is therefore applicable to a wide range of solution chemistry and biochemical processes. In the **activated complex theory** of reactions, it is supposed that as two reactants approach, their potential energy rises and reaches a maximum, as illustrated by the reaction profile in Fig. 7.19. This maximum corresponds to the formation of an **activated complex**. Unlike in collision theory, the activated complex is supposed to have a definite composition and a loose structure. It can be pictured as a cluster of atoms that is poised to pass on to products or to collapse back into the reactants from which it was formed (Fig. 7.20): an activated complex is not a reaction intermediate that can be isolated and studied like ordinary molecules. The concept of an activated complex is applicable to reactions in solutions as well as to the gas phase, because we can think of the activated complex as perhaps involving any solvent molecules that may be present.

Initially only the reactants A and B are present. As the reaction event proceeds, A and B come into contact, distort, and begin to exchange or discard atoms. The potential energy rises to a maximum, and the cluster of atoms that corresponds to the region close to the maximum is the activated complex. The potential energy falls as the atoms rearrange in the cluster, and reaches a value characteristic of the products. The climax of the reaction is at the peak of the potential energy. Here, two reactant molecules have come to such a degree of closeness and distortion that a small further distortion will send them in the direction of products. This crucial configuration is called the **transition state** of the reaction. Although some molecules entering the transition state might revert to reactants, if they pass through this configuration it is probable that products will emerge from the encounter.

As an example, consider the approach of an H atom to an F_2 molecule. For simplicity, we imagine the approach as occurring along the F—F bond direction. At great distances the potential energy is the sum of the potential energies of H and F_2. When H and F_2 are so close that their orbitals start to overlap, the F—F bond begins to stretch and a bond begins to form between H and the nearer F. The H atom comes closer, the F—F bond lengthens, the H—F bond shortens and strengthens, and the atoms enter the range of locations characteristic of the activated complex. There comes a stage when the cluster of three atoms that constitutes the activated complex has maximum potential energy and is poised at the transition state. An infinitesimal compression of the H—F bond and a stretch of F—F takes the complex through the transition state. Distances further along the reaction coordinate represent stages at which the H—F bond forms more fully and the F—F bond breaks. Motion along the reaction coordinate from left to right therefore represents the progress of H and F_2 through these configurations. Whether or not a colliding H atom and F_2 molecule actually crosses the potential barrier depends on the kinetic energy the molecules have initially, because they must be able to climb the barrier and attain the transition state.

In an actual reaction, H atoms approach F_2 molecules from all angles, and the specification of the reaction coordinate is a subtle problem; it is even more subtle for a reaction taking place in solution, for then the surrounding solvent molecule may be involved in the formation of the activated complex. We shall therefore regard the reaction coordinate simply as an indication of the distortions in the reactant molecules (and the surrounding medium, if that is relevant) as the activated complex is formed, the critical transition state is reached, and the product molecules emerge. At the transition state, motion along the reaction coordinate corresponds to some complicated collective vibration-like motion of all the atoms in the complex (and the motion of the solvent molecules).

In a simple form of activated complex theory, it is supposed that the activated complex is in equilibrium with the reactants, and that its abundance in the reaction mixture can be expressed in terms of an equilibrium constant, which is normally denoted $K^{\ddagger}$:

$$\text{reactants} \rightleftharpoons \text{activated complex} \qquad K^{\ddagger} = \frac{[\text{activated complex}]}{[\text{reactants}]}$$

Then, if we suppose that the rate at which products are formed is proportional to the concentration of the activated complex, we can write

$$\text{rate of formation of products} \propto [\text{activated complex}]$$
$$\propto K^{\ddagger}[\text{reactants}]$$

Now, by comparing this expression with the form of the rate law

$$\text{rate of formation of products} = k[\text{reactants}]$$

we see that the rate constant k is proportional to the equilibrium constant $K^{\ddagger}$ for the formation of the activated complex. We have already seen that an equilibrium constant may be expressed in terms of the standard reaction

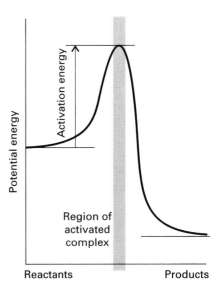

Fig. 7.19 The same type of graph as in Fig. 7.15 represents the reaction profile that is considered in activated complex theory. The activation energy is the potential energy of the activated complex relative to that of the reactants.

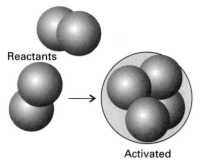

Reactants

Activated complex

Fig. 7.20 In the activated complex theory of chemical reactions, two reactants encounter each other (either in a gas phase collision or as a result of diffusing together through a solvent), and if they have sufficient energy, form an activated complex. The activated complex is depicted here by a relatively loose cluster of atoms that may undergo rearrangement into products. In an actual reaction, only some atoms—those at the actual reaction site—might be significantly loosened in the complex, the bonding of the others remaining almost unchanged. This would be the case for CH_3 groups attached to a carbon atom that was undergoing substitution.

Gibbs energy, which in this case is the **activation Gibbs energy**, $\Delta^{\ddagger}G$, for the formation of the activated complex from the reactants. It follows from eqn (3.24) that

$$K^{\ddagger} = e^{-\Delta^{\ddagger}G/RT} \qquad (18)$$

and therefore, by using

$$\Delta^{\ddagger}G = \Delta^{\ddagger}H - T\Delta^{\ddagger}S \qquad (19)$$

that

$$k \propto e^{-(\Delta^{\ddagger}H - T\Delta^{\ddagger}S)/RT}$$
$$\propto e^{\Delta^{\ddagger}S/R}e^{-\Delta^{\ddagger}H/RT} \qquad (20)$$

This expression has the form of the Arrhenius expression, eqn (14), if we identify the **enthalpy of activation**, $\Delta^{\ddagger}H$, with the activation energy and the **entropy of activation**, $\Delta^{\ddagger}S$, with the pre-exponential factor (more precisely, with $R \ln A$).

The advantage of activated complex theory over collision theory is that it is applicable to reactions in solution as well as in the gas phase. It also gives some clue to the calculation of the steric factor P, for the orientation requirements are carried in the entropy of activation. Thus, if there are strict orientation requirements (for example, in the approach of a substrate molecule to an enzyme), then the entropy of activation will be strongly negative (representing a decrease in disorder when the activated complex forms), and the pre-exponential factor will be small. In practice, it is occasionally possible to estimate the sign and magnitude of the entropy of activation and hence to estimate the rate constant. The general importance of activated complex theory is that it shows that even a complex series of events—not only a collisional encounter in the gas phase—displays Arrhenius-like behaviour, and that the concept of activation energy (and its consequences, such as eqn (15) for the effect of temperature on the rate constant) is applicable.

Exercise E7.11

In a particular reaction in water, it is proposed that two ions of opposite charge come together to form an electrically neutral activated complex. Is the contribution of the solvent to the entropy of activation likely to be positive or negative?

[*Answer*: positive, as H_2O is less organized around the neutral species]

Catalysis

Raising the temperature is one way of accelerating a reaction. Another way would be to find a means of lowering the activation energy, for then at the same temperature a higher proportion of molecules would be able to pass over the activation barrier. The height of the activation barrier for a given reaction path is, however, outside our control: it is determined by the electronic structures of the reactants and the arrangement of atoms in the activated complex. To change an activation barrier, we have to provide another route for the reaction—another reaction mechanism.

A **catalyst**, a species that increases the reaction rate but is not itself consumed in a reaction, acts by providing an alternative reaction path with a lower activation energy (Fig. 7.21). A catalyst that is in the same phase as the reactants (for example, is dissolved in the same solvent) is called a **homogeneous catalyst**, and one that is in a different phase (most commonly a solid introduced into a gas-phase reaction) is called a **heterogeneous catalyst**. Many industrial processes make use of heterogeneous catalysts, which include platinum, rhodium, and various metal oxides, but increasingly attention is turning to homogeneous catalysts, partly because they are easier to cool.

A strong acid can act as a homogenous catalyst for some reactions, and its action illustrates the general principle of catalysis, that a new reaction pathway is being provided. For instance, a strong acid can donate a proton to an organic species, and the resulting cation (the conjugate acid of the organic compound) may have a lower activation energy for reaction with another reactant (Fig. 7.22). A metal acts as a heterogeneous catalyst for certain gas-phase reactions by providing a surface to which a reactant can attach by the process of **chemisorption**, the formation of chemical bonds to a surface. For example, hydrogen molecules may attach as atoms to a nickel surface, and these atoms react much more readily with another species (such as a hydrocarbon) than the original molecules (Fig. 7.23). The chemisorption step therefore results in a reaction pathway with a lower activation energy than in the absence of the catalyst. Modern homogeneous catalysts include complexes of rhodium or palladium, which can form bonds to organic molecules, and enable them to undergo rearrangements at temperatures far below those required in the absence of a catalyst (Fig. 7.24). These complexes can be regarded as the limit of a metal surface, for their active site is just a single metal atom.

Nature stumbled on catalysts long before chemists painstakingly fabricated them. Natural catalysts are the protein molecules called **enzymes**. Despite the complexity of these large molecules, the central mode of action is the same as we have described: the enzyme provides a reaction pathway with a low activation energy, and hence accelerates the reaction for which it has evolved. We examine the kinetics of enzyme reactions in Section 7.7.

Accounting for the rate laws

We now move on to the second stage of the analysis of kinetic data, their explanation in terms of a postulated **reaction mechanism**, the sequence

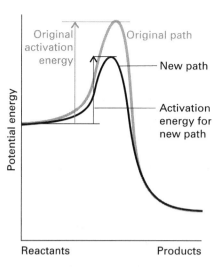

Fig. 7.21 A catalyst provides an alternative reaction pathway with a lower activation energy than the uncatalysed reaction.

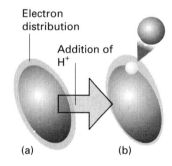

Fig. 7.22 In acid catalysis, the attachment of a proton to a species may so distort the electron distribution that subsequent attack by another reactant may be facilitated.

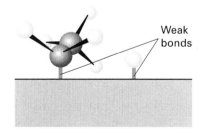

Fig. 7.23 In a reaction catalysed by a surface, a reactant chemisorbed to the surface may react with a hydrogen atom formed by the chemisorption of hydrogen.

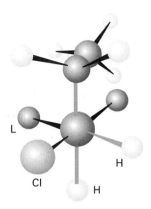

Fig. 7.24 Certain metal complexes may also act as highly localized analogues of solid surfaces: an organic material may bind to the metal atom, and hence be prepared for a subsequent reaction step.

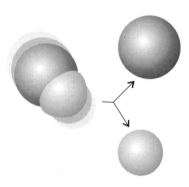

Fig. 7.25 In a unimolecular elementary reaction, an energetically excited species decomposes without further interaction with other species present in the system.

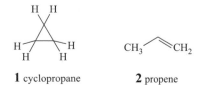

1 cyclopropane   **2** propene

of elementary molecular events that lead from the reactants to the products.

7.6 Elementary reactions

Many reactions occur in a series of steps called **elementary reactions**, each of which involves only one or two molecules. We shall denote an elementary reaction by writing its chemical equation without displaying the physical state of the species, as in

$$H + Br_2 \longrightarrow HBr + Br$$

We have already used this convention in some of the reactions discussed earlier in the chapter. This equation signifies that a specific H atom attacks a specific Br_2 molecule to produce a molecule of HBr and a Br atom. Ordinary chemical equations merely summarize the overall stoichiometry of the reaction and do not imply any specific mechanism.

The **molecularity** of an elementary reaction is the number of molecules coming together to react. In a **unimolecular reaction** a single molecule shakes itself apart or its atoms into a new arrangement (Fig. 7.25). An example is the isomerization of cyclopropane to propene (**1** → **2**). The radioactive decay of nuclei (for example, the emission of a β particle from the nucleus of a tritium atom, which is used in mechanistic studies to follow the course of particular groups of atoms) is 'unimolecular' in the sense that a single nucleus shakes itself apart. In a **bimolecular reaction**, two molecules collide and exchange energy, atoms, or groups of atoms, or undergo some other kind of change, as in the reaction between H and F_2 or between H and Br_2 (Fig. 7.26). It is important to distinguish molecularity from order:

- The order of a reaction is an empirical quantity, and is obtained by inspection of the experimentally determined rate law.
- The molecularity of a reaction refers to an individual *elementary* reaction that has been postulated as a step in a proposed mechanism.

Many substitution reactions in organic chemistry (for instance, S_N2 nucleophilic substitutions) are bimolecular and involve an activated complex that is formed from two reactant species.

The rate law of an elementary reaction (but not of an overall reaction in general) can be written down from its chemical equation. Thus, the rate law of a unimolecular elementary reaction is first-order in the reactant:

$$A \longrightarrow products \qquad rate = k[A] \qquad (21)$$

A unimolecular reaction is first-order because the number of A molecules that decay in a short interval is proportional to the number available to decay. For example, ten times as many decay in the same interval when there are initially 1000 A molecules than when there are only 100 present.

Therefore, the rate of decomposition of A is proportional to its concentration.

The rate law of an elementary bimolecular reaction is second order:

$$A + B \longrightarrow products \qquad rate = k[A][B] \qquad (22)$$

A bimolecular reaction is second order because its rate is proportional to the rate at which the reactants meet, which is proportional to both their concentrations. Therefore, if we believe (or simply postulate) that a reaction is a single step, bimolecular process, we can write down the rate law (and then, ideally, go on to test it).

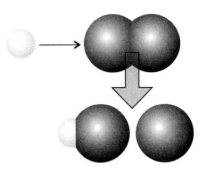

Fig. 7.26 In a bimolecular elementary reaction, two species are involved in the process.

The interpretation of an experimentally determined rate law is full of pitfalls, partly because a simple rate law can also result from a complex reaction scheme. We shall see below how to string simple steps together into a mechanism and how to arrive at the corresponding rate law. For the present we emphasize that:

If the reaction is an elementary bimolecular process, then it has second-order kinetics; however, if the kinetics are second order, then the reaction could be bimolecular but might be complex.

The postulated mechanism can be explored only by detailed detective work on the system, and by investigating whether side products or intermediates appear during the course of the reaction. Detailed analysis of this kind was one of the ways, for example, in which the reaction

$$H_2(g) + I_2(g) \longrightarrow 2\,HI(g)$$

was shown to proceed by a complex mechanism after many years during which people had accepted on good, but insufficiently meticulous evidence, that it was a fine example of a simple bimolecular reaction

$$H_2 + I_2 \longrightarrow 2\,HI$$

in which atoms exchanged partners during a collision.

7.7 The formulation of rate laws

A rate law is an experimentally determined aspect of a reaction. Once it has been determined, the next step is often to propose a reaction mechanism that is consistent with it. In this section, we describe the second of these two steps.

We introduce the technique by considering the rate law for the gas-phase oxidation of nitrogen monoxide, NO, which is found experimentally to be third-order overall:

$$2\,NO(g) + O_2(g) \longrightarrow 2\,NO_2(g)$$
$$\text{rate of formation of } NO_2 = k[NO]^2[O_2] \qquad (23)$$

261

That the reaction is third-order overall accounts for the relatively slow oxidation of nitrogen monoxide in the atmosphere (as in the formation of nitrogen oxide pollutants, NO_x) in the absence of other processes, because the rate depends on the square of the concentration of NO, which is very small if the concentration itself is small.

One explanation of the observed reaction order might be that the reaction is a single termolecular (three-molecule) elementary step; but a termolecular process is the simultaneous collision of three particles, and such collisions occur very infrequently. Therefore, although termolecular collisions may contribute, the rate of reaction by this mechanism is so slow that another mechanism usually dominates. Indeed, the additional observation that the reaction rate decreases as the temperature is raised points to a complex reaction mechanism, because simple reactions always go faster at higher temperatures.

The following mechanism has been proposed:
1. Two NO molecules combine to form a dimer:

$$NO + NO \longrightarrow N_2O_2, \qquad \text{rate of formation of } N_2O_2 = k_a[NO]^2$$

This step is plausible, because NO is an odd-electron species, and two molecules can form a covalent bond when they meet. That the N_2O_2 dimer is also known in the solid makes the suggestion plausible: it is often a good strategy to decide whether a proposed intermediate is the analogue of a known compound.
2. The N_2O_2 dimer decomposes into NO molecules:

$$N_2O_2 \longrightarrow NO + NO, \qquad \text{rate of decomposition of } N_2O_2 = k_a'[N_2O_2]$$

This step, the reverse of Step 1, is a unimolecular decay: the dimer shakes itself apart. We shall adopt the convention in which the rate constant of a reverse reaction is marked with a prime (as in k_a for the forward reaction and k_a' for its reverse).
3. Alternatively, an O_2 molecule collides with the dimer and results in the formation of NO_2:

$$N_2O_2 + O_2 \longrightarrow NO_2 + NO_2$$
$$\text{rate of consumption of } N_2O_2 = k_b[N_2O_2][O_2]$$

The rate at which NO_2 is formed in this step is

$$\text{rate of formation of } NO_2 = 2k_b[N_2O_2][O_2]$$

The 2 appears in the rate law because two NO_2 molecules are formed in each reaction event, and so the concentration of NO_2 increases at twice the rate that the concentration of N_2O_2 decays.

The steady-state approximation

Now we proceed to derive the rate law on the basis of this proposed mechanism. The rate of formation of product comes directly from Step 3:

$$\text{rate of formation of } NO_2 = 2k_b[N_2O_2][O_2]$$

However, this expression is not an acceptable overall rate law because it is expressed in terms of the intermediate N_2O_2: *an acceptable rate law for an overall reaction is expressed solely in terms of the species that appear in the overall reaction.* Therefore, we need to find an expression for the concentration of N_2O_2. To do so, we consider the net rate of formation of the intermediate, the difference between its rates of formation and decay. Because N_2O_2 is formed by Step 1 but decays by Steps 2 and 3, its net rate of formation is

$$\text{net rate of formation of } N_2O_2 = k_a[NO]^2 - k_a'[N_2O_2] - k_b[N_2O_2][O_2]$$

At this stage we introduce the **steady-state approximation**, in which it is supposed that the concentrations of all intermediates remain constant and small throughout the reaction (except right at the beginning and right at the end). For our purposes, we identify the intermediate (which, in general, is any species that does not appear in the overall reaction but which has been invoked in the mechanism) as N_2O_2, and write

$$\text{net rate of formation of } N_2O_2 = 0$$

It follows from the expression above that

$$k_a[NO]^2 - k_a'[N_2O_2] - k_b[N_2O_2][O_2] = 0$$

This equation can be rearranged to give an equation for the concentration of N_2O_2:

$$[N_2O_2] = \frac{k_a[NO]^2}{k_a' + k_b[O_2]}$$

It follows that the rate of formation of NO_2 is

$$\text{rate of formation of } NO_2 = 2k_b[N_2O_2][O_2] = \frac{2k_ak_b[NO]^2[O_2]}{k_a' + k_b[O_2]} \quad (24)$$

At this stage, the rate law is more complex than the observed law, but resembles it. The two expressions become identical if we suppose that the rate of decomposition of the dimer is much greater than its rate of reaction with oxygen, for then

$$k_a'[N_2O_2] \gg k_b[N_2O_2][O_2]$$

or, after cancelling the $[N_2O_2]$,

$$k_a' \gg k_b[O_2]$$

When this condition is satisfied, the denominator in the overall rate law can be approximated by k_a' alone, and we conclude that

$$\text{rate of formation of } NO_2 \approx \left(\frac{2k_a k_b}{k_a'}\right)[NO]^2[O_2] \qquad (25)$$

which has the observed overall third-order form, eqn (23). Moreover, we can identify the observed rate constant as the following combination of rate constants for the elementary reactions:

$$k = \frac{2k_a k_b}{k_a'} \qquad (26)$$

The proposed mechanism is consistent with the anomalous temperature dependence, because although each one of the elementary rate constants increases with temperature, if k_a' increases more rapidly than $k_a k_b$ increases, then the overall rate constant k will *decrease* with increasing temperature and the reaction will go more slowly. That k_a' has a strong temperature dependence is consistent with the mechanism, for the decomposition step, which relies on the dissociation of the dimer into NO molecules, can be expected to have a high activation energy and, as we have seen, a high activation energy implies that a reaction has a rate that depends strongly on temperature. The general conclusion is that, although the rate constants of elementary reactions almost invariably increase with increasing temperature, the observed rate constant, being a composite of several elementary rate constants, may either increase or decrease with temperature. In practice, most composite rate constants increase with temperature, so the rates of most reactions do increase with temperature.

Exercise E7.12

An alternative mechanism that may apply when the concentration of O_2 is high and that of NO is low is one in which the first step is $NO + O_2 \longrightarrow NO\!-\!O_2$ and its reverse, followed by $NO\!-\!O_2 + NO \longrightarrow NO_2 + NO_2$. Confirm that this mechanism also leads to the observed rate law when the concentration of NO is low.

[*Answer*:
$$\text{rate} = 2k_a k_b[NO]^2[O_2]/(k_a' + k_b[NO]) \approx (2k_a k_b/k_a')[NO]^2[O_2]]$$

The rate-determining step

The mechanism of oxidation of nitrogen monoxide can be used to introduce another important concept. Suppose that the rate of Step 3 is very fast, so that k_a' may be neglected relative to $k_b[O_2]$ in eqn (24). (One way to achieve this condition is to increase the concentration of O_2 in the reaction mixture.) Then eqn (24) simplifies to

$$\text{rate of formation of } NO_2 \approx \frac{2k_a k_b [NO]^2 [O_2]}{k_b [O_2]} = 2k_a [NO]^2 \qquad (27)$$

Now the reaction is second order in NO, and the concentration of O_2 does not appear in the rate law. The explanation is that the rate of reaction of N_2O_2 is so great (on account of the high concentration of O_2 in the system), that as soon as any N_2O_2 is formed, it reacts. Therefore, the rate of formation of NO_2 is determined by the rate at which N_2O_2 is formed, which is a bimolecular, second-order elementary process. The formation of N_2O_2 in this mechanism, and in the presence of a high concentration of O_2, is an example of a **rate-determining step**, the slowest step in a reaction mechanism, which controls the rate of the overall reaction. The rate-determining step is like a slow ferry crossing between two fast highways: the overall rate at which traffic can reach its destination is determined by the rate at which it can make the ferry crossing.

When the concentration of O_2 is reduced to the point that eqn (25) is applicable, the rate determining step in the reaction becomes the slow reaction of N_2O_2 with the scarce O_2 molecules. Now the rate of the overall reaction is determined by the value of k_b, the rate constant of the slowest step. It also depends on the ratio of the rate constants for the fast forward and reverse steps, k_a and k_a'. The latter dependence can be explained by considering the extreme case in which these two reactions are so fast compared with the rate determining step that they reach a state of dynamic equilibrium. The two rates are then equal, and by setting

$$k_a [NO]^2 = k_a' [N_2O_2]$$

we can deduce that the ratio of rate constants is equal to the equilibrium constant for the formation of the intermediate:

$$K = \frac{[N_2O_2]}{[NO]^2} = \frac{k_a}{k_a'}$$

Therefore, the rate constant in eqn (26) can be written

$$k = 2K k_b \qquad (28)$$

The first factor K effectively determines the concentration of the reaction intermediate and the second, the rate constant for the slow step, the rate at which that intermediate forms products.

That the rate determining step is the N_2O_2 formation reaction when the O_2 concentration is high, but is the rate of the reaction of N_2O_2 when the O_2 concentration is low, emphasizes that the rate determining step is not necessarily a fixed quantity, but may switch from one elementary step to another as the conditions are changed.

The Michaelis–Menten mechanism of enzyme action

Another example of a reaction in which an intermediate is formed is the **Michaelis–Menten mechanism** of enzyme action. The rate of an enzyme catalysed reaction in which a substrate S is converted into products P,

$$S \xrightarrow{\text{E}} P$$

is found to depend on the concentration of the enzyme E even though the enzyme undergoes no net change, and a typical rate law was given at the start of the chapter:

$$\text{rate of formation of product} = \frac{k[\text{E}][\text{S}]}{[\text{S}] + K_M} \tag{29}$$

The proposed mechanism (with all species in an aqueous environment) is

Step 1: The bimolecular formation of a combination, ES, of the enzyme and the substrate:

$$\text{E} + \text{S} \longrightarrow \text{ES} \qquad \text{rate of formation of ES} = k_a[\text{E}][\text{S}]$$

Step 2: The unimolecular decomposition of the complex:

$$\text{ES} \longrightarrow \text{E} + \text{S} \qquad \text{rate of decomposition of ES} = k_a'[\text{ES}]$$

Step 3: The unimolecular formation of products and the release of the enzyme from its combination with the substrate:

$$\text{ES} \longrightarrow \text{P} + \text{E} \qquad \text{rate of formation of P} = k_b[\text{ES}]$$
$$\text{rate of consumption of ES} = k_b[\text{ES}]$$

We seek the rate law for the rate of formation of product, which according to Step 3 is

$$\text{rate of formation of P} = k_b[\text{ES}]$$

As shown in the following *Justification*, the resulting rate law is

$$\text{rate of formation of P} = k[\text{E}]_0, \qquad \text{with } k = \frac{k_b[\text{S}]}{[\text{S}] + K_M} \tag{30}$$

where the **Michaelis constant**, K_M, is

$$K_M = \frac{k_a' + k_b}{k_a} \tag{31}$$

and the concentration of the complex ES is

$$[ES] = \frac{[E][S]}{K_M} \qquad (32)$$

Justification

To make progress, we need to know the concentration of the intermediate ES (which cannot occur in the overall rate law). In accord with the steady-state approximation, we set up an expression for the net rate of formation of ES (allowing for its formation in Step 1 and its removal in Steps 2 and 3), and then set that net rate equal to zero:

net rate of formation of ES $= k_a[E][S] - k_a'[ES] - k_b[ES] = 0$

It follows that

$$[ES] = \frac{k_a[E][S]}{k_a' + k_b}$$

However, there is now a small complication: [E] and [S] are the molar concentrations of the free enzyme and free substrate, and if $[E]_0$ is the total concentration of enzyme, then

$$[E] + [ES] = [E]_0$$

Because only a little enzyme is added, the free substrate concentration is almost the same as the total substrate concentration, and we can ignore the fact that [S] differs slightly from $[S] + [ES]$. Therefore,

$$[ES] = \frac{k_a([E]_0 - [ES])[S]}{k_a' + k_b}$$

which rearranges to

$$[ES] = \frac{k_a[E]_0[S]}{k_a' + k_b + k_a[S]}$$

It follows that the rate of formation of product is

rate of formation of P $= k[E]_0$,   with $k = \dfrac{k_b[S]}{[S] + K_M}$ $\qquad (33)$

and K_M given by eqn (31).

According to eqn (30), the rate of enzymolysis is first-order in the enzyme concentration, but the effective rate constant k depends on the concentration of substrate. When $[S] \gg K_M$, the effective rate constant is equal to k_b, and the rate law in eqn (30) reduces to

$$\text{rate of formation of P} = k_b[E]_0 \qquad (34)$$

The rate is independent of the concentration of S because there is so much substrate present that it remains at effectively the same concentration even though products are being formed. Moreover, the rate is a maximum, and $k_b[E]_0$ is called the **maximum velocity**, v_{max}, of the enzymolysis:

$$v_{max} = k_b[E]_0 \qquad (35)$$

The constant k_b is called the **maximum turnover number**. The rate determining step is Step 3, because there is ample ES present (because S is so abundant), and the rate is determined by the rate at which ES reacts to form the product. It follows from eqn (30) that the reaction rate v at a general substrate composition is related to the maximum velocity by

$$v = \frac{[S]}{[S] + K_M} \times v_{max} \qquad (36)$$

Equation (36) is the basis of the analysis of enzyme kinetic data by using a **Lineweaver–Burk plot**, a graph of $1/v$ (the reciprocal of the reaction rate) against $1/[S]$ (the reciprocal of the substrate concentration). If we take the reciprocal of both sides of eqn (36) it becomes

$$\frac{1}{v} = \frac{[S] + K_M}{v_{max}[S]} = \frac{1}{v_{max}} + \left(\frac{K_M}{v_{max}}\right)\frac{1}{[S]} \qquad (37)$$

Because this expression has the form

$$y = \text{intercept} + \text{slope} \times x$$

with $y = 1/v$ and $x = 1/[S]$, if $1/v$ is plotted against $1/[S]$, a straight line should be obtained. The slope of the straight line is K_M/v_{max} and the extrapolated intercept at $1/[S] = 0$ is equal to $1/v_{max}$ (Fig. 7.27). Therefore, the intercept can be used to find v_{max}, and then that value combined with the slope to find the value of K_M. Alternatively, note that the extrapolated intercept with the horizontal axis (where $1/v = 0$) occurs at $1/[S] = -1/K_M$.

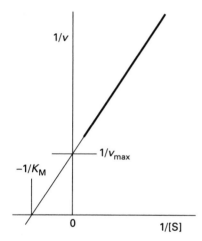

Fig. 7.27 A Lineweaver–Burk plot is used to analyse kinetic data on enzyme-catalysed reactions. The reciprocal of the rate of formation of products ($1/v$) is plotted against the reciprocal of the substrate concentration ($1/[S]$). All the data points (which typically lie in the black region of the line) correspond to the same overall enzyme concentration, $[E]_0$. The intercept of the extrapolated (grey) straight line with the horizontal axis is used to obtain the Michaelis constant, K_M. The intercept with the vertical axis, is used to determine $v_{max} = k_b[E]_0$, and hence k_b. The slope may also be used, for it is equal to K_M/v_{max}.

Exercise E7.13

Show that a plot of v against $v/[S]$ is an alternative route to the value of K_M.

[*Answer*: $v = v_{max} - K_M \times (v/[S])$]

Enzyme inhibition

The action of an enzyme may be partially suppressed by the presence of a foreign substance, which is called an **inhibitor**. An inhibitor, which we denote I, may be a poison that has been administered (perhaps accidentally) to the organism, or it may be a substance that is naturally present in a cell, and which is a component of the regulatory mechanism of the cell. The use of the techniques we have been developing for investigating the mode of action of enzymes can be illustrated by the problem of deciding whether the inhibition of an enzyme is competitive or noncompetitive. In **competitive inhibition** the inhibitor competes for the active site and so reduces the ability of the enzyme to bind the substrate (Fig. 7.28). In **noncompetitive inhibition** the inhibitor does not compete for the active site; instead, it attaches to another part of the enzyme molecule, thereby distorting it and reducing its ability to bind the substrate (Fig. 7.29). We shall now show that the two kinds of inhibition can be distinguished by making use of kinetic data on the rate of enzyme action.

First, consider competitive inhibition. We suppose that the inhibitor molecule I is in equilibrium with the complex EI it forms when it is bound to the active site:

$$EI \rightleftharpoons E + I \qquad K_I = \frac{[E][I]}{[EI]} \tag{38}$$

The rate of formation of product turns out to be

$$\text{rate of formation of P} = \frac{k_b[S][E]_0}{[S] + (1 + [I]/K_I)K_M} \tag{39}$$

If we compare this expression with eqn (30), we see that the role of the inhibitor is to modify K_M; therefore, in a Lineweaver–Burk plot, the slope and the intercept with the horizontal axis change as [I] is changed, but the intercept with the vertical axis (the value of $v_{max} = k_b[E]_0$) remains unchanged (Fig. 7.30).

Justification

The only change between this case and the case of no inhibition, is that some of the enzyme is uselessly linked to the inhibitor, so the total concentration of enzyme is

$$[E]_0 = [E] + [ES] + [EI]$$

in place of $[E] + [ES]$ in the former calculation. It remains the case that

$$[ES] = \frac{[E][S]}{K_M}, \qquad \text{rate of formation of P} = k_b[ES]$$

The first of these expressions can be combined with

$$[EI] = \frac{[E][I]}{K_I}$$

to write

$$[E]_0 = [E] + \frac{[E][S]}{K_M} + \frac{[E][I]}{K_I}$$

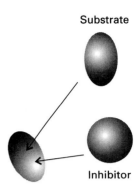

Fig. 7.28 In competitive inhibition, both the substrate (the egg shape) and the inhibitor compete for the active site, and reaction ensues only is the substrate is successful in attaching there.

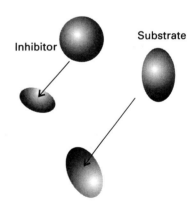

Fig. 7.29 In one version of noncompetitive inhibition, the substrate and the inhibitor attach to distant sites of the enzyme molecule, and a complex in which they are both attached (IES) does not lead to the formation of product.

269

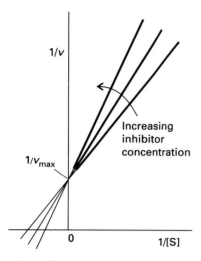

Fig. 7.30 A Lineweaver–Burk plot may be used to distinguish competitive and noncompetitive inhibition kinetically. An indication that competitive inhibition is occurring is that the intercept with the vertical axis does not move as the concentration of inhibitor is increased.

It follows that

$$[E] = \frac{[E]_0}{1 + [S]/K_M + [I]/K_I}$$

from which it follows that

$$[ES] = \frac{[E][S]}{K_M} = \frac{[S][E]_0}{[S] + K_M(1 + [I]/K_I)}$$

When substituted into the rate equation for the formation of product, we obtain eqn (39).

Now consider noncompetitive inhibition. We suppose that the inhibitor is in equilibrium with a bound state IE, but that the site occupied by I is not the active site for the attachment of S (which is why we write it IE and not EI, to suggest the use of a site that is distant from the one used to form ES). Moreover, because I and S are not in competition for the same site, I may also bind to the complex ES to give a complex that we shall denote IES:

$$IES \rightleftharpoons I + ES \qquad K_I' = \frac{[I][ES]}{[IES]} \tag{40}$$

We suppose that although I and S may both bind to E, the enzyme can bring about a change in S only if I is not present. (The presence of I in IE allows S to bind, but so affects the structure of the enzyme that the latter cannot carry out its function.) Therefore, only ES can give rise to products; IES cannot. In this scenario, the rate of formation of product turns out to follow the rate law

$$\text{rate of formation of P} = \frac{k_b[S][E]_0}{([S] + K_M)(1 + [I]/K_I)} \tag{41}$$

In a Lineweaver–Burk plot with different values of [I], the straight lines now pass through a common intercept with the horizontal axis (because $1/v = 0$ at $1/[S] = -1/K_M$ independent of the inhibitor properties), but the slope and the intercept with the vertical axis both increase as the concentration of inhibitor is increased (Fig. 7.31).

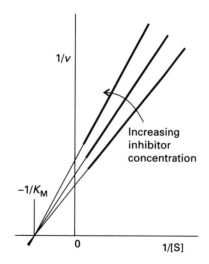

Fig. 7.31 When there is noncompetitive inhibition, the intercept with the horizontal axis does not move as the concentration of inhibitor is increased.

Justification

The total concentration of enzyme is now given by the expression

$$[E]_0 = [E] + [ES] + [IE] + [IES]$$

The concentrations of the complexes are given by the same equations as before, but with the addition of the equilibrium for the double complex IES:

$$[ES] = \frac{[E][S]}{K_M} \qquad [IE] = \frac{[I][E]}{K_I} \qquad [IES] = \frac{[I][ES]}{K_I'}$$

From now on we shall suppose that the equilibrium between I and its bound state with the complex is independent of whether S is attached to the enzyme (because the sites are so distant), and hence set K_I' equal to K_I. With this simplification in mind, we can express the total concentration of enzyme as

$$
\begin{aligned}
[E]_0 &= \frac{K_M[ES]}{[S]} + [ES] + \frac{K_M[I][ES]}{K_I[S]} + \frac{[I][ES]}{K_I} \\
&= \frac{[ES]}{[S]}\left(K_M + [S] + \frac{K_M[I]}{K_I} + \frac{[I][S]}{K_I} \right) \\
&= \frac{[ES]}{[S]}\left(1 + \frac{[I]}{K_I} \right)(K_M + [S])
\end{aligned}
$$

This expression can be rearranged into an equation for [ES]:

$$[ES] = \frac{[S][E]_0}{([S] + K_M)(1 + [I]/K_I)}$$

The rate of formation of product is proportional to [ES] (because we are assuming that IES is inactive), and so at this stage we can use the same expression for the rate as in the noninhibited case,

$$\text{rate of formation of P} = k_b[ES]$$

When the expression for [ES] that we have just calculated is substituted into that rate equation, we obtain eqn (41).

Now we can see how the distinction between competitive and noncompetitive inhibition can be made. A series of Lineweaver–Burk plots are made for different inhibitor concentrations; if the plots resemble those in Fig. 7.30, then the inhibition is competitive. On the other hand, if the plots resemble those in Fig. 7.31, then the inhibition is noncompetitive.

7.8 Unimolecular reactions

A number of gas phase reactions follow first-order kinetics, as in the isomerization of cyclopropane mentioned earlier, when the strained triangular molecule bursts apart into an acyclic alkene:

$$cyclo\text{-}C_3H_6 \longrightarrow CH_3CH{=}CH_2 \qquad \text{rate} = k[cyclo\text{-}C_3H_6]$$

The problem with explaining first-order rate laws is based on the expectation that a molecule acquires enough energy to react as a result of its collisions with other molecules. The cyclopropane molecule, for example, needs energy to overcome the activation barrier to breaking a carbon–carbon bond and reorganizing the arrangement of hydrogen atoms. Collisions, though, are simple bimolecular events, and so how can they result in a first-order rate law? First-order gas-phase reactions are widely called **unimolecular reactions** because the rate-determining step is an elementary unimolecular reaction in which the reactant molecule changes into the product. This term must be used with caution, however, because the composite mechanism has bimolecular as well as unimolecular steps.

The first successful explanation of unimolecular reactions is ascribed to Frederick Lindemann in 1921.[†] The **Lindemann mechanism** is as follows:

Step 1. A reactant molecule A becomes energetically excited (denoted A^*) by collision with another A molecule:

$$A + A \longrightarrow A^* + A \qquad \text{rate of formation of } A^* = k_a[A]^2$$

Step 2. The energized molecule might lose its excess energy by collision with another molecule:

$$A^* + A \longrightarrow A + A \qquad \text{rate of deactivation of } A^* = k_a'[A^*][A]$$

Step 3. Alternatively, the excited molecule might shake itself apart (as may happen with vibrationally excited cyclopropane) and form products P. That is, it might undergo the unimolecular decay

$$A^* \longrightarrow P \qquad \text{rate of formation of } P = k_b[A^*]$$
$$\text{rate of consumption of } A^* = k_b[A^*]$$

If the unimolecular step, Step 3, is slow enough to be the rate-determining step, the overall reaction will have first-order kinetics, as observed. We can demonstrate this explicitly by applying the steady-state approximation to the net rate of formation of A^*:

$$\text{net rate of formation of } A^* = k_a[A]^2 - k_a'[A^*][A] - k_b[A^*] = 0$$

This equation solves to

$$[A^*] = \frac{k_a[A]^2}{k_b + k_a'[A]}$$

[†]K.J. Laidler, in his *Chemical kinetics* (Harper and Row, 1987), gives an interesting historical summary of the origin of the mechanism. Apparently, Lindemann sketched the mechanism at a meeting and published a brief note; almost simultaneously a young Danish doctoral student, J.A. Christiansen, published his Ph.D. thesis in which the same mechanism was proposed and developed in much greater detail. The 'Lindemann–Christiansen mechanism' would therefore appear to be a fairer name than the conventional 'Lindemann mechanism'.

It follows that the rate law for the formation of products is

$$\text{rate of formation of P} = k_b[A^*] = \frac{k_a k_b[A]^2}{k_b + k_a'[A]} \qquad (42)$$

At this stage the rate law is not first order in A. However, if the rate of deactivation by (A^*, A) collisions is much greater than the rate of unimolecular decay, in the sense that $k_a'[A^*][A] \gg k_b[A^*]$, which corresponds to

$$k_a'[A] \gg k_b \qquad (43)$$

then we can neglect k_b in the denominator of the rate law and obtain

$$\text{rate of formation of P} = k[A], \quad \text{with } k = \frac{k_a k_b}{k_a'} \qquad (44)$$

Equation (44) is a first-order rate law, as we set out to show.

Exercise E7.14

Suppose that an inert gas M is present and dominates the excitation of A and de-excitation of A^*. Devise the rate law for the formation of products.

$$[\text{rate} = k_a k_b[A][M]/(k_b + k_a'[M])]$$

Chain reactions

Many gas-phase reactions and liquid-phase polymerization reactions are **chain reactions**, reactions in which an intermediate produced in one step generates a reactive intermediate in a subsequent step, then that intermediate generates another reactive intermediate, and so on.

7.9 The structure of chain reactions

The intermediates responsible for the propagation of a chain reaction are called **chain carriers**. In a radical chain reaction the chain carriers are radicals. Ions may also propagate chains, and in nuclear fission the chain carriers are neutrons.

The first chain carriers are formed in the **initiation step** of the reaction. For example, Cl atoms are formed by the dissociation of Cl_2 molecules either as a result of vigorous intermolecular collisions in a **thermolysis reaction** or as a result of absorption of a photon in a **photolysis reaction**. The chain carriers produced in the initiation step attack other reactant molecules in the **propagation steps**, and each attack gives rise to a new chain carrier. An example is the attack of a methyl radical on ethane:

$$\cdot CH_3 + CH_3CH_3 \longrightarrow CH_4 + \cdot CH_2CH_3$$

The dot signifies the unpaired electron and marks the radical. In some cases the attack results in the production of more than one chain carrier. An example of such a **branching step** is

$$\cdot O \cdot + H_2O \longrightarrow HO \cdot + HO \cdot$$

where the attack of one O atom on an H_2O molecule forms two $\cdot OH$ radicals.[†]

The chain carrier might attack a product molecule formed earlier in the reaction. Because this attack decreases the net rate of formation of product, it is called a **retardation step**. For example, in a photochemical reaction in which HBr is formed from H_2 and Br_2, an H atom might attack an HBr molecule, leading to H_2 and Br:

$$\cdot H + HBr \longrightarrow H_2 + \cdot Br$$

Retardation does not end the chain, because one radical ($\cdot H$) gives rise to another ($\cdot Br$), but it does deplete the concentration of the product. Elementary reactions in which radicals combine and end the chain are called **termination steps**, as in

$$CH_3CH_2 \cdot + \cdot CH_2CH_3 \longrightarrow CH_3CH_2CH_2CH_3$$

In an **inhibition step**, radicals are removed other than by chain termination, such as by reaction with the walls of the vessel or with foreign radicals:

$$CH_3CH_2 \cdot + \cdot R \longrightarrow CH_3CH_2R$$

The NO molecule has an unpaired electron and is a very efficient chain inhibitor. The observation that a gas-phase reaction is quenched when NO is introduced is a good indication that a radical chain mechanism is in operation.

7.10 The rate laws of chain reactions

A chain reaction often leads to a complicated rate law (but not always). As a first example, consider the thermal reaction of H_2 with Br_2. The overall reaction and the observed rate law are

$$H_2(g) + Br_2(g) \longrightarrow 2\,HBr(g)$$

$$\text{rate of formation of HBr} = \frac{k[H_2][Br_2]^{3/2}}{[Br_2] + k'[HBr]} \tag{45}$$

[†] In the notation to be introduced in Chapter 8, an O atom has the configuration $[He]2s^22p^4$, with two unpaired electrons.

The complexity of the rate law suggests that a complicated mechanism is involved, and the following radical chain mechanism has been proposed:

Step 1. Initiation:

$$Br_2 \longrightarrow 2\,Br\cdot \qquad \text{rate of consumption of } Br_2 = k_a[Br_2]$$

At low pressures this elementary reaction is bimolecular and second-order in Br_2.

Step 2. Propagation:

$$Br\cdot + H_2 \longrightarrow HBr + H\cdot \qquad \text{rate} = k_b[Br][H_2]$$
$$H\cdot + Br_2 \longrightarrow HBr + Br\cdot \qquad \text{rate} = k_b'[H][Br_2]$$

In this and the following steps, 'rate' means either the rate of formation of one of the products or the rate of consumption of one of the reactants. We shall specify the species only if the rates differ.

Step 3. Retardation:

$$H\cdot + HBr \longrightarrow H_2 + Br\cdot \qquad \text{rate} = k_c[H][HBr]$$

Step 4. Termination:

$$Br\cdot + \cdot Br + M \longrightarrow Br_2 + M \qquad \text{rate of formation of } Br_2 = k_d[Br]^2$$

The third body M, a molecule of an inert gas, removes the energy of recombination; the constant concentration of M has been absorbed into the rate constant k_d. Other possible termination steps include the recombination of H atoms to form H_2 and the combination of H and Br atoms; however, it turns out that only Br atom recombination is important.

Now we establish the rate law for the reaction. The experimental rate law is expressed in terms of the rate of formation of product, HBr, so we start by writing an expression for its net rate of formation. Because HBr is formed in Step 2 (by both reactions) and consumed in Step 3,

$$\begin{aligned} &\text{net rate of formation of HBr} \\ &\quad = k_b[Br][H_2] + k_b'[H][Br_2] - k_c[H][HBr] \end{aligned} \tag{46}$$

To make progress, we need the concentrations of the intermediates Br and H. Therefore, we set up the expressions for their net rate of formation and apply the steady state assumption to both:

$$\text{net rate of formation of H} = k_b[Br][H_2] - k_b'[H][Br_2] - k_c[H][HBr] = 0$$
$$\begin{aligned} \text{net rate of formation of Br} &= 2k_a[Br_2] - k_b[Br][H_2] + k_b'[H][Br_2] \\ &\quad + k_c[H][HBr] - 2k_d[Br]^2 = 0 \end{aligned}$$

The steady-state concentrations of the intermediates are found by solving these two equations and are

$$[Br] = \left(\frac{k_a[Br_2]}{k_d}\right)^{1/2}$$

$$[H] = \frac{k_b(k_a/k_d)^{1/2}[Br_2]^{1/2}}{k_b'[Br_2] + k_c[HBr]}$$

When we substitute these concentrations into eqn (46) we obtain

$$\text{rate of formation of HBr} = \frac{2k_b(k_a/k_d)^{1/2}[H_2][Br_2]^{3/2}}{[Br_2] + (k_c/k_b')[HBr]} \qquad (47)$$

This equation has the same form as the empirical rate law, and we can identify the two empirical rate coefficients as

$$k = 2k_b\left(\frac{k_a}{k_d}\right)^{1/2} \qquad k' = \frac{k_c}{k_b'} \qquad (48)$$

We can conclude that the proposed mechanism is at least consistent with the observed rate law. Additional support for the mechanism would come from the detection of the proposed intermediates (by spectroscopy), and the measurement of individual rate constants for the elementary steps and confirming that they correctly reproduced the observed composite rate constants.

7.11 Explosions

A **thermal explosion** is due to the rapid increase of reaction rate with temperature. If the energy released in an exothermic reaction cannot escape, the temperature of the reaction system rises, and the reaction goes faster. The acceleration of the rate results in a faster rise of temperature, and so the reaction goes even faster... catastrophically fast. A **chain-branching explosion** may occur when there are chain branching steps in a reaction, for then the number of chain carriers grows exponentially and the rate of reaction may cascade into an explosion.

An example of both types of explosion is provided by the reaction between hydrogen and oxygen:

$$2\,H_2(g) + O_2(g) \longrightarrow 2\,H_2O(g)$$

Although the net reaction is very simple, the mechanism is very complex and has not yet been fully elucidated. It is known that a chain reaction is involved, and that the chain carriers include $\cdot H$, $\cdot O\cdot$, $\cdot OH$, and $\cdot O_2H$. Some steps are:

$$\text{Initiation :} \quad H_2 + O_2 \longrightarrow \cdot O_2H + \cdot H$$
$$\text{Propagation :} \quad O_2 + \cdot H \longrightarrow \cdot O \cdot + \cdot OH \quad \text{(branching)}$$
$$\cdot O \cdot + H_2 \longrightarrow \cdot OH + \cdot H \quad \text{(branching)}$$
$$H_2 + \cdot OH \longrightarrow \cdot H + H_2O$$

The two branching steps can lead to a chain-branching explosion.

The occurrence of an explosion depends on the temperature and pressure of the system, and the explosion regions for the reaction are shown in Fig. 7.32. At very low pressures, the system is outside the explosion region and the mixture reacts smoothly. At these pressures the chain carriers produced in the branching steps can reach the walls of the container where they combine (with an efficiency that depends on the composition of the walls). Increasing the pressure of the mixture (along the broken line in the illustration) takes the system through the **lower explosion limit** (provided that the temperature is greater than about 730 K). The mixture then explodes because the chain carriers react before reaching the walls and the branching reactions are explosively efficient. The reaction is smooth when the pressure is above the **upper explosion limit**. The concentration of molecules in the gas is then so great that the radicals produced in the branching reaction combine in the body of the gas, and gas-phase reactions such as $O_2 + \cdot H \longrightarrow \cdot O_2H$ can occur. Recombination reactions like this are facilitated by three-body collisions, because the third body (M) can remove the excess energy and allow the formation of a bond:

$$O_2 + \cdot H + M \longrightarrow \cdot O_2H + M^*$$

The radical $\cdot OH_2$ is relatively unreactive and can reach the walls, where it is removed. At low pressures three-particle collisions are unimportant and recombination is much slower. At higher pressures, when three-particle collisions are important, the explosive propagation of the chain by the radicals produced in the branching step is partially quenched because $\cdot O_2H$ is formed in place of $\cdot O \cdot$ and $\cdot OH$. If the pressure is increased to above the third explosion limit the reaction rate increases so much that a thermal explosion occurs.

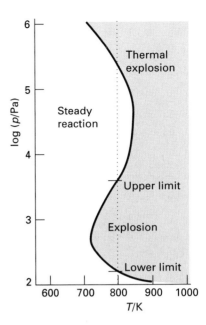

Fig. 7.32 The explosion limits of the reaction between hydrogen and oxygen. In the explosive regions, the reaction proceeds explosively when heated homogeneously.

7.12 Photochemical reactions

Many reactions can be initiated by the absorption of light. The most important of all are the photochemical processes that capture the Sun's radiant energy. Some of these reactions lead to the heating of the atmosphere during the daytime by absorption in the ultraviolet region as a result of reactions like those depicted in Fig. 7.33. Others include the absorption of red and blue light by chlorophyll and the subsequent use of the energy to bring about the synthesis of carbohydrates from carbon dioxide and water. Without photochemical processes the world would be simply a warm, sterile, rock.

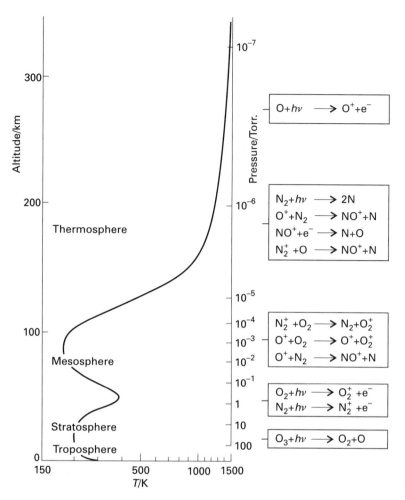

Fig. 7.33 The temperature profile through the atmosphere and some of the reactions that occur. The temperature peak at about 50 km is due to the absorption of solar radiation by the O_2 and N_2 reactions.

Quantum yield

A molecule acquires enough energy to react by absorbing photons. The **Stark–Einstein law** states that one photon is absorbed by each molecule responsible for the primary photochemical process. The law is valid under normal conditions (when the radiant intensity is not very high), but is not valid when high intensity lasers are the source of radiation, for they produce beams of light of such high photon density that a molecule may absorb more than one photon before undergoing reaction. Even when the law is obeyed, it leaves open the possibility that an excited molecule does not lead to products: there are many ways in which the excitation may be lost other than by dissociation or ionization. We therefore speak of the **primary quantum yield**, ϕ, the number of reactant molecules producing specified primary products (atoms or ions, for instance) for each photon absorbed.

The primary product of photon absorption—a radical, a photoexcited molecule, or an ion—may be successful in initiating a process that leads to products. Now we have to be aware that one successfully excited molecule might initiate the consumption of more than one reactant molecule. We therefore need to introduce the **overall quantum yield**, Φ, which is the number of reactant molecules that react for each photon absorbed. In the photolysis of HI, for example, the processes are

$$HI + h\nu \longrightarrow H + I$$
$$H + HI \longrightarrow H_2 + I$$
$$2I \longrightarrow I_2$$

The overall quantum yield is 2 because the absorption of one photon leads to the destruction of two HI molecules. In a photochemically initiated chain reaction, Φ may be very large, and values of about 10^4 are common. In such cases the chain reaction acts as a chemical amplifier of the initial absorption step.

Example *Using the quantum yield*
The overall quantum yield for the formation of ethene from 4-heptanone with 313 nm light is 0.21. How many molecules of 4-heptanone per second, and what chemical amount per second, are destroyed when the sample is irradiated with a 50 W, 313 nm source under conditions of total absorption?

Strategy
We need to calculate the number of photons emitted by the lamp per second; all are absorbed (by assertion); the number of molecules destroyed per second is the number of photons absorbed multiplied by the overall quantum yield Φ. The number of photons emitted by the source per second is the power (joules per second) divided by the energy of a single photon ($E = h\nu$, with $\nu = c/\lambda$).

Solution
The energy of a photon of wavelength 313 nm is

$$E = \frac{hc}{\lambda} = \frac{(6.626\,08 \times 10^{-34}\,\text{J s}) \times (2.997\,92 \times 10^8\,\text{m s}^{-1})}{313 \times 10^{-9}\,\text{m}}$$
$$= 6.35 \times 10^{-19}\,\text{J}$$

A 50 W ($50\,\text{J s}^{-1}$) source therefore generates photons at a rate

$$\text{rate of photon production} = \frac{50\,\text{J s}^{-1}}{6.35 \times 10^{-19}\,\text{J}}$$
$$= 7.9 \times 10^{19}\,\text{s}^{-1}$$

The number of 4-heptanone molecules destroyed per second is therefore 0.21 times this quantity, or

$$\text{rate of 4-heptanone consumption} = 0.21 \times (7.9 \times 10^{19}\,\text{s}^{-1})$$
$$= 1.7 \times 10^{19}\,\text{s}^{-1}$$

Division by the Avogadro constant gives

$$\text{rate of 4-heptanone consumption} = \frac{1.7 \times 10^{19}\,\text{s}^{-1}}{6.022\,14 \times 10^{23}\,\text{mol}^{-1}}$$
$$= 2.8 \times 10^{-5}\,\text{mol s}^{-1}$$

Exercise E7.15

The overall quantum yield for another reaction at 290 nm is 0.30. For what length of time must irradiation with a 100 W source continue in order to destroy 1.0 mol of molecules?

[*Answer*: 3.8 h]

Photochemical rate laws

As an example of how to incorporate the photochemical activation step into a mechanism, consider the photochemical activation of the reaction

$$H_2(g) + Br_2(g) \longrightarrow 2\,HBr(g)$$

In place of the first step in the thermal reaction we have

$$Br_2 + h\nu \longrightarrow 2\,Br \qquad \text{rate of formation of Br} = I_{abs}$$

where I_{abs} is the number of photons of the appropriate frequency absorbed divided by the time interval and the volume. It follows that I_{abs} should take the place of $k_a[Br_2]$ in the thermal reaction scheme, and so from eqn (47) we can write

$$\text{rate of formation of HBr} = \frac{2k_b(1/k_d)^{1/2}[H_2][Br_2]\sqrt{I_{abs}}}{[Br_2] + (k_c/k_b')[HBr]} \qquad (49)$$

Although the details of this expression are complicated, the essential prediction is clear: the reaction rate should depend on the square root of the absorbed light intensity. This prediction is confirmed experimentally.

EXERCISES

7.1 The rate of formation of C in the reaction $2A + B \longrightarrow 3C + 2D$ is $2.2\,mol\,L^{-1}\,s^{-1}$. State the rates of formation and consumption of A, B, and D.

7.2 The rate law for the reaction in Exercise 7.1 was reported as rate $= k[A][B][C]$. What are the units of k?

7.3 If the rate laws are expressed with (a) concentrations in numbers of molecules per metre cubed (molecules m^{-3}), (b) pressures in kilopascals, what are the units of the second-order and third-order rate constants?

7.4 The rate constant for the first-order decomposition of N_2O_5 in the reaction $2N_2O_5(g) \longrightarrow 4NO_2(g) + O_2(g)$ is $k = 3.38 \times 10^{-5}\,s^{-1}$ at $25\,°C$. What is the half-life of N_2O_5? What will be the total pressure, initially 500 Torr for the pure N_2O_5 vapour, (a) 10 s, (b) 10 min after initiation of the reaction?

7.5 In a study of the alcohol dehydrogenase catalysed oxidation of ethanol, the molar concentration of ethanol decreased in a first-order reaction from $220\,mmol\,mol^{-1}$ to $56.0\,mmol\,mol^{-1}$ in $1.22 \times 10^4\,s$. What is the rate constant of the reaction?

7.6 The elimination of carbon dioxide from pyruvate ions by a decarboxylase enzyme was monitored by measuring the partial pressure of the gas as it was formed. In one experiment, the partial pressure increased from zero to 100 Pa in 522 s in a first-order reaction. What is the rate constant of the reaction?

7.7 In the study of a second-order gas phase reaction, it was found that the molar concentration of a reactant fell from $220\,mmol\,L^{-1}$ to $56.0\,mmol\,L^{-1}$ in $1.22 \times 10^4\,s$. What is the rate constant of the reaction?

7.8 Carbonic anhydrase is a zinc-based enzyme that catalyses the conversion of carbon dioxide to carbonic acid. In an experiment to study its effect, it was found that the molar concentration of carbon dioxide in solution decreased from $220\,mmol\,mol^{-1}$ to $56.0\,mmol\,mol^{-1}$ in $1.22 \times 10^4\,s$. What is the rate constant of the first-order reaction?

7.9 The formation of NOCl from NO in the presence of a large excess of chlorine is pseudosecond order in NO. In an experiment to study the reaction, the partial pressure of NOCl increased from zero to 100 Pa in 522 s. What is the rate constant of the reaction?

7.10 A number of reactions that take place on the surfaces of catalysts are zero order in the reactant. One example is the decomposition of ammonia on hot tungsten. In one experiment, the partial pressure of ammonia decreased from 21 kPa to 10 kPa in 770 s. (a) What is the rate constant for the zero-order reaction? (b) How long will it take all the ammonia to disappear?

7.11 The half-life of pyruvic acid in the presence of an aminotransferase enzyme (which converts it to alanine) was found to be 221 s. How long will it take for the concentration of pyruvic acid to fall to one-sixty-fourth of its initial value in this first-order reaction?

7.12 The half life for the (first-order) radioactive decay of ^{14}C is 5730 y (it emits β rays with an energy of 0.16 MeV). An archaeological sample contained wood that had only 69 per cent of the ^{14}C found in living trees. What is its age?

7.13 One of the hazards of nuclear explosions is the generation of ^{90}Sr and its subsequent incorporation in place of calcium in bones. This nuclide emits β rays of energy 0.55 MeV, and has a half-life of 28.1 y. Suppose $1.00\,\mu g$ was absorbed by a newly born child. How much will remain after (a) 19 y, (b) 75 y if none is lost metabolically?

7.14 The second-order rate constant for the reaction

$$CH_3COOC_2H_5(aq) + OH^-(aq) \longrightarrow$$
$$CH_3CO_2^-(aq) + CH_3CH_2OH(aq)$$

is $0.11\,L\,mol^{-1}\,s^{-1}$. What is the concentration of ester after (a) 15 s, (b) 15 min when ethyl acetate is added to sodium hydroxide so that the initial concentrations are $[NaOH] = 0.055\,mol\,L^{-1}$ and $[CH_3COOC_2H_5] = 0.150\,mol\,L^{-1}$?

7.15 A reaction $2A \longrightarrow P$ has a second-order rate law with $k = 1.24\,mL\,mol^{-1}\,s^{-1}$. Calculate the time required for the concentration of A to change from $0.260\,mol\,L^{-1}$ to $0.026\,mol\,L^{-1}$.

7.16 The composition of a liquid phase reaction $2A \longrightarrow B$ was followed spectrophotometrically with the following results:

t/min	0	10	20	30	40	∞
$[B]/(mol\,L^{-1})$	0	0.089	0.153	0.200	0.230	0.312

Determine the order of the reaction and its rate constant.

7.17 A rate constant is $1.78 \times 10^{-4}\,\mathrm{L\,mol^{-1}\,s^{-1}}$ at $19\,°C$ and $1.38 \times 10^{-3}\,\mathrm{L\,mol^{-1}\,s^{-1}}$ at $37\,°C$. Evaluate the Arrhenius parameters of the reaction.

7.18 The activation energy for the decomposition of benzene diazonium chloride is $99.1\,\mathrm{kJ\,mol^{-1}}$. At what temperature will the rate be 10 per cent greater than its rate at $25\,°C$?

7.19 Which reaction responds more strongly to changes of temperature, one with an activation energy of $52\,\mathrm{kJ\,mol^{-1}}$ or one with an activation energy of $25\,\mathrm{kJ\,mol^{-1}}$?

7.20 The rate constant of a reaction increases by a factor of 1.23 when the temperature is increased from $20\,°C$ to $27\,°C$. What is the activation energy of the reaction?

7.21 Food rots about 40 times more rapidly at $25\,°C$ than when it is stored at $4\,°C$. Estimate the overall activation energy for the processes responsible for its decomposition.

7.22 Suppose that the rate constant of a reaction *decreases* by a factor of 1.23 when the temperature is increased from $20\,°C$ to $27\,°C$. How should you report the activation energy of the reaction?

7.23 The enzyme urease catalyses the reaction in which urea is hydrolysed to ammonia and carbon dioxide. The half-life of urea in the pseudofirst-order reaction for a certain amount of urease doubles when the temperature is lowered from $20\,°C$ to $10\,°C$ and the Michaelis constant is largely unchanged. What is the activation energy of the reaction?

7.24 The activation energy of the first-order decomposition of dinitrogen oxide into N_2 and O is $251\,\mathrm{kJ\,mol^{-1}}$. The half-life of the reactant is $6.5\,\mathrm{Ms}$ ($1\,\mathrm{Ms} = 10^6\,\mathrm{s}$) at $455\,°C$. What will it be at $550\,°C$?

7.25 The reaction $2\,H_2O_2(aq) \longrightarrow 2\,H_2O(l) + O_2(g)$ is catalysed by Br^- ions. If the mechanism is:

$$H_2O_2(aq) + Br^-(aq) \longrightarrow$$
$$H_2O(l) + BrO^-(aq) \quad \text{(slow)}$$
$$BrO^-(aq) + H_2O_2(aq) \longrightarrow$$
$$H_2O(l) + O_2(g) + Br^-(aq) \quad \text{(fast)}$$

give the order of the reaction with respect to the various participants.

7.26 The reaction mechanism

$$A_2 \longrightarrow 2\,A \quad \text{(fast)}$$
$$A + B \longrightarrow P \quad \text{(slow)}$$

involves an intermediate A. Deduce the rate law for the formation of P.

7.27 Consider the following mechanism for renaturation of a double helix from its strands A and B:

$$A + B \longrightarrow \text{unstable helix} \quad \text{(fast)}$$
$$\text{unstable helix} \longrightarrow \text{stable double helix} \quad \text{(slow)}$$

Derive the rate equation for the formation of the double helix and express the rate constant of the renaturation reaction in terms of the rate constants of the individual steps.

7.28 The enzyme-catalysed conversion of a substrate at $25\,°C$ has a Michaelis constant of $0.045\,\mathrm{mol\,L^{-1}}$. The rate of the reaction is $1.15 \times 10^{-3}\,\mathrm{mol\,L^{-1}\,s^{-1}}$ when the substrate concentration is $0.110\,\mathrm{mol\,L^{-1}}$. What is the maximum velocity of this enzymolysis?

7.29 Find the condition for which the reaction rate of an enzymolysis that follows Michaelis–Menten kinetics is half its maximum value.

7.30 Consider the following mechanism for the thermal decomposition of R_2:

$$
\begin{array}{lll}
(1) & R_2 \longrightarrow 2\,R \\
(2) & R + R_2 \longrightarrow P_B + R' \\
(3) & R' \longrightarrow P_A + R \\
(4) & 2\,R \longrightarrow P_A + P_B
\end{array}
$$

where R_2, P_A, and P_B are stable hydrocarbons and R and R' are radicals. Find the dependence of the rate of decomposition of R_2 on the concentration of R_2.

7.31 Refer to Fig. 7.32 and determine the pressure range for branching chain explosion in the hydrogen–oxygen reaction at (a) $700\,K$, (b) $800\,K$, and (c) $900\,K$.

7.32 In a photochemical reaction $A \longrightarrow 2\,B + C$, the overall quantum yield with $500\,nm$ light is $2.1 \times 10^2\,\mathrm{mol\,einstein^{-1}}$, where $1\,\mathrm{einstein} = 1\,\mathrm{mol}$ photons. After exposure of $300\,\mathrm{mmol}\,A$ to the light, $2.15\,\mathrm{mmol}\,B$ is formed. How many photons were absorbed by A?

7.33 In an experiment to measure the quantum efficiency of a photochemical reaction, the absorbing substance was exposed to $490\,nm$ light from a $100\,W$ source for 45 minutes. The intensity of the transmitted light was 35 per cent of the intensity of the incident light. As a result of irradiation, $0.297\,\mathrm{mol}$ of the absorbing substance decomposed. Find the quantum efficiency.

7.34 The condensation reaction of acetone, $(CH_3)_2CO$ (propanone), in aqueous solution is catalysed by bases, B, which react reversibly with acetone to form the carbanion $C_3H_5O^-$. The carbanion then reacts with a molecule of acetone to give the product. A simplified version of the mechanism is

$$(1) \quad AH + B \longrightarrow BH^+ + A^-$$
$$(2) \quad A^- + BH^+ \longrightarrow AH + B$$
$$(3) \quad A^- + HA \longrightarrow product$$

where AH stands for acetone and A^- its carbanion. Use the steady state approximation to find the concentration of the carbanion and derive the rate equation for the formation of the product.

7.35 Consider the acid-catalysed reaction

$$HA + H^+ \overset{(1)}{\underset{(2)}{\rightleftharpoons}} HAH^+ \qquad (fast)$$

$$HAH^+ + B \overset{(3)}{\longrightarrow} BH^+ + AH \qquad (slow)$$

Deduce the rate law and show that it can be made independent of the specific term $[H^+]$.

7.36 Consider the following chain mechanism:

$$(1) \quad AH \longrightarrow A\cdot + H\cdot$$
$$(2) \quad A\cdot \longrightarrow B\cdot + C$$
$$(3) \quad AH + B\cdot \longrightarrow A\cdot + D$$
$$(4) \quad A\cdot + B\cdot \longrightarrow P$$

Identify the initiation, propagation, and termination steps, and use the steady-state approximation to deduce that the decomposition of AH is first order in AH.

7.37 The rate, v, of an enzyme-catalysed reaction was measured when various amounts of substrate S were present and the concentration of enzyme was $12.5\ \mu mol\ L^{-1}$. The following results were obtained:

$[S]/(mmol\ L^{-1})$	1.0	2.0	3.0	4.0	5.0
$v/(\mu mol\ L^{-1}\ s^{-1})$	1.1	1.8	2.3	2.6	2.9

Determine the Michaelis–Menten constant, the maximum velocity of the reaction, and the maximum turnover number of the enzyme.

7.38 The following results were obtained when the rate of an enzymolysis was monitored (a) without inhibitor, (b) with inhibitor at a concentration of $18\ \mu mol\ L^{-1}$.

$[S]/(10^{-4}\ mol\ L^{-1})$	1.0	3.0	7.0	12.0	18.0
$v/(\mu mol\ L^{-1}\ s^{-1})$ (a)	0.49	0.95	1.3	1.5	1.6
(b)	0.27	0.52	0.71	0.81	0.86

Is the inhibition competitive or noncompetitive?

Atomic structure

It is impossible to understand the majority of modern explanations in chemistry without a familiarity with the principal concepts of quantum mechanics. The same is true of virtually all the spectroscopic techniques that are now so central to investigations of composition and structure. Present day techniques for studying chemical reactions have progressed to the point where the information is so detailed that quantum mechanics has to be used in its interpretation. And, of course, the very currency of chemistry—the electronic structures of atoms and molecules—cannot be discussed without making use of quantum mechanical concepts.

The role—indeed, the existence—of quantum mechanics was realized only during the twentieth century. It was once thought that the motion of atomic and subatomic particles could be expressed in terms of the laws of classical mechanics introduced in the seventeenth century by Isaac Newton (see *Further information 3*), for these laws were very successful at explaining the motion of planets and everyday objects such as pendulums and projectiles. However, towards the end of the nineteenth century, experimental evidence accumulated showing that classical mechanics failed when it was applied to very small particles, such as individual atoms, nuclei, and electrons, and it took until 1926 to identify the appropriate concepts and equations for describing them.

The failures of classical physics

In this section, we shall see how it came to be realized that classical mechanics had shortcomings that were particularly serious when applied to systems in which only small energies were being transferred. To appreciate these shortcomings, we need to know that classical physics is based on two presumptions:

1. A particle travels in a **trajectory**, a path with a precise position and a precise velocity at each instant.
2. Any type of motion can be excited to a state of arbitrary energy.

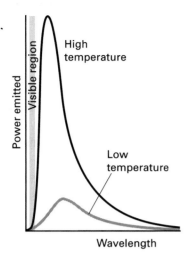

Fig. 8.1 The power emitted by a black body at two temperatures. Note how the power increases in the visible region (the tinted range of wavelengths) as the temperature is raised, and how the peak maximum moves to shorter wavelengths. The total power (the area under the curve) increases as the temperature is increased (as T^4).

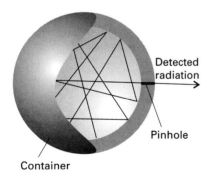

Fig. 8.2 A good approximation to a black-body radiator is a pinhole in a heated cavity. Any radiation inside is reflected many times and comes to thermal equilibrium with the walls at a temperature T. The pinhole allows some radiation to escape and be detected.

These two presumptions agree with everyday experience. For example, a pendulum swings with a precise oscillating motion and can be made to oscillate with any energy simply by pulling it back to an arbitrary angle and then letting it swing freely. We can predict its position and the speed at which it is swinging at every instant. Everyday experience, however, does not extend to a familiarity with the behaviour of individual atoms and subatomic particles, and careful experiments of the type described below have shown that the laws of classical mechanics fail to account for the observed behaviour of very small particles. Classical mechanics is in fact only an *approximate* description of the motion of particles, and the approximation is invalid when it is applied to submicroscopic particles, such as molecules, atoms, and electrons.

8.1 Black-body radiation

By **electromagnetic radiation** we mean a wave of electric and magnetic fields travelling through space. Such waves are generated by the acceleration of electric charge, as in the oscillating motion of electrons in the antenna of a radio transmitter. We do not need to know many of the properties of electromagnetic radiation, but those we do need to understand for this section are summarized in *Further information 6*. A hot object emits electromagnetic radiation because its atoms and electrons are ceaselessly being accelerated into and out of different states of motion. At high temperatures, an appreciable proportion of the radiation is in the visible region of the spectrum, and a higher proportion of short-wavelength blue light is generated as the temperature is raised. We observe this behaviour when an iron bar glowing red hot becomes white hot when heated further, because then more blue light mixes into the red light and changes the perceived colour to white. The precise dependence is illustrated in Fig. 8.1, which shows how the energy output varies with wavelength at a series of temperatures. The curves are those of an ideal emitter called a **black body**, which is an object capable of emitting and absorbing all frequencies of radiation uniformly. A good approximation to a black body is a pinhole in a container, because the radiation leaking out of the hole has been absorbed and re-emitted inside so many times that it has come to thermal equilibrium with the walls (Fig. 8.2), and the temperature of the radiation-filled space inside the container is the same as that of the walls.

The experimental observations

We shall consider the **energy density** of the radiation in a cavity, the total energy in the cavity divided by its volume, and in particular the contribution to the total energy density from radiation of different wavelengths. Figure 8.1 shows two main features. The first is that shorter wavelengths contribute more to the energy density as the temperature is raised. As a result, the perceived colour shifts towards the blue, as already mentioned. An analysis of the data led Wilhelm Wien (in 1893) to summarize this shift by a statement now known as **Wien's displacement law**:

$$T\lambda_{\max} = \text{constant} \tag{1}$$

where the value of the constant is 0.29 cm K. In this expression, λ_{max} is the wavelength of the maximum contribution to the energy density when the temperature is T. The displacement law implies that as T increases, λ_{max} decreases enough to preserve the same value of $T\lambda_{\text{max}}$. One interesting application of Wien's displacement law is to the estimation of the temperatures of stars, and other inaccessible hot objects, from the intensity profile of the light they emit, for the intensity of each wavelength emitted is proportional to the energy density at that wavelength. For example, the maximum emission of the Sun occurs at $\lambda_{\text{max}} \approx 490$ nm, so its surface temperature must be close to

$$T = \frac{0.29 \times 10^{-2}\,\text{m K}}{490 \times 10^{-9}\,\text{m}} = 5.9 \times 10^3\,\text{K}$$

or about 6000 K.

Exercise E8.1

Estimate the wavelength at the maximum energy output of an incandescent lamp if the filament is at 3000 °C

[*Answer*: 890 nm]

The second feature of black-body radiation had been noticed in 1879 by Josef Stefan, who considered the sharp rise in the **emittance**, M, the power emitted (the rate of energy output) divided by the surface area, as the temperature is raised. He established what is now called the **Stefan–Boltzmann law**:

$$M = aT^4 \qquad (2)$$

with $a = 56.7\,\text{nW m}^{-2}\,\text{K}^{-4}$ (where $1\,\text{nW} = 10^{-9}\,\text{W}$). This law implies that each square centimetre of the surface of a black body at 1000 K radiates about 5.7 W when all wavelengths are taken into account, but it radiates $3^4 = 81$ times that power (460 W) when the temperature is increased by a factor of 3, to 3000 K. The law is the basis of seeking as high a temperature as possible for an incandescent lamp, for then the emission is as strong as possible.

Exercise E8.2

Suppose technological advances made it possible to produce a ceramic material that could be used as a filament at 3800 °C instead of 3000 °C. By what factor would the power output of a lamp that used the new material increase?

[*Answer*: 2.4]

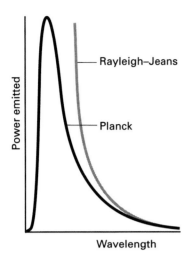

Fig. 8.3 Theoretical attempts to account for black-body radiation. The Rayleigh–Jeans law leads to an infinite energy density at short wavelengths and gives rise to the ultraviolet catastrophe. The Planck distribution is in excellent agreement with experiment.

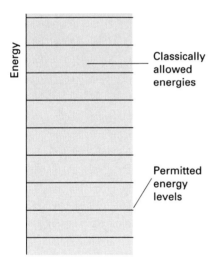

Fig. 8.4 According to classical physics, an oscillator (including the oscillators that correspond to vibrations of the electromagnetic field and correspond to radiation of a particular frequency) can have any energy (as depicted by the tinted range of energies). Planck's proposal implied that an oscillator could be excited only in discrete steps, for it can possess only certain energies (those depicted by the horizontal lines in the illustration).

The attempted classical interpretation

The physicist Lord Rayleigh studied black-body radiation from a classical viewpoint. In his day (at the end of the nineteenth century), electromagnetic radiation was regarded as waves in a jellylike 'ether'. If the ether could oscillate at a certain frequency ν, then radiation of that frequency would be present in it. Rayleigh took the view that the ether could oscillate with any frequency and so waves could exist in it of any wavelength. He calculated the contribution to the energy density in a wavelength range of width $\delta\lambda$ at any wavelength. With minor help from James Jeans, Rayleigh arrived at the **Rayleigh–Jeans law**:

$$\text{energy density in the wavelength range } \lambda \text{ to } \lambda + \delta\lambda = \frac{8\pi kT}{\lambda^4}\delta\lambda \qquad (3)$$

The power emitted in the wavelength range $\delta\lambda$ is proportional to this energy density.

Unfortunately (for Rayleigh, Jeans, and classical physics), although the Rayleigh–Jeans formula is quite successful at long wavelengths (low frequencies), it fails badly at short wavelengths (high frequencies). Thus, as λ decreases, the energy density in the cavity, and hence the power emitted per unit area, increases towards infinity and never passes through a maximum (Fig. 8.3). The equation therefore predicts that oscillations of very short wavelength (high frequency, corresponding to ultraviolet light, X-rays, and even γ-rays) are strongly excited even at room temperature. So, according to classical physics, every time you strike a match, you blast the surroundings with γ rays! This absurd result is called the **ultraviolet catastrophe**.

Planck's resolution: quantization

The German physicist Max Planck studied black-body radiation from the viewpoint of thermodynamics, in which he was an expert. In 1900 he found that he could account for the characteristics of black-body radiation by proposing that the energy of each electromagnetic oscillator is limited to discrete values and cannot be varied arbitrarily. Thus, the oscillation of the electromagnetic field that corresponds to yellow light, for instance, could be stimulated only if a certain energy was provided. This limitation of the energy of an object to discrete values is called the **quantization of energy**. Specifically, Planck proposed that the energy of an oscillator of frequency ν is restricted to an integral multiple of the quantity $h\nu$, where h is a fundamental constant now known as the **Planck constant**:

$$E = nh\nu \qquad \text{where } h = 6.626 \times 10^{-34}\,\text{J s} \qquad (4)$$

with $n = 0, 1, 2, \cdots$ (Fig. 8.4). To use this expression, we need to note that $1\,\text{Hz} = 1\,\text{s}^{-1}$, so $1\,\text{Hz} \times 1\,\text{s} = 1$.

Exercise E8.3

What is the minimum energy that can be used to excite an oscillator corresponding to yellow light (of frequency 5.2×10^{14} Hz)?

[*Answer*: 3.4×10^{-19} J]

When Planck calculated the energy density in the wavelength range λ to $\lambda + \delta\lambda$ on the basis of his quantization postulate, he obtained

energy density in the wavelength range λ to $\lambda + \delta\lambda$

$$= \frac{8\pi hc}{\lambda^5} \left(\frac{1}{e^{hc/\lambda kT} - 1} \right) \delta\lambda \qquad (5)$$

The crucial difference between this formula and the Rayleigh–Jeans law in eqn (3) is the exponential in the denominator, which has the effect of making the energy density approach zero as λ approaches zero,[†] and hence eliminates the ultraviolet catastrophe. The graph predicted by this Planck distribution is shown in Fig. 8.3, it compares very favourably with the experimental curve shown in Fig. 8.1.

Exercise E8.4

Calculate the ratio of the energy densities predicted by the Planck and Rayleigh–Jeans formulae for yellow (580 nm) light at 1000 K.

[*Answer*: Rayleigh/Planck $= 2.4 \times 10^9$]

The physical reason why Planck's quantization hypothesis is successful is as follows. The atoms in the walls of the black body undergo thermal motion, and this motion excites the oscillators of the electromagnetic field. According to classical mechanics, all the electromagnetic oscillators are excited, even those of very high frequency, and the corresponding wavelengths of radiation are emitted, including radiation of very short wavelength. According to quantum mechanics, however, the oscillators are excited only if they can acquire an energy of at least $h\nu$. This minimum energy is too large for the walls to supply in the case of the high-frequency oscillators, so the latter remain unexcited. The effect of quantization is to quench the contribution from the high-frequency oscillators, for they cannot be excited with the

[†] As λ approaches zero, $hc/\lambda kT$ approaches infinity and $e^{hc/\lambda kT}$ approaches infinity too; therefore the denominator in the Planck formula becomes infinite and so the expression itself approaches zero.

energy available, and hence the very short wavelength radiation is not emitted.

The Planck distribution accounts quantitatively for the Stefan–Boltzmann and Wien laws. Thus, when the area under the graph in Fig. 8.3 is calculated (to obtain the total energy density over the entire wavelength range), the resulting expression is proportional to T^4, in accord with the Stefan–Boltzmann law. Similarly, when the wavelength λ_{max} that corresponds to the maximum point of the curve is calculated, its position is found to be inversely proportional to the temperature, which is in agreement with Wien's law.

8.2 Heat capacities

Heat capacities were first introduced in Section 2.4, where we saw that they are the constant of proportionality, C, between the rise in temperature, ΔT, of a sample and the heat, q, needed to bring that increase about:

$$q = C\Delta T$$

We can suspect that there are similarities between black-body radiation and heat capacities: the former involves examining how energy is taken up by the oscillations of the electromagnetic field, whereas the latter involves examining how energy is taken up by the oscillation of atoms about their mean positions in the solid.

On the basis of some somewhat slender experimental evidence, the French scientists Pierre-Louis Dulong and Alexis-Thérèse Petit had proposed in 1819 that the molar heat capacity of all monatomic solids—such as metals—was equal (in modern units) to about $24\,\mathrm{J\,K^{-1}\,mol^{-1}}$. Classical physics was able to account for this value quite readily, for if it is assumed that the atomic oscillators can be excited to any energy, then the predicted molar heat capacity is $3R$, where R is the gas constant, and $3R = 24\,\mathrm{J\,K^{-1}\,mol^{-1}}$.

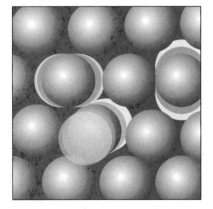

Fig. 8.5 An atom in a solid can oscillate about its position in three perpendicular directions, and when a solid is heated, the vigour of the motion increases. This illustration depicts a view of one layer of atoms in a solid and shows the oscillations of some of the atoms.

Justification

The explanation of Dulong and Petit's law takes three steps. The first is to note that if a solid consists of N atoms, then because each atom can oscillate in any of three perpendicular directions (Fig. 8.5), the solid is equivalent to a collection of $3N$ oscillators. The second step makes use of a conclusion from classical physics known as the **equipartition theorem**, which implies that, at a temperature T, the average energy of an oscillator is kT, where k is the **Boltzmann constant**. The Boltzmann constant is equal to the gas constant R divided by the Avogadro constant N_A, so it follows that $R = N_A \times k$. It now follows from these two steps that the total energy of the N vibrating atoms is $3N \times kT$. The energy per mole of atoms is therefore $3N_A kT$, or $3RT$, because the number of atoms per mole is N_A. In the final step, we note that when the temperature of the sample increases by ΔT, the molar energy increases by $3R\Delta T$. This increase in energy must be supplied as heat from the surroundings, so the heat required to raise the temperature by ΔT is $3R\Delta T$. It follows by comparison with the definition of heat capacity ($q = C\Delta T$) that $C = 3R$.

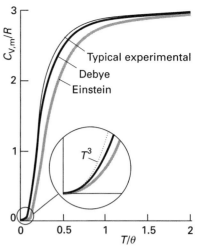

Fig. 8.6 Experimental low-temperature heat capacities and theoretical predictions. The Dulong and Petit law predicts no falling off at low temperatures. The Einstein formula predicts the general temperature dependence quite well, but is everywhere too low. Debye's modification of Einstein's calculation gives very good agreement with experiment. The inset shows a magnified view of the graphs close to $T = 0$. The Debye curve is proportional to T^3 in this region.

The apparent success of classical mechanics in accounting for observed heat capacities was short-lived, for when technological advances made it possible to measure heat capacities at low temperatures, all substances were found to have values significantly lower than $24\,\text{J K}^{-1}\,\text{mol}^{-1}$, and at very low temperatures the heat capacity was found to approach zero (Fig. 8.6). Even some quite common substances were found to have molar heat capacities well below the expected value even at room temperature: the value for diamond, for instance, is only $6.1\,\text{J K}^{-1}\,\text{mol}^{-1}$ at $25°C$.

To account for these new observations, Einstein took the view that each atom oscillates about its mean position with a single frequency ν. He then borrowed Planck's hypothesis, and asserted that the only permitted energy of any oscillating atom is an integral multiple of $h\nu$ (exactly as for electromagnetic oscillators). On the basis of this model, Einstein was able to deduce the expression

$$C = 3Rf^2 \qquad f = \frac{(h\nu/kT)e^{h\nu/2kT}}{e^{h\nu/kT} - 1} \qquad (6)$$

This expression is plotted as a function of temperature in Fig. 8.6. We see that it does indeed predict that the heat capacity of a solid should decrease as the temperature is lowered, and that it approaches zero as the temperature approaches zero.

The physical reason for the success of the Einstein model is that at low temperatures there is only enough energy available for a few atoms to be able to oscillate. Because so few atoms can be involved in taking up energy, the solid is incapable of absorbing heat readily and consequently its heat capacity is low. At higher temperatures there is enough energy available for all the oscillators to become active: all $3N$ oscillators contribute, and the heat capacity approaches its classical value of $3R$.

A more refined approach to the calculation was carried out by the Dutch physicist Peter Debye. He allowed for the atoms to oscillate with a range of frequencies rather than the single frequency supposed by Einstein. The graph of his more complicated expression is similar to Einstein's, but the numerical agreement with the experimental data is better (see Fig. 8.6). An important practical conclusion from Debye's calculation is that at low temperatures, the heat capacity of a solid is expected to be proportional to T^3. This dependence, which is called the **Debye T^3 law**, is used to extrapolate measurements of heat capacities to $T = 0$ in the experimental determination of entropies (Section 3.2).

8.3 Waves as particles and particles as waves

So far, we have seen that two observations—on the electromagnetic field and the heat capacities of solids—have led to the overthrow of the classical view that oscillators can have any energy. We shall now see how three other experimental observations upset another central concept of classical physics, the distinction between waves and particles.

The photoelectric effect

Planck's view that an electromagnetic oscillator of frequency ν can possess only the energies $0, h\nu, 2h\nu, \cdots$ inspired a new view of the nature of electromagnetic radiation. Instead of thinking of radiation of a given frequency as the excitement of the electromagnetic field to one of its permitted states of oscillation at that frequency, we can think of that radiation as consisting of $0, 1, 2, \cdots$ particles, each particle having an energy $h\nu$. When there is only one such particle present, the energy of the radiation is $h\nu$, when there are two particles of that frequency, their total energy is $2h\nu$, and so on. These particles are now called **photons**.[†] According to the photon picture of radiation, a ray of light of frequency ν consists of a stream of photons, each one having an energy $h\nu$. As the intensity of the ray is increased, the number of photons increases, but each one continues to have the energy $h\nu$. An intense beam of monochromatic (single-frequency) radiation consists of a dense stream of photons; a weak beam of radiation of the same frequency consists of a relatively small number of the same type of photons.

Example *Calculating the number of photons*
Calculate the number of photons emitted by a 100 W yellow lamp in 10.0 s. Take the wavelength of yellow light as 560 nm and assume 100 per cent efficiency.

Strategy
The total energy emitted by a lamp in a given interval is its power multiplied by the time interval of interest (1 J = 1 W s). The number of photons emitted in that time is therefore the total energy divided by the energy of one photon. We calculate the energy of a single photon from the formula $E = h\nu$. For this calculation we need to know that the wavelength and frequency are related by $\nu = c/\lambda$.

Solution
The electromagnetic energy emitted by the lamp (if all the energy it consumes is converted into radiation of a single frequency) is

$$\text{total energy} = (100\,\text{W}) \times (10.0\,\text{s}) = 1.00 \times 10^3\,\text{J}$$

[†]The name was coined by the chemist G.N. Lewis, whom we meet again in Chapter 9.

Each photon has an energy

$$E = h\nu = \frac{hc}{\lambda} = \frac{(6.626 \times 10^{-34}\,\text{J s}) \times (2.998 \times 10^8\,\text{m s}^{-1})}{560 \times 10^{-9}\,\text{m}}$$
$$= 3.55 \times 10^{-19}\,\text{J}$$

The number of photons required to carry away the total energy we have calculated is therefore

$$N = \frac{\text{total energy}}{\text{energy of one photon}} = \frac{1.00 \times 10^3\,\text{J}}{3.55 \times 10^{-19}\,\text{J}}$$
$$= 2.82 \times 10^{21}$$

Exercise E8.5

How many 1000 nm photons does a 1 mW infrared rangefinder emit in 0.1 s?

[*Answer*: 5×10^{14}]

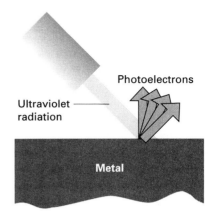

Fig. 8.7 The experimental arrangement to demonstrate the photoelectric effect. A beam of ultraviolet radiation is used to irradiate a patch of the surface of a metal, and electrons are ejected from the surface if the frequency of the radiation is above a threshold value that depends on the metal.

Evidence that confirmed the view that radiation can be interpreted as a stream of particles comes from the **photoelectric effect**, the ejection of electrons from metals when they are exposed to ultraviolet radiation (Fig. 8.7). The characteristics of the photoelectric effect are as follows:

1. No electrons are ejected, regardless of the intensity of the radiation, unless its frequency exceeds a threshold value characteristic of the metal.
2. The kinetic energy of the ejected electrons varies linearly with the frequency of the incident radiation but is independent of its intensity.
3. Even at low light intensities, electrons are ejected immediately if the frequency is above the threshold value.

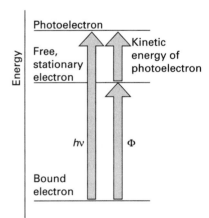

Fig. 8.8 In the photoelectric effect, an incoming photon brings a definite quantity of energy, $h\nu$. It collides with an electron close to the surface of the metal target, and transfers its energy to it. The difference between the work function, Φ, and the energy $h\nu$ appears as the kinetic energy of the ejected electron.

These observations strongly suggest an interpretation of the photoelectric effect in which an electron is ejected in a collision with a particle-like projectile, provided the projectile carries enough energy to expel the electron from the metal. If we suppose that the projectile is a photon of energy $h\nu$, where ν is the frequency of the radiation, then the conservation of energy requires that the kinetic energy of the electron (which is equal to $\frac{1}{2}m_e v^2$, when the speed of the electron is v) should be equal to the energy supplied by the photon less the energy Φ required to remove the electron from the metal (Fig. 8.8):

$$\tfrac{1}{2}m_e v^2 = h\nu - \Phi \tag{7}$$

The quantity Φ (uppercase phi) is called the **work function** of the metal.

When $h\nu < \Phi$, photoejection (the ejection of electrons by light) cannot occur because the photon supplies insufficient energy to expel the electron: this conclusion is consistent with observation (1). Equation (7) predicts that the kinetic energy of an ejected electron should vary linearly with the frequency, in agreement with observation (2). When a photon collides with an electron, it gives up all its energy, and so we should expect electrons to appear as soon as the collisions begin, provided the photons carry sufficient energy: this conclusion agrees with observation (3). Thus, the photoelectric effect is strong evidence for the existence of photons. Indeed, an analysis like this earned Einstein his Nobel Prize.

The diffraction of electrons

The photoelectric effect shows that light has certain properties of particles. Although contrary to the long-established wave theory of light, a similar view had been held before, but discarded. No significant scientist, however, had taken the view that matter is wavelike. Nevertheless, experiments carried out in 1925 forced people to even that conclusion. The crucial experiment was performed by the American physicists Clinton Davisson and Lester Germer, who observed the diffraction of electrons by a crystal (Fig. 8.9). **Diffraction** is the interference between waves caused by an object in their path, and results in a series of bright and dark fringes where the waves are detected. It is a typical characteristic of waves.

The Davisson–Germer experiment, which has since been repeated with other particles (including molecular hydrogen), shows clearly that 'particles' have wavelike properties. We have also seen that 'waves' have particle-like properties. Thus we are brought to the heart of modern physics. *When examined on an atomic scale, the concepts of particle and wave melt together, particles taking on the characteristics of waves, and waves the characteristics of particles.* This joint wave–particle character of matter and radiation is called **wave–particle duality**.

As these concepts emerged there was an understandable confusion about how to combine both aspects of matter into a single description. Some progress towards coordinating wave and particle properties was made by Louis de Broglie when, in 1924, he suggested that any particle travelling with a linear momentum p (the product of its mass and velocity, $p = mv$) should have (in some sense) a wavelength given by the **de Broglie relation**:

$$\lambda = \frac{h}{p} \tag{8}$$

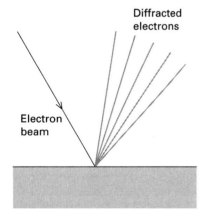

Diffracted electrons

Electron beam

Fig. 8.9 In the Davisson–Germer experiment, a beam of electrons was directed on a single crystal of nickel, and the scattered electrons showed a variation in intensity with angle that corresponded to the pattern that would be expected if the electrons had a wave character and were diffracted by the layers of atoms in the solid.

The de Broglie relation implies that the wavelength of a particle should decrease as its speed increases (Fig. 8.10). It also implies that, for a given speed, heavy particles should have shorter wavelengths than lighter particles. Equation (8) was confirmed by the Davisson–Germer experiment, for the wavelength it predicts for the electrons they used in their experiment agrees with the details of the diffraction pattern they observed. We shall build on the relation, and understand it more, in the next section.

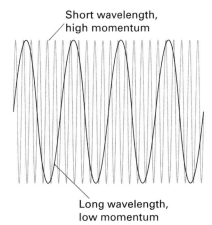

Short wavelength, high momentum

Long wavelength, low momentum

Fig. 8.10 According to the de Broglie relation, a particle with low momentum has a long wavelength whereas a particle with high momentum has a short wavelength. A high momentum can result either from a high mass or from a high velocity (because $p = mv$). Macroscopic objects have such large masses that, even if they are travelling very slowly, their wavelengths are undetectably short.

Example *Estimating the de Broglie wavelength*
Estimate the wavelength of electrons that have been accelerated from rest through a potential difference of 1.00 kV.

Strategy
To use the de Broglie relation, we need to know the linear momentum, $p = m_e v$, of the accelerated electron. We would know that if we knew its kinetic energy $E_K = \frac{1}{2} m_e v^2$, because we can combine these two expressions and rearrange them into $p = \sqrt{2 m_e E_K}$. The kinetic energy acquired by an electron that is accelerated from rest by falling through a potential difference $\mathcal{V}$ is $e\mathcal{V}$, where e is the magnitude of its charge (see *Further information 3*), so we can write $E_K = e\mathcal{V}$ and obtain $p = \sqrt{2 m_e e \mathcal{V}}$. It follows that the de Broglie wavelength is

$$\lambda = \frac{h}{\sqrt{2 m_e e \mathcal{V}}}$$

At this stage, all we need do is to substitute the data and use the relations $1\,\mathrm{C\,V} = 1\,\mathrm{J}$ and $1\,\mathrm{J} = 1\,\mathrm{kg\,m^2\,s^{-2}}$.

Solution
Substituting the data and the fundamental constants (from inside the front cover) gives

$$\lambda =$$

$$\frac{6.626 \times 10^{-34}\,\mathrm{J\,s}}{\{2 \times (9.110 \times 10^{-31}\,\mathrm{kg}) \times (1.602 \times 10^{-19}\,\mathrm{C}) \times (1.00 \times 10^{3}\,\mathrm{V})\}^{1/2}}$$
$$= 3.88 \times 10^{-11}\,\mathrm{m}$$

The wavelength of 38.8 pm is comparable to typical bond lengths in molecules (about 100 pm). Electrons accelerated in this way are used in the technique of electron diffraction for the determination of molecular structure.

Exercise E8.7

Calculate the wavelength of an electron in a 10 MeV particle accelerator ($1\,\mathrm{MeV} = 10^6\,\mathrm{eV}$).

[*Answer*: 0.39 pm]

8.4 Atomic and molecular spectra

The most directly compelling evidence for the quantization of energy comes from the frequencies of radiation absorbed or emitted by atoms and molecules. We shall only mention this point here, and leave it for a more complete treatment later (Chapter 11). A typical atomic emission spectrum is shown in Fig. 8.11 and a typical molecular absorption spectrum is shown in Fig. 8.12. The obvious feature of both is that radiation is emitted at a series of discrete frequencies. This observation can be understood if the energy of the atoms or molecules is also confined to discrete values, for then energy can be discarded only in packets (Fig. 8.13). For example, if the energy of an atom decreases by ΔE, then the energy is carried away as a photon of frequency $\nu = \Delta E/h$, and radiation of that frequency, a so-called **spectroscopic line**, appears in the spectrum.

Classical mechanics utterly failed in its attempts to account for the existence of discrete spectroscopic lines, just as it failed to account for the other experiments described above. Such total failure showed that the basic concepts of classical mechanics were false. A new mechanics, quantum mechanics, had to be devised to take its place.

The dynamics of microscopic systems

We shall take the de Broglie relation as our starting point, and abandon the classical concept of particles moving along trajectories. From now on, we shall adopt the quantum mechanical view that a particle is spread through space like a wave. As for a wave in water where the water accumulates in some places but is low in others, there are regions where the particle is more likely to be found than others. To describe this distribution, we introduce the concept of **wavefunction**, ψ (psi), in place of the trajectory, and then set up a scheme for calculating and interpreting ψ. To a very crude first approximation, a wavefunction can be visualized as a blurred version of a trajectory (Fig. 8.14); however, we shall refine this picture in the following sections.

8.5 The Schrödinger equation

In 1926, the Austrian physicist Erwin Schrödinger proposed an equation for finding the wavefunction of any system. The **Schrödinger equation** for a particle of mass m moving in one dimension with energy E is

$$-\frac{\hbar^2}{2m}\frac{d^2\psi}{dx^2} + V\psi = E\psi \tag{9}$$

In this expression, V, which may depend on the position x of the particle, is the potential energy; $\hbar$ (which is read as h-bar) is a convenient modification of Planck's constant:

$$\hbar = \frac{h}{2\pi} = 1.05459 \times 10^{-34}\,\text{J s}$$

The fact that the Schrödinger equation is a differential equation should not cause too much consternation. We shall not need to solve it explicitly, and the rare cases where we shall need to show the explicit forms of its solution will involve mathematical functions no more complicated than $\sin x$ and e^{-x}.

Justification

The form of the Schrödinger equation can be justified to a certain extent by the following remarks. Consider a region where the potential energy is zero. Then the equation simplifies to

$$-\frac{\hbar^2}{2m}\frac{d^2\psi}{dx^2} = E\psi \tag{10}$$

and a solution is

$$\psi = \sin kx \qquad k = \sqrt{\frac{2mE}{\hbar^2}}$$

as may be verified by substitution of the solution into both sides of the equation and using

$$\frac{d\sin kx}{dx} = k\cos kx \qquad \frac{d\cos kx}{dx} = -k\sin kx$$

The function $\sin kx$ is a wave of wavelength $\lambda = 2\pi/k$, as can be seen by comparing $\sin kx$ with the standard form of a harmonic wave of wavelength λ, which is $\sin(2\pi x/\lambda)$ (Fig. 8.15). Next, we note that the energy of the particle is entirely kinetic (because $V = 0$ everywhere), and so

$$E = \tfrac{1}{2}mv^2 = \frac{(mv)^2}{2m} = \frac{p^2}{2m}$$

Because the energy is related to k by

$$E = \frac{k^2\hbar^2}{2m}$$

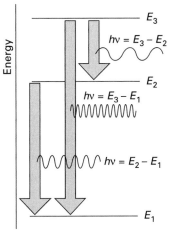

Fig. 8.13 Spectral lines can be accounted for if we assume that a molecule emits a photon as it changes between discrete energy levels. High-frequency radiation is emitted when the two states involved in the transition are widely separated in energy; low-frequency radiation is emitted when the two states are close in energy.

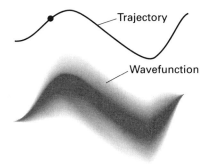

Fig. 8.14 According to classical mechanics, a particle may have a well defined trajectory, with a precisely specified position and momentum at each instant (as represented by the precise path in the diagram). According to quantum mechanics, a particle cannot have a precise trajectory; instead, there is only a probability that it may be found at a specific location at any instant. The wavefunction that determines its probability distribution is a kind of blurred version of the trajectory. Here, the wavefunction is represented by areas of shading: the darker the area, the greater the probability of finding the particle there.

297

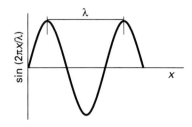

Fig. 8.15 The wavelength of a harmonic wave of the form $\sin(2\pi x/\lambda)$.

it follows from a comparison of the two equations that

$$p = k\hbar$$

Therefore, the linear momentum is related to the wavelength of the wavefunction by

$$p = \frac{2\pi}{\lambda} \times \frac{h}{2\pi} = \frac{h}{\lambda}$$

which is de Broglie's relation. We see, in the case of freely moving particles, that the Schrödinger equation has led to an experimentally verified conclusion.

One feature of the solution of the Schrödinger equation (and which is common to all differential equations), is that an infinite number of possible solutions are allowed *mathematically*. For instance, if $\sin x$ is a solution of a particular Schrödinger equation, then so too is $a \sin bx$, where a and b are arbitrary constants. (This property is easily verified by substituting $\psi = a \sin bx$ into the equation in the *Justification* above, and checking that it is a solution for all values of a and b.) However, it turns out that only some of these solutions are acceptable *physically*. To be acceptable, a solution must satisfy certain constraints called **boundary conditions** (Fig. 8.16). Suddenly, we are at the heart of quantum mechanics: the fact that only some solutions are acceptable, together with the fact that each solution corresponds to a characteristic value of E, implies that only certain values of the energy are acceptable. That is, *when the Schrödinger equation is solved subject to the boundary conditions that the solutions must satisfy, we find that the energy of the system is quantized.* Planck and his immediate successors had to postulate the quantization of energy for each system they considered: now we see that quantization is an automatic feature of a single equation, the Schrödinger equation, that is universally applicable to all systems. Later in this chapter we shall see exactly which energies are allowed in a variety of systems, the most important of which (for chemistry) are atoms.

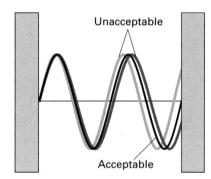

Fig. 8.16 Although an infinite number of solutions of the Schrödinger equation exist, not all of them are physically acceptable. Acceptable wavefunctions have to satisfy certain boundary conditions, that vary from system to system. In the example shown here, where the particle is confined between two impenetrable walls, the only acceptable wavefunctions are those that fit between the walls (like the vibrations of a stretched string). Because each wavefunction corresponds to a characteristic energy, and the boundary conditions rule out many solutions, only certain energies are permissible.

The Born interpretation

Before we demonstrate the role of boundary conditions, it will be helpful to understand the physical significance of a wavefunction. The interpretation of ψ that is widely used is based on a suggestion made by the German physicist Max Born. He made use of an analogy with the wave theory of light, in which the square of the amplitude of an electromagnetic wave is interpreted as its intensity and therefore (in quantum terms) as the number of photons present. The **Born interpretation** asserts that:

The probability of finding a particle in a small (strictly, infinitesimal) region of volume δV is proportional to $\psi^2 \times \delta V$.

That is, wherever the square of ψ is large, there is a high probability of finding the particle. Wherever the square of ψ is small, there is only a small chance of finding the particle. The density of shading in Fig. 8.17 represents this **probabilistic interpretation**, an interpretation that accepts that we can make predictions only about the probability of finding a particle somewhere, in contrast to classical physics, which claims to be able to predict *precisely* that a particle will be at a given point on its path at a given instant.

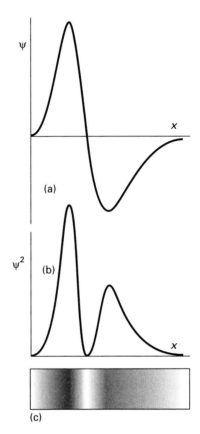

Example *Interpreting a wavefunction*
The wavefunction of an electron in the lowest energy state of a hydrogen atom is proportional to e^{-r/a_0}, with $a_0 = 52.9\,\text{pm}$ and r the distance from the nucleus (Fig. 8.18). Calculate the relative probabilities of finding the electron inside a small volume δV located at (a) the nucleus, (b) a distance a_0 from the nucleus.

Strategy
The probability is given by $\psi^2 \times \delta V$ evaluated at the specified location. The volume of interest is so small (even on the scale of the atom) that we can ignore the variation of ψ within it and write

$$\text{probability} \propto \psi^2 \times \delta V$$

with ψ^2 evaluated at the point in question.

Solution
(a) At the nucleus, $r = 0$, and so there $\psi^2 \propto 1.0$ and

$$\text{probability} \propto 1.0 \times \delta V$$

(b) At a distance $r = a_0$ in an arbitrary (but definite) direction, $\psi^2 \propto e^{-2} = 0.14$ and

$$\text{probability} \propto 0.14 \times \delta V$$

Therefore, the ratio of probabilities is $1.0/0.14 = 7.1$. It is more probable (by a factor of 7.1) that the electron will be found at the nucleus than in the same volume element located at a distance a_0 from the nucleus.

Fig. 8.17 (a) A wavefunction does not have a direct physical interpretation. However, (b) its square (its square modulus if it is complex) tells us the probability of finding a particle at each point. The probability implied by the wavefunction shown here is depicted by the density of shading in (c).

Exercise E8.8

The wavefunction for the lowest energy state in the ion He^+ is proportional to e^{-2r/a_0}. Repeat the calculation for this ion. Any comment?

[*Answer*: 55; a more compact wavefunction]

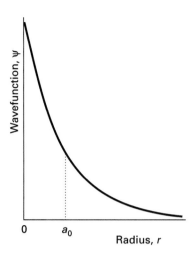

Fig. 8.18 The wavefunction for an electron in the ground state of a hydrogen atom is an exponentially decaying function of the form e^{-r/a_0}, where a_0 is the Bohr radius, a combination of fundamental constants with a value close to 53 pm.

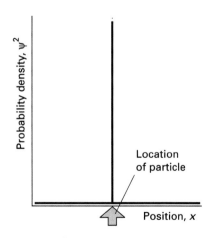

Fig. 8.19 The wavefunction for a particle with a well defined position is a sharply spiked function that has zero amplitude everywhere except at the particle's position.

The uncertainty principle

We have seen that, according to the de Broglie relation, a wave of constant wavelength (that is, the wavefunction $\sin 2\pi x/\lambda$) corresponds to a particle with a definite linear momentum $p = h/\lambda$. However, a wave does not have a definite location at a single point in space, so we cannot speak of the precise position of the particle if it has a definite momentum. Indeed, because a sine wave spreads throughout the whole of space we cannot say *anything* about the location of the particle: because the wave spreads everywhere, the particle may be found anywhere in the whole of space. This statement is one half of the **uncertainty principle** proposed by Werner Heisenberg in 1927, in one of the most celebrated results of quantum mechanics:

It is impossible to specify simultaneously, with arbitrary precision, both the momentum and the position of a particle.

Before discussing the principle further, we must establish the other half, that if we know the position of a particle exactly, then we can say nothing about its momentum. If the particle is at a definite location, then its wavefunction must be nonzero there and zero everywhere else (Fig. 8.19). Such a wavefunction can be simulated by forming a **superposition** of many wavefunctions; that is, by adding together the amplitudes of a large number of sine functions (Fig. 8.20). This procedure is successful because the amplitudes of the waves add together at one location to give a nonzero total amplitude, but cancel everywhere else. In other words, we can create a sharply localized wavefunction by adding together wavefunctions corresponding to many different wavelengths, and therefore, by the de Broglie relation, of many different linear momenta.

The superposition of a few sine functions gives a broad, ill-defined wavefunction, but as the number of functions increases the wavefunction becomes sharper because of the more complete interference between the positive and negative regions of the components. When an infinite number of components is used, the wavefunction is a sharp, infinitely narrow spike like that in Fig. 8.19, which corresponds to perfect localization of the particle. Now the particle is perfectly localized, but at the expense of discarding all information about its momentum.

The quantitative version of the position-momentum uncertainty relation is

$$\Delta p \Delta x \geq \tfrac{1}{2}\hbar \tag{11}$$

The quantity Δp is the 'uncertainty' in the linear momentum and Δx is the uncertainty in position (which is proportional to the width of the peak in Fig. 8.20).[†] Equation (11) expresses quantitatively the fact that the more

---

[†]Strictly, the uncertainty in momentum is the root mean square (r.m.s.) deviation of the momentum from its mean value, $\Delta p = \sqrt{\langle p^2 \rangle - \langle p \rangle^2}$, where the angle brackets denote mean values. Likewise, the uncertainty in position is the r.m.s. deviation in the mean value of position, $\Delta x = \sqrt{\langle x^2 \rangle - \langle x \rangle^2}$.

closely the location of a particle is specified (the smaller the value of Δx), then the greater the uncertainty in its momentum (the larger the value of Δp) parallel to that coordinate, and vice versa (Fig. 8.21).

It must be understood that the uncertainty principle applies to location and momentum *along the same axis*. It is silent on location on one axis and momentum along a perpendicular axis. The restrictions it implies are summarized in Table 8.1.

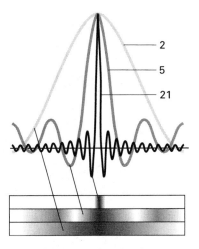

Fig. 8.20 The wavefunction for a particle with an ill-defined location can be regarded as the sum (superposition) of several wavefunctions of different wavelength that interfere constructively in one place but destructively elsewhere. As more waves are used in the superposition, the location becomes more precise at the expense of uncertainty in the particle's momentum. An infinite number of waves are needed to construct the wavefunction of a perfectly localized particle.

Example *Using the uncertainty principle*
The speed of a certain projectile of mass 1.0 g is known to within $1.0 \times 10^{-6}\,\mathrm{m\,s^{-1}}$. What is the minimum uncertainty in its position along its line of flight?

Strategy
We can estimate Δp from $m\Delta v$, where Δv is the uncertainty in the speed; then we use eqn (11) to estimate the minimum uncertainty in position, Δx, where x is the direction in which the projectile is travelling.

Solution
The uncertainty in position is

$$\Delta x \geq \frac{\hbar}{2m\Delta v} = \frac{1.054 \times 10^{-34}\,\mathrm{J\,s}}{2 \times (1.0 \times 10^{-3}\,\mathrm{kg}) \times (1.0 \times 10^{-6}\,\mathrm{m\,s^{-1}})}$$
$$= 5.3 \times 10^{-26}\,\mathrm{m}$$

This degree of uncertainty is completely negligible for all practical purposes. However, when the mass is that of an electron, the same uncertainty in speed implies an uncertainty in position far larger than the diameter of an atom, and so the concept of a trajectory, the simultaneous possession of a precise position and momentum, is untenable.

Table 8.1 Constraints of the uncertainty principle

Variable 1: Variable 2	x	y	z	p_x	p_y	p_z
x	✓	✓	✓	✗	✓	✓
y	✓	✓	✓	✓	✗	✓
z	✓	✓	✓	✓	✓	✗
p_x	✗	✓	✓	✓	✓	✓
p_y	✓	✗	✓	✓	✓	✓
p_z	✓	✓	✗	✓	✓	✓

Observables that can be determined simultaneously with arbitrary precision are marked with a ✓.

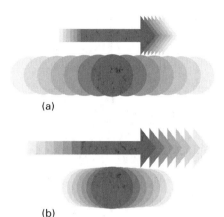

(a)

(b)

Fig. 8.21 A representation of the content of the uncertainty principle. The range of locations of a particle is shown by the circles, and the range of momenta by the arrows. In (a), the position is quite uncertain, and the range of momenta is small. In (b), the location is much better defined, and now the momentum of the particle is quite uncertain.

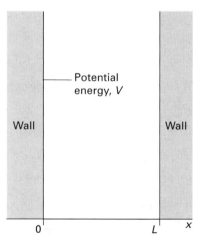

Fig. 8.22 A particle in a one-dimensional region with impenetrable walls at either end. Its potential energy is zero between $x = 0$ and $x = L$ and rises abruptly to infinity as soon as the particle touches either wall.

Exercise 8.9

Estimate the minimum uncertainty in the speed of an electron in a hydrogen atom (taking its diameter as 100 pm).

[*Answer*: 500 km s^{-1}]

The uncertainty principle summarizes the difference between classical and quantum mechanics. Classical mechanics supposed, falsely as we now know, that the position and momentum of a particle could be specified simultaneously with arbitrary precision. However, quantum mechanics shows that position and momentum are **complementary**, that is, not simultaneously specifiable, and that we have to make a choice: we can specify position at the expense of momentum, or momentum at the expense of position.

8.6 Applications of quantum mechanics

We shall now illustrate some of the concepts that have been introduced. We shall describe two concrete examples. One is translational motion (motion in a straight line), and the other is rotation (motion in a circle).

Translational motion: the particle in a box

First, we consider the translational motion of a particle that can travel in a straight line in one dimension (along the x axis) but is confined between two walls separated by a distance L. We shall suppose that this **particle in a box** has mass m and can travel freely from $x = 0$ to $x = L$. The potential energy of the particle is zero inside the box but rises abruptly to infinity at the walls (Fig. 8.22).

The boundary conditions for this system are the requirement that each acceptable wavefunction of the particle must fit inside the box exactly, like the vibrations of a violin string (as in Fig. 8.16). It follows that the wavelength, λ, of the permitted wavefunctions must be one of the values

$$\lambda = 2L, L, \frac{2L}{3}, \frac{2L}{4}, \cdots$$

In general, the wavelength must be one of the values given by the expression

$$\lambda = \frac{2L}{n} \qquad n = 1, 2, \cdots$$

Each wavefunction is a sine wave with one of these wavelengths; therefore, because a sine wave of wavelength λ is proportional to $\sin 2\pi x/\lambda$, the permitted wavefunctions are

$$\psi_n = N \sin \frac{n\pi x}{L} \qquad n = 1, 2, \cdots \tag{12}$$

The constant N is called the **normalization constant**. It is chosen so that the total probability of finding the particle inside the box is 1. For a particle in a box, $N = (2/L)^{1/2}$.

Justification

According to the Born interpretation, the probability of finding the particle in an infinitesimal region of length dx is equal to $\psi^2\, dx$. Therefore, the total probability of finding the particle in a region between $x = 0$ and $x = L$ is the sum (integral) of all the contributions from each infinitesimal region in that range:

$$\int_0^L \psi^2\, dx = 1$$

Substitution of the form of the wavefunction gives

$$N^2 \int_0^L \sin^2 \frac{n\pi x}{L}\, dx = 1$$

Because

$$\int \sin^2 ax\, dx = \tfrac{1}{2}x - \frac{\sin 2ax}{4a} + \text{constant}$$

it follows that

$$N^2 \times \tfrac{1}{2}L = 1$$

and therefore that

$$N = \sqrt{\frac{2}{L}}$$

It is now a simple matter to find the permitted energy levels because the only contribution to the energy is the kinetic energy of the particle (the potential energy is zero everywhere inside the box.) First, we note that it follows from the de Broglie relation, eqn (8), that the only values of the linear momentum that are acceptable are

$$p = \frac{h}{\lambda} = \frac{nh}{2L} \qquad n = 1, 2, \cdots$$

Then, because the kinetic energy of a particle of momentum p and mass m is $E = p^2/2m$, it follows that the permitted energies of the particle are

$$E_n = \frac{n^2 h^2}{8mL^2} \qquad n = 1, 2, \cdots \qquad (13)$$

As we see in eqns (12) and (13), the energies and wavefunctions of a particle in a box are labelled with the number n. A **quantum number**, of which n is an example, is an integer (in certain cases, as we shall see, a half-integer) that labels the state of the system. As well as acting as a label, a quantum number specifies certain physical properties of the system: in the present example, n specifies the energy of the particle through eqn (13).

The permitted energies of the particle are shown in Fig. 8.23 together with the shapes of the wavefunctions for $n = 1$ to 5. All the wavefunctions except the one of lowest energy ($n = 1$) possess points called **nodes** where the function passes through zero. The number of nodes in the wavefunctions shown in the illustration increases from 0 (for $n = 1$) to 6 (for $n = 7$), and is $n - 1$ for a particle in a box in general. The points at the edges of the box where $\psi = 0$ are not nodes, because the wavefunction does not pass *through* zero there. It is a general feature of quantum mechanics that as the number of nodes in a wavefunction increases, then the energy increases too.

The particle in a box introduces another important general feature of quantum mechanics. Because the quantum number n cannot be zero (for this system), the lowest energy that the particle may possess is not zero, as would be allowed by classical mechanics, but $E_1 = h^2/8mL^2$ (the energy when $n = 1$). This lowest, irremovable energy is called the **zero-point energy**. The existence of a zero-point energy is consistent with the uncertainty principle because, if a particle is confined to a finite region, its location is not

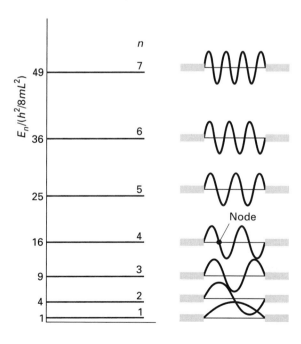

Fig. 8.23 The allowed energy levels and the corresponding (sine wave) wave functions for a particle in a box. Note that the energy levels increase as n^2, and so their spacing increases as n increases. Each wavefunction is a standing wave, and successive functions possess one more half wave and have a correspondingly shorter wavelength.

completely indefinite; consequently its momentum cannot be specified precisely as zero, and therefore its kinetic energy cannot be precisely zero either. The zero-point energy is not a special, mysterious kind of energy: it is simply the last remnant of energy that a particle cannot give up. For a particle in a box it can interpreted as the energy arising from a ceaseless fluctuating motion of the particle between the two confining walls of the box.

The energy difference between adjacent levels is

$$\Delta E = E_{n+1} - E_n = (2n + 1)\frac{h^2}{8mL^2} \qquad (14)$$

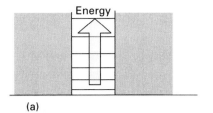

(a)

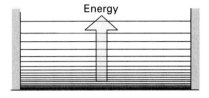

(b)

Fig. 8.24 (a) A narrow box has widely spaced energy levels; (b) a wide box has closely spaced energy levels. (In each case, the separations depend on the mass of the particle too.)

This expression shows that the difference decreases as the length L of the box increases, and that it becomes zero when the walls are infinitely far apart (Fig. 8.24). Atoms and molecules free to move in laboratory-sized vessels may therefore be treated as though their translational energy is not quantized, because L is so large. The expression also shows that the separation decreases as the mass of the particle increases. Particles of macroscopic mass (like balls and planets, and even minute specks of dust) behave as though their translational motion is unquantized. Both these conclusions are true in general:

1. The larger the size of the system, the less important are the effects of quantization.
2. The greater the mass of the particle, the less important are the effects of quantization.

Exercise E8.10

Consider an electron that is a part of a conjugated polyene (such as a carotene molecule) of length 2.0 nm. What is the energy in electron-volts ($1\,eV = 1.602 \times 10^{-19}$ J) to excite it from the level with $n = 5$ to the next higher level?

[*Answer*: 1.0 eV]

Rotational motion: the particle on a ring

The discussion of translational motion focused on linear momentum. When we turn to rotational motion we have to focus on the analogous *angular* momentum. The **angular momentum** of a particle that is travelling on a circular path of radius r is defined as

$$\text{angular momentum} = p \times r \qquad (15)$$

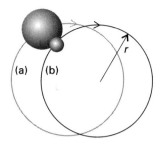

Fig. 8.25 A particle travelling on a circular path has a moment of inertia I that is given by mr^2. (a) This heavy particle has a large moment of inertia about the central point; (b) this light particle is travelling on a path of the same radius, but it has a smaller moment of inertia. The moment of inertia plays a role in circular motion that is the analogue of the mass for linear motion: a particle with a high moment of inertia is difficult to accelerate into a given state of rotation, and requires a strong braking force to stop its rotation.

where p is its linear momentum (the product of its mass and velocity, $p = mv$) at any instant. A particle that is travelling at high speed in a circle has a higher angular momentum than a particle of the same mass travelling more slowly. An object with a high angular momentum (like a flywheel) requires a strong braking force (more precisely, a strong torque) to bring it to a standstill.

To see what quantum mechanics tells us about rotational motion, we consider a particle of mass m moving in a horizontal circular path of radius r. The energy of the particle is entirely kinetic because the potential energy is zero everywhere, and we can therefore write $E = p^2/2m$. By using eqn (15), this energy can be expressed in terms of the angular momentum as

$$E = \frac{(\text{angular momentum})^2}{2mr^2}$$

The quantity mr^2 is called the **moment of inertia** of the particle about the centre of its path, and denoted I: a heavy particle in a path of large radius has a large moment of inertia (Fig. 8.25). It follows that the energy of the particle is

$$E = \frac{(\text{angular momentum})^2}{2I} \qquad (16)$$

Now we use the de Broglie relation to see that the energy of rotation is quantized. To do so, we express the angular momentum in terms of the wavelength of the particle:

$$\text{angular momentum} = p \times r = \frac{h}{\lambda} \times r$$

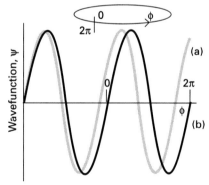

Fig. 8.26 Two solutions of the Schrödinger equation for a particle on a ring. The circumference has been opened out into a straight line; the points at $\phi = 0$ and 2π are identical. The solution in (a) is unacceptable because it has different values after each circuit, and so interferes destructively with itself. The solution in (b) is acceptable because it reproduces itself on successive circuits.

Suppose for the moment that λ can take an arbitrary value. In that case, the amplitude of the wavefunction depends on the angle as shown in Fig. 8.26. When the angle increases beyond 2π (that is, 360°), the wavefunction continues to change, but for an arbitrary wavelength it gives rise to a different amplitude at each point, and the interference between the waves on successive circuits cancels the amplitude of the wave on its previous circuit. Thus, this particular arbitrary wave cannot survive in the system. An acceptable solution is obtained only if the wavefunction reproduces itself on successive circuits. The acceptable wavefunctions have wavelengths that match after each circuit, and therefore have wavelengths that are given by the expression

$$\lambda = \frac{2\pi r}{n} \qquad n = 0, 1, 2, \cdots$$

(The value $n = 0$, which gives an infinite wavelength, corresponds to a uniform amplitude.) It follows that the permitted energies are

$$E_n = \frac{(hr/\lambda)^2}{2I} = \frac{(hn/2\pi)^2}{2I} = \frac{n^2\hbar^2}{2I}$$

with $n = 0, \pm 1, \pm 2, \cdots$.

Two points need to be made about the expression for the energy before we use it. One is that a particle can travel either clockwise or counterclockwise around a ring. These different directions are represented by positive and negative values of n: the energy depends on n^2, so the difference in sign does not affect the energy. Second, it is conventional in the discussion of rotational motion (for reasons that will become clear later) to denote the quantum number by m_l in place of n. Therefore, the final expression for the energy levels is

$$E_{m_l} = \frac{m_l^2 \hbar^2}{2I} \qquad m_l = 0, \pm 1, \pm 2, \cdots \qquad (17)$$

These energy levels are drawn in Fig. 8.27.

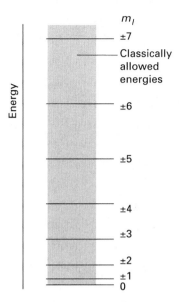

Fig. 8.27 The energy levels of a particle that can move on a circular path. Classical physics allowed the particle to travel with any energy (as represented by the continuous grey band); quantum mechanics, however, allows only discrete energies. Each energy level, other than the one with $m_l = 0$, is doubly degenerate, because the particle may rotate either clockwise or counterclockwise with the same energy.

Exercise E8.11

The moment of inertia of an HCl molecule is 2.6×10^{-47} kg m^2. What is the minimum energy needed to start it rotating in a plane?

[*Answer*: 2.1×10^{-22} J]

As we have remarked, the occurrence of m_l as its square in the expression for the energy means that two states of motion, such as that with $m_l = +1$ and $m_l = -1$, both correspond to the same energy. Such a condition, where more than one state has the same energy, is called **degeneracy**. All the states with $|m_l| > 0$ are doubly degenerate because two states correspond to the same energy for each value of $|m_l|$. The state with $m_l = 0$, the lowest energy state of the particle, is **nondegenerate**, meaning that only one state has a particular energy (in this case, zero). The origin of the degeneracy is easy to identify: a particle may travel either clockwise or counterclockwise on the circle, and the different signs of m_l correspond to the two opposite directions of motion. The state with $m_l = 0$ is nondegenerate because the particle is stationary, so the question of its direction of travel does not arise.

An important additional conclusion is that the angular momentum of the particle is quantized. This conclusion is an example of an observable other than the energy that is confined to discrete values. In the present case, we can use the relation between angular momentum and linear momentum (angular momentum $= p \times r$), and between linear momentum and the allowed wavelengths of the particle ($\lambda = $ circumference/an integer $= 2\pi r/m_l$), to conclude that the angular momentum of a particle around an axis is confined to the values

$$\text{angular momentum} = pr = \frac{hr}{\lambda} = \frac{hr}{(2\pi r/m_l)} = \frac{m_l h}{2\pi}$$

307

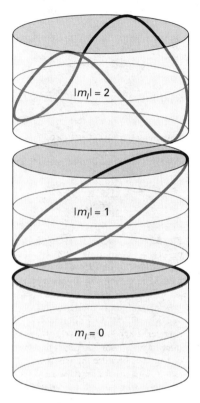

Fig. 8.28 The wavefunction of a particle on a ring. As the wavelength shortens, the angular momentum grows in steps of $\hbar$.

That is, the angular momentum of the particle around the axis is confined to the values

$$\text{angular momentum around axis} = m_l \hbar \qquad (18)$$

with $m_l = 0, \pm 1, \pm 2, \cdots$. Positive values of m_l correspond to counterclockwise rotation (as seen from above) and negative values correspond to clockwise rotation (Fig. 8.28). The quantized motion can be thought of in terms of the rotation of a bicycle wheel that can rotate only with a discrete series of angular momenta, so that as the wheel is accelerated, the angular momentum jerks from the values 0 (when the wheel is stationary) to $\hbar, 2\hbar, \cdots$ but can have no intermediate value.

A final point concerning the rotational motion of a particle is that it does not have a zero-point energy: m_l may take the value 0, so E may be zero. This conclusion is also consistent with the uncertainty principle. Although the particle is certainly between the angles 0 and 360° on the ring, that range is equivalent to not knowing anything about where it is on the ring. Consequently, the angular momentum may be specified exactly, and a value of zero is possible. When the angular momentum is zero precisely, the energy of the particle is also zero precisely.

The structures of atoms

We now have enough background to move on to the major purpose of this chapter: the discussion of the electronic structures of atoms. When considering the structures of atoms we need to distinguish between hydrogenic atoms and many-electron atoms. A **hydrogenic atom** is a one-electron atom or ion of general atomic number Z. Hydrogenic atoms include H, He$^+$, Li^{2+}, C^{5+}, and even U^{91+} (such very highly ionized atoms may be found in the outer regions of stars). A **many-electron atom** is an atom or ion that has more than one electron. Many-electron atoms include all neutral atoms other than H (for instance, helium, with its two electrons, is a many-electron atom in this sense). Hydrogenic atoms, and H in particular, are important because the Schrödinger equation can be solved for them and their structures can be discussed exactly. They also provide a set of concepts that are used to describe the structures of many-electron atoms and (as we shall see in the next chapter) the structures of molecules too.

8.7 The experimental basis: the spectra of hydrogenic atoms

When an electric discharge is passed through gaseous hydrogen, the H_2 molecules dissociate. The energetically excited H atoms that are produced emit electromagnetic radiation of discrete frequencies as they discard energy and return to their **ground state**, their state of lowest energy (Fig. 8.29). The

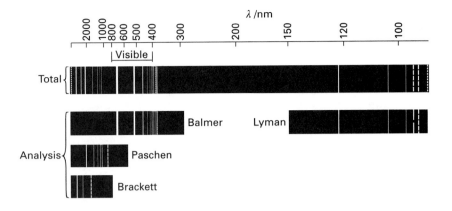

Fig. 8.29 The spectrum of atomic hydrogen. The spectrum is shown at the top, and is analysed into overlapping series below. The Balmer series lies largely in the visible region.

record of wavenumbers, frequencies, or wavelengths[†] of the radiation emitted is called the **emission spectrum** of the atom. In its earliest form, the radiation was detected photographically as a series of lines (the focused image of the slit that the light was sampled through), and the components of radiation present in a spectrum are still widely referred to as spectroscopic 'lines'.

The first important contribution to understanding this spectrum was made by the Swiss schoolteacher Johann Balmer, who pointed out in 1885 that (in modern terms) the wavenumbers of the light in the visible region of the electromagnetic spectrum fit the expression

$$\tilde{\nu} = \mathcal{R}_H \left(\frac{1}{2^2} - \frac{1}{n^2} \right) \qquad n = 3, 4, \cdots$$

The lines described by this formula are now called the **Balmer series** of the spectrum. Later, another set of lines was discovered in the ultraviolet region of the spectrum, and is called the **Lyman series**. Yet another set was discovered in the infrared region when detectors became available for that region, and is called the **Paschen series**. With this additional information available, the Swedish spectroscopist Johannes Rydberg noted (in 1890) that all the lines were described by the expression

$$\tilde{\nu} = \mathcal{R}_H \left(\frac{1}{n_1^2} - \frac{1}{n_2^2} \right) \qquad n_1 = 1, 2, \cdots \qquad n_2 = n_1 + 1, n_1 + 2, \cdots \qquad (19)$$

with $\mathcal{R}_H = 109\,677\,\text{cm}^{-1}$, a quantity now called the **Rydberg constant** for hydrogen. The first five series of lines then correspond to n_1 taking the values 1 (Lyman), 2 (Balmer), 3 (Paschen), 4 (Brackett), and 5 (Pfund).

The existence of discrete spectroscopic lines strongly suggests that the energy of the electron in the hydrogen atom is quantized. Thus, because the total energy is conserved when a **transition**, a change of state, occurs from

[†]The essential properties of electromagnetic radiation are summarized in *Further information 6*.

one energy level to another, when an atom changes its energy by ΔE, the difference must be carried away as a photon of frequency ν (as was illustrated in Fig. 8.13), where

$$\Delta E = h\nu \qquad (20)$$

This relation is called the **Bohr frequency condition**. It follows that we can expect to observe discrete lines if only certain energy states of an electron in an atom are permitted.

8.8 The interpretation: the structures of hydrogenic atoms

The quantum mechanical description of the structure of a hydrogenic atom is based on Rutherford's **nuclear model**, in which the atom is pictured as a central nucleus of charge Ze and one electron. To derive the details of the structure of the atom, we have to set up and solve the Schrödinger equation for the atom, using for the potential energy the **Coulomb potential energy** of the interaction between the nucleus of charge $+Ze$ and the electron of charge $-e$:

$$V = -\frac{Ze^2}{4\pi\varepsilon_0 r} \qquad (21)$$

where ε_0 is the **vacuum permittivity** (a fundamental constant, see inside front cover) and e is the elementary electric charge. The resulting equation can be solved, but we shall only summarize the results.

The energy levels

When the appropriate boundary conditions are imposed, we obtain the physically acceptable solutions, the wavefunctions of the electron in the atom and their corresponding energies. Because only certain wavefunctions are allowed by the boundary conditions, only certain energies can occur, which is qualitatively in accord with the spectroscopic evidence. These allowed energy levels are given by the expression

$$E_n = -\frac{hcZ^2\mathcal{R}}{n^2} \qquad hc\mathcal{R} = \frac{\mu e^4}{32\pi^2\varepsilon_0^2\hbar^2} \qquad (22)$$

The quantity μ is the **reduced mass**:

$$\mu = \frac{m_e m_N}{m_e + m_N} \qquad (23)$$

where m_e is the mass of the electron and m_N is the mass of the nucleus. For all except the most precise considerations, the mass of the nucleus is so much bigger than the mass of the electron that the latter may be neglected in the denominator, and then

$$\mu \approx \frac{m_e m_N}{m_N} = m_e \qquad (24)$$

In other words, to a good approximation the reduced mass can be set equal to the mass of the electron, m_e. The constant $\mathcal{R}$ is numerically identical to the experimental Rydberg constant $\mathcal{R}_H$ when m_N is not ignored but set equal to the mass of the nucleus.

The quantum number n that appears in eqn (22) is called the **principal quantum number**. It is confined to the values

$$n = 1, 2, \cdots$$

and so on without limit. We can calculate the energy of the electron in the atom by substituting the value of n into eqn (22). The resulting energy levels are depicted in Fig. 8.30. Note how they are widely separated at low values of n, but then converge as n increases. At low values of n the electron is confined close to the nucleus by the pull between opposite charges, and the energy levels resemble those of a narrow box. At high values of n, when the electron has such a high energy that it can travel out to large distances, the energy levels are close together, like those of a particle in a large box.

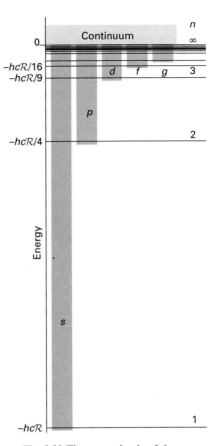

Fig. 8.30 The energy levels of the hydrogen atom. The energies are relative to a proton and an infinitely distant, stationary electron. The degeneracy of the orbitals with the same value of n but different values of l (that is, the degeneracy of orbitals in different subshells but the same shell of the atom) is a special characteristic of one-electron, hydrogenic atoms.

Exercise E8.12

The shortest wavelength transition in the Paschen series occurs at 821 nm in hydrogen; at what wavelength does it occur in Li^{2+}? [*Hint*: Think about the variation of energies with atomic number Z.]

[*Answer*: $\frac{1}{9} \times 821$ nm $= 91.2$ nm]

All the energies given by eqn (22) are negative, which signifies that an electron that is part of the atom has a lower energy than when it is not part of the atom. The zero of energy (which occurs at $n = \infty$) corresponds to the infinitely widely separated and stationary electron and nucleus. The state of lowest energy, the ground state of the atom, is the one with $n = 1$ (the lowest permitted value of n and hence the largest negative value of the energy), and the energy of this state is

$$E_1 = -hc\mathcal{R}$$

The negative sign means that the ground state lies $hc\mathcal{R}$ below the energy of the infinitely separated electron and nucleus. The first excited state of the atom, the state with $n = 2$, lies at

$$E_2 = -\frac{1}{4}hc\mathcal{R}$$

This energy level is $\frac{3}{4}hc\mathcal{R}$ above the ground state (see Fig. 8.30).

The empirical expression for the spectroscopic lines observed in the spectrum of hydrogen is now easy to explain: when a transition occurs, an electron jumps from an energy level with one quantum number (n_2) to a level with a lower energy (with quantum number n_1) and as a result its energy changes by

$$\Delta E = \frac{hc\mathcal{R}}{n_2^2} - \frac{hc\mathcal{R}}{n_1^2}$$

This energy is carried away by a photon of energy $hc\tilde{\nu}$, and by equating this energy to ΔE, we immediately obtain eqn (19).

In a spectroscopic transition, an electron moves between the energy levels of an atom. However, it is possible to remove an electron completely from an atom and eqn (22) can be used to account for the minimum energy required, which is called the **ionization energy**, I. For a hydrogen atom, the ionization energy is the energy required to raise the electron from the ground state (with $n = 1$ and energy $E_1 = -hc\mathcal{R}_H$) to the state corresponding to complete removal of the electron (the state with $n = \infty$ and zero energy). Therefore, the energy that must be supplied is

$$I = hc\mathcal{R}_H = 2.179 \times 10^{-18}\,\text{J}$$

which corresponds to $1312\,\text{kJ}\,\text{mol}^{-1}$ or $13.59\,\text{eV}$.

Exercise E8.13

Predict the ionization energy of He^+ given that the ionization energy of H is $13.59\,\text{eV}$. *Hint*: Decide how the energy of the ground state varies with Z.

[*Answer*: $54.36\,\text{eV}$]

The wavefunctions

The one-electron wavefunctions of hydrogenic atoms are called **orbitals**. The name expresses something less definite than the 'orbits' of classical mechanics. An electron that is described by a particular orbital is said to 'occupy' that orbital. To specify an orbital in this three-dimensional system, we have to give the values of three quantum numbers: the principal quantum number n, which we have already met, and two others. One is the **azimuthal quantum number**, l, which is confined to the values

$$l = 0, 1, 2, \cdots, n - 1$$

For reasons that will become clear, the azimuthal quantum number is also called the *orbital angular momentum quantum number*. Note that for a given value of n, there are n allowed values of l. The other is the **magnetic quantum number**, m_l, which is confined to the values

$$m_l = l, l-1, l-2, \cdots, -l$$

Note that for a given value of l, there are $2l + 1$ values of m_l (for example, when $l = 3$, m_l may have any of the seven values $+3, +2, +1, 0, -1, -2, -3$). It follows from these restrictions that there is only one orbital with $n = 1$, because when $n = 1$ the only value that l can have is 0, and that in turn implies that m_l can have only the value 0. Likewise, there are four orbitals with $n = 2$, because l can take the values 0 and 1, and in the latter case m_l can have the three values $+1$, 0, and -1. In general, there are n^2 orbitals with a given value of n.

Although we need all three quantum numbers to specify a given orbital, we have already seen that in a hydrogenic atom (and *only* in hydrogenic atoms), the energy depends only on the principal quantum number n. Therefore, all orbitals of the same value of n but different values of l and m_l have the same energy. This degeneracy is the reason why all orbitals with the same value of n are said to belong to the same **shell** of the atom. It is common to refer to successive shells by the letters K (1), L (2), M (3), and N (4). Thus, all four orbitals of the shell with $n = 2$ form the L shell of the atom.

Orbitals with the same value of n but different values of l form the **subshells** of a given shell. These subshells are generally denoted by the letters $s, p, \cdots$, using the correspondence s (for $l = 0$), p ($l = 1$), d ($l = 2$), f ($l = 3$), and then alphabetically, $g, h, \cdots$, with the omission of i. In practice, only the s, p, d, and f subshells are of any importance. For $n = 1$, there is only one subshell, the one with $l = 0$. When $n = 2$, there are two subshells, namely the $2s$ subshell (with $l = 0$) and the $2p$ subshell (with $l = 1$). The general pattern of the first three shells and their subshells are listed in Fig. 8.31. In a hydrogenic atom, all the subshells of a given shell correspond to the same energy (because, as we have seen, the energy depends on n and not on l).

Each subshell contains $2l + 1$ individual orbitals (corresponding to the $2l + 1$ values of m_l for each value of l). Thus, in any given subshell, the number of orbitals is as given in Fig. 8.31. An s subshell ($l = 0$) consists of only one orbital ($m_l = 0$), which is called an **s orbital**. A p subshell ($l = 1$) consists of three **p orbitals** (corresponding to $m_l = +1, 0, -1$). An electron that occupies an s orbital is called an **s electron**. Similarly, we can speak of $p, d, \cdots$ electrons according to the orbitals they occupy.

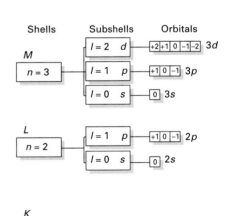

Fig. 8.31 The structures of atoms are described in terms of shells of electrons that are labelled by the principal quantum number n, and a series of n subshells of these shells, with each subshell of a shell being labelled by the quantum number l. Each subshell consists of $2l + 1$ orbitals.

Exercise E8.14

How many orbitals are there in a shell with $n = 5$?

[*Answer*: 25]

s orbitals

The mathematical form of the wavefunction for a $1s$ orbital ($n = 1, l = 0, m_l = 0$) for a hydrogen atom is

$$\psi = \frac{1}{\sqrt{\pi a_0^3}} e^{-r/a_0} \tag{25}$$

We need to see what it tells us about the probable location of an electron that has this wavefunction, which is the case if the atom is in its ground state. First, we note that the wavefunction depends only on the radius, r, of the point of interest and is independent of angle (the latitude and longitude of the point), so the orbital has the same amplitude at all points of the same radius regardless of direction in space relative to the nucleus. Because the probability of finding an electron is proportional to the square of the wavefunction, we now know that the electron will be found with the same probability in any direction (for a given distance from the nucleus). We summarize this angular independence by saying that a $1s$ orbital is **spherically symmetrical**.

Next, we note that ψ depends on distance from the nucleus (Fig. 8.32), and that its amplitude decays exponentially from a maximum value of $1/(\pi a_0^3)^{1/2}$ at the nucleus (at $r = 0$). It follows from the exponentially decaying form of the $1s$ orbital that, in the ground state of a hydrogenic atom, the most probable point at which the electron will be found is at the nucleus itself. A method of depicting the probability of finding the electron at each point in space is to represent ψ^2 by the density of shading in a diagram (Fig. 8.33(a)). A simpler procedure is to show only the **boundary surface**, the shape that captures about 90 per cent of the electron probability. For the $1s$ orbital, the boundary surface is a sphere (Fig. 8.33(c)).

Exercise E8.15

What is the probability of finding the electron in a volume of $1\,\text{pm}^3$ centred on the nucleus in a hydrogen atom?

[*Answer*: 2.2×10^{-6}, one observation in 455 000]

We often need to know the probability that an electron will be found at a given distance from a nucleus regardless of its angular position (Fig. 8.34). As we show in the *Justification* below, the probability that an electron in an s orbital will be found between a radius r and a slightly greater radius $r + \delta r$ is calculated by evaluating

$$\text{probability} = 4\pi r^2 \psi^2 \times \delta r \tag{26}$$

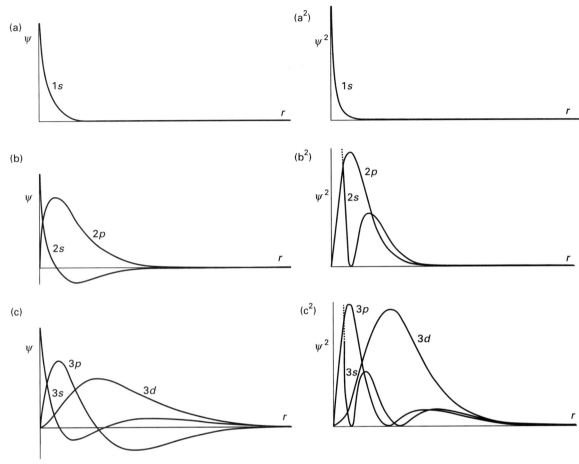

Fig. 8.32 (a,b,c) The radial wavefunctions of the first few states of the hydrogen atom and (a², b², c²) the probability densities. Note that the *s*-orbitals have a nonzero and finite value at the nucleus. The vertical scales are different in each case.

The factor $4\pi r^2 \psi^2$ multiplying the width of the region is called the **radial distribution function**, and by examining its shape we can see how the probability varies as the distance from the nucleus increases. The principal point is that, because r^2 increases as r increases but ψ^2 decreases exponentially, it follows that the radial distribution function goes through a maximum. The location of the maximum marks the most probable radius (not point) at which the electron will be found. For a $1s$ orbital of hydrogen, the maximum occurs at the **Bohr radius**, a_0, which is at 53 pm from the nucleus. An analogy that might help to fix the significance of the radial distribution function for an electron is the corresponding distribution for the population of the Earth regarded as a perfect sphere. The radial distribution function is zero at the centre of the Earth and for the next 6400 km (to the surface of the planet), when it peaks sharply and then rapidly decays again to zero. It remains virtually zero for all radii more than about 10 km above the surface. Almost all the population will be found very close to $r = 6400$ km, and it is not relevant that people are dispersed over a very wide range of latitudes and longitudes. The small probabilities of finding people above and below

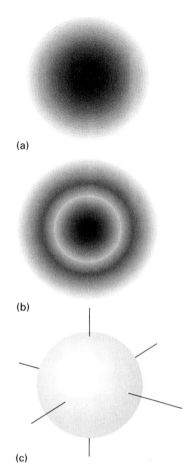

Fig. 8.33 Representations of the first two *s* hydrogenic atomic orbitals, (a) 1*s*, (b) 2*s*, in terms of the electron densities (as represented by the density of shading) and (c) the boundary surface within which there is a 90 per cent probability of finding the electron. The boundary surfaces are the same for both orbitals, except that the surface for the 2*s* orbital has a larger radius.

6400 km anywhere in the world corresponds to the population that happens to be down mines or living in places as high as Denver or Tibet at the time.

Justification

The mathematical expression for the radial distribution function is found by considering the probability of finding the electron in a spherical shell of radius r and thickness δr. The volume of such a shell is its surface area, $4\pi r^2$, multiplied by its thickness, δr, and is therefore $4\pi r^2 \delta r$. According to the Born interpretation, the probability of finding an electron inside a small volume of magnitude δV is given by the value of $\psi^2 \times \delta V$. Therefore, interpreting δV as the volume of the shell, we obtain $\psi^2 \times 4\pi r^2 \delta r$ for the probability of finding the particle anywhere in its volume.

A 2*s* orbital (an orbital with $n = 2$, $l = 0$, and $m_l = 0$) is also spherical, and so its boundary surface is a sphere. Because a 2*s* orbital spreads further out from the nucleus than a 1*s* orbital—because the electron it describes has more energy to climb away from the nucleus—its boundary surface is a sphere of larger radius. It also differs from a 1*s* orbital in its radial dependence (Fig. 8.33(b)), for although the wavefunction has a nonzero value at the nucleus (like all *s* orbitals), it passes through zero before commencing its exponential decay towards zero at large distances. The fact that the wavefunction passes through zero everywhere at a certain radius is summarized by saying that the orbital has a **radial node**: there is zero probability of finding the electron at the location of a radial node. A 3*s* orbital has two radial nodes, a 4*s* orbital has three radial nodes.

p and *d* orbitals

All ***p* orbitals** (orbitals with $l = 1$) have a double-lobed appearance like that shown in Fig. 8.35. The two lobes are separated by a **nodal plane** that cuts through the nucleus. There is zero probability of finding an electron on this plane, and therefore (because the nucleus lies in the plane) zero probability of finding the electron at the nucleus.

The exclusion of the electron from the nucleus is a common feature of all *p* orbitals. To understand its origin, we need to know that the value of the quantum number l tells us the magnitude of the angular momentum of the electron around the nucleus (in classical terms, how rapidly it is circulating around the nucleus) through the expression

$$\text{angular momentum} = \sqrt{l(l+1)} \times \hbar \qquad (27)$$

Because $l = 1$, the angular momentum of an electron in a *p* orbital is $\sqrt{2}\hbar$. On the other hand, for an *s* orbital, the orbital angular momentum is zero (because $l = 0$), and in classical terms the electron does not circulate around the nucleus. Therefore, whereas a p electron is flung away from the nucleus by the centrifugal force arising from its motion, an s electron is not. The same centrifugal effect appears in all orbitals with angular momentum (those for which $l > 0$), such as d orbitals and f orbitals, and all such orbitals have nodes at the nucleus.

A p shell consists of three individual orbitals ($m_l = +1, 0, -1$). The three orbitals are normally represented by their boundary surfaces, as depicted in Fig. 8.35. The p_x orbital has a symmetrical double-lobed shape directed along the x-axis, and similarly the p_y and p_z orbitals are directed along the y and z axes, respectively. As n increases, the p orbitals become bigger (for the same reason as s orbitals), and acquire a more complex nodal structure. However, their boundary surfaces retain the double-lobed shape shown in the illustration.

Orbitals with $l = 2$ are called **d orbitals**. There are five d orbitals in each shell (for $n \geq 3$), each one corresponding to one of the values $m_l = +2, +1, 0, -1$ and -2. The five boundary surfaces of the orbitals (which are similar for the d orbitals of other shells) are shown in Fig. 8.36.

The physical significance of the quantum number m_l can now be explained. It indicates the component of the electron's orbital angular momentum around an arbitrary axis passing through the nucleus. Positive values of m_l correspond to clockwise motion seen from below and negative values correspond to counterclockwise motion. The larger the value of $|m_l|$, the higher the angular momentum around the arbitrary axis. In fact,

$$\text{component of angular momentum} = m_l \hbar$$

An s electron has $m_l = 0$, and has no angular momentum about any axis. A p electron can rotate clockwise about an axis ($m_l = -1$) with an angular momentum $\hbar$, clockwise (as seen from below) with the same angular momentum about that axis ($m_l = +1$), or not at all ($m_l = 0$) about that axis (but about other axes instead). An electron in the d subshell can rotate with five different amounts of angular momentum about an arbitrary axis ($+2\hbar, +\hbar, 0, -\hbar, -2\hbar$).

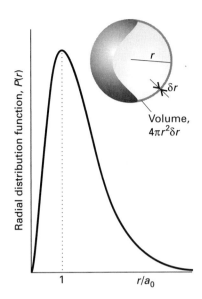

Fig. 8.34 The radial distribution function gives the probability that the electron will be found anywhere in a shell of radius r regardless of angle. The graph shows the output from an imaginary shell-like detector of variable radius and fixed thickness δr.

8.9 Electron spin

To complete the description of the state of a hydrogenic atom, we need to introduce one more concept, that of electron **spin**. The spin of an electron is an intrinsic angular momentum that every electron possesses and which cannot be changed or eliminated (just like its mass or its charge). The name 'spin' is evocative of a ball spinning on its axis, and (so long as it is treated with caution) this classical interpretation can be used to help to visualize the motion. However, in fact spin is a purely quantum mechanical phenomenon and has no classical counterpart, so the analogy must be used with care.

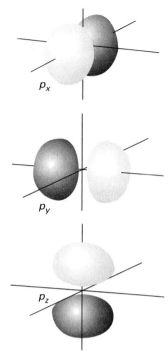

Fig. 8.35 The boundary surfaces of *p* orbitals. A nodal plane passes through the nucleus and separates the two lobes of each orbital. The light and dark areas denote regions of opposite sign of the wavefunction.

We shall make use of two properties of electron spin:

1. Electron spin is described by a quantum number s (the analogue of l for orbital angular momentum), but s is fixed at the single value $\frac{1}{2}$ for all electrons at all times.
2. The direction of the spin angular momentum can be clockwise or counterclockwise; these two states are distinguished by the quantum number m_s which can take the values $+\frac{1}{2}$ or $-\frac{1}{2}$, but no other values.

An electron with $m_s = +\frac{1}{2}$ is called an *α* **electron** and denoted by the symbol ↑; an electron with $m_s = -\frac{1}{2}$ is called a *β* **electron** and denoted by the symbol ↓.

The existence of electron spin was confirmed by an experiment performed by Otto Stern and Walther Gerlach in 1921, who shot a beam of silver atoms through an inhomogeneous magnetic field (Fig. 8.37). A silver atom has 47 electrons, and (for reasons that will become clear later) 23 of the spins are ↑ and 23 spins are ↓; the one remaining spin may be either ↑ or↓. Because the angular momenta of the ↑ and ↓ spins cancel each other, the atom behaves as if it had the spin of a single electron. The idea behind the Stern–Gerlach experiment was that a rotating, charged body—in this case an electron—behaves like a magnet and interacts with the applied field. The magnetic field pushes or pulls the electron according to the orientation of the electron's spin, and so the initial beam of atoms should split into two beams, one corresponding to atoms with ↑ spin and the other to atoms with ↓ spin. This result was observed.

Other fundamental particles also have characteristic spins. For example, protons and neutrons are spin-$\frac{1}{2}$ particles (that is, for them $s = \frac{1}{2}$) and so invariably spin with a single, irremovable angular momentum. Because the

Fig. 8.36 The boundary surfaces of *d* orbitals. Two nodal planes in each orbital intersect at the nucleus and separate the four lobes of each orbital. The light and dark areas denote regions of opposite sign of the wavefunction.

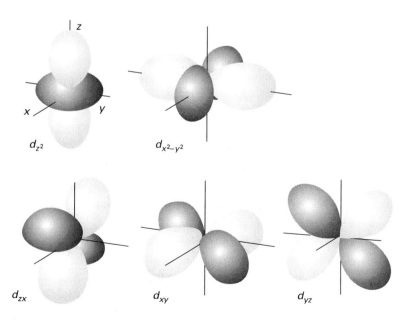

318

masses of a proton and a neutron are so much greater than the mass of an electron, yet they all have the same spin angular momentum, the classical picture of proton and neutron spin would be of particles spinning much more slowly than an electron. Some elementary particles have $s = 1$, and so have a higher intrinsic angular momentum than an electron. For our purposes the most important spin-1 particle is the photon. It is a very deep feature of nature, that the particles from which matter is built have half-integral spin (such as electrons and quarks, all of which have $s = \frac{1}{2}$) whereas the particles that transmit forces between these particles, and so bind them together into entities like nuclei, atoms, and planets, all have integral spin (such as $s = 1$ for the photon, which transmits the electromagnetic interaction between charged particles).

8.10 Spectral transitions and selection rules

We have already seen that when an electron makes a transition from an orbital in a shell with a principal quantum number n_2 into an orbital of a shell with principal quantum number n_1, the excess energy is emitted as a photon and contributes to one of the spectral lines given by the Rydberg formula, eqn (19). We can think of the sudden change in the distribution of the electron as it changes its spatial distribution from one orbital to another orbital as jolting the electromagnetic field into oscillation, and that oscillation corresponds to the generation of a photon of light.

It turns out, however, that not all transitions between all available orbitals are possible. For example, it is not possible for an electron in a $3d$ orbital to make a transition to a $1s$ orbital. Transitions are classified as either **allowed**, if they can contribute to the spectrum, or **forbidden**, if they cannot. The forbidden or allowed character of a transition can be traced to the role of the photon spin, which we mentioned above. When a photon, with its one unit of angular momentum, is generated in a transition, the angular momentum of the electron must change by one unit to compensate for the angular momentum carried away by the photon. That is, the angular momentum must be conserved—neither created nor destroyed—just as linear momentum is conserved in collisions. Thus, an electron in a d orbital with $l = 2$ cannot make a transition into an s orbital with $l = 0$ because the photon cannot carry away enough angular momentum. Similarly, an s electron cannot make a transition to another s orbital, because then there is no change in the electron's angular momentum to make up for the angular momentum carried away by the photon.

A **selection rule** is a statement about which spectroscopic transitions are allowed. They are derived (for atoms) by identifying the transitions that conserve angular momentum when a photon is emitted or absorbed. The selection rules for hydrogenic atoms are

$$\Delta l = \pm 1 \qquad \Delta m_l = 0, \pm 1$$

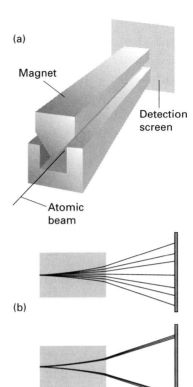

Fig. 8.37 (a) The experimental arrangement for the Stern–Gerlach experiment: the magnet is the source of an inhomogeneous field. (b) The classically expected result, when the orientations of the electron spins can take all angles. (c) The observed outcome using silver atoms, when the electron spins can adopt only two orientations ($\uparrow$ and $\downarrow$).

The principal quantum number n can change by any amount consistent with the Δl for the transition because it does not relate directly to the angular momentum.

Example *Using the selection rules*
To what orbitals may a 4*d* electron make spectroscopic transitions?

Strategy
We apply the selection rules, principally the rule concerning l, and identify the orbital to which the transition may occur.

Solution
Because $l = 2$, the final orbital must have $l = 1$ or 3. Thus, an electron may make a transition from a 4*d* orbital to any *np* orbital (subject to $\Delta m_l = 0, \pm 1$) and to any *nf* orbital (subject to the same rule). However, it cannot undergo a transition to any other orbital, so a transition to any *ns* orbital or another *nd* orbital is forbidden.

Exercise E8.16

To what orbitals may a 4*s* electron make spectroscopic transitions?

[*Answer*: *np* orbitals only]

Selection rules enable us to construct a **Grotrian diagram** (Fig. 8.38), which is a diagram that summarizes the energies of the states and the allowed transitions between them. The thicknesses of the transition lines in the diagram indicate in a general way their relative intensities in the spectrum. The intensities may also be calculated from the wavefunctions of the two states, but we shall not deal with this aspect here.

The structures of many-electron atoms

The Schrödinger equations for many-electron atoms are extremely complicated because all the electrons interact with one another. Even for a He atom, with its two electrons, no mathematical expression for the orbitals and energies can be given, and we are forced to make approximations.

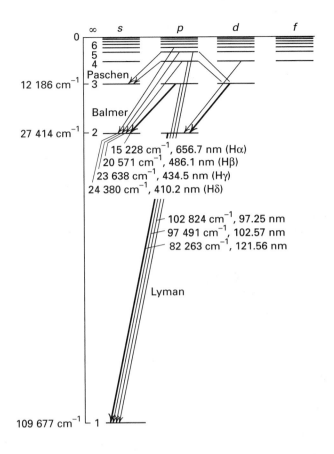

0

∞   s   p   d   f

6
5
4
Paschen
12 186 cm^{-1} — 3

Balmer
27 414 cm^{-1} — 2

15 228 cm^{-1}, 656.7 nm (Hα)
20 571 cm^{-1}, 486.1 nm (Hβ)
23 638 cm^{-1}, 434.5 nm (Hγ)
24 380 cm^{-1}, 410.2 nm (Hδ)

102 824 cm^{-1}, 97.25 nm
97 491 cm^{-1}, 102.57 nm
82 263 cm^{-1}, 121.56 nm

Lyman

109 677 cm^{-1} — 1

8.11 The orbital approximation

The actual wavefunction of a many-electron atom, if it could be calculated, would be a very complicated function of the coordinates of all the electrons. However, in the **orbital approximation**, we suppose that a reasonable first approximation to this unknown exact wavefunction is obtained by thinking of each electron as occupying its 'own' orbital, and writing

$$\psi = \psi(1)\psi(2)\cdots \tag{28}$$

where $\psi(1)$ is the wavefunction of electron 1, $\psi(2)$ that of electron 2, and so on. We can think of the individual orbitals as resembling the hydrogenic orbitals, but with nuclear charges that are modified by the presence of all the other electrons in the atom. This description is only approximate, but it is a useful model for discussing the properties of atoms, and is the starting point for more sophisticated descriptions of atomic structure.

The orbital approximation allows us to express the electronic structure of an atom by reporting its **configuration**, the list of occupied orbitals (usually, but not necessarily, in its ground state). For example, because the ground state of a hydrogen atom consists of a single electron in a 1s orbital, we report its configuration as $1s^1$ (read 'one-s-one'). A helium atom has two electrons. We can imagine forming the atom by adding the electrons in succession to the orbitals of the bare nucleus (of charge $2e$). The first electron occupies a $1s$ hydrogenic orbital, but because $Z = 2$, the orbital is more compact than in H itself. The second electron joins the first in the same $1s$ orbital, and so the electron configuration of the ground state of He is $1s^2$ (read 'one-s-two').

The Pauli principle

Lithium, with $Z = 3$, has three electrons. Two of its electrons occupy a $1s$ orbital drawn even more closely than in He around the more highly charged nucleus. The third electron, however, does not join the first two in the $1s$ orbital because a $1s^3$ configuration is forbidden by the **Pauli exclusion principle**. This fundamental quantum mechanical principle states the following:

> No more than two electrons may occupy any given orbital, and if two electrons do occupy one orbital, then their spins must be paired.

Electrons with **paired spins**, denoted ↑↓, have zero net spin angular momentum because the spin angular momentum of one electron is cancelled by the spin of the other (one electron has $m_s = +\frac{1}{2}$ and the other has $m_s = -\frac{1}{2}$, and their sum is 0). The exclusion principle is the key to understanding the structures of complex atoms, to chemical periodicity, and to molecular structure. It was proposed by the Austrian Wolfgang Pauli in 1924 when he was trying to account for the absence of some lines in the spectrum of helium.

Lithium's third electron cannot enter the $1s$ orbital because that orbital is already full: we say the K shell is **complete** and that the two electrons form a **closed shell**. Because a similar closed shell occurs in the He atom, we denote it [He]. The third electron is excluded from the K shell and must occupy the next available orbital, which is one with $n = 2$ and hence belonging to the L shell. However, we now have to decide whether the next available orbital is the $2s$ orbital or a $2p$ orbital, and therefore whether the lowest energy configuration of the atom is $[\mathrm{He}]2s^1$ or $[\mathrm{He}]2p^1$.

Penetration and shielding

Unlike in hydrogenic atoms, the $2s$ and $2p$ orbitals (and, in general, all orbitals of a given shell) are not degenerate in many-electron atoms. For reasons we shall now explain, s electrons generally lie lower in energy than p electrons of a given shell, and p electrons lie lower than d electrons.

An electron in a many-electron atom experiences a Coulombic repulsion from all the other electrons present. When the electron is at a distance r from the nucleus, the repulsion it experiences is the same as would be generated by a point negative charge located on the nucleus. The magnitude of this charge is equal to the charge of all the other electrons within a sphere of radius r (Fig. 8.39). The effect of the point negative charge is to lower the full charge of the nucleus from Ze to $Z_{\mathrm{eff}}e$, the **effective nuclear**

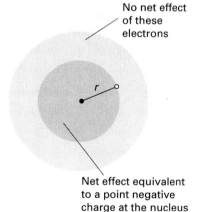

No net effect of these electrons

Net effect equivalent to a point negative charge at the nucleus

Fig. 8.39 An electron at a distance r from the nucleus experiences a Coulombic repulsion from all the electrons within a sphere of radius r and which is equivalent to a point negative charge located on the nucleus. The effect of the point charge is to reduce the apparent nuclear charge of the nucleus from Ze to $Z_{\mathrm{eff}}e$.

charge. To express the fact that an electron experiences a nuclear charge that has been modified by the other electrons present, we say that the electron experiences a **shielded** nuclear charge. The electrons do not actually 'block' the full Coulombic attraction of the nucleus: the effective charge is simply a way of expressing the net outcome of the nuclear attraction and the electronic repulsions in terms of a single equivalent charge at the centre of the atom.

The effective nuclear charges experienced by s and p electrons are different because the electrons have different wavefunctions (Fig. 8.40). An s electron has a greater **penetration** through inner shells than a p electron in the sense that it is more likely to be found close to the nucleus than a p electron of the same shell (the p orbital, remember, has a node passing through the nucleus). As a result of this penetration, an s electron experiences less shielding than a p electron and it therefore experiences a larger Z_{eff}. Consequently, by the combined effects of penetration and shielding, an s electron is more tightly bound than a p electron of the same shell. Similarly, a d electron penetrates less than a p electron of the same shell, and it therefore experiences more shielding and an even smaller Z_{eff}.

The consequence of penetration and shielding is that, in general, the energies of orbitals in the same shell of a many-electron atom lie in the order

$$s < p < d < f$$

The individual orbitals of a given subshell (such as the three p orbitals of the p subshell) remain degenerate because they all have the same radial characteristics and so experience the same effective nuclear charge.

We can now complete the Li story. Because the shell with $n = 2$ consists of two nondegenerate subshells, with the $2s$ orbital lower in energy than the three $2p$ orbitals, the third electron occupies the $2s$ orbital. This arrangement results in the ground state configuration $1s^2 2s^1$, or $[He]2s^1$. It follows that we can think of the structure of the atom as consisting of a central nucleus surrounded by a complete helium-like shell of two $1s$ electrons, and around that a more diffuse $2s$ electron. The electrons in the outermost shell of an atom in its ground state are called the **valence electrons** because they are largely responsible for the chemical bonds that the atom forms (and, as we shall see, the extent to which an atom can form bonds is called its 'valence'). Thus, the valence electron in Li is a $2s$ electron, and lithium's other two electrons belong to its core, where they take little part in bond formation.

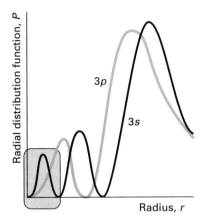

Fig. 8.40 An electron in an s orbital (here a $3s$ orbital) is more likely to be found close to the nucleus than an electron in a p orbital of the same shell. Hence it experiences less shielding and is more tightly bound.

The building-up principle

The extension of the procedure used for H, He, and Li to other atoms is called the **building-up principle** (or the *Aufbau* principle, from the German word for building up). The building-up principle specifies an order of occupation of atomic orbitals that reproduces the experimentally determined ground state configurations of neutral atoms.

We imagine the bare nucleus of atomic number Z, and then feed into the available orbitals Z electrons one after the other. The first rule of the building-up principle is:

The order of occupation of orbitals is

$1s$ $2s$ $2p$ $3s$ $3p$ $4s$ $3d$ $4p$ $5s$ $4d$ $5p$ $6s$ $5d$ $4f$ $6p$ $\cdots$

and, according to the Pauli exclusion principle, each orbital may accommodate up to two electrons.

This order of occupation is approximately the order of energies of the individual orbitals, because in general the lower the energy of the orbital, the lower the total energy of the atom as a whole when that orbital is occupied. An s orbital is complete as soon as two electrons are present in it. Each of the three p orbitals of a shell can accommodate two electrons, so a p subshell is complete as soon as six electrons are present in it. A d subshell, which consists of five orbitals, can accommodate up to ten electrons.

As an example, consider a carbon atom. Because $Z = 6$ for carbon, there are six electrons to accommodate. Two enter and fill the $1s$ orbital, two enter and fill the $2s$ orbital, leaving two electrons to occupy the orbitals of the $2p$ subshell. Hence its ground configuration is $1s^2 2s^2 2p^2$, or more succinctly $[\text{He}]2s^2 2p^2$, with $[\text{He}]$ the helium-like $1s^2$ core. However, it is possible to be more precise. On electrostatic grounds, we can expect the last two electrons to occupy different $2p$ orbitals, for they will then be further apart on average and repel each other less than if they were in the same orbital. Thus, one electron can be thought of as occupying the $2p_x$ orbital and the other the $2p_y$ orbital, and the lowest energy configuration of the atom is $[\text{He}]2s^2 2p_x^1 2p_y^1$. The same rule applies whenever degenerate orbitals of a subshell are available for occupation. Thus, another rule of the building-up principle is:

Electrons occupy different orbitals of a given subshell before doubly occupying any one of them.

It follows that a nitrogen atom ($Z = 7$) has the configuration $[\text{He}]2s^2 2p_x^1 2p_y^1 2p_z^1$. Only when we get to oxygen ($Z = 8$) is a $2p$ orbital doubly occupied, giving the configuration $[\text{He}]2s^2 2p_x^2 2p_y^1 2p_z^1$. An additional point arises when electrons occupy degenerate orbitals (such as the three $2p$ orbitals) singly, as they do in C, N, and O, for there is then no requirement that their spins should be paired. We need to know whether the lowest energy is achieved when the electron spins are the same (both ↑, for instance, denoted ↑↑, if there are two electrons in question, as in C) or when they are paired (↑↓), This problem is resolved by **Hund's rule**:

In its ground state, an atom adopts a configuration with the greatest number of unpaired electrons.

The explanation of Hund's rule is complicated, but it reflects the quantum mechanical property of **spin correlation**, that electrons in different orbitals with parallel spins have a tendency to stay well apart and hence repel each other less.[†] We can now conclude that in the ground state of a C atom, the two $2p$ electrons have the same spin, that all three $2p$ electrons in an N atom have the same spin, and that the two electrons that singly occupy different $2p$ orbitals in an O atom have the same spin (the two in the $2p_x$ orbital are necessarily paired).

Neon, with $Z = 10$, has the configuration $[He]2s^2 2p^6$, which completes the L shell. This closed-shell configuration is denoted $[Ne]$, and acts as a core for subsequent elements. The next electron must enter the $3s$ orbital and begin a new shell, and so a Na atom, with $Z = 11$, has the configuration $[Ne]3s^1$. Like lithium with the configuration $[He]2s^1$, sodium has a single s electron outside a complete core.

Exercise E8.17

Deduce the ground state electron configuration of sulfur.

[*Answer*: $[Ne]3s^2 3p_x^2 3p_y^1 3p_z^1$]

This analysis has brought us to the origin of chemical periodicity. The L shell is completed by eight electrons, and so the element with $Z = 3$ (Li) should have similar properties to the element with $Z = 11$ (Na). Likewise, Be ($Z = 4$) should be similar to Mg ($Z = 12$), and so on up to the noble gases He ($Z = 2$), Ne ($Z = 10$), and Ar ($Z = 18$).

The occupation of d orbitals

Argon has complete $3s$ and $3p$ subshells, and as the $3d$ orbitals are high in energy, the atom effectively has a closed-shell configuration. Indeed, the $4s$ orbitals are so lowered in energy by their ability to penetrate close to the nucleus that the next electron (for potassium) occupies a $4s$ orbital rather than a $3d$ orbital and the K atom resembles a Na atom. The same is true of a Ca atom, which has the configuration $[Ar]4s^2$. However, at this point, the $3d$ orbitals become comparable in energy to the $4s$ orbitals (Fig. 8.41), and they commence to be filled.

Ten electrons can be accommodated in the five $3d$ orbitals, which accounts for the electron configurations of scandium to zinc. However, the building-up principle has less clear-cut predictions about the ground state configurations of these elements because electron–electron repulsions are comparable to the energy difference between the $4s$ and $3d$ orbitals, and a simple analysis no longer works. At gallium, the energy of the $3d$ orbitals has fallen so far below those of the $4s$ and $4p$ orbitals that they (the full $3d$ orbitals) can be largely ignored, and the building-up principle can be used

---

[†]The effect of spin correlation is to allow the atom to shrink slightly, so the electron–nucleus interaction is improved when the spins are parallel.

Fig. 8.41 Energy levels of many-electron atoms in the periodic table. The inset shows a magnified view close to $Z = 20$.

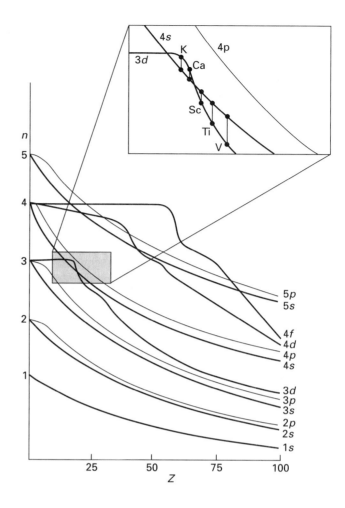

in the same way as in preceding periods. Now the $4s$ and $4p$ subshells constitute the valence shell, and the period terminates with krypton. Because 18 electrons have intervened since argon, this period is the first **long period** of the periodic table. The existence of the **d-block elements** (the 'transition metals') reflects the stepwise occupation of the $3d$ orbitals, and the subtle shades of energy differences along this series gives rise to the rich complexity of inorganic (and bioinorganic) d-metal chemistry. A similar intrusion of the f orbitals in Periods 6 and 7 accounts for the existence of the **f block** of the periodic table (the lanthanides and actinides).

The configurations of cations and anions

The configurations of cations of elements in the s, p, and d blocks of the periodic table are derived by removing electrons from the ground-state configuration of the neutral atom in a specific order. First, we remove p electrons (if any are present), then s electrons, and then as many d electrons as are necessary to achieve the stated charge. For instance, because the configuration of Fe is $[Ar]3d^64s^2$, an Fe^{3+} cation has the configuration $[Ar]3d^5$.

The configurations of anions are derived by continuing the building-up procedure and adding electrons to the neutral atom until the configuration of the next noble gas has been reached. Thus, the configuration of an O^{2-} ion is achieved by adding two electrons to $[He]2s^22p^4$, giving $[He]2s^22p^6$, the configuration of Ne.

Exercise E8.18

Give the electron configurations of (a) a Cu^{2+} ion and (b) an S^{2-} ion.

[*Answer*: (a) $[Ar]3d^9$, (b) $[Ne]3s^23p^6$]

8.12 Periodic trends in atomic properties

The periodic recurrence of analogous ground state electron configurations as the atomic number increases accounts for the periodic variation in the properties of atoms. Here we concentrate on two aspects of atomic periodicity: atomic radius and ionization energy.

Atomic radius

Atomic radii are defined in terms of the internuclear distances between bonded atoms in solids and molecules. They are of great significance in chemistry, for the size of an atom is one of the most important properties for determining how many chemical bonds an element can form. Moreover, the size and shape of a molecule depend on the sizes of the atoms of which it is composed, and molecular shape and size is a crucial aspect of a molecule's biological function. Atomic radius also has an important technological aspect, because the similarity of the atomic radii of the d-block elements is the main reason why they can be blended together to form so many different alloys, particularly varieties of steel. If there is one single attribute of an element that determines its chemical properties (either directly, or indirectly through the variation of other properties), then it is atomic radius.

In general, atomic radii decrease from left to right across a period and increase down each group (Fig. 8.42). The decrease across a period can be traced to the increase in nuclear charge, which draws the electrons in closer to the nucleus. The increase in nuclear charge is partly cancelled by the increase in the number of electrons, but one electron does not fully shield one nuclear charge, so the increase in nuclear charge dominates. The increase in atomic radius down a group (despite the increase in nuclear charge) is

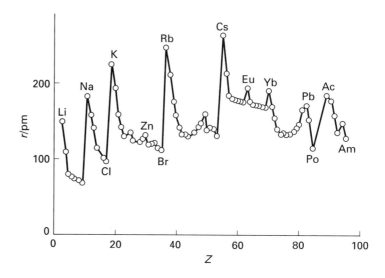

Fig. 8.42 The variation of radius through the periodic table. Note the contraction of radii following the lanthanides in Period 6 (following Yb, ytterbium).

explained by the fact that the valence shells of successive periods correspond to higher principal quantum numbers. That is, successive periods correspond to the start and then completion of successive (and more distant) shells of the atom that surround each other like the successive layers of an onion, and the need to occupy a more distant shell leads to a larger atom despite the increased nuclear charge.

A modification of the increase down a group is encountered in Period 6, for the radii of the atoms late in the *d* block and in the following regions of the *p* block are not as large as would be expected by simple extrapolation down the group. The reason can be traced to the fact that in Period 6 the *f* orbitals are occupied. An *f* electron is a very inefficient shielder of nuclear charge (for reasons connected with its radial extension), and as the atomic number increases from La to Yb, there is a considerable contraction in radius. By the time the *d* block resumes (at lutetium, Lu), the poorly shielded but considerably increased nuclear charge has drawn in the surrounding electrons, and the atoms are compact. They are so compact, that the metals in this region of the periodic table (iridium to lead) are very dense. The reduction in radius below that expected by extrapolation from preceding periods is called the **lanthanide contraction**.

Ionization energy

The minimum energy necessary to remove an electron from a many-electron atom is its **first ionization energy**, I_1. The **second ionization energy**, I_2, is the minimum energy needed to remove a second electron (from the singly-charged cation). The variation of the first ionization energy through the periodic table is shown in Fig. 8.43 and some numerical values are given in Table 2.3. The ionization energy of an element plays a central role in

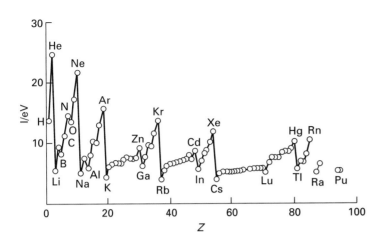

Fig. 8.43 The periodic variation of the first ionization energies of the elements.

determining the ability of its atoms to participate in bond formation (for bond formation, as we shall see in Chapter 9, is a consequence of the relocation of electrons from one atom to another). After atomic radius, it is the most important property for determining an element's chemical characteristics.

Lithium has a low first ionization energy: its outermost electron is well-shielded from the nucleus by the core ($Z_{eff} = 1.3$ compared with $Z = 3$) and it is easily removed. Beryllium has a higher nuclear charge than lithium, and its outermost electron (one of the two $2s$ electrons) is more difficult to remove: its ionization energy is larger. The ionization energy decreases between beryllium and boron because in the latter the outermost electron occupies a $2p$ orbital and is less strongly bound than if it had been a $2s$ electron. The ionization energy increases between boron and carbon because the latter's outermost electron is also $2p$ and the nuclear charge has increased. Nitrogen has a still higher ionization energy because of the further increase in nuclear charge.

There is now a kink in the curve because the ionization energy of oxygen is lower than would be expected by simple extrapolation. At oxygen a $2p$ orbital must become doubly occupied, and the electron–electron repulsions are increased above what would be expected by simple extrapolation along the row. (The kink is less pronounced in the next row, between phosphorus and sulfur, because their orbitals are more diffuse.) The values for oxygen, fluorine, and neon fall roughly on the same line, the increase of their ionization energies reflecting the increasing attraction of the nucleus for the outermost electrons.

The outermost electron in sodium is $3s$. It is far from the nucleus, and the latter's charge is shielded by the compact, complete neon-like core. As a result, the ionization energy of sodium is substantially lower than that of neon. The periodic cycle starts again along this row, and the variation of the ionization energy can be traced to similar reasons.

EXERCISES

8.1 Incandescent lamps are a common feature of everyday life. Calculate the power radiated by a $5.0\,\text{cm} \times 2.0\,\text{cm}$ section of the surface of a hot body at 3000 K.

8.2 A photodetector produces $0.68\,\mu\text{W}$ when exposed to radiation of wavelength 245 nm. How many photons does it detect per second?

8.3 A diffraction experiment requires the use of electrons of wavelength 550 pm. Calculate the velocity of the electrons.

8.4 Calculate the linear momentum of photons of wavelength (a) 725 nm, (b) 75 pm, (c) 20 m.

8.5 Recall (Section 1.1) that pressure is force divided by area, and that force is rate of change of momentum. Suppose that you designed a spacecraft to work by photon pressure. The sail was a completely absorbing fabric of area $1.0\,\text{km}^2$ and you directed a red laser beam of wavelength 650 nm on to it from a base on the Moon. What is (a) the force, (b) the pressure exerted by the radiation on the sail? (c) Suppose the mass of the spacecraft was 1.0 kg. Given that, after a period of acceleration from standstill, $speed = (force/mass) \times time$, how long would it take for the craft to accelerate to a speed of $1.0\,\text{m s}^{-1}$?

8.6 The energy required for the ionization of a certain atom is $3.44 \times 10^{-18}\,\text{J}$. the absorption of a photon of unknown wavelength ionizes the atom and ejects an electron with velocity $1.03 \times 10^{6}\,\text{m s}^{-1}$. Calculate the wavelength of the incident radiation.

8.7 The speed of a certain proton is $3.5 \times 10^{5}\,\text{m s}^{-1}$. If the uncertainty in its momentum is 0.0100 per cent, what uncertainty in its location must be tolerated?

8.8 A hydrogen atom, treated as a point mass, is confined to an infinite one-dimensional square well of width 1.0 nm. How much energy does it have to give up to fall from the level with $n = 2$ to the lowest energy level?

8.9 The pores in zeolite catalysts are so small that quantum mechanical effects on the distribution of atoms and molecules within them can be significant. Calculate the location in a box of length L at which the probability of a particle being found is 50 per cent of its maximum probability when $n = 1$.

8.10 The blue solution formed when an alkali metal dissolves in liquid ammonia consists of the metal cations and electrons trapped in a cavity formed by ammonia molecules. (a) Calculate the spacing between the levels with $n = 4$ and $n = 5$ of an electron in a one-dimensional box of length 5.0 nm. (b) What is the wavelength of the radiation emitted when the electron makes a transition between the two levels?

8.11 Calculate the energy per photon and the energy per mole of photons for radiation of wavelength (a) 600 nm (red), (b) 550 nm (yellow), (c) 400 nm (violet), (d) 200 nm (ultraviolet), (e) 150 pm (X-ray), (f) 1.0 cm (microwave).

8.12 How fast would a particle of mass 1.0 g need to travel to have the same linear momentum as a photon of radiation of wavelength 300 nm?

8.13 A certain lamp emits blue light of wavelength 350 nm. How many photons does it emit each second if its power is (a) 1.00 W, (b) 100 W?

8.14 An FM radio transmitter broadcasts at 98.4 MHz with a power of 45 kW. How many photons does it generate per second?

8.15 The peak of the Sun's emission occurs at about 480 nm; estimate the temperature of its surface.

8.16 The work function for metallic caesium is 2.14 eV. Calculate the kinetic energy and the speed of the electrons ejected by light of wavelength (a) 750 nm, (b) 250 nm.

8.17 Calculate the size of the quantum involved in the excitation of (a) an electronic motion of frequency $1.0 \times 10^{15}\,\text{Hz}$, (b) a molecular vibration of period $2.0 \times 10^{-14}\,\text{s}$, (c) a pendulum of period 0.50 s. Express the results in joules and in kilojoules per mole.

8.18 Calculate the de Broglie wavelength of (a) a mass of 1.0 g travelling at $1.0\,\text{m s}^{-1}$, (b) the same, travelling at $1.00 \times 10^{5}\,\text{km s}^{-1}$, (c) a He atom travelling at $1000\,\text{m s}^{-1}$ (a typical speed at room temperature).

8.19 Calculate the de Broglie wavelength of an electron accelerated from rest through a potential difference of (a) 1.00 V, (b) 1.00 kV, (c) 100 kV. *Hint*

The electron is accelerated to a kinetic energy equal to $e\mathcal{V}$.

8.20 Calculate the minimum uncertainty in the speed of a ball of mass 500 g that is known to be within 5.0 μm of a certain point on a bat.

8.21 What is the minimum uncertainty in the position of a bullet of mass 5.0 g that is known to have a speed somewhere between $350.00\,000\,1\,\mathrm{m\,s^{-1}}$ and $350.00\,000\,0\,\mathrm{m\,s^{-1}}$?

8.22 An electron is confined to a linear region with a length of the same order as the diameter of an atom (c. 100 pm). Calculate the minimum uncertainties in its position and speed.

8.23 In an X-ray photoelectron experiment, a photon of wavelength 150 pm ejects an electron from the inner shell of an atom and it emerges with a speed of $2.24 \times 10^7\,\mathrm{m\,s^{-1}}$. Calculate the binding energy of the electron.

8.24 The Planck distribution, eqn (5), gives the energy of electromagnetic radiation in the wavelength range $\delta\lambda$ at the wavelength λ. Calculate the energy density in the range 650 nm to 655 nm inside a cavity of volume $10\,\mathrm{cm^3}$ when its temperature is (a) 25 °C, (b) 3000 °C.

8.25 The wavelength of the emission maximum from a small pinhole in an electrically heated container was determined at a series of temperatures, and the results are given below. Deduce a value for Planck's constant.

$\theta/°C$	1000	1500	2000	2500	3000	3500
$\lambda_{max}/\mathrm{nm}$	2181	1600	1240	1035	878	763

8.26 Calculate the wavelength of the line with $n = 5$ in the Balmer series of the spectrum of atomic hydrogen.

8.27 The frequency of one of the lines in the Paschen series of the spectrum of atomic hydrogen is 2.7415×10^{15} Hz. Identify the principal quantum number of the upper state in the transition.

8.28 One of the terms of the H atom is at $27\,414\,\mathrm{cm^{-1}}$. What is (a) the wavenumber, (b) the energy of the term with which it combines to produce light of wavelength 486.1 nm?

8.29 When ultraviolet radiation of wavelength 58.4 nm from a helium lamp is directed on to a sample of krypton, electrons are ejected with a speed of $1.59 \times 10^6\,\mathrm{m\,s^{-1}}$. Calculate the ionization energy of krypton.

8.30 Given that a 3s-orbital is proportional to $(6 - 6\rho + \rho^2)\mathrm{e}^{-\rho/2}$, with $\rho = 2Zr/3a_0$, locate its radial nodes.

8.31 The wavefunction of one of the d orbitals is proportional to $\sin\theta\cos\theta$. At what angles does it have nodal planes?

8.32 What is the orbital angular momentum (as multiples of $\hbar$) of an electron in the orbitals (a) 1s, (b) 3s, (c) 3d, (d) 2p, (e) 3p? Give the numbers of angular and radial nodes in each case.

8.33 State the orbital degeneracy of the levels in the hydrogen atom that have energy (a) $-hc\mathcal{R}_H$, (b) $-\frac{1}{9}hc\mathcal{R}_H$, and (c) $-\frac{1}{49}hc\mathcal{R}_H$.

8.34 At what radius does the probability of finding an electron in a small volume located at a point in the ground state of an H atom fall to 25 per cent of its maximum value?

8.35 At what radius in the H atom does the radial distribution function of the ground state have (a) 25 per cent, (b) 10 per cent of its maximum value?

8.36 What is the probability of finding an electron anywhere in one lobe of a p orbital given that it occupies the orbital?

8.37 Which of the following transitions are allowed in the normal electronic emission spectrum of an atom: (a) $2s \rightarrow 1s$, (b) $2p \rightarrow 1s$, (c) $3d \rightarrow 2p$, (d) $5d \rightarrow 2s$, (e) $5p \rightarrow 3s$?

8.38 How many electrons can occupy subshells with the following values of l: (a) 0, (b) 3, (c) 5.

8.39 Give the electron configurations of the ground states of the first 18 elements in the periodic table.

8.40 The 'Humphreys series' is another group of lines in the spectrum of atomic hydrogen. It begins at 12 368 nm and has been traced to 3281.4 nm. (a) What are the transitions involved? (b) What are the wavelengths of the intermediate transitions?

8.41 At what wavelength would you expect the longest wavelength transition of the Humphreys series to occur in He^+? *Hint.* The energy levels of hydrogenic atoms and ions are proportional to Z^2.

8.42 A series of lines in the spectrum of atomic hydrogen lies at 656.46 nm, 486.27 nm, 434.17 nm, and 410.29 nm. What is the wavelength of the next line in the series? What is the ionization energy of the atom when it is in the lower state of the transitions?

8.43 The Li^{2+} ion is hydrogenic and has a Lyman series at $740\,747\,cm^{-1}$, $877\,924\,cm^{-1}$, $925\,933\,cm^{-1}$, and beyond. Show that the energy levels are of the form $-hc\mathcal{R}_{Li}/n^2$ and find the value of $\mathcal{R}_{Li}$ for this ion. Go on to predict the wavenumbers of the two longest-wavelength transitions of the Balmer series of the ion and find the ionization energy of the ion.

8.44 If we lived in a four-dimensional world, there would be one s orbital, four p orbitals, and nine d orbitals in their respective subshells. (a) Suggest what form the periodic table might take for the first 24 elements. (b) Which elements (using their current names) would be noble gases?

The chemical bond

The chemical bond is central to all aspects of chemistry. Reactions make them and break them, and the structures of solids and individual molecules depend on them. The physical properties of individual molecules and of bulk samples of matter also stem in large part from the shifts in electron density that take place when atoms form bonds to one another. The theory of the origin of the numbers, strengths, and three-dimensional arrangements of chemical bonds between atoms is called **valence theory**, and is the topic of this chapter. Valence theory is an attempt to explain the properties of molecules ranging from the smallest, such as why N_2 is so inert that it acts as a diluent for the aggressive oxidizing power of atmospheric oxygen, to the most complex molecules known, such as the function of protein molecules as enzymes and the molecular biology of DNA. The description of chemical bonding has become highly developed through the use of computers, and it is now possible to compute details of the electron distribution in molecules of almost any complexity. However, much can also be achieved in terms of a simple qualitative understanding of bond formation, and that is the focus of this chapter.

As should be well known from introductory chemistry, chemists generally distinguish between two types of bond:

- An **ionic bond** is a chemical bond formed by the *transfer* of electrons from one atom to another and the consequent attraction between the ions so formed.
- A **covalent bond** is a chemical bond formed when two atoms *share* a pair of electrons.

The character of a covalent bond, which we concentrate on in this chapter, was identified by G.N. Lewis (in 1916, before quantum mechanics was fully developed). We shall assume that Lewis's ideas are familiar, but for convenience they are reviewed in *Further information 8*. In this chapter we develop the modern theory of chemical bond formation in terms of the quantum mechanical properties of electrons and place Lewis's ideas in a modern context. We shall see that an ionic bond can be regarded as a limiting type of covalent bond in which the sharing of electrons has given way to the possession by one atom of both electrons of the bond. However, there are certain aspects of ionic solids (solids in which ionic bonding is a good description) that we shall treat separately in Chapter 10.

Lewis's original theory was unable to account for the shapes adopted by molecules; and shape is a crucial aspect of the properties of a molecule. For instance, it is central to an enzyme's action. The most elementary (but quite successful) explanation of the shapes adopted by molecules is the **valence-shell electron pair repulsion model** (VSEPR model) in which it is supposed that the shape of a molecule is determined by the repulsions between electron pairs in the valence shell. We shall suppose that this model is already familiar; however, a brief outline is included in *Further information 9*. Once again, the purpose of this chapter is to extend these elementary arguments and indicate some of the contributions that quantum theory has made to understanding why a molecule adopts its characteristic shape.

All theories of molecular structure make the same simplification at the outset. Whereas the Schrödinger equation for a hydrogen atom can be solved exactly, an exact solution is not possible for any molecule because even the simplest molecule consists of three particles (two nuclei and one electron). The **Born–Oppenheimer approximation** is therefore adopted, in which it is supposed that the nuclei, being so much heavier than an electron, move relatively slowly and may be treated as stationary while the electrons move around them. We can therefore think of the nuclei as being fixed at arbitrary locations, and then solve the Schrödinger equation for the wave-function of the electrons alone. The approximation is quite good for molecules in their electronic ground states, for calculations suggest that (in classical terms) the nuclei in H_2 move through only about 1 pm while the electron speeds through 1000 pm. These figures suggest that the error of assuming that the nuclei are stationary is small.

The Born–Oppenheimer approximation allows us to select an internuclear separation in a diatomic molecule, and (in principle) to solve the Schrödinger equation for the electrons for that nuclear separation. Then we can choose a different separation and repeat the calculation, and so on. In this way we can explore how the energy of the molecule varies with bond length (and, in more complicated molecules, with angles too), and obtain a **molecular potential energy curve**. A typical example of such a curve is illustrated in Fig. 9.1. It is called a *potential* energy curve because the kinetic energy of the nuclei (which are supposed stationary) is ignored. Once the curve has been calculated, we can identify the **equilibrium bond length**, the internuclear separation at the minimum of the curve, and D_e, the depth of the minimum below the energy of the infinitely widely separated atoms. In the next chapter we shall see that the narrowness of the well in the potential energy curve is an indication of the stiffness of the bonds.

There are two major approaches to the calculation of molecular structure, valence bond (VB) theory and molecular orbital (MO) theory. Almost all modern computational work makes use of MO theory, and we shall concentrate on it in this chapter. Valence bond theory, though, has left its imprint on the language of chemistry, and it is important to know the quantum mechanical significance of terms such as hybridization and resonance, for these concepts are still widely used in qualitative discussions of molecular structure. We begin, therefore, with a brief account of VB theory with a view to establishing the concepts that it has introduced into chemistry.

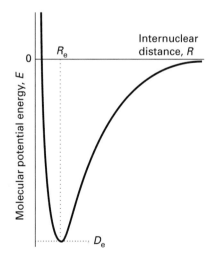

Fig. 9.1 A molecular potential energy curve. The equilibrium bond length R_e corresponds to the energy minimum.

Valence bond theory

In **valence bond theory**, a bond is regarded as forming when an electron in one atomic orbital pairs its spin with that of an electron supplied by another atomic orbital (Fig. 9.2). However, for us to understand what these words really mean, we have to examine the wavefunction for the two electrons that form the bond.

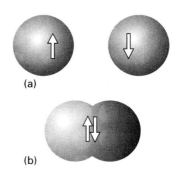

Fig. 9.2 In the valence bond theory, a σ-bond is formed when two electrons in orbitals on neighbouring atoms pair and the orbitals merge to form a cylindrical electron cloud.

9.1 Diatomic molecules

We shall begin by considering the simplest possible chemical bond, the one in molecular hydrogen. Suppose the two atoms (which we label A and B) are far apart, then we can be confident that electron 1 is on atom A and electron 2 is on atom B, and write the overall wavefunction for the two electrons as

$$\psi = \psi_{\text{H}1s\text{A}}(1)\psi_{\text{H}1s\text{B}}(2)$$

When the two atoms are at their bonding distance, it may still be true that electron 1 is on A and electron 2 is on B. However, an equally likely arrangement is for electron 1 to be on B and electron 2 to be on A, in which case the wavefunction would be

$$\psi = \psi_{\text{H}1s\text{A}}(2)\psi_{\text{H}1s\text{B}}(1)$$

Whenever two possibilities are equally likely, the rules of quantum mechanics tell us to **superimpose**, or add together, the two corresponding wavefunctions. Therefore, the wavefunction for the two electrons in a hydrogen molecule is

$$\psi(\text{H}\text{—}\text{H}) = \psi_{\text{H}1s\text{A}}(1)\psi_{\text{H}1s\text{B}}(2) + \psi_{\text{H}1s\text{A}}(2)\psi_{\text{H}1s\text{B}}(1) \qquad (1)$$

with the two locations of the electrons contributing equally. This wavefunction is the valence-bond wavefunction for the bond in molecular hydrogen. It expresses the idea that we cannot keep track of either electron, and that their distributions blend together. For technical reasons stemming from the Pauli exclusion principle, this wavefunction can exist only if the two electrons it describes have opposite spins (that is, are paired), so the merging of orbitals that gives rise to a bond is accompanied by the pairing of the two electrons that contribute to it. Bonds do not form *because* electrons tend to pair: bonds are *allowed* to form by the electrons pairing their spins.

The distribution of the electrons is determined by the wavefunction ψ, and because it is built from the merging of H1s orbitals, we can expect the overall distribution to be sausage-shaped (as in Fig. 9.2). A VB wavefunction with cylindrical symmetry around the internuclear axis is called a σ **bond**. It is so called because, when viewed along the bond, it resembles a pair of electrons in an s orbital (and σ, sigma, is the Greek equivalent of s). All VB wavefunctions are constructed in a similar way, by using the atomic orbitals available on the participating atoms.In general, therefore, for orbitals that we can symbolize A and B on atoms A and B respectively, a VB wavefunction for an A—B bond is

$$\psi(\mathrm{A}-\mathrm{B}) = A(1)B(2) + A(2)B(1) \qquad (2)$$

The energy of a molecule for a given internuclear separation is calculated by substituting the VB wavefunction into the Schrödinger equation for the molecule with the nuclei at a fixed separation R, and solving the equation for the energy E. When the wavefunction in eqn (2) is used in this way and the energy is plotted against R, a curve like the one shown in Fig. 9.1 is obtained. The energy falls below that of two separated H atoms as the two atoms are brought together and each electron becomes free to migrate to the other atom. However, the energy reduction that follows from this process is counteracted by an increase in energy from the Coulombic repulsion between the two positively charged nuclei, which has the form[†]

$$V_{\mathrm{nuc,nuc}} \propto \frac{1}{R} \qquad (3)$$

This positive contribution to the energy becomes large as R becomes small, and the total energy curve passes through a minimum and then climbs to a strongly positive value as the two nuclei are pressed together.

A similar description can be applied to more complicated molecules with more that one electron able to contribute to the bonding. For example, to construct the valence bond description of N_2, we consider the valence electron configuration of each atom:

$$\mathrm{N}\ 2s^2 2p_x^1 2p_y^1 2p_z^1$$

It is conventional to take the z axis to be the internuclear axis, so we can imagine each atom as having a $2p_z$ orbital pointing towards a $2p_z$ orbital on the other atom, with the $2p_x$ and $2p_y$ orbitals perpendicular to the axis (Fig. 9.3). Each of these p orbitals is occupied by one electron, so we can think of the molecule as forming by the merging of matching orbitals on neighbouring atoms and the accompanying pairing of the electrons that occupy them. A cylindrically symmetric σ bond can form by the merging of the opposing $2p_z$ orbitals and the pairing of the two electrons. However, the remaining p orbitals cannot merge to give σ bonds as they do not have cylindrical symmetry around the internuclear axis. Instead, the $2p_x$ orbitals merge and the two electrons pair to form a **π bond**, and the $2p_y$ orbitals similarly merge and their electrons pair to form another π bond. In general, a π bond arises from the merging of two p orbitals that approach side-by-side and the pairing of the electrons that they contain. A π bond is so called because, viewed along

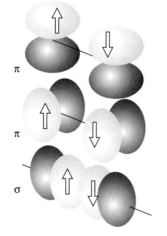

Fig. 9.3 The bonds in N_2 are built by allowing the electrons in the N2p orbitals to pair. However, only one orbital on each atom can form a σ bond: the orbitals perpendicular to the axis form π bonds.

[†]More precisely, for two nuclei with atomic numbers Z_A and Z_B, the contribution to the potential energy is

$$V_{\mathrm{nuc,nuc}} = \frac{Z_A Z_B e^2}{4\pi\varepsilon_0 R}$$

where ε_0 is the vacuum permittivity (see inside the front cover).

the internuclear axis, it resembles a pair of electrons in a p orbital (and π is the Greek equivalent of p). It follows that the overall bonding pattern in N_2 is a σ bond plus two π bonds (Fig. 9.4), which is consistent with the Lewis structure : N≡N : in which the atoms are linked by a triple bond.

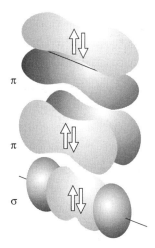

Fig. 9.4 The electrons in the $2p$ orbitals of two neighbouring N atoms merge to form σ and π bonds. The electrons in the $N2p_z$ orbitals pair to form a bond of cylindrical symmetry. Electrons in the $N2p$ orbitals that lie perpendicular to the axis also pair form two π bonds.

Exercise E9.1

Describe the valence-bond ground state of a Cl_2 molecule.

[*Answer*: one $\sigma(Cl3p_z, Cl3p_z)$ bond]

9.2 Polyatomic molecules

The concept of the merging of singly occupied orbitals to form a bond consisting of two paired spins can be extended readily to polyatomic species. Each σ bond in a polyatomic molecule is formed by the merging of orbitals with cylindrical symmetry about the internuclear axis and the pairing of the spins of the electrons they contain. Likewise, π bonds are formed by pairing electrons that occupy atomic orbitals of the appropriate symmetry. A simple description of the electronic structure of H_2O should make this clear.

The valence electron configuration of an O atom is $2s^2 2p_x^2 2p_y^1 2p_z^1$. The two unpaired electrons in the $O2p$ orbitals can each pair with an electron in a $H1s$ orbital, and each combination results in the formation of a σ bond (each bond has cylindrical symmetry about the respective O—H internuclear distance). Because the $2p_y$ and $2p_z$ orbitals lie at $90°$ to each other, the two σ bonds they form also lie at $90°$ to each other (Fig. 9.5). We predict, therefore, that H_2O should be an angular molecule, which it is. However, the model predicts a bond angle of $90°$, whereas the actual bond angle is $104°$.

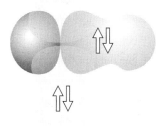

Exercise E9.2

Give a valence bond description of NH_3, and predict the bond angle of the molecule on the basis of this description. The experimental bond angle is $107°$.

[*Answer*: three $\sigma(N2p, H1s)$ bonds; $90°$]

Fig. 9.5 The bonding in an H_2O molecule can be pictured in terms of the pairing of an electron belonging to one H atom with an electron in an $O2p$ orbital; the other bond is formed likewise, but using a perpendicular $O2p$ orbital. The predicted bond angle is $90°$, which is in poor agreement with the experimental bond angle ($104°$).

While broadly correct, valence bond theory seems to have two deficiencies. One is the poor estimate it provides for the bond angle in H_2O (and other molecules, such as NH_3). Indeed, the theory appears to make worse predictions than the qualitative VSEPR model, which predicts HOH and HNH bond angles of slightly less than $109°$ in H_2O and NH_3, respectively. The second major deficiency is its apparent inability to account for carbon's tetravalence: the ground state configuration of C is $2s^2 2p_x^1 2p_y^1$, which suggests that a carbon atom should be capable of forming only two bonds, not four. However, both deficiencies are overcome by the combined consequences of **promotion**, the excitation of an electron to an orbital of higher energy, and **hybridization**, the blending together of the orbitals of the excited atom.

Promotion

The promotion of an electron is its excitation to an orbital of higher energy as a bond is formed. The promotion is worthwhile if the energy it requires can be more than recovered in the greater strength or number of bonds that can be formed as a result. In carbon, for example, the promotion of a $2s$ electron to a $2p$ orbital leads to the configuration $2s^1 2p_x^1 2p_y^1 2p_z^1$, with four unpaired electrons in separate orbitals. These electrons may pair with four electrons in orbitals provided by four other atoms (such as four H$1s$ orbitals if the molecule is CH_4), and as a result the atom can form four σ bonds. Although energy is required to promote the $2s$ electron, it is more than recovered by the atom's ability to form four bonds in place of the two bonds of the unpromoted atom. We can now also see why tetravalent carbon is so common: its promotion energy is quite small because the promoted electron leaves a doubly occupied $2s$ orbital and enters a vacant $2p$ orbital, hence significantly relieving the electron–electron repulsion it experiences in the former.

Hybridization

Our description of the bonding in CH_4 (and its homologues) is still incomplete because it appears to imply the presence of three σ bonds of one type (formed from the merging of H$1s$ and C$2p$ orbitals) and a fourth σ bond of a distinctly different type (formed from the merging of H$1s$ and C$2s$). It is well known, however, that all four bonds in methane are exactly equivalent both in terms of their chemical properties and their physical properties (their lengths, strengths, and stiffnesses).

This problem is overcome by another technical feature of quantum mechanics: the same electron density distribution can be described in different ways. In this case, the electron density distribution in the promoted atom can be described *either* as arising from four electrons in one s and three p orbitals, *or* as arising from four electrons in four different mixtures of these orbitals. Mixtures (more formally, linear combinations) of atomic orbitals *on the same atom* are called **hybrid orbitals**. We can picture them as formed by interference between the waves corresponding to the C$2s$ and C$2p$ orbitals. The origin of the hybridization can be appreciated by thinking of the four original atomic orbitals, which are waves centred on a nucleus, as being like ripples spreading from a single point on the surface of a lake: the waves interfere destructively and constructively in different regions, and give rise to four new shapes. The specific linear combinations that give rise to four equivalent hybrid orbitals are

$$h_1 = s + p_x + p_y + p_z \qquad h_2 = s - p_x - p_y + p_z$$
$$h_3 = s - p_x + p_y - p_z \qquad h_4 = s + p_x - p_y - p_z$$

As a result of the constructive and destructive interference between the positive and negative regions of the component orbitals, each hybrid orbital has a large lobe pointing towards one corner of a regular tetrahedron (Fig. 9.6). Because each hybrid is built from one s orbital and three p orbitals, it is called an sp^3 **hybrid orbital**.

It is now easy to see how the valence bond description of the methane molecule leads to a tetrahedral molecule containing four equivalent C—H bonds. It is energetically favourable (in the end, after bonding has been taken into account) for the carbon atom to undergo promotion. The promoted configuration has a distribution of electrons that is equivalent to one electron occupying each of four tetrahedral hybrid orbitals. Each hybrid orbital of the promoted atom contains a single unpaired electron; a hydrogen $1s$ electron can pair with each one, giving rise to a σ bond pointing in a tetrahedral direction. Because each sp^3 hybrid orbital has the same composition, all four σ bonds are identical apart from their orientation in space (Fig. 9.7).

Hybridization can also be used to describe the structure of an ethene molecule and the torsional rigidity of double bonds. An ethene molecule is planar, with HCH and HCC bond angles close to 120°. To reproduce this σ-bonding structure, we think of each C atom as being promoted to a $2s^1 2p^3$ configuration. However, instead of using all four orbitals to form hybrids, we form sp^2 **hybrid orbitals** by allowing the s orbital and *two* p orbitals to interfere constructively and destructively. As shown in Fig. 9.8, the three hybrid orbitals

$$h_1 = s + \sqrt{2}\,p_x \qquad h_2 = s + \sqrt{\tfrac{3}{2}}\,p_x - \sqrt{\tfrac{1}{2}}\,p_y \qquad h_3 = s - \sqrt{\tfrac{3}{2}}\,p_x - \sqrt{\tfrac{1}{2}}\,p_y$$

lie in a plane and point towards the corners of an equilateral triangle. The third $2p$ orbital ($2p_z$) is not included in the hybridization, and its axis is perpendicular to the plane in which the hybrids lie.

The structure of an ethene molecule can now be described as follows. The sp^2-hybridized C atoms each form three σ bonds by spin pairing with either the h_1 hybrid of the other C atom or with H$1s$ electrons. The σ framework therefore consists of bonds at 120° to each other. When the two CH$_2$ groups lie in the same plane, the two electrons in the unhybridized $2p_z$ orbitals can pair and form a π bond (Fig. 9.9). The formation of this π bond locks the framework into the planar arrangement, for any rotation of one CH$_2$ group relative to the other leads to a weakening of the π bond (and consequently an increase in energy of the molecule).

A similar description applies to a linear ethyne (acetylene) molecule, H—C≡C—H. Now the carbon atoms are sp **hybridized**, and the σ bonds are formed using hybrid atomic orbitals of the form

$$h_1 = s + p_z \qquad h_2 = s - p_z$$

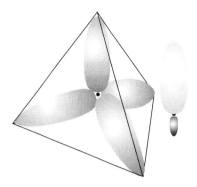

Fig. 9.6 The $2s$ and three $2p$ orbitals of a carbon atom hybridize, and the resulting hybrid orbitals point towards the corners of a regular tetrahedron.

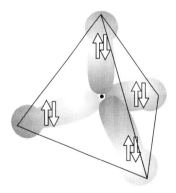

Fig. 9.7 The valence bond description of the structure of CH$_4$. Each σ bond is formed by the pairing of an electron in an H$1s$ orbital with an electron in one of the hybrid orbitals shown in Fig. 9.6. The resulting molecule is regular tetrahedral.

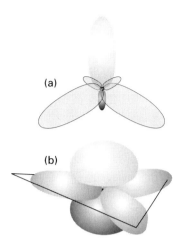

Fig. 9.8 (a) Trigonal planar hybridization is obtained when an *s* and two *p* orbitals are hybridized. The three lobes lie in a plane and make an angle of 120° to each other. (b) The remaining *p* orbital in the valence shell of an *sp²*-hybridized atom lies perpendicular to the plane of the three hybrids.

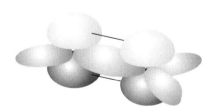

Fig. 9.9 The valence bond description of the structure of a carbon–carbon double bond, as in ethene. The electrons in the two *sp²* hybrids that point towards each other pair and form a σ bond. Electrons in the two *p* orbitals that are perpendicular to the plane of the hybrids pair, and form a π bond. The electrons in the remaining hybrid orbitals are used to form bonds to other atoms (in ethene itelf, to H atoms).

These two orbitals lie along the *z*-axis. The electrons in them pair either with an electron in the corresponding hybrid orbital on the other C atom or with an electron in the H1*s* orbitals. Electrons in the two remaining *p* orbitals on each atom, which are perpendicular to the molecular axis, pair to form two perpendicular π bonds (as in Fig. 9.10).

Other hybridization schemes, particularly those involving *d* orbitals, are often invoked to account for (or at least be consistent with) other molecular geometries (Table 9.1). The hybridization of *N* atomic orbitals always results in the formation of *N* hybrid orbitals. For example, sp^3d^2 hybridization results in six equivalent hybrid orbitals pointing towards the corners of a regular octahedron. This octahedral hybridization scheme is sometimes invoked to account for the structure of octahedral molecules, such as SF_6.

Exercise E9.3

Describe the bonding in a PCl_5 molecule in VB terms.

[*Answer*: Five σ bonds formed from sp^3d hybrids]

The 'pure' schemes in Table 9.1 are not the only possibilities: it is possible to form hybrid orbitals with intermediate proportions of atomic orbitals. For example, as more *p*-orbital character is included in an *sp*-hybridization scheme, the hybridization changes towards sp^2 and the angle between the hybrids changes continuously from 180° for pure *sp* hybridization to 120° for pure sp^2 hybridization. If the proportion of *p* character continues to be increased (by reducing the admixture of *s* orbital), then the hybrids eventually become pure *p* orbitals at an angle of 90° to each other. Now we can account for the structure of H_2O, with its bond angle of 104°. Each O—H σ bond is formed from an O atom hybrid orbital with a composition that lies between pure *p* (which would lead to a bond angle of 90°) and pure sp^2 (which would lead to a bond angle of 120°). The actual bond angle and hybridization adopted are found by calculating the energy of the molecule as the bond angle is varied, and looking for the angle at which the energy is a minimum.

Table 9.1 Hybrid orbitals

Number	Shape	Hybridization
2	Linear	sp
3	Trigonal planar	sp^2
4	Tetrahedral	sp^3
5	Bipyramidal	sp^3d
6	Octahedral	sp^3d^2

Resonance

Another term introduced by VB theory into chemistry is **resonance**, the superposition of different electron distributions in the same nuclear framework. To understand what this means, consider the VB description of a purely covalently bonded HCl molecule, which could be written

$$\psi_{\text{cov}} = \psi_{\text{H1}s}(1)\psi_{\text{Cl2}p_z}(2) + \psi_{\text{H1}s}(2)\psi_{\text{Cl2}p_z}(1)$$

We have supposed that the bond is formed by the spin pairing of electrons in the hydrogen $1s$ orbital and the chlorine $2p_z$ orbital. However, there is something wrong with this description: it allows electron 1 to be on the H atom when electron 2 is on the Cl atom, and vice versa, but it does not allow *both* electrons to be on the Cl atom simultaneously. On physical grounds, we might expect the covalent character of HCl to be only a partial description of the molecule: because the Cl atom is so electronegative, we ought to expect a polar molecule with H^+Cl^- playing a role in its description. The wavefunction for this ionic structure is

$$\psi_{\text{ion}} = \psi_{\text{Cl2}p_z}(1)\psi_{\text{Cl2}p_z}(2) \tag{4}$$

However, this wavefunction alone is unrealistic, because HCl is not an ionic species. The wavefunction for the molecule is a *superposition* of the covalent and ionic descriptions, and we write

$$\psi = \psi_{\text{cov}} + \lambda\psi_{\text{ion}} \tag{5}$$

The parameter λ is found by using the **variation theorem**:

> If an arbitrary wavefunction is used to calculate the energy, then the value calculated is never less than the true energy.

The arbitrary wavefunction is called a **trial wavefunction**. The theorem implies that if we vary the coefficients (in this case, λ) in the trial wavefunction until we achieve the lowest energy, then those coefficients and the corresponding wavefunction will be the best available. This approach, expressing a wavefunction as the superposition of wavefunctions corresponding to a variety of structures, is called **resonance**; in this case, it is called specifically **ionic–covalent resonance**. The interpretation of the wavefunction, which is called a **resonance hybrid**, is that if we were to inspect the molecule, then the proportion of the time that it would be found with an ionic structure is proportional to λ^2. Resonance is not a flickering between the contributing states: it is a blending of their characteristics, much as a mule is a blend of a horse and a donkey.

One of the most familiar examples of resonance is in the VB description of benzene, where the wavefunction of the molecule is written as a superposition of the wavefunctions of the two Kekulé structures (**1**) and (**2**):

$$\psi = \psi_{\text{Kek1}} + \psi_{\text{Kek2}}$$

Fig. 9.10 The electronic structure of ethyne (acetylene). The electrons in the two sp hybrids on each atom pair to form σ bonds either with the other C atom or with an H atom. The remaining two unhybridized $2p$ orbitals on each atom are perpendicular to the axis: the electrons in corresponding orbitals on each atom pair to form two π bonds.

1 Kekulé structure 1

2 Kekulé structure 2

The two contributing structures have identical energies, so they contribute equally to the superposition. The effect of resonance in this case is to distribute double-bond character around the ring and to make all carbon–carbon bonds equivalent. Because the wavefunction is improved by allowing resonance, it follows from the variation theorem that the energy of the molecule is lowered relative to either Kekulé structure alone: this lowering is called the **resonance stabilization** of the molecule, and is largely responsible for the unusual stability of aromatic rings. Resonance always lowers the energy (because it improves the wavefunction), and the lowering is greatest when the contributing structures have similar energies.

Molecular orbitals

In **molecular orbital theory**, it is accepted that electrons should not be regarded as belonging to particular bonds but should be treated as spreading throughout the entire molecule: every electron contributes to the strength of every bond. This theory has been more fully developed than valence bond theory and provides the language that is widely used in modern discussions of bonding in small inorganic molecules, d-metal complexes, and solids. To introduce it, we shall follow the same strategy as in Chapter 8 where the one-electron hydrogen atom was taken as the fundamental species for discussing atomic structure, and then developed into a description of complex atoms. In this chapter we use the simplest molecule of all, the one-electron hydrogen molecule-ion, H_2^+, to introduce the essential features of bonding, and then use it as a guide to the structures of more complex systems. These applications include diatomic molecules, polyatomic molecules, and finally solids consisting of effectively infinite numbers of atoms.

9.3 Linear combinations of atomic orbitals

The wavefunctions obtained by solving the Schrödinger equation for the H_2^+ ion are called **molecular orbitals**. A molecular orbital is like an atomic orbital, but spreads over all the nuclei in the molecule. As for an atomic orbital, the probability of finding an electron at a particular location is proportional to the square of the wavefunction at that point. So, where the molecular orbital has a large amplitude, the electron that occupies it has a high probability of being found. Where the molecular orbital is zero (at one of its nodes), there is zero probability of finding the electron.

Exact mathematical formulas for the molecular orbitals of H_2^+ may be obtained (within the Born–Oppenheimer approximation), but they are very complicated functions and do not give much insight into the form of the orbitals, the contributions to the energy, or the form that orbitals are likely to take for more complex polyatomic molecules. Therefore, we shall adopt a simpler procedure that, while more approximate, gives more insight. The approximation we shall describe is used almost universally in modern MO computations.

First, we note that the wavefunction of the electron in the ground state of H_2^+ can be expected to resemble a $1s$ orbital centred on nucleus A when it is close to nucleus A, and to resemble a $1s$ orbital centred on nucleus B when it is close to nucleus B. This resemblance suggests that an approximate form for the molecular orbital should be

$$\psi = \psi_{A1s} + \psi_{B1s} \tag{6}$$

where ψ_{A1s} is a $1s$ orbital centred on A and ψ_{B1s} is the corresponding orbital centred on B. When the electron is close to A its distance from B is large, the wavefunction ψ_{B1s} is small, and therefore ψ is almost pure ψ_{A1s}, as we require. Similarly, ψ is almost pure ψ_{B1s} close to B. The technical term for a sum of the kind in eqn (6) is a **linear combination of atomic orbitals** (LCAO), and we shall use that name from now on. An approximate molecular orbital formed from a linear combination of atomic orbitals is called an **LCAO-MO**.

The form of the orbital constructed in eqn (6) is shown in Fig. 9.11. It is called a σ **orbital** because it resembles an s orbital when viewed along the axis:[†] in general, any molecular orbital with cylindrical symmetry about the internuclear axis is called a σ orbital. Because it is the σ orbital of lowest energy (as we shall see), it is labelled 1σ. An electron that occupies a σ orbital is called a σ **electron**. In the ground state of the H_2^+ ion, there is a single 1σ electron, and so we report the configuration of the molecule-ion as $1\sigma^1$.

We can see by examining the molecular orbital the origin of the lowering of energy that is responsible for the formation of the bond. The two atomic orbitals are like waves centred on neighbouring nuclei. In the internuclear region, the amplitudes interfere constructively (Fig. 9.12), and the wavefunction has an enhanced amplitude there. Because the amplitude is increased, by the Born interpretation there is an increased probability of finding the electron between the two nuclei, where it is in a good position to interact strongly with both of them. Hence the energy of the molecule is lower than that of the separate atoms, where each electron can interact strongly with only one nucleus. In elementary MO theory, *the bonding effect of an electron that occupies a molecular orbital is ascribed to its accumulation in the internuclear region as a result of the constructive interference of the contributing atomic orbitals.*

9.4 Bonding and antibonding orbitals

A 1σ orbital is an example of a **bonding molecular orbital**, a molecular orbital which, if occupied, contributes to the strength of a bond between

[†]More precisely, it is so called because an electron that occupies a σ orbital has zero orbital angular momentum around the internuclear axis, just as an s electron has zero orbital angular momentum around a nucleus.

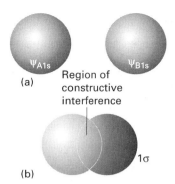

Fig. 9.11 The formation of a bonding molecular orbital (a σ orbital). Two H1s orbitals come together, and where they overlap, interfere constructively and give rise to an enhanced amplitude in the internuclear region. The resulting orbital has cylindrical symmetry about the internuclear axis. When it is occupied by two paired electrons, to give the configuration σ^2, we have a σ bond.

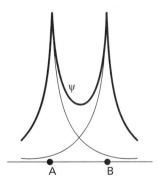

Fig. 9.12 The wavefunction along the internuclear axis. Note that there is an enhancement of amplitude between the nuclei, so there is an increased probability of finding the bonding electrons in that region.

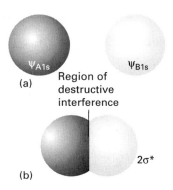

(a)

Region of
destructive
interference

(b)

$2\sigma^*$

Fig. 9.13 The formation of an antibonding molecular orbital (a σ^* orbital). Two H1s orbitals come together, and where they overlap with opposite signs (as depicted by different shades of grey), interfere destructively and give rise to a decreased amplitude in the internuclear region. There is a nodal plane exactly half way between the nuclei, on which any electrons that occupy the orbital will not be found.

two atoms. As in VB theory, we can substitute the wavefunction in eqn (6) into the Schrödinger equation for the molecule-ion with the nuclei at a fixed separation R and solve for the energy. The molecular potential energy curve obtained by plotting the energy against R is very similar to the one drawn in Fig. 9.1 The energy of the molecule decreases as R is decreased from large values because the electron is increasingly likely to be found in the internuclear region as the two atomic orbitals interfere more effectively. However, at small separations, there is too little space between the nuclei for significant accumulation of electron density there. In addition, the nucleus–nucleus repulsion $V_{\mathrm{nuc,nuc}}$ given in eqn (3) becomes large as the distance R is decreased. As a result, after an initial decrease, at small internuclear separations the energy of the molecule rises and the potential energy curve passes through a minimum. Calculations on H_2^+ give the equilibrium bond length as 130 pm and the bond dissociation energy as 171 kJ mol^{-1}; the experimental values are 106 pm and 250 kJ mol^{-1}, and so this simple LCAO-MO description of the molecule, while inaccurate, is not absurdly wrong.

It is an important feature of MO theory that when N atomic orbitals are used to construct an LCAO, we obtain N different molecular orbitals. So far, we have considered the LCAO in which the two atomic orbitals are added together. However, there is a second MO that can be constructed from the same two atomic orbitals. It is formed by taking the difference between the two orbitals:

$$\psi' = \psi_{\mathrm{A}1s} - \psi_{\mathrm{B}1s}$$

Because ψ' is cylindrically symmetrical around the internuclear axis it is also a σ orbital, and is denoted 2σ (Fig. 9.13). When it is used in the Schrödinger equation, we find that an electron that occupies it has a higher energy than when it is in the other σ orbital (the 1σ orbital), and, indeed, it has a higher energy than when it is in either of the atomic orbitals.

Exercise E9.4

Show that the molecular orbital written above is zero on a plane cutting through the internuclear axis at its mid point. Take each atomic orbital to be of the form e^{-r}.

The origin of the high energy of 2σ can be traced to the existence of a **nodal plane**, a plane on which the wavefunction passes through zero, lying half way between the nuclei and perpendicular to the internuclear axis (Fig. 9.13). The two atomic orbitals cancel on this plane as a result of destructive interference, because they have opposite signs, and there is zero probability of finding the electron on this plane. In drawings like that in Figs 9.11 and

9.13, we represent overlap of orbitals with the same sign (as in the formation of 1σ) by shading of the same tint; the overlap of orbitals of opposite sign (as in the formation of 2σ) is represented by one orbital of a light tint (or white) and another orbital of a dark tint.

The 2σ orbital is an example of an **antibonding orbital**, an orbital that, if occupied, decreases the strength of a bond between two atoms. Antibonding orbitals are sometimes marked with an asterisk, as in $2\sigma^*$. The antibonding character of the 2σ orbital is partly a result of the exclusion of the electron from the internuclear region and its relocation outside the bonding region. Because of its location, the electron helps to pull the nuclei apart rather than helping to pull them together. An antibonding orbital is often slightly more strongly antibonding than the corresponding bonding orbital is bonding. This is partly because, although the 'gluing' and 'anti-gluing' effects of the electron are similar, the nuclei repel each other in both cases, and this repulsion pushes both levels up in energy. We shall display the relative energies of the atomic orbitals and the bonding and antibonding molecular orbitals that they are used to form in a **molecular orbital energy level diagram** like that shown in Fig. 9.14. Note that it is not quite symmetrical about the energy of the atomic orbitals.

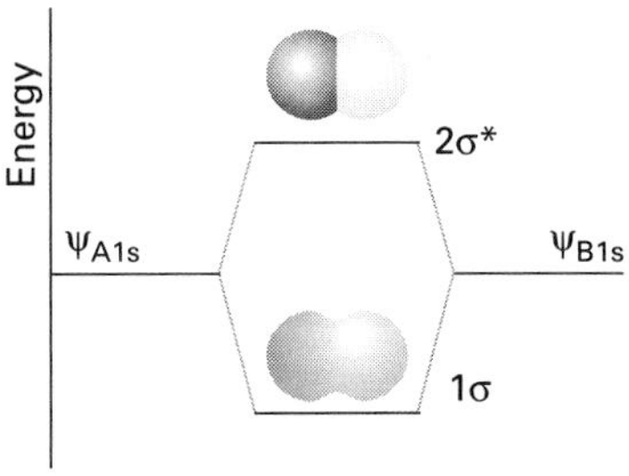

Fig. 9.14 A molecular orbital energy level diagram for orbitals constructed from $(1s,1s)$-overlap, the separation of the levels corresponding to the equilibrium bond length.

9.5 The structures of diatomic molecules

In Chapter 8 we used the hydrogenic atomic orbitals and the building-up principle to deduce the ground electronic configurations of many-electron atoms. We can adopt an analogous procedure for many-electron diatomic molecules (such as H_2 with two electrons and even Br_2 with 70), but using the H_2^+ molecular orbitals as a basis. As we shall illustrate in the following sections, first with H_2 and then with heavier molecules, the general procedure is as follows:

1. Construct molecular orbitals by forming linear combinations of all suitable valence atomic orbitals supplied by the atoms (the meaning of 'suitable' will be explained shortly); N atomic orbitals result in N molecular orbitals.
2. Accommodate the valence electrons supplied by the atoms so as to achieve the lowest overall energy subject to the constraint of the Pauli exclusion principle, that no more than two electrons may occupy a single orbital (and then must be paired).
3. As for atoms, if more than one molecular orbital of the same energy is available, add the electrons to each individual orbital before doubly occupying any one orbital (because that minimizes electron-electron repulsions).
4. Take note of Hund's rule (Section 8.11), that if electrons occupy different degenerate orbitals, then they do so with parallel spins.

The following sections show how these rules are used in practice.

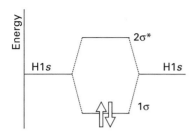

Fig. 9.15 The ground electronic configuration of H_2 is obtained by accommodating the two electrons in the lowest available orbital (the bonding orbital).

Exercise E9.5

How many molecular orbitals can be built from the valence shell orbitals in O_2?

[*Answer*: 8]

Hydrogen and helium molecules

The first step in the discussion of H_2, the simplest many-electron diatomic molecule, is to build the molecular orbitals. Because each H atom of H_2 contributes a $1s$ orbital (as in H_2^+), we can form the 1σ and $2\sigma^*$ bonding and antibonding orbitals from them, as we have seen already. At the equilibrium internuclear separation these orbitals will have the energies that can be represented schematically by the horizontal lines in Fig. 9.15.

There are two electrons to accommodate (one from each atom), and both can enter the 1σ orbital by pairing their spins. The ground state configuration is therefore $1\sigma^2$, and the atoms are joined by a bond consisting of an electron pair in a bonding σ orbital. These two electrons bind the two nuclei together more strongly and closely than the single electron in H_2^+, and the bond length is reduced from 106 pm to 74 pm. A pair of electrons in a σ orbital is called a σ **bond**, and is very similar to the σ bond of valence bond theory (the two differ in certain details of the electron distribution between the two atoms joined by the bond). We can conclude that *the importance of an electron pair in bonding stems from the fact that two is the maximum number of electrons that can enter each bonding molecular orbital*. Electrons do not 'want' to pair: they pair because in that way they are able to occupy a low-energy orbital.

A similar argument shows why helium is a monatomic gas. Consider a hypothetical He_2 molecule. Each He atom contributes a $1s$ orbital to the linear combination used to form the molecular orbitals, and so 1σ and $2\sigma^*$ molecular orbitals can be constructed. They differ in detail from those in H_2 because the $He1s$ orbitals are more compact, but the general shape is the same, and for qualitative discussions we can use the same molecular orbital energy level diagram as for H_2. Because each atom provides two electrons, there are four electrons to accommodate. Two can enter the 1σ orbital, but then it is full (by the Pauli exclusion principle) and the next two must enter the antibonding $2\sigma^*$ orbital (Fig. 9.16). The ground electronic configuration of He_2 is therefore

$$He_2 \; 1\sigma^2 2\sigma^{*2}$$

Because an antibonding orbital is slightly more antibonding than a bonding orbital is bonding, the He_2 molecule has a higher energy than the separated atoms, and so it does not form. Hence, two ground-state He atoms do not form bonds to each other, and helium is a monatomic gas.

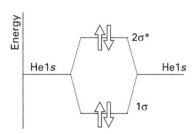

Fig. 9.16 The ground electronic configuration of the four-electron molecule He_2 has two bonding electrons and two antibonding electrons. It has a higher energy than the separated atoms, and so He_2 is unstable relative to two He atoms.

Example *Judging the stability of diatomic molecules*
Decide whether Li_2 is likely to exist on the assumption that only the valence *s* orbitals contribute to its molecular orbitals.

Strategy
Decide what molecular orbitals can be formed from the available valence orbitals, rank them in order of energies, then feed in the electrons supplied by the valence orbitals of the atoms. Judge whether there is a net bonding or net antibonding effect between the atoms.

Solution
Each molecular orbital is built from $2s$ atomic orbitals, which give one bonding and one antibonding combination (1σ and $2\sigma^*$, respectively). Each Li atom supplies one valence electron; the two electrons fill the 1σ orbital, to give the configuration $1\sigma^2$, which is bonding.

Exercise E9.6

Is LiH likely to exist if the Li atom uses only its $2s$ orbital for bonding?

[*Answer*: yes, $(Li2s, H1s)\sigma^2$]

Period 2 diatomic molecules

We shall now see how the concepts we have introduced apply to other **homonuclear diatomic molecules**, which are diatomic molecules formed from identical atoms, such as N_2 and Cl_2, and diatomic ions such as O_2^{2-}. In line with the building-up procedure, we first consider the molecular orbitals that may be formed and do not (at this stage) trouble about how many electrons are available.

The atomic orbitals available for the construction of linear combinations are:

- The **core orbitals**, the orbitals of the inner, closed shells.
- The **valence orbitals**, the orbitals of the valence shell.
- The **virtual orbitals**, the orbitals of the atom that are unoccupied in its ground state.

In elementary treatments (but not in modern sophisticated treatments) the core orbitals are ignored as being too compact to have significant overlap with orbitals on other atoms. The virtual orbitals are ignored on the grounds that they are too high in energy to participate in bonding. Therefore, molecular orbitals are formed using only the valence orbitals.

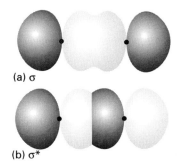

(a) σ

(b) σ*

Fig. 9.17 (a) The interference leading to the formation of a σ-bonding orbital and (b) the corresponding antibonding orbital when two p orbitals overlap along an internuclear axis.

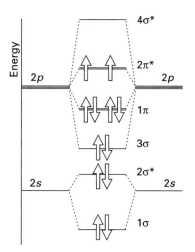

Fig. 9.18 A typical molecular orbital energy level diagram for Period 2 homonuclear diatomic molecules. the valence atomic orbitals are drawn in the columns on the left and the right; the molecular orbitals are shown in the middle. Note that the π orbitals form doubly degenerate pairs. The sloping lines joining the molecular orbitals to the atomic orbitals show the principal composition of the molecular orbitals. This diagram is suitable for O_2 and F_2.

In Period 2, the valence orbitals are $2s$ and $2p$. Suppose first that we consider these two types of orbital separately. Then the $2s$ orbitals on each atom overlap to form bonding and antibonding combinations that we shall denote 1σ and $2\sigma^*$, respectively. Likewise, the two $2p_z$ orbitals (by convention, the internuclear axis is denoted the z axis) have cylindrical symmetry around the internuclear axis, and hence may participate in σ-orbital formation to give the bonding and antibonding combinations 3σ and $4\sigma^*$, respectively (Fig. 9.17). The resulting energy levels of the σ orbitals are shown in the MO energy level diagram in Fig. 9.18. Now, strictly we should not consider the s and p_z orbitals separately, because they both have cylindrical symmetry, and both of them can contribute to σ-orbital formation. Therefore, we should combine all four orbitals together to form four σ molecular orbitals, each one of the form

$$\psi = c_1\psi_{2s}(A) + c_2\psi_{2s}(B) + c_3\psi_{2p_z}(A) + c_4\psi_{2p_z}(B)$$

The four coefficients, c_i, which represent the different contributions that each atomic orbital makes to the overall molecular orbital, must be found by solving the Schrödinger equation. However, in practice, the two lowest energy combinations of this kind are very similar to the combinations 1σ and $2\sigma^*$ of $2s$ orbitals that we have described, and the two highest energy combinations are very similar to the 3σ and $4\sigma^*$ combinations of $2p_z$ orbitals. In each case there will be small differences: the $2\sigma^*$ orbital, for instance, will be contaminated by some $2p_z$ character and the 3σ orbital will be contaminated by some $2s$ character, and their energies will be slightly shifted from where they would be if we considered only the 'pure' combinations. Nevertheless, the changes are not great, and we can continue to think of 1σ and $2\sigma^*$ as forming one bonding and antibonding pair, and of 3σ and $4\sigma^*$ as forming another pair. The four orbitals are shown in the centre column of Fig. 9.18. There is no guarantee that $2\sigma^*$ and 3σ will be in the exact location shown in the illustration, and the locations shown in Fig. 9.19 are found in some molecules.

Now consider the $2p_x$ and $2p_y$ orbitals of each atom, which are perpendicular to the internuclear axis and may overlap broadside-on. This overlap may be constructive or destructive, and results in a bonding and an antibonding **π orbital**, which we label 1π and $2\pi^*$, respectively. The notation π is the analogue of p in atoms, for when viewed along the axis of the molecule, a π orbital looks like a p orbital (Fig. 9.20).[†] The two $2p_x$ orbitals overlap to give a bonding and an antibonding π orbital, as do the two $2p_y$ orbitals too. The two bonding combinations have the same energy; likewise, the two antibonding combinations have the same energy. Hence, each π energy level is doubly degenerate and consists of two distinct orbitals. Two electrons in a π orbital constitute a **π bond**: such a bond resembles a π bond of valence bond theory, but the details of the electron distribution are slightly different.

[†]More precisely, an electron in a π orbital has one unit of orbital angular momentum about the internuclear axis.

The relative order of the σ and π orbitals in a molecule cannot be predicted readily (it varies with the energy separation between the $2s$ and $2p$ orbitals of the atom), and in some molecules the order shown in Fig. 9.18 applies, whereas others have the order shown in Fig. 9.19. The change in order can be seen in Fig. 9.21, which shows the calculated energy levels for the Period 2 homonuclear diatomic molecules. A useful rule is that, for neutral molecules, the order shown in Fig. 9.18 is valid for O_2 and F_2, whereas the order shown in Fig. 9.19 is valid for the preceding elements of the period.

Symmetry and overlap

One central feature of molecular orbital theory can now be addressed. We have seen that s and p_z orbitals may contribute to the formation of σ orbitals, and that p_x and p_y orbitals may contribute to π orbitals. However, we never have to consider orbitals formed by the overlap of s and p_x orbitals (or p_y orbitals). When building molecular orbitals, it is necessary to consider linear combinations only of atomic orbitals of the same symmetry with respect to the internuclear axis. Because an s orbital has cylindrical symmetry around the internuclear axis, but a p_x orbital does not, the two atomic orbitals cannot contribute to the same molecular orbital. The reason for this distinction based on symmetry can be understood by considering the overlap between an s orbital and a p_x orbital (Fig. 9.22): although there is constructive interference between the two orbitals on one side of the axis, there is an exactly compensating amount of destructive interference on the other side of the axis, and the net overlap (and hence the net bonding or antibonding effect) is zero.

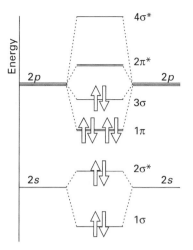

Fig. 9.19 A typical molecular orbital energy level diagram for Period 2 homonuclear diatomic molecules up to and including N_2.

Justification

The extent to which two orbitals overlap is measured by the **overlap integral**, S:

$$S = \int \psi_A^* \psi_B \, d\tau$$

where the integration is over all space. If the atomic orbital ψ_A on A is small wherever the orbital ψ_B on B is large, or vice versa, then the product of their amplitudes is everywhere small and the integral—the sum of these products—is small (Fig. 9.23(a)). If ψ_A and ψ_B are simultaneously large in some region of space, then S may be large (Fig. 9.23(b)). If the two atomic orbitals are identical (for example, $1s$ orbitals on the same nucleus), $S = 1$. For two $1s$ orbitals at the equilibrium bond length in H_2^+, $S = 0.59$, which is an unusually large value. Typical values for orbitals with $n = 2$ are in the range 0.2 to 0.3.

Now consider the arrangement in (Fig. 9.23(c)) in which an s orbital overlaps a p_x orbital of a different atom.

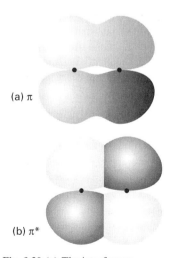

(a) π

(b) π^*

Fig. 9.20 (a) The interference leading to the formation of a π-bonding orbital and (b) the corresponding antibonding orbital.

Fig. 9.21 The variation of the orbital energies of Period 2 homonuclear diatomic molecules.

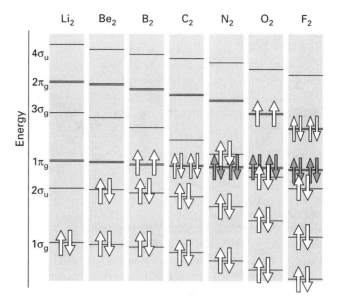

At some point r the product $\psi_A^* \psi_B$ may be large. However, there is a point r' where $\psi_A^* \psi_B$ has exactly the same magnitude but an opposite sign. When the integral is evaluated, these two contributions are added together and cancel. For every point in the upper half of the diagram, there is a point in the lower half that cancels it, and so $S = 0$. Therefore, there is no net overlap between the s and p orbitals in this arrangement.

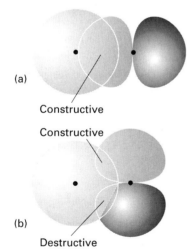

(a) Constructive

Constructive

(b) Destructive

Fig. 9.22 Overlapping s and p orbitals. (a) End-on overlap leads to nonzero overlap and to the formation of an axially symmetric σ orbital. (b) Broadside overlap leads to no net accumulation or reduction of electron density and does not contribute to bonding.

We now have the criteria for selecting atomic orbitals from which molecular orbitals are to be built:

1. Use all available valence orbitals from both atoms (in polyatomic molecules, from all the atoms).
2. Classify the atomic orbitals as having σ and π symmetry with respect to the internuclear axis, and build σ and π orbitals from all atomic orbitals of a given symmetry.
3. If there are N_σ atomic orbitals of σ symmetry, then N_σ σ molecular orbitals can be built with progressively higher energy from strongly bonding to strongly antibonding.
4. If there are N_π atomic orbitals of π symmetry, then N_π π molecular orbitals can be built with progressively higher energy from strongly bonding to strongly antibonding. The π orbitals occur in doubly degenerate pairs.

As a general rule, the energy of each species of orbital (σ or π) increases with the number of internuclear nodes. The lowest energy orbital of a given species has no internuclear nodes and the highest energy orbital has a nodal plane between each pair of neighbouring atoms (Fig. 9.24).

Example *Assessing the contribution of d orbitals*
Can *d* orbitals contribute to σ and π orbitals in diatomic molecules?

Strategy
We need to assess the symmetry of *d* orbitals with respect to the internuclear *z* axis: orbitals of the same symmetry can contribute to a given molecular orbital.

Solution
A d_{z^2} orbital has cylindrical symmetry around *z* and so can contribute to σ orbitals. The d_{zx} and d_{yz} orbitals have π symmetry with respect to the axis (Fig. 9.25), and so can contribute to π orbitals.

Exercise E9.7

Sketch the 'δ orbitals' (orbitals that resemble *d* orbitals when viewed along the internuclear axis) that may be formed by the remaining two *d* orbitals (and which contribute to bonding in some *d*-metal cluster compounds).

[*Answer*: see Fig. 9.25]

The electronic structures of homonuclear diatomic molecules

Figures 9.18 and 9.19 show the general layout of the valence-shell atomic orbitals of Period 2 atoms on the left and right of the MO energy level diagram. The lines in the middle are an indication of the energies of the molecular orbitals that can be formed by overlap of atomic orbitals. From the eight valence shell orbitals (four from each atom), we can form eight molecular orbitals: four are σ orbitals and four, in two pairs, are doubly degenerate π orbitals. With the orbitals established, the ground state electron configurations of the molecules are derived by adding the appropriate number of electrons to the orbitals and following the building-up rules. Charged species (such as the peroxide ion, O_2^{2-}, and C_2^+) need either more or fewer electrons (for anions and cations, respectively) than the neutral molecules.

We shall illustrate the procedure with N_2, which has ten valence electrons; for this molecule we use Fig. 9.19. The first two electrons pair, enter, and fill the 1σ orbital; the next two enter and fill the $2\sigma^*$ orbital. Six electrons remain. There are two 1π orbitals, and so four electrons can be accommodated in them. The two remaining electrons enter the 3σ orbital. The ground state configuration of N_2 is therefore

$$N_2 \, 1\sigma^2 2\sigma^{*2} 1\pi^4 3\sigma^2$$

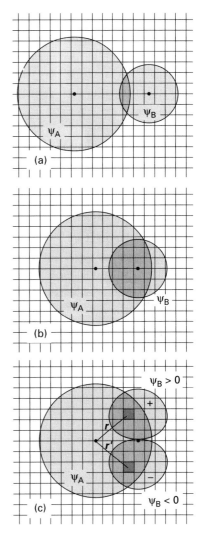

Fig. 9.23 A schematic representation of the contributions to the overlap integral. (a) $S \approx 0$ because the orbitals are far apart and their product is always small. (b) S is large (but less than 1) because the product $\psi_A \psi_B$ is large over a substantial region. (c) $S = 0$ because the positive region of overlap is exactly cancelled by the negative region.

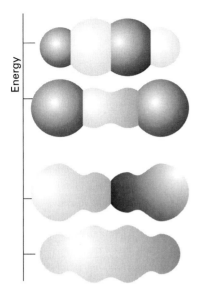

Fig. 9.24 A schematic representation of the four molecular orbitals that can be formed from four *s* orbitals in a chain of four atoms. The lowest energy combination (the bottom diagram) is formed from atomic orbitals with the same sign, and there are no internuclear nodes. The next higher orbital has one node (at the centre of the molecule). The next higher orbital has two internuclear nodes, and the uppermost, highest energy orbital, has three internuclear nodes, one between each neighbouring pair of atoms, and is fully antibonding. The sizes of the spheres reflect the contributions of each atom to the molecular orbital; the shading represents different signs.

This configuration is also depicted in Fig. 9.19.

The strength of a bond in a molecule is the net outcome of the bonding and antibonding effects of the occupied orbitals. The **bond order**, b, is defined as the difference between the number of occupied bonding orbitals between the atoms of interest and the number of occupied antibonding orbitals. In a diatomic molecule

$$b = \tfrac{1}{2}(n - n^*) \tag{7}$$

where n is the number of electrons in bonding orbitals and n^* is the number of electrons in antibonding orbitals. Each electron pair in a bonding orbital increases the bond order by 1 and each pair in an antibonding orbital decreases it by 1. For H_2, $b = 1$, corresponding to a single bond between the two atoms: this bond order is consistent with the Lewis structure H—H for the molecule. In He_2, which has equal numbers of bonding and antibonding electrons (with $n = 2$ and $n^* = 2$), the bond order is $b = 0$, and there is no bond. In N_2, 1σ, 3σ, and 1π are bonding orbitals, and $n = 2 + 2 + 4 = 8$; however, $2\sigma^*$ (the antibonding partner of 1σ) is antibonding, so $n^* = 2$ and the bond order of N_2 is $b = \tfrac{1}{2}(8 - 2) = 3$. This value is consistent with the Lewis structure :N≡N:, in which there is a triple bond between the two atoms.

The bond order is a useful parameter for discussing the characteristics of bonds, because it correlates with bond length, and the greater the bond order between atoms of a given pair of atoms, the shorter the bond. The bond order also correlates with bond strength, and the greater the bond order, the greater the strength. The high bond order of N_2 is consistent with its high dissociation energy (942 kJ mol^{-1}).

Example *Writing the electron configuration of a diatomic molecule*
Write the ground state electron configuration of O_2 and calculate the bond order.

Strategy
Decide which MO energy level diagram to use (Fig. 9.18 or Fig. 9.19), and count the valence electrons. Then accommodate these electrons by using the building-up principle.

Solution
Because the molecule is oxygen, the molecular orbital energy level diagram in Fig. 9.18 should be used. There are 12 valence electrons to accommodate. The first 10 electrons recreate the N_2 configuration (with a reversal of the order of the 3σ and 1π orbitals); the remaining two electrons must occupy the $2\pi^*$ orbitals. The configuration and bond order are therefore

$$O_2 \; 1\sigma^2 2\sigma^{*2} 3\sigma^2 1\pi^4 2\pi^{*2}$$

This configuration is also depicted in Fig. 9.18. Because 1σ, 3σ, and 1π are bonding and $2\sigma^*$ and $2\pi^*$ are antibonding, the bond order is $b = \frac{1}{2}(8-4) = 2$. This bond order accords with the classical view that oxygen has a double bond.

Exercise E9.8

Write the electron configuration of F_2 and deduce its bond order.

[*Answer*: $1\sigma^2 2\sigma^{*2} 3\sigma^2 1\pi^4 2\pi^{*4}$, $b = 1$]

We see from the example above that the electron configuration of O_2 is

$$O_2 \quad 1\sigma^2 2\sigma^{*2} 3\sigma^2 1\pi^4 2\pi^{*2}$$

According to the building-up principle, the two $2\pi^*$ electrons in O_2 will occupy different orbitals: one will enter the $2\pi^*$ orbital formed by overlap of $2p_x$ and the other will enter its degenerate partner, the $2\pi^*$ orbital formed from overlap of the $2p_y$ orbitals. Because the two electrons occupy different orbitals, they will have parallel spins ($\uparrow\uparrow$). Therefore, we can predict that an O_2 molecule will be magnetic because the magnetic effects of the two unpaired spins do not cancel. Specifically, oxygen is predicted to be a **paramagnetic** substance, a substance that is drawn into a magnetic field. Most substances (those with paired electron spins) are **diamagnetic**, and are pushed out of a magnetic field. That O_2 is in fact a paramagnetic gas is a striking confirmation of the superiority of the molecular orbital description of the molecule over the Lewis description (which requires all the electrons to be paired). The property of paramagnetism is utilized to monitor the oxygen content of incubators by measuring the magnetism of the gases they contain.

An F_2 molecule has two more electrons than an O_2 molecule and the configuration and bond order are

$$F_2 \quad 1\sigma^2 2\sigma^{*2} 3\sigma^2 1\pi^4 2\pi^{*4} \qquad b = 1$$

We conclude that F_2 is a singly-bonded molecule, in agreement with its Lewis structure F—F. The low bond order is consistent with the low dissociation energy of F_2 (154 kJ mol^{-1}). A hypothetical Ne_2 would have two further electrons:

$$Ne_2 \quad 1\sigma^2 2\sigma^{*2} 3\sigma^2 1\pi^4 2\pi^{*4} 4\sigma^{*2} \qquad b = 0$$

(The orbital $4\sigma^*$ is the antibonding partner of 3σ.) The bond order of zero agrees with the monatomic character of neon.

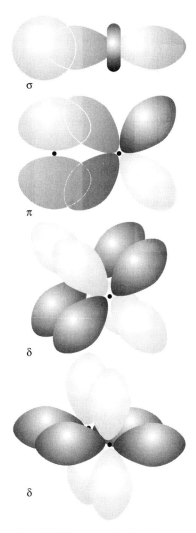

σ

π

δ

δ

Fig. 9.25 The types of molecular orbital to which *d* orbitals can contribute. The σ and π combinations can be formed with *s*, *p*, and *d* orbitals of the appropriate symmetry, but the δ orbitals can be formed only by the *d* orbitals of the two atoms.

The chemical bond

Example *Judging the relative bond strengths of molecules and ions*
Judge whether N_2^+ is likely to have a larger or smaller dissociation energy than N_2.

Strategy
Because a species with the larger bond order is likely to have the larger dissociation energy, we should compare their electronic configurations, and assess their bond orders.

Solution
From Fig. 9.19,

$$N_2 \ 1\sigma^2 2\sigma^{*2} 1\pi^4 3\sigma^2 \qquad b = 3$$
$$N_2^+ \ 1\sigma^2 2\sigma^{*2} 1\pi^4 3\sigma^1 \qquad b = 2.5$$

Because the cation has the smaller bond order, we expect it to have the smaller dissociation energy. The experimental dissociation energies are $945 \ kJ \ mol^{-1}$ for N_2 and $842 \ kJ \ mol^{-1}$ for N_2^+.

Exercise E9.9

Which can be expected to have the higher dissociation energy, F_2 or F_2^+?

[*Answer*: F_2^+]

Parity

We have seen a little of the importance of the symmetry of atomic orbitals in the construction of molecular orbitals. Symmetry plays a role in many discussions of molecular orbitals themselves, particularly in the interpretation of the electronic transitions that give rise to molecular spectra (Chapter 11).

The classification of molecular orbitals as σ and π is one aspect of molecular (as distinct from atomic) symmetry that we have already seen, for σ and π orbitals differ in their symmetry about the internuclear axis. However, there is another aspect of molecular symmetry that applies in homonuclear diatomic molecules rather than to diatomic molecules in general. The orbitals of homonuclear diatomic molecules may also be classified according to their **parity**, their behaviour under the process called **inversion** (Fig. 9.26). To decide on the parity of an orbital, we consider an arbitrary point in a homonuclear diatomic, and note the sign of the orbital. Then we imagine travelling through the centre of the molecule and out to the corresponding point on the other side. If the orbital has the same sign (the same shading) at that point, then it has **even parity** and is denoted g (from *gerade*,

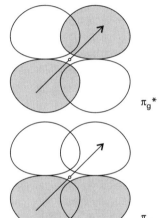
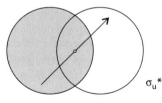
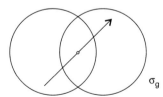

Fig. 9.26 The parity of an orbital is even (g) if its amplitude is unchanged under inversion in the centre of symmetry of the molecule, but odd (u) if the amplitude changes sign. Heteronuclear diatomic molecules do not have a centre of inversion, and so the g,u classification is irrelevant.

354

the German word for even). If the orbital has opposite sign, then it has **odd parity** and is denoted u (from *ungerade*, uneven). When we deal with diatomic molecules, the parity designation applies only to homonuclear species. Heteronuclear diatomic molecules (such as HCl) do not have inversion symmetry of the kind that we have described: such molecules do not have a 'centre'.

The diagram in Fig. 9.26 shows that a bonding σ orbital has even parity; so it is written σ_g; an antibonding σ orbital has odd parity and is written σ_u^*. A bonding π orbital has odd parity and is denoted π_u and an antibonding π orbital has even parity, denoted π_g^*. It follows that the full-dress version of the electron configuration of the ground state of N_2 is

$$N_2\ 1\sigma_g^2 2\sigma_u^{*2} 1\pi_u^4 3\sigma_g^2$$

This detailed specification of a configuration is needed (in the main) only when discussing electronic transitions and molecular selection rules.

Exercise E9.10

Write the full electron configuration of the ground state of the F_2 molecule.

[*Answer*: $1\sigma_g^2 2\sigma_u^{*2} 3\sigma_g^2 1\pi_u^4 2\pi_g^{*4}$]

9.6 Heteronuclear diatomic molecules

A **heteronuclear diatomic molecule** is a diatomic molecule formed from atoms of two different elements, such as CO and HCl. The electron distribution in the covalent bond between the atoms is not symmetrical between the atoms because it is energetically favourable for the electron pair to be found closer to one atom rather than the other. This imbalance results in a **polar bond**, which is a covalent bond in which the electron pair is shared unequally by the two atoms.

Electronegativity

The atom that draws the bonding electron pair to it more strongly is called the more **electronegative** atom. A numerical scale of electronegativity (Table 9.2) was originally formulated by Linus Pauling, who based it on considerations of bond dissociation energy, E:

$$|\chi_A - \chi_B| = 0.102 \times \sqrt{\Delta/\text{kJ mol}^{-1}}$$
$$\Delta = E(A\text{—}B) - \tfrac{1}{2}\{E(A\text{—}A) + E(B\text{—}B)\} \tag{8}$$

Table 9.2 Electronegativities of the main-group elements

H						
2.1						
Li	Be	B	C	N	O	F
1.0	1.5	2.0	2.5	3.0	3.5	4.0
Na	Mg	Al	Si	P	S	Cl
0.9	1.2	1.5	1.8	2.1	2.5	3.0
K	Ca	Ga	Ge	As	Se	Br
0.8	1.0	1.6	1.8	2.0	2.4	2.8
Rb	Sr	In	Sn	Sb	Te	I
0.8	1.0	1.7	1.8	1.9	2.1	2.5
Cs	Ba	Tl	Pb	Bi	Po	
0.7	0.9	1.8	1.8	1.9	2.0	

A somewhat simpler definition was proposed by Robert Mulliken, who related the electronegativity, χ, to the ionization energy, I, and the electron affinity, E_{ea}, of the element:

$$\chi = \tfrac{1}{2}(I + E_{ea}) \tag{9}$$

This relation is plausible, because an atom that has a high electronegativity is likely to be one that has a high ionization energy (so that it is unlikely to lose electrons to another atom in the molecule) and a high electron affinity (so that it is energetically favourable for an electron to move towards it). The Mulliken electronegativities are broadly in line with the Pauling electronegativities. Electronegativities show a periodicity, and the elements with the highest electronegativities are those close to fluorine in the periodic table.

The location of the bonding electron pair close to one atom in a heteronuclear molecule results in that atom having a net negative charge, which is called a **partial negative charge** and denoted $\delta-$. There is a compensating partial positive charge $\delta+$ on the other atom. In a typical heteronuclear diatomic molecule, the more electronegative element has the partial negative charge and the more electropositive element has the partial positive charge.

Exercise E9.11

Predict the (weak) polarity of a C—H bond.

[*Answer*: $^{\delta-}C—H^{\delta+}$]

Polar covalent bonds

A **polar covalent bond** consists of two electrons in an orbital of the form

$$\psi = c_A \psi_A + c_B \psi_B \qquad (10)$$

with unequal coefficients. The proportion of the atomic orbital ψ_A in the bond is c_A^2 and that of ψ_B is c_B^2. A **nonpolar bond**, a covalent bond in which the electron pair is shared equally between the two atoms and there are zero partial charges on each atom, has $c_A^2 = c_B^2$. A pure ionic bond, in which one atom has obtained virtually sole possession of the electron pair (as in Cs^+F^-, to a first approximation), has one coefficient zero (so that A^+B^- would have $c_A^2 = 0$ and $c_B^2 = 1$. A general feature of molecular orbitals between dissimilar atoms is that the atomic orbital with the lower energy (that belonging to the more electronegative atom) makes the larger contribution to the lowest energy molecular orbital. The opposite is true of the highest (most antibonding) orbital, for which the principal contribution comes from the atomic orbital with higher energy (the less electronegative atom). A schematic representation of this point is shown in Fig. 9.27.

These features of polar bonds can be illustrated by considering HF. The general form of the molecular orbitals of HF is

$$\psi = c_H \psi_H + c_F \psi_F \qquad (11)$$

where ψ_H is an H1s orbital and ψ_F is an F2p_z orbital. Because the ionization energy of a hydrogen atom is 13.6 eV, we know that the energy of the H1s orbital is -13.6 eV (as usual, the zero of energy is the infinitely separated electron and proton, Fig. 9.28). Similarly, from the ionization energy of fluorine, which is 18.6 eV, we know that the energy of the F2p_z orbital is -18.6 eV, about 5 eV lower than the H1s orbital. It follows that the bonding σ orbital in HF is mainly F2p_z and the antibonding σ orbital is mainly H1s orbital in character. The two electrons in the bonding orbital are most likely to be found in the F2p_z orbital, so there is a partial negative charge on the F atom and a partial positive charge on the H atom.

A systematic way of finding the coefficients in the linear combinations is to use the variation theorem (Section 9.2) and to look for the values of the coefficients that result in the lowest energy. For example, when the variation principle is applied to an H_2 molecule, the calculated energy is lowest when the two H1s orbitals contribute equally to a bonding orbital; however, when the principle is applied to an HF molecule, the lowest energy is obtained for the orbital

$$\psi = 0.33\psi_{H1s} + 0.94\psi_{F2p_z}$$

We see that indeed the F2p_z orbital does make the greater contribution to the bonding σ orbital.

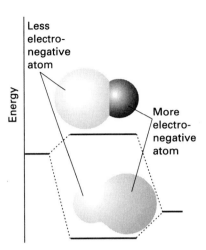

Fig. 9.27 A schematic representation of the relative contributions of atoms of different electronegativities to bonding and antibonding molecular orbitals. In the bonding orbital, the more electronegative atom makes the greater contribution (represented by the larger sphere), and the electrons of the bond are more likely to be found on that atom. The opposite is true of an antibonding orbital. A part of the reason why an antibonding orbital is of high energy is that the electrons that occupy it are likely to be found on the more electropositive atom.

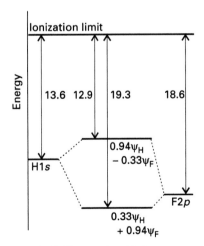

Fig. 9.28 The atomic orbital energy levels of H and F atoms and the molecular orbitals they form. The bonding orbital has predominantly F atom character and the antibonding orbital has predominantly H atom character.

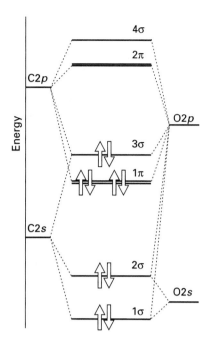

Fig. 9.29 The molecular orbital energy level diagram for CO.

What percentage of its time does a σ electron in HF spend in a $F2p_z$ orbital?

[*Answer*: 88 per cent $(= (0.94)^2 \times 100$ per cent)]

A schematic diagram for CO is shown in Fig. 9.29. It illustrates a number of points we have made. The ground configuration is

$$CO\ 1\sigma^2 2\sigma^2 1\pi^4 3\sigma^2$$

(Note that the parity designation is inapplicable because the molecule is heteronuclear.) The lowest energy orbitals are predominantly of O character as that is the more electronegative element. The **highest occupied molecular orbital** (HOMO) is 3σ, which is a largely nonbonding orbital centred on C, and so the two electrons that occupy it can be regarded as a lone pair on the C atom. The **lowest unoccupied molecular orbital** (LUMO) is 2π, which is largely a doubly degenerate orbital of $2p$ character on carbon. This combination of a lone pair orbital on C and a pair of empty π orbitals also largely on C is at the root of the importance of carbon monoxide in d-block chemistry, because it enables it to form an extensive series of carbonyl complexes by a combination of electron donation from the 3σ orbital and electron acceptance into the 2π orbitals.

9.7 The structures of polyatomic molecules

The bonds in polyatomic molecules are built in the same way as in diatomic molecules, the only difference being that we use more atomic orbitals to construct the molecular orbitals, and these molecular orbitals spread over the entire molecule. In general, a molecular orbital is a linear combination of all the atomic orbitals of all the atoms in the molecule. In H_2O, for instance, the atomic orbitals are the two $H1s$ orbitals, the $O2s$ orbital, and the three $O2p$ orbitals (if we consider only the valence shell). From these six atomic orbitals we can construct six molecular orbitals that spread over all three atoms. The molecular orbitals differ in energy, the lowest energy, most strongly bonding orbitals having the least number of nodes between adjacent atoms, and the highest energy, most strongly antibonding orbitals having the greatest numbers of nodes between neighbouring atoms.

According to MO theory, the bonding influence of a single electron pair is distributed over all the atoms, and each electron pair (the maximum number of electrons that can occupy any single molecular orbital) helps to bind all the atoms together. In the LCAO-MO approximation, each molecular orbital is modelled as a linear combination of atomic orbitals, with atomic orbitals contributed by all the atoms in the molecule. No attempt is made to construct bonds between individual pairs of atoms, and all s and p orbitals are included in the construction of molecular orbitals without resorting to the artifice of hybridization. Thus, a typical molecular orbital in H_2O constructed from $H1s$ orbitals (denoted s_A and s_B) and $O2s$ and $O2p$ orbitals (denoted s_O and p_O) will have the composition

$$\psi = c_1 s_A + c_2 s_O + c_3 p_O + c_4 s_B \qquad (12)$$

Because four atomic orbitals are being used to form the molecular orbital, there will be four possible molecular orbitals: the lowest energy (most bonding) orbital will have no internuclear nodes and the highest energy (most antibonding) orbital will have a node between each pair of neighbouring nuclei.

As an illustration, consider the molecular orbitals that may be formed from the p orbitals perpendicular to the molecular plane of benzene, C_6H_6. Because there are six such orbitals, it is possible to form six molecular orbitals of the form

$$\psi = c_1 p_1 + c_2 p_2 + c_3 p_3 + c_4 p_4 + c_5 p_5 + c_6 p_6 \qquad (13)$$

The lowest energy, most strongly bonding orbital has no internuclear nodes, and has the form[†]

$$\psi = p_1 + p_2 + p_3 + p_4 + p_5 + p_6$$

and is illustrated in Fig. 9.30. It is strongly bonding because the constructive interference between neighbouring p orbitals results in a good accumulation of electron density between the nuclei (but slightly off the internuclear axis, as in the π bonds of diatomic molecules). The most antibonding orbital has the form

$$\psi = p_1 - p_2 + p_3 - p_4 + p_5 - p_6$$

The alternation of signs in the linear combination results in destructive interference between neighbours, and the molecular orbital has a nodal plane between each pair of neighbours, as shown in the illustration. The remaining four molecular orbitals are more difficult to establish by qualitative arguments, but they have the form shown in Fig. 9.30, and lie in energy between the most bonding and most antibonding orbitals. Note that the four intermediate orbitals form two doubly degenerate pairs, one net bonding and the other net antibonding.

The energies of the six π molecular orbitals in benzene may be calculated by solving the Schrödinger equation, and are also shown in the molecular orbital energy level diagram (Fig. 9.31). There are six electrons to be accommodated (one is supplied by each C atom), and they occupy the lowest

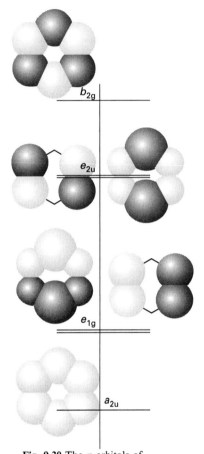

Fig. 9.30 The π orbitals of benzene. The lowest energy orbital is fully bonding between neighbouring atoms but the uppermost orbital is fully antibonding. The two pairs of doubly degenerate molecular orbitals have an intermediate number of internuclear nodes. As usual, light and dark shading represents different signs of the wavefunction. The orbitals have opposite signs below the plane of the ring.

[†]We are ignoring normalization factors, for clarity. In this and the following case it would be $1/\sqrt{6}$ if we ignored overlap.

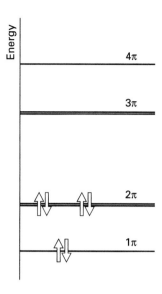

Energy

4π

3π

2π

1π

Fig. 9.31 The π molecular orbital energy level diagram for benzene, and the configuration in its ground state.

Fig. 9.32 When six electrons occupy the three lowest orbitals, the resulting overall electron density is the same at each atom: this uniformity is represented by the double-doughnut electron cloud.

three orbitals. The resulting electron distribution is like a double doughnut (Fig. 9.32). It is an important feature of the configuration that the only molecular orbitals occupied have a net bonding character, for this is one contribution to the stability (in the sense of low energy) of the benzene molecule. It may be helpful to note the similarity between the molecular orbital energy level diagram for benzene and that for N_2 (see Fig. 9.19): the strong bonding in benzene is echoed in the strong bonding in nitrogen.

A feature of the molecular orbital description of benzene is that each molecular orbital spreads either all round or partially round the C_6 ring. That is, π bonding is **delocalized**, and each electron pair helps to bind together several or all of the C atoms. The delocalization of bonding influence is a primary feature of molecular orbital theory, and we shall encounter it in its extreme when we come to consider the electronic structures of solids.

The band theory of solids

Modern chemistry is deeply concerned with the properties of solids. Apart from their intrinsic usefulness for construction, modern solids have made possible the semiconductor revolution and recent advances in ceramics have given rise to the hope that we may now be on the verge of a superconductor revolution. Solids are used widely in chemical industry as catalysts, where the details of their action often depends on the details of their electronic structure, particularly at the surfaces where the reaction takes place. Advances in understanding of electron mobility in solids are also useful in biology, where electron transport processes are responsible for many biochemical processes, particularly photosynthesis and respiration.

For our purposes, we shall concentrate on the simplest types of solid. We shall distinguish two types of solid according to the variation of their electrical conductivity with temperature. A **metallic conductor** is a substance with a conductivity that decreases as the temperature is raised. Metallic conductors include the metallic elements, their alloys, and graphite. A **semiconductor** is a substance with a conductivity that increases as the temperature is raised. Semiconductors include silicon, diamond, and gallium arsenide. A semiconductor generally has a lower conductivity than that typical of metals, but the magnitude of the conductivity is not the criterion of the distinction. It is conventional to classify semiconductors with very low electrical conductivities as **insulators**. We shall use the latter term, but it should be appreciated that it is one of convenience rather than one of fundamental significance. We shall not consider **superconductors**, which are substances that conduct electricity with zero resistance.

9.8 The formation of bands

The electronic structures of solids can be treated in a uniform way in terms of molecular orbital theory. Indeed, it is possible to regard a solid as just another system in which electron delocalization (like that in benzene, but on a much bigger scale) is responsible for bonding and properties. In a solid, atom after atom lies in a three-dimensional array and electrons take part in bonding that spreads throughout the sample. We shall consider a one-dimensional solid initially, which consists of a single, infinitely long line of atoms, each one having one s orbital available for forming molecular orbitals. We can construct the LCAO-MOs of the solid by adding atoms to a line and then find the electronic structure by using the building-up principle.

One atom contributes one s orbital at a certain energy (Fig. 9.33(a)). When a second atom is brought up it overlaps the first, and forms a bonding and antibonding orbital (Fig. 9.33(b)). The third atom overlaps its nearest neighbour (and only slightly the next-nearest), and three molecular orbitals are formed from these three atomic orbitals (Fig. 9.33(c)). The fourth atom leads to the formation of a fourth molecular orbital (Fig. 9.33(d)). At this stage we can begin to see that the general effect of bringing up successive atoms is to spread the range of energies covered by the molecular orbitals, and also to fill in the range of energies with more and more orbitals (one more for each additional atom). When N atoms have been added to the line, there are N molecular orbitals covering a band of finite width. When N is infinitely large, the difference between neighbouring energy levels is infinitely small, but the band still has finite width. This band consists of N different molecular orbitals, the lowest-energy orbital being fully bonding, and the highest-energy orbital being fully antibonding between adjacent atoms (Fig. 9.34).

The band formed from overlap of s orbitals is called the **s band**. If the atoms have p orbitals available, then the same procedure leads to a **p band** (as in the upper half of Fig. 9.34). If the atomic p orbitals lie higher in energy than the s orbitals, then the p band lies higher than the s band, and there may be a **band gap**, a range of energies for which no orbitals exist.

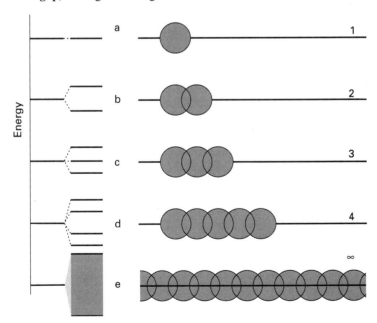

Fig. 9.33 The formation of a band of N molecular orbitals by successive addition of N atoms to a line. Note that the band remains of finite width, and although it looks continuous when N is large, it consists of N different orbitals.

Fig. 9.34 The overlap of *s* orbitals gives rise to an *s* band, and the overlap of *p*-orbitals gives rise to a *p* band. In this case the *s* and *p* orbitals of the atoms are so widely spaced that there is a band gap. In many cases the separation is less, and the bands overlap.

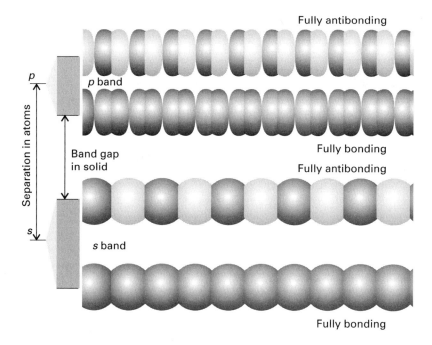

Fully antibonding

p band

Fully bonding

Fully antibonding

Band gap in solid

s band

Fully bonding

9.9 The occupation of orbitals

Now consider the electronic structure of a solid formed from atoms each able to contribute one electron (for example, the alkali metals). There are N atomic orbitals and therefore N molecular orbitals squashed into an apparently continuous band. There are N electrons to accommodate. Because the orbitals in the bands are so close together, electrons can be excited out of them by thermal motion of the atoms. The possibility of thermal excitation is a complication we can avoid by considering the solid at $T = 0$ when there is no such motion and all the electrons occupy the lowest available orbitals.

At $T = 0$, only the lowest $\frac{1}{2}N$ molecular orbitals are occupied (Fig. 9.35), and the highest occupied molecular orbital is called the **Fermi level**. However, unlike in the discrete molecules we have considered so far, there are empty orbitals just above and very close in energy to the Fermi level, and so it requires hardly any energy to excite the uppermost electrons. Some of the electrons are therefore very mobile, and give rise to electrical conductivity.

The electrical conductivity of a metallic solid decreases with increasing temperature even though more electrons are excited into empty orbitals. This apparent paradox is resolved by noting that the increase in temperature causes more vigorous thermal motion of the atoms, so collisions between the moving electrons and an atom are more likely. That is, the electrons are scattered out of their paths through the solid, and are less efficient at transporting charge.

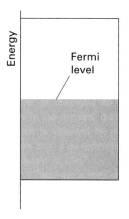

Fig. 9.35 When N electrons occupy a band of N orbitals, it is only half full and the electrons near the Fermi level (the top of the filled levels) are mobile.

9.10 Insulators and semiconductors

When each atom provides two electrons, the $2N$ electrons fill the N orbitals of the s band. The Fermi level now lies at the top of the band (at $T = 0$), and there is a gap before the next band begins (Fig. 9.36(a)). As the temperature is increased, electrons populate the empty orbitals of the upper band (Fig. 9.36(b)). They are now mobile, and the solid is an electric conductor. In fact, it is a semiconductor, because the electrical conductivity depends on the number of electrons that are promoted across the gap, and that increases as the temperature is raised. If the gap is large, though, very few electrons will be promoted at ordinary temperatures, and the conductivity will remain close to zero, giving an insulator. Thus, the conventional distinction between an insulator and a semiconductor is related to the size of the band gap and is not absolute like the distinction between a metal (incomplete bands at $T = 0$) and a semiconductor (full bands at $T = 0$).

Another method of increasing the number of charge carriers and enhancing the semiconductivity of a solid is to implant foreign atoms into an otherwise pure material. If these **dopants** can trap electrons as indium or gallium atoms can in silicon), then they withdraw electrons from the filled band, leaving holes which allow the remaining electrons to move (Fig. 9.37(a)). This doping procedure gives rise to **p-type semiconductivity**, the p indicating that the holes are positive relative to the electrons in the band. Alternatively, a dopant might carry excess electrons (for example, phosphorus atoms introduced into germanium), and these additional electrons occupy otherwise empty bands, giving **n-type semiconductivity** (Fig. 9.37(b)), where n denotes the negative charge of the carriers. The preparation of doped but otherwise ultrapure materials was described in Section 4.7.

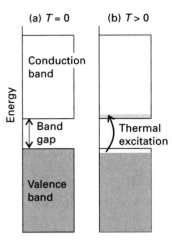

Fig. 9.36 (a) When $2N$ electrons are present, the band is full and the material is an insulator at $T = 0$. (b) At temperatures above $T = 0$, electrons populate the levels of the conduction band at the expense of the valence band, and the solid is a semiconductor.

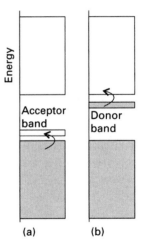

Fig. 9.37 (a) A dopant with fewer electrons than its host can form a narrow band that accepts electrons from the valence band. The holes in the valence band are mobile, and the substance is a p-type semiconductor. (b) A dopant with more electrons than its host forms a narrow band that can supply electrons to the conduction band. The electrons it supplies are mobile, and the substance is an n-type semiconductor.

EXERCISES

9.1 Give the ground-state electron configurations of (a) Li_2, (b) Be_2, and (c) C_2.

9.2 Give the ground-state electron configurations of (a) H_2^-, (b) N_2, and (c) O_2. For heteronuclear diatomic molecules, a good first approximation is that the energy level diagram is much the same as for homonuclear diatomic molecules.

9.3 Three biologically important diatomic species, either because they promote or inhibit life, are (a) CO, (b) NO, and (c) CN^-. The first binds to haemoglobin, the second is a neurotransmitter, and the third interrupts the electron-transfer chain. Their biochemical action is a reflection of their orbital structure. Deduce their ground-state electron configurations.

9.4 From the ground-state electron configurations of B_2 and C_2, predict which molecule should have the greater dissociation energy.

9.5 Some chemical reactions proceed by the initial loss or transfer of an electron to a diatomic species. Which of the molecules N_2, NO, O_2, C_2, F_2, and CN would you expect to be stabilized by (a) the addition of an electron to form AB^-, (b) the removal of an electron to form AB^+?

9.6 The existence of compounds of the noble gases was once a great surprise and stimulated a great deal of theoretical work. Sketch the molecular orbital energy level diagram for XeF and deduce its ground-state electron configurations. Is XeF likely to have a shorter bond length than XeF^+?

9.7 Where it is appropriate, give the parity of (a) $2\pi^*$ in F_2, (b) 3σ in NO, (c) 1δ in Tl_2, (d) $2\delta^*$ in Fe_2.

9.8 Give the parities of the first four levels of a particle-in-a-box wavefunctions.

9.9 State the parities of the six π-orbitals of benzene (see Fig. 9.30).

9.10 Two important diatomic molecules for the welfare of humanity are NO and N_2: the former is both a pollutant and a neurotransmitter, and the latter is the ultimate source of the nitrogen of proteins and other biomolecules. Use the electron configurations of NO and N_2 to predict which is likely to have the shorter bond length.

9.11 Put the following species in order of increasing bond length: F_2^-, F_2, F_2^+.

9.12 Show that the sp^2 hybrid orbital $(s + \sqrt{2}p)/\sqrt{3}$ is normalized to 1 if the s and p orbitals are normalized to 1.

9.13 Find another sp^2 hybrid orbital that is orthogonal to (has zero overlap with) the hybrid orbital in the preceding problem.

9.14 Normalize the molecular orbital $\psi_{sA} + \lambda\psi_{sB}$ in terms of the parameter λ and the overlap integral S.

9.15 Which of the following species are expected to be linear: (a) CO_2, (b) NO_2, (c) NO_2^+, (d) NO_2^-, (e) SO_2, (f) H_2O, (g) H_2O_2? Give reasons in each case.

9.16 Use the VSEPR model to predict the shapes of (a) H_2S, (b) SF_6, (c) XeF_4, (d) SF_4.

9.17 Construct the molecular orbital energy level diagrams of (a) ethene (ethylene) and (b) ethyne (acetylene) on the basis that the molecules are formed from the appropriately hybridized CH_2 or CH fragments.

9.18 Predict the electronic configurations of (a) the benzene anion, (b) the benzene cation. Estimate the π-bond energy in each case.

9.19 Many of the colours of vegetation are due to electronic transitions in conjugated π-electron systems. In the free-electron molecular orbital (FEMO) theory, the electrons in a conjugated molecule are treated as independent particles in a box of length L. Sketch the form of the two occupied orbitals in butadiene predicted by this model and predict the minium excitation energy of the molecule. The tetraene $CH_2{=}CHCH{=}CHCH{=}CHCH{=}CH_2$ can be treated as a box of length $8R$, where $R = 140\,pm$ (as in this case, an extra half bond-length is often added at each end of the box). Calculate the minimum excitation energy of the molecule and sketch the HOMO and LUMO.

Cohesion and structure

Contents

The origin of cohesion

Fluids

Crystal structure

Natural biopolymers

Atoms and molecules with complete valence shells are still able to interact with one another: they can exercise attractive forces over the range of several atomic diameters, and they can repel one another when pressed together. These residual forces are highly important. They account, for instance, for the condensation of gases and the structures of molecular solids. All organic liquids and solids, ranging from small molecules like benzene to essentially infinite molecules like cellulose and the polymers from which fabrics are made, are bound together by the forces of cohesion that we explore in this chapter. These forces are also responsible for the structural organization of biological macromolecules, for they twist the long polypeptide chains of proteins into characteristic shapes and then pin them together in the arrangement essential to their function.

To include all the interactions that are responsible for the cohesion of condensed phases means that we also have to include the Coulombic interaction between ions that is responsible for the existence of ionic solids. Although these interactions are usually regarded as versions of valence forces, it is appropriate to include these compounds here because closed-shell species are involved; but, of course, these species are now charged ions, and conventionally are regarded as participating in ionic bonding.

The second major question that we confront in this chapter is the determination of the structures of condensed phases and macromolecules. How is it possible to locate atoms in a crystal, to measure their separations, and to determine how they stack together? How can we determine the highly complex arrangements of atoms in a protein, which arises from the interplay of the valence forces responsible for bonds and the non-valence forces responsible for twisting and packing the bonded chains? The principal technique for investigating the arrangements of atoms in condensed phases, primarily crystalline solids, is X-ray diffraction, and we shall see a little of its basis in this chapter.

The origin of cohesion

The term **van der Waals forces** denotes the interactions between closed-shell molecules. The attractive contributions to these forces include the interactions between the partial electric charges of polar molecules. Van der Waals forces also include the repulsive interactions that are responsible for the prevention of the complete collapse of matter to densities as high as those characteristic of atomic nuclei. The repulsive interactions arise from the Pauli principle and the exclusion of electrons from regions of space where the orbitals of closed-shell species overlap.

10.1 Lattice enthalpy

A measure of the strength of binding in a solid, the analogue of the dissociation energy of individual molecules, is the **lattice enthalpy**, $\Delta H_L^{\ominus}$. The lattice enthalpy is defined as the molar enthalpy change accompanying the complete separation of the entities that compose the solid (such as ions if the solid is ionic, and molecules if the solid is molecular) under standard conditions. For example, the lattice enthalpy of an ionic solid such as calcium chloride, $CaCl_2$, is the standard molar enthalpy change accompanying the process

$$CaCl_2(s) \longrightarrow Ca^{2+}(g) + 2\,Cl^-(g)$$

and the lattice enthalpy of a molecular solid, such as ice, is the standard molar enthalpy of the transformation

$$H_2O(s) \longrightarrow H_2O(g)$$

Because the lattice enthalpy is invariably a positive quantity, it is normally reported without its $+$ sign. The lattice enthalpy of a molecular solid is the same as its standard enthalpy of sublimation. The lattice enthalpy of a metal is its enthalpy of atomization.

The greater the lattice enthalpy, the stronger the bonding within the solid, because more energy must be supplied as heat to drive the ions or molecules apart and form the gas. Some experimental lattice enthalpies are given in Table 10.1. Most lattice enthalpies of ionic solids are of the order of $10^3\,kJ\,mol^{-1}$, but there are considerable variations. The lattice enthalpies of ionic solids are significantly larger than those of most molecular solids, which is a sign that ion–ion forces are much stronger than other forces of cohesion. For example, the lattice enthalpy of solid carbon dioxide is only $25\,kJ\,mol^{-1}$, and even the value for water (in which there are strong hydrogen bonds) is only $50\,kJ\,mol^{-1}$. A feature to note from the data in the table is

Table 10.1 Lattice enthalpies, $\Delta H_L^{\ominus}/kJ\,mol^{-1}$

LiF	1037	LiCl	852	LiBr	815	LiI	761
NaF	926	NaCl	786	NaBr	752	NaI	705
KF	821	KCl	717	KBr	689	KI	649
MgO	3850	CaO	3461	SrO	3283	BaO	3114
MgS	3406	CaS	3119	SrS	2974	BaS	2832
Al_2O_3	15.9×10^3						

that ionic solids composed of small, highly charged ions (such as Al_2O_3) have very high lattice enthalpies. Another feature is that molecular solids in which there is hydrogen bonding also have higher lattice enthalpies than solids in which hydrogen bonding cannot occur.

The determination of lattice enthalpy

Lattice enthalpies are determined experimentally by using a **Born–Haber cycle**, which is a cycle (a closed path) of steps that includes lattice formation as one stage. The value of the lattice enthalpy—the only unknown in a well-chosen cycle—is found from the requirement that the sum of the enthalpy changes round a complete cycle is zero (because enthalpy is a state property). A typical cycle for an ionic compound has the form shown in Fig. 10.1. Some of the enthalpy changes required to complete the cycle are in a slightly disguised form. For example, the enthalpy of atomization of a metallic element is in fact its enthalpy of sublimation,

$$M(s) \longrightarrow M(g)$$

and the enthalpy of atomization of a diatomic gaseous element is in fact its enthalpy of dissociation:

$$X_2(g) \longrightarrow 2\,X(g)$$

The following example illustrates how the cycle is used. Care should be taken to use data for a single temperature.

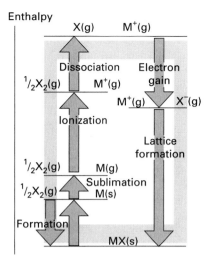

Fig. 10.1 The Born–Haber cycle for the determination of one of the unknown enthalpies, most commonly the lattice enthalpy. Upward pointing arrows denote positive changes in enthalpy; downward pointing arrows denote negative enthalpy changes. All the steps in the cycle correspond to the same temperature.

Example *Using a Born–Haber cycle to determine a lattice enthalpy*
Calculate the lattice enthalpy of KCl(s) using a Born–Haber cycle and the information given in the margin, which are all for 25 °C.

Strategy
First, draw the cycle, showing the atomization of the elements, their ionization, and the formation of the solid lattice; then complete the cycle (for the step solid compound ⟶ original elements) by using the enthalpy of formation. The sum of enthalpy changes round the cycle is zero, so include the numerical data and set the sum of all the terms equal to zero; then solve the equation for the one unknown (the lattice enthalpy).

Solution
The required cycle is shown in Fig. 10.2. The first step is the sublimation(atomization) of solid potassium:

$$\Delta H^{\ominus}/\text{kJ mol}^{-1}$$
$$K(s) \longrightarrow K(g) \qquad +89$$

Chlorine atoms are formed by dissociation of Cl_2:

$$\tfrac{1}{2}Cl_2(g) \longrightarrow Cl(g) \qquad +122$$

Step	$\Delta H^{\ominus}/\text{kJ mol}^{-1}$
Sublimation of K(s)	+89
Ionization of K(g)	+418
Dissociation of Cl_2(g)	+244
Electron attachment to Cl(g)	−349
Formation of KCl(s)	−348

367

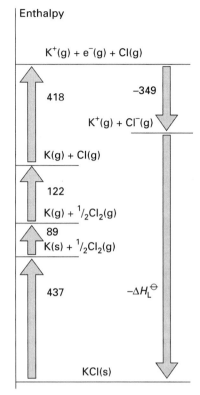

Enthalpy

K⁺(g) + e⁻(g) + Cl(g)

418

−349

K⁺(g) + Cl⁻(g)

K(g) + Cl(g)

122

K(g) + ¹/₂Cl₂(g)

89

K(s) + ¹/₂Cl₂(g)

437

$-\Delta H_L^{\ominus}$

KCl(s)

Fig. 10.2 The Born–Haber cycle for the calculation of the lattice enthalpy of potassium chloride. The sum of the enthalpy changes around the cycle is zero. The numerical values are in kilojoules per mole.

Now, potassium ions are formed by ionization of the gas-phase atoms:

$$K(g) \longrightarrow K^+(g) + e^-(g) \qquad +418$$

and chloride ions are formed from the chlorine atoms:

$$Cl(g) + e^-(g) \longrightarrow Cl^-(g) \qquad -349$$

The solid is now formed:

$$K^+(g) + Cl^-(g) \longrightarrow KCl(s) \qquad -\Delta H_L^{\ominus}$$

and the cycle is completed by decomposing KCl(s) into its elements:

$$KCl(s) \longrightarrow K(s) + \tfrac{1}{2}Cl_2(g) \qquad +437$$

(the reverse of its formation). The sum of the enthalpy changes is $-\Delta H_L^{\ominus} + 719\,\text{kJ mol}^{-1}$; however, the sum must be equal to zero, so $\Delta H_L^{\ominus} = 719\,\text{kJ mol}^{-1}$.

Exercise E10.1

Calculate the lattice enthalpy of magnesium bromide from the data in the margin and the information in Appendix 1.

Coulombic contributions to lattice enthalpies

Our next task is to account for the variation in lattice enthalpies, and in particular to account for the fact that, for ionic solids, the lattice enthalpy is high for small, highly charged ions. The dominant interaction in an ionic lattice is the Coulombic interactions between ions, which is far stronger than any other attractive interaction, so we concentrate on that.

The starting point is the Coulombic potential energy for the interaction of two ions

$$V_{12} = \frac{(z_1 e) \times (z_2 e)}{4\pi\varepsilon_0 r_{12}} \tag{1}$$

where ε_0 is the vacuum permittivity (see inside the front cover). This expression applies to ions of charge numbers z_1 and z_2 (with cations having positive charge numbers and anions negative charge numbers) separated by a distance r_{12}. To calculate the total potential energy of all the ions in a crystal, we have to sum this expression over all the pairs of ions in the solid. Nearest neighbours (which have opposite signs) contribute a large negative (attractive) term, second-nearest neighbours (which have the same sign) contribute a slightly weaker positive (repulsive) term, and so on (Fig. 10.3). The overall result, however, is that there is a net attraction between the cations and anions and a favourable (negative) contribution to the energy of the solid. For example, in a uniformly spaced one-dimensional

Step	$\Delta H^{\ominus}/\text{kJ mol}^{-1}$
Sublimation of Mg(s)	+148
Ionization of Mg(g) to Mg²⁺(g)	+2187
Dissociation of Br₂(g)	+31
Electron attachment to Br(g)	−325

line of alternating cations and anions with $z_1 = +z$ and $z_2 = -z$, the total Coulombic interaction of one ion with all the others is equal to

$$V = -\frac{z^2 e^2}{4\pi\varepsilon_0 d} \times 2\ln 2 \qquad (2)$$

where d is the distance between neighbouring ions in the crystal. The negative sign tells us that the potential energy of the ion is lower in this hypothetical crystal than in a gas of widely separated ions. Although this calculation is based on an unrealistic model of a crystal, it is already showing the features we are seeking to explain: the energy of the ion is greatly lowered if the charge numbers (z) of the ions are high and their diameters are small (so that d is small).

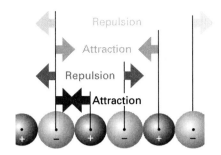

Fig. 10.3 There are alternating positive and negative contributions to the potential energy of a crystal lattice on account of the repulsions between ions of like charge and attractions of ions of opposite charge. The total potential energy is negative, but the sum might converge quite slowly.

Justification

Consider a line of alternating cations and anions extending in an infinite direction to the left and right of the ion of interest. The Coulombic energy of interaction with the ions on the right is the following sum of terms, where the negative terms represent attractions between ions of charge opposite to that of the ion of interest and the positive terms represent repulsions between ions of the same charge:

$$V = \frac{1}{4\pi\varepsilon_0} \times \left(-\frac{z^2 e^2}{d} + \frac{z^2 e^2}{2d} - \frac{z^2 e^2}{3d} + \frac{z^2 e^2}{4d} - \cdots \right)$$

$$= -\frac{z^2 e^2}{4\pi\varepsilon_0 d} \times \left(1 - \tfrac{1}{2} + \tfrac{1}{3} - \tfrac{1}{4} + \cdots \right) = -\frac{z^2 e^2}{4\pi\varepsilon_0} \times \ln 2$$

This calculation has used the mathematical relation

$$\ln 2 = 1 - \tfrac{1}{2} + \tfrac{1}{3} - \tfrac{1}{4} + \cdots$$

The interaction of the ion of interest with the ions to its left is the same, so the total potential energy of interaction is twice this expression, as given in eqn (2).

When the calculation is repeated for more realistic, three-dimensional arrays of ions it is also found that the potential energy depends on the charge numbers of the ions and the value of a single parameter d, which may be taken as the distance between the centres of nearest neighbours:

$$V = \frac{N_A e^2}{4\pi\varepsilon_0} \left(\frac{z_1 z_2}{d} \right) \mathcal{A} \qquad (3)$$

where $\mathcal{A}$ is a numerical constant called the **Madelung constant**. Because the charge number of cations is positive and that of anions is negative, the product $z_1 z_2$ is negative. Therefore, V is also negative, which corresponds

369

Table 10.2 Madelung constants

Structural type	$\mathcal{A}$
Caesium chloride	1.763
Fluorite	2.519
Rock salt	1.748
Rutile	2.408

to a lowering in potential energy relative to the gas of widely separated ions. The value of the Madelung constant for a single line of ions is $2\ln 2 = 1.386\cdots$, as we have already seen. The computed values of the Madelung constant for a variety of lattices are given in Table 10.2.

So far, we have considered only the Coulombic interaction between ions. However, even neutral atoms repel each other when they are pressed together and their wavefunctions overlap, and we have to take this repulsive force into account. These additional repulsions work against the net Coulombic attraction between ions, and so they raise the energy of the solid. When their effect is taken into account, it turns out that the lattice enthalpy is given by the **Born–Meyer equation**:

$$\Delta H_{\mathrm{L}}^{\ominus} = |z_1 z_2| \frac{N_{\mathrm{A}} e^2}{4\pi\varepsilon_0 d}\left(1 - \frac{d^*}{d}\right)\mathcal{A} \tag{4}$$

where d^* is an empirical parameter that is often taken as 34.5 pm (because that value is found to give reasonable agreement with experiment). The modulus signs ($|\cdots|$) mean that we should remove any minus sign from the product of z_1 and z_2, which results in a positive value for the lattice enthalpy. The details of this expression are not important: what is important is that because it shows that $\Delta H_{\mathrm{L}}^{\ominus} \propto z_1 z_2/d$, the expression confirms the experimental results that the lattice enthalpy increases with increasing charge number of the ions and with decreasing ionic radius. The second conclusion follows from the fact that the smaller the ionic radii, the smaller the value of d.

Exercise E10.2

Which can be expected to have the greater lattice enthalpy, magnesium oxide or strontium oxide?

[*Answer*: MgO]

10.2 Permanent and induced electric dipole moments

To understand the origin of the attractive forces between *uncharged* species, we need to understand the origin of the electric dipole moment of a molecule. An **electric dipole** consists of two charges q and $-q$ separated by a distance l. The value of ql called the **electric dipole moment**, μ. We shall represent dipole moments by an arrow pointing from the negative charge to the positive charge.

Because a dipole moment is the product of a charge and a length, the SI unit to use is coulomb-metre (C m). However, it is often much more convenient to report it in **debye**, D, where

$$1\,\mathrm{D} = 3.336 \times 10^{-30}\,\mathrm{C\,m}$$

for then experimental values for molecules are close to 1. The unit is named after Peter Debye, the Dutch pioneer of the study of dipole moments of molecules. The dipole moment of a pair of charges e and $-e$ separated by 100 pm is 1.6×10^{-29} C m, corresponding to 4.8 D. Dipole moments of small molecules are typically smaller than that, at about 1 D.

Polar molecules

A **polar molecule** is a molecule with a permanent electric dipole moment arising from the partial charges on its atoms (Section 9.6). **Nonpolar molecules**, molecules that have zero permanent electric dipole moment, may temporarily acquire a dipole moment in an electric field as a result of the distortion of their electronic distributions and the locations of their nuclei. Similarly, polar molecules may have their existing dipole moments modified temporarily by the applied field. The importance of electric dipole moments in chemistry stems from the fact that their partial charges may interact with the charges of ions (and contribute to solvation) and with the partial charges on other molecules (and contribute to molecular cohesion).

All heteronuclear diatomic molecules are polar because the difference in electronegativities of their two atoms results in nonzero partial charges. Typical dipole moments are 1.08 D for HCl and 0.42 D for HI (Table 10.3). A *very* approximate relation between the dipole moment and the difference in Pauling electronegativities (Table 9.2) of the two atoms, $\Delta\chi$, is

$$\mu/D \approx \Delta\chi \qquad (5)$$

Table I0.3 Dipole moments (μ) and polarizability volumes (α')

	μ/D	$\alpha'/10^{-30}$ m^3
Ar	0	1.66
CCl$_4$	0	10.5
C$_6$H$_6$	0	10.4
H$_2$	0	0.819
H$_2$O	1.85	1.48
NH$_3$	1.47	2.22
HCl	1.08	2.63
HBr	0.80	3.61
HI	0.42	5.45

The polarizability volumes are mean values over all orientations of the molecule.

Exercise E10.3

Estimate the electric dipole moment of an HBr molecule.

[*Answer*: 0.7 D; experimental 0.80 D]

Because it attracts the electrons more strongly, the more electronegative atom is usually the negative end of the dipole. However, there are exceptions, particularly when antibonding orbitals are occupied. Thus the dipole moment of CO is very small (0.12 D) but the negative end of the dipole is on the C atom even though the O atom is more electronegative. This apparent paradox is resolved as soon as we realize that antibonding orbitals are occupied in CO (see Fig. 9.29), and because electrons in antibonding orbitals tend to be found closer to the less electronegative atom, they can give rise to a negative partial charge on that atom. If the negative partial charge arising from this charge density is larger than the partial charge arising from the electrons in bonding orbitals, then the net effect will be a small negative partial charge on the less electronegative atom.

1 Ozone, O_3

2 Carbon dioxide, CO_2

Molecular symmetry is of the greatest importance in deciding whether a polyatomic molecule is polar or not. Indeed, molecular symmetry is more important than the question of whether or not the atoms in the molecule belong to the same element. Homonuclear polyatomic molecules may be polar if they have low symmetry because the atoms may be in different environments and hence carry different partial charges. For instance, the angular molecule ozone, O_3 (**1**), is a homonuclear polyatomic molecule, but it is polar because the central O atom is different from the outer two (it is bonded to two atoms, they are bonded only to one) and the electron density on the central O atom is different from that on the two outer O atoms. Heteronuclear polyatomic molecules may be nonpolar if they have high symmetry because individual bond dipoles may then cancel. The heteronuclear *linear* triatomic molecule CO_2, for example, is nonpolar because, although there are partial charges on all three atoms, the dipole moment associated with the OC bond points in the opposite direction to the dipole moment associated with the CO bond, and the two cancel (**2**).

Exercise E10.4

Use the **VSEPR** model to judge whether ClF_3 is polar or nonpolar.

[*Answer*: Polar]

To a first approximation, it is possible to resolve the dipole moment of a polyatomic molecule into contributions from various groups of atoms in the molecule and the directions in which these individual contributions lie (Fig. 10.4). Thus, *p*-dichlorobenzene is nonpolar by symmetry on account of the cancellation of two equal but opposing C—Cl moments (exactly as in carbon dioxide). *o*-Dichlorobenzene has a dipole moment which is approximately the resultant of two monochlorobenzene dipole moments arranged at 60° to each other. This technique of 'vector addition' can be applied with fair success to other series of related molecules, and the resultant μ_{res} of two equal dipole moments μ that make an angle θ to each other (**3**) is approximately

$$\mu_{res} \approx 2\mu \cos \tfrac{1}{2}\theta \qquad (6)$$

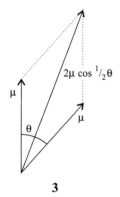

3

Exercise E10.5

Estimate the ratio of the electric dipole moments of *ortho* and *meta* disubstituted benzenes.

[*Answer*: $\mu(\text{ortho})/\mu(\text{meta}) = 1.7$]

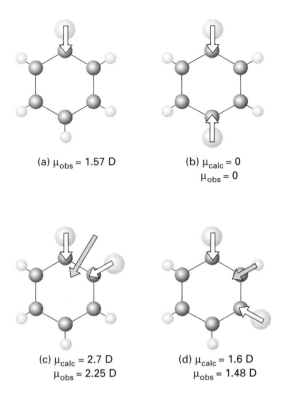

Fig. 10.4 The dipole moments of the dichlorobenzene isomers can be obtained approximately by vectorial addition of two chlorobenzene dipole moments (1.57 D).

(a) $\mu_{obs} = 1.57$ D

(b) $\mu_{calc} = 0$
$\mu_{obs} = 0$

(c) $\mu_{calc} = 2.7$ D
$\mu_{obs} = 2.25$ D

(d) $\mu_{calc} = 1.6$ D
$\mu_{obs} = 1.48$ D

Fig. 10.4 The dipole moments of the dichlorobenzene isomers can be obtained approximately by vectorial addition of two chlorobenzene dipole moments (1.57 D).

Interactions between dipoles

The potential energy of a point dipole μ_1 in the presence of a charge q_2 in the orientation shown in Fig. 10.5 is

$$V \propto \frac{\mu_1 q_2}{r^2} \tag{7}$$

This interaction energy decreases more rapidly with distance than that between two point charges (as r^2 rather than r) because, from the viewpoint of the single charge, the partial charges of the point dipole seem to merge and cancel as the distance r increases. Likewise, the interaction energy between two point dipoles μ_1 and μ_2 in the orientation shown in Fig. 10.6 is

$$V \propto \frac{\mu_1 \mu_2}{r^3} \tag{8}$$

This potential energy decreases even more rapidly because the charges of both dipoles seem to merge as the separation of the dipoles increases.

The potential energy of interaction between two polar molecules is a complicated function of the angle between them. However, when the two dipoles are parallel, as in Fig. 10.7, the potential energy is simply

$$V \propto \frac{\mu_1 \mu_2}{r^3} \times f \qquad \text{where } f = 1 - 3\cos^2\theta \tag{9}$$

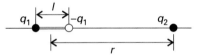

Fig. 10.5 The array of charges used to calculate the interaction between a dipole and a collinear point charge.

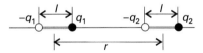

Fig. 10.6 The array of charges used to calculate the interaction between two collinear electric dipoles.

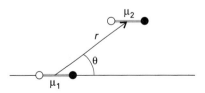

Fig. 10.7 The array of charges used to calculate the interaction between two parallel electric dipoles.

(a)

(b)

Fig. 10.8 A dipole–dipole interaction. When a pair of molecules can adopt all relative orientations with equal probability, the favourable orientations (a) and the unfavourable ones (b) cancel, and the average interaction is zero. In an actual fluid, the interactions in (a) slightly predominate.

The factor f takes into account how the like or opposite charges come closer to one another as the relative orientation of the dipoles is changed. The potential energy is negative (attractive) when $\theta < 54.7°$ because opposite charges are then closer than like charges. It is positive (repulsive) when $\theta > 54.7°$ because then like charges are then closer than unlike charges. The *average* potential energy of interaction between polar molecules that are freely rotating in a fluid is zero because the attractions and repulsions cancel. However, because the potential energy of a dipole near another dipole depends on their relative orientations, the molecules do not in fact rotate completely freely even in a gas. As a result, the lower energy orientations are marginally favoured, and so there is a nonzero interaction between polar molecules (Fig. 10.8). The detailed calculation of the average interaction energy is quite complicated, but the form of the final answer is very simple:

$$V = -\frac{C}{r^6} \qquad \text{where } C \propto \frac{(\mu_1 \mu_2)^2}{T} \qquad (10)$$

The important features of this expression are the dependence of the average interaction energy on the inverse *sixth* power of the separation and its inverse dependence on the temperature. The temperature-dependence reflects the way that the greater thermal motion overcomes the mutual orienting effects of the dipoles at higher temperatures. At 25 °C the average interaction energy for pairs of molecules with $\mu = 1\,\text{D}$ is about $-1.4\,\text{kJ mol}^{-1}$ when the separation is 0.3 nm. This energy should be compared with the average molar kinetic energy of $\frac{3}{2}RT = 3.7\,\text{kJ mol}^{-1}$ at the same temperature: the two are not very dissimilar, but they are both much less than the energies involved in the making and breaking of chemical bonds.

Induced dipole moments

So far, we have considered interactions between charged species (an ion interacting with another ion) and between species that have no net charge but which have nonzero dipole moments (two polar molecules interacting, or a polar molecule like water interacting with an ion). Now we move on to consider the interaction between molecules, at least one of which has neither a net charge nor a permanent dipole moment. An example might be a polar group on a polypeptide interacting with a Xe atom that is being used as an anaesthetic.

Although a molecule might not have a permanent electric dipole moment (methane, for example, does not), it may acquire a temporary **induced dipole moment**, μ^*, as a result of the influence of an electric field (perhaps a field generated by a nearby ion or polar molecule). The charge of the ion or the partial charge of the permanent dipole distorts the electron distribution of the polarizable molecule, and gives rise to an electric dipole in it. The magnitude of the induced dipole moment is proportional to the strength of the applied electric field, $\mathcal{E}$, and we write

$$\mu^* = \alpha \mathcal{E} \qquad (11)$$

where the constant α is the **polarizability** of the molecule. The larger the polarizability of the molecule, the greater the distortion that is caused by a given electric field. If the molecule has few electrons, they are tightly controlled by the nuclear charges and the polarizability of the molecule is low. If the molecule contains large atoms with electrons some distance from the nucleus, the nuclear control is less, the electron distribution is flabbier, and the polarizability is greater. The polarizability depends on the orientation of the molecule with respect to the field unless the molecule is tetrahedral (such as CCl_4), octahedral (such as SF_6), or icosahedral (C_{60}, buckminsterfullerene). Atoms, tetrahedral, octahedral, and icosahedral molecules have isotropic (orientation-independent) polarizabilities; all other molecules have anisotropic (orientation-dependent) polarizabilities.

Some polarizabilities are reported in Table 10.3 as 'polarizability volumes'. The polarizability has the same units as (dipole moment)/(electric field strength), or $(Cm)/(Vm^{-1})$, which rearranges to $CV^{-1}m^2$. Because that collection of units is awkward, α is usually converted to the **polarizability volume**, α', by using the relation

$$\alpha' = \frac{\alpha}{4\pi\varepsilon_0} \tag{12}$$

The polarizability volume has the dimensions of volume (hence its name) and is comparable in magnitude to the volume of a molecule.[†]

Exercise E10.6

What strength of electric field is required to induce an electric dipole moment of $1.0\,\mu D$ in a molecule of polarizability volume $1.1 \times 10^{-31}\,m^3$ (like CCl_4)?

[*Answer*: $2.7\,kV\,cm^{-1}$]

A polar molecule with dipole moment μ_1 can induce a dipole moment in a polarizable molecule (which may itself be either polar or nonpolar) because the partial charges of the polar molecule give rise to an electric field that distorts the second molecule. That induced dipole interacts with the permanent dipole of the first molecule, and the two are attracted together (Fig. 10.9). The formula for the interaction energy is

$$V = -\frac{C}{r^6} \qquad \text{where } C \propto \mu_1^2\alpha_2 \tag{13}$$

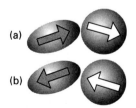

(a)

(b)

Fig. 10.9 A dipole–induced-dipole interaction. The induced dipole (light arrows) follows the changing orientation of the permanent dipole (dark arrow).

[†]When using older compilations of data, it is useful to note that polarizability volumes expressed in cubic centimetres (cm^3) have the same numerical values as the 'polarizabilities' reported using c.g.s. electrical units, and so the tabulated values previously called 'polarizabilities' can be used directly.

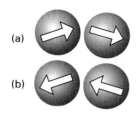

(a)

(b)

Fig. 10.10 In the dispersion interaction, an instantaneous dipole on one molecule induces a dipole on another molecule, and the two dipoles then interact to lower the energy. The directions of the two instantaneous dipoles are correlated, and, although they occur in different orientations at different instants, the interaction does not average to zero.

where α_2 is the polarizability of molecule 2. The negative sign shows that the interaction is attractive. For a molecule with $\mu = 1\,\mathrm{D}$ (such as HCl) near a molecule of polarizability volume $\alpha' = 1.0 \times 10^{-31}\,\mathrm{m}^3$ (such as benzene, Table 10.3) the average interaction energy is about $-0.8\,\mathrm{kJ\,mol}^{-1}$ when the separation is 0.3 nm.

Dispersion interactions

Finally, we consider the interactions between species that have neither a net charge nor a permanent electric dipole moment (such as two Xe atoms in a gas). We know that uncharged, nonpolar species can interact because they form condensed phases, such as benzene, liquid hydrogen, and liquid xenon.

The interaction between nonpolar species arises from the *transient* dipoles that they possess as a result of fluctuations in the instantaneous positions of their electrons (Fig. 10.10). Suppose, for instance, that the electrons in one molecule flicker into an arrangement that results in partial positive and negative charges and thus gives it an instantaneous dipole moment μ_1. While it exists, this dipole can polarize the other molecule and induce in it an instantaneous dipole moment μ_2. The two dipoles attract each other and the potential energy of the pair is lowered. Although the first molecule will go on to change the size and direction of its dipole (perhaps within as short a time as 10^{-16} s), the second will follow it; that is, the two dipoles are *correlated* in direction like two meshing gears, with a positive partial charge on one molecule appearing close to a negative partial charge on the other molecule, and vice versa. Because of this correlation of the relative positions of the partial charges, and their resulting attractive interaction, the attraction between the two instantaneous dipoles does not average to zero. Instead, it gives rise to a **dispersion interaction**. Polar molecules also interact by a dispersion interaction as well as by dipole–dipole interactions.

The strength of the dispersion interaction depends on the polarizability of the first molecule because the magnitude of the instantaneous dipole moment μ_1 depends on the looseness of the control that the nuclear charge has over the outer electrons. If the control is loose, the electron distribution can undergo relatively large fluctuations; moreover, if the control is loose, then it can also respond strongly to applied electric fields and hence have a high polarizability. It follows that a high polarizability is a sign of large fluctuations in local charge density. The strength also depends on the polarizability of the second molecule, for that polarizability determines how readily a dipole can be induced in molecule 2 by molecule 1. We therefore expect $V \propto \alpha_1 \alpha_2$. The actual calculation of the dispersion interaction is quite involved, but a reasonable approximation to the interaction energy is given by the **London formula**:

$$V = -\frac{C}{r^6} \qquad \text{where } C \propto \alpha_1 \alpha_2 \left(\frac{I_1 I_2}{I_1 + I_2}\right) \tag{14}$$

where I_1 and I_2 are the ionization energies of the two molecules (Table 2.3). Once again, the interaction turns out to be proportional to the inverse sixth power of the separation. For two CH_4 molecules, $V \approx -5\,kJ\,mol^{-1}$ when $r = 0.3\,nm$.

Hydrogen bonding

The strongest intermolecular interaction arises from the formation of a **hydrogen bond**, in which a hydrogen atom lies between two strongly electronegative atoms and binds them together. The bond is denoted $X{-}H \cdots Y$, with X and Y being nitrogen, oxygen, or fluorine. Unlike the other interactions we have considered, hydrogen bonding is not universal, but is restricted to molecules that contain these atoms.

The most elementary description of the formation of a hydrogen bond is that it is the result of a Coulombic interaction between the partly exposed positive charge of a proton bound to an electron-withdrawing X atom (in the fragment X—H) and the negative charge of a lone pair on the second atom Y:

$$^{\delta-}X{-}H^{\delta+} \cdots : Y^{\delta-}$$

A slightly more sophisticated version of this description is to regard hydrogen bond formation as the formation of a Lewis acid–base complex in which the partly exposed proton of the X—H group is the Lewis acid and :Y, with its lone pair, is the Lewis base:

$$X{-}H + :Y \rightarrow X{-}H \cdots Y$$

Molecular orbital theory provides an alternative description that is more in line with the concept of delocalized bonding and the ability of an electron pair to bind more than one pair of atoms (Section 9.7). Thus, if the X—H bond is regarded as formed from the overlap of an orbital on X (which we denote $X\sigma$) and a hydrogen $1s$ orbital (H1s), and the lone pair on Y occupies an orbital on Y (the orbital $Y\sigma$), then when the two molecules are close together, we can build three molecular orbitals from the three basis orbitals:

$$\psi = c_1\psi_{X\sigma} + c_2\psi_{H1s} + c_3\psi_{Y\sigma}$$

One of the molecular orbitals is bonding, one almost nonbonding, and the third antibonding (Fig. 10.11). These three orbitals need to accommodate four electrons (two from the original X—H bond and two from the lone pair of Y), and they may do so if two enter the bonding orbital and two enter the nonbonding orbital. Because the antibonding orbital remains empty, the net effect is a lowering of energy.

Hydrogen bond formation, which has a typical strength of the order of $20\,kJ\,mol^{-1}$, dominates all other van der Waals interactions when it can occur. It accounts for the rigidity of molecular solids such as sucrose and ice, the secondary structure of proteins (the formation of helices and sheets of polypeptide chains), the low vapour pressure of liquids such as water, and their high viscosities and surface tensions. Hydrogen bonding also

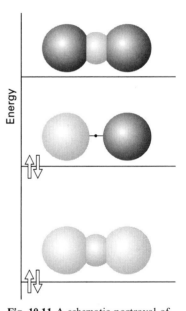

Fig. 10.11 A schematic portrayal of the molecular orbitals that can be formed from an X, H, and Y orbital and which gives rise to an $XH \cdots Y$ hydrogen bond. The lowest energy combination is fully bonding, the next nonbonding, and the uppermost is antibonding. The antibonding orbital is not occupied by the electrons provided by the XH bond and the :Y lone pair, so the configuration shown may result in a net lowering of energy in certain cases (namely when the X and Y atoms are N, O, or F).

377

Table 10.4 Interaction potential energies

Interaction type	Distance dependence of potential energy	Typical energy/ kJ mol^{-1}	Comment
Ion–ion	$1/r$	250	Only between ions
Ion-dipole	$1/r^2$	15	
Dipole-dipole	$1/r^3$	2	Between stationary polar molecules
	$1/r^6$	0.3	Between rotating polar molecules
London (dispersion)	$1/r^6$	2	Between all types of molecules

The energy of a hydrogen bond X—H $\cdots$ Y is typically 20 kJ mol^{-1} and occurs on contact for X, Y = N, O, or F.

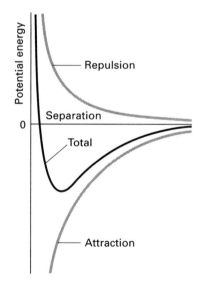

Fig. 10.12 The general form of an intermolecular potential energy curve (the graph of the potential energy of two closed shell species as the distance between them is changed). The attractive (negative) contribution has a long range, but the repulsive (positive) interaction increases more sharply once the molecules come into contact. The overall potential energy is shown by the heavy line.

contributes to the solubility in water of species such as ammonia and compounds containing hydroxyl groups and to the hydration of anions. In this last case, even ions such as Cl$^-$ and HS$^-$ can participate in hydrogen bond formation with water, for their charge enables them to interact with the hydroxylic protons of H$_2$O.

The strengths and distance dependences of the attractive forces that we have considered so far are summarized in Table 10.4.

10.3 The total interaction

The total attractive interaction energy between rotating molecules that cannot participate in hydrogen bonding is the sum of the contributions from the dipole–dipole, dipole–induced dipole, and dispersion interactions. Only the dispersion interaction contributes if both molecules are nonpolar. All three interactions vary as the inverse sixth power of the separation, and so we may write

$$V = -\frac{C}{r^6} \qquad (15)$$

where C is a coefficient that depends on the identity of the molecules.

When molecules are squeezed together (for instance, during the impact of a collision, under the force exerted by a weight pressing on a substance, or simply as a result of the attractive forces drawing the molecules together), repulsive terms become important and begin to dominate the attractive forces (Fig. 10.12). These repulsive interactions arise in large measure from the Pauli exclusion principle, which forbids pairs of electrons being in the same region of space. The repulsions increase steeply with decreasing separation in a way that can be deduced only by very extensive, complicated molecular structure calculations. In many cases, however, progress can be made by using a greatly simplified representation of the potential energy, where the details are ignored and the general features expressed by a few adjustable parameters.

One such approximation is the **hard-sphere potential**, in which it is assumed that the potential energy rises abruptly to infinity as soon as the particles come within some separation σ (Fig. 10.13):

$$V = \begin{cases} \infty & \text{for } r \leq \sigma \\ 0 & \text{for } r > \sigma \end{cases} \qquad (16)$$

This very simple potential is surprisingly useful for assessing a number of properties. Another widely used approximation is to express the short-range repulsive potential energy as inversely proportional to a high power of r:

$$V = +\frac{C^*}{r^n} \qquad (17)$$

where C^* is another constant (the star signifies repulsion). Typically, n is set equal to 12, in which case the repulsion dominates the $1/r^6$ attractions strongly at short separations because then $C^*/r^{12} \gg C/r^6$. The sum of the repulsive interaction with $n = 12$ and the attractive interaction given by eqn (14) is called the **Lennard-Jones (12,6) potential**. It is normally written in the form

$$V = 4\varepsilon\left\{ \left(\frac{\sigma}{r}\right)^{12} - \left(\frac{\sigma}{r}\right)^{6} \right\} \qquad (18)$$

and is drawn in Fig. 10.14. The two parameters are now ε, the depth of the well, and σ, the separation at which $V = 0$. Some typical values are listed in Table 10.5. Although the (12,6)-potential has been used in many calculations, there is plenty of evidence to show that $1/r^{12}$ is a very poor representation of the repulsive potential, and that an exponential form $e^{-r/\sigma}$ is superior. An exponential function is more faithful to the exponential decay of atomic wavefunctions at large distances, and hence to the distance-dependence of the overlap that is responsible for repulsion.

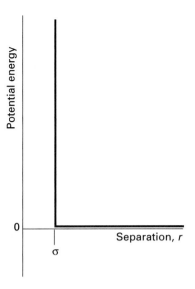

Fig. 10.13 The true intermolecular potential can be modelled in a variety of ways. One of the simplest is this hard-sphere potential, in which there is no potential energy of interaction until the two molecules are separated by a distance σ, when the potential energy rises abruptly to infinity as the impenetrable hard spheres repel each other.

Table 10.5 Lennard-Jones parameters for the (12,6) potential

	$\varepsilon/\text{kJ mol}^{-1}$	σ/pm
Ar	128	342
Br_2	536	427
C_6H_6	454	527
Cl_2	368	412
H_2	34	297
He	11	258
Xe	236	406

Exercise E10.7

At what separation does the minimum of the potential energy curve occur for a Lennard-Jones potential?

[*Answer*: $r = 2^{1/6}\sigma$]

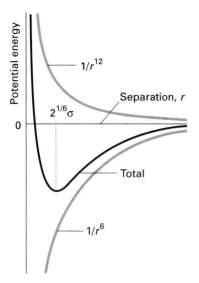

Fig. 10.14 The Lennard-Jones potential is another approximation to the true intermolecular potential energy curves. It models the attractive component by a contribution that is proportional to $1/r^6$, and the repulsive component by a contribution that is proportional to $1/r^{12}$. Specifically, these choices result in the Lennard-Jones (12,6)-potential. Although there are good theoretical reasons for the former, there is plenty of evidence to show that $1/r^{12}$ is only a very poor approximation to the repulsive part of the curve.

Fluids

Although intermolecular forces might just seem like the left-overs after atoms have gone about their main business of forming chemical bonds, they are of crucial importance for the structure and functioning of the world. At the lowest level of importance, they are responsible for the imperfections of real gases. However, imperfections are a sign that molecules can cohere, and when the temperature is low enough for the kinetic energy of the molecules to have been reduced sufficiently, a gas will condense to a liquid and then freeze to a solid.

10.4 Real gases

As we saw in Section 1.4, the equation of state of a perfect gas is

$$\frac{pV_\mathrm{m}}{RT} = 1$$

(where $V_\mathrm{m} = V/n$). For real gases, when intermolecular forces are not negligible, this equation of state is modified to

$$\frac{pV_\mathrm{m}}{RT} = 1 + \frac{B}{V_\mathrm{m}} + \frac{C}{V_\mathrm{m}^2} + \cdots \tag{19}$$

where B is the second virial coefficient and C is the third virial coefficient; both coefficients depend on the temperature. Because the virial coefficients are consequences of the intermolecular forces, by measuring them we can determine the strengths of the forces.

The second virial coefficient can be calculated quite easily when the intermolecular potential energy depends only on the separation r of the particles, as in the dispersion interaction between noble-gas atoms and non-polar molecules. For example, if the intermolecular potential is modelled by the hard-sphere potential specified in eqn (16), then it turns out that

$$B = \tfrac{2}{3}\pi N_\mathrm{A}\sigma^3 \tag{20}$$

and the larger the molecules (the larger the value of σ), the greater the second virial coefficient. The evaluation of B for a more realistic intermolecular potential, such as the Lennard-Jones potential, is more difficult, but nevertheless can be carried out numerically on a computer for a choice of the parameters that appear in eqn (18). These parameters are varied until the calculated value of B agrees with the experimental value. The values in Table 10.5 were calculated in this way.

Exercise 10.8

The second virial coefficient of helium is $12.0\,\mathrm{cm^3\,mol^{-1}}$ at $0\,^\circ\mathrm{C}$; estimate the radius of the atom assuming that the atom can be modelled as a hard sphere.

[*Answer*: 212 pm]

10.5 Liquids

The starting point for the discussion of gases is the totally chaotic distribution of the molecules of a perfect gas. The starting point for the discussion of solids (as we shall see) is the well ordered structure of perfect crystals. The liquid state is between these two extremes: there is some structure and some disorder. The particles of a liquid are held together by intermolecular forces, but their kinetic energies are comparable to their potential energies. As a result, although the molecules are not free to escape completely from the bulk, the whole structure is very mobile. The flow of molecules is like a crowd of spectators leaving a stadium.

The best description of the average locations of the particles in the liquid is in terms of the **pair distribution function**, g. This function is defined so that $g\delta r$ is the probability that a molecule will be found in the range δr at a distance r from another.[†] It follows that if g passes through a maximum at a radius of, for instance, 0.5 nm, then the most probable distance (regardless of direction) at which a second molecule will be found will be at 0.5 nm from the first molecule.

In a crystal, g is a periodic array of sharp spikes, representing the certainty (in the absence of defects and thermal motion) that particles lie at definite locations. This regularity continues out to large distances (to the edge of the crystal, billions of molecules away), and so we say that crystals have **long-range order**. When the crystal melts, the long-range order is lost and wherever we look at long distances from a given particle there is equal probability of finding a second particle. Close to the first particle, though, there may be a remnant of order (Fig. 10.15). Its nearest neighbours might still adopt approximately their original positions, and even if they are displaced by newcomers the new particles might adopt their vacated positions. It may still be possible to detect, on average, a sphere of nearest neighbours at a distance r_1, and perhaps beyond them a sphere of next-nearest neighbours at r_2. The existence of this **short-range order** means that g can be expected to have a broad but pronounced peak at r_1, a smaller and broader peak at r_2, and perhaps some more structure beyond that.

The shape of the pair distribution function can be determined by X-ray diffraction, for g can be extracted from the diffuse diffraction pattern

[†]Recall the analogous quantity used to describe the distance of an electron from an atom, Section 8.8.

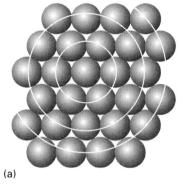

(a)

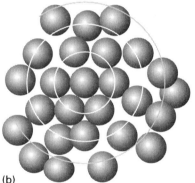

(b)

Fig. 10.15 (a) In a perfect crystal at $T = 0$, the distribution of molecules (or ions) is highly regular, and the pair distribution function shows a series of sharp peaks showing the regular organization of rings of neighbours around any selected central molecule or ion. (b) In a liquid, there remain some elements of structure close to each molecule, but the further the distance, the less the correlation. The pair distribution function now shows a pronounced (but broadened) peak corresponding to the nearest neighbours of the molecule of interest (which are only slightly more disordered than in the solid), and a suggestion of a peak for the next ring of molecules, but little structure at greater distances.

Cohesion and structure

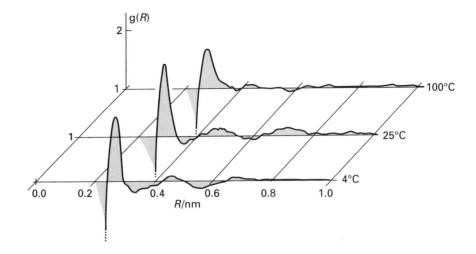

Fig. 10.16 The experimentally determined radial distribution function of the oxygen atoms in liquid water at three temperatures. Note the expansion as the temperature is raised.

characteristic of liquid samples in much the same way as a crystal structure is obtained from X-ray diffraction of crystals (see Section 10.11). The shells of local structure shown in the example in Fig. 10.16 (for water) are unmistakable. Closer analysis shows that any given H_2O molecule is surrounded by other molecules at the corners of a tetrahedron, similar to the arrangement in ice (Fig. 10.17). The form of g at 100 °C shows that the intermolecular forces (in this case, largely hydrogen bonds) are strong enough to affect the local structure right up to the boiling point.

10.6 Molecular motion in liquids

In a gas, a molecule flies through empty space, and may traverse dozens of molecular diameters before colliding with another molecule. In a liquid, a molecule is surrounded by neighbours, and it can move only a fraction of a diameter, perhaps because its neighbours move aside momentarily, before colliding. Molecular motion in liquids is a series of short steps, with incessantly changing directions, like people in an aimless, milling crowd.

The process of migration by means of a random jostling motion though a liquid is called **diffusion**. We think of the motion of the molecule

Fig. 10.17 A fragment of the crystal structure of ice. Each O atom is at the centre of a tetrahedron of four O atoms at a distance of 276 pm. The central O atom is attached by two short OH bonds to two H atoms and by two relatively long O⋯H bonds to two neighbouring H_2O molecules. Overall, the structure consists of planes of hexagonal puckered rings of H_2O molecules (like the chair form of cyclohexane).

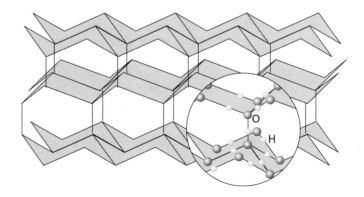

as occurring in a series of short jumps in random directions, a so-called **random walk** (Fig. 10.18). If there is an initial concentration gradient in the liquid (for instance, a solution may have a high concentration of solute in one region), then the rate at which the molecules spread out and make the solution homogeneous is proportional to the concentration gradient, and we write

$$\text{rate of diffusion} = D \times \text{concentration gradient}$$

The coefficient D is called the **diffusion coefficient**: if it is large, then molecules diffuse rapidly. Some values are given in Table 10.6. Diffusion coefficients are of the greatest importance for discussing the spread of pollutants in lakes and through the atmosphere. In both cases, the spread of pollutant may be assisted by bulk motion of the fluid as a whole (as when a wind blows in the atmosphere). This motion is called **convection**. Because diffusion is often a slow process, we speed up the spread of solute molecules by inducing convection by stirring a fluid or turning on an extractor fan.

We can use the diffusion coefficient to describe the probable location of a given molecule. In this case, we can even consider the diffusion of a solvent molecule through a pure solvent, when D is called the **self-diffusion coefficient**. Although a molecule undergoing a random walk may take many steps in a given time, it has only a small probability of being found far from its starting point because some of the steps lead it away from the starting point but others lead it back (Fig. 10.19). The net distance travelled in a time t from the starting point is measured by the **root mean square distance**, d, with

$$d = \sqrt{2Dt} \qquad (21)$$

Thus, the net distance increases only as the square root of the time, so for a particle to be found twice as far (on average) from its starting point, we must wait twice as long. Likewise, the average distance is proportional to the square root of the diffusion coefficient.

Exercise E10.9

The self-diffusion coefficient of H_2O in water is $2.26 \times 10^{-9}\ \mathrm{m^2\,s^{-1}}$ at $25\,^{\circ}\mathrm{C}$. How long does it take for an H_2O molecule to travel (a) 1.0 cm, (b) 2.0 cm from its starting point in a sample of unstirred water?

[*Answer*: (a) 6.1 h, (b) 25 h]

If we think of a molecule as moving through a fluid in a random walk, then we can expect there to be a relation between the diffusion coefficient and the

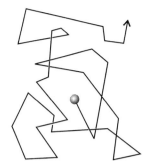

Fig. 10.18 One possible path of a random walk in three dimensions. In this general case, the step length is also a random variable.

Table 10.6 Diffusion coefficients at $25\,^{\circ}\mathrm{C}$, $D/10^{-9}\ \mathrm{m^2\,s^{-1}}$

H_2O in water	2.26
Ar in tetrachloromethane	3.63
CH_3OH in water	1.58
$C_{12}H_{22}O_{11}$ (sucrose) in water	0.522
NH_2CH_2COOH in water	0.673
O_2 in tetrachloromethane	3.82

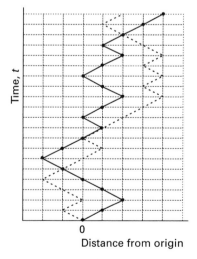

Fig. 10.19 The directions of a typical sequence of steps in a one-dimensional random walk. Note how the walker might still be close to the origin after many steps. Two walks are shown here. The equations in the text refer to an average outcome of many such walks.

rate at which the molecule takes its steps and the distance of each step. The actual relation is called the **Einstein–Smoluchowski equation**, and is

$$D = \frac{\lambda^2}{2\tau} \tag{22}$$

where λ is the length of each step and τ is the time each one takes. This equation tells us that a molecule that takes long steps in a short time has a high diffusion coefficient, which is what we should expect. However, we see that D is proportional to the *square* of the step length, which is more difficult to anticipate by a physical argument.

Exercise E10.10

Suppose an H_2O molecule moves through one molecular diameter (about 200 pm) each time it takes a step in a random walk. What is the time for each step at 25 °C?

[*Answer*: 9 ps]

The diffusion coefficient increases with temperature because an increase in temperature enables a molecule to escape more easily from the attractive forces exerted by its neighbours. The speed of the random walk is increased, which corresponds to a decrease in the time τ in the Einstein–Smoluchowski equation. If we suppose that the rate $(1/\tau)$ of the random walk follows an Arrhenius temperature dependence with an activation energy E_a, then the diffusion coefficient can be expected to follow the relation

$$D \propto e^{-E_a/RT} \tag{23}$$

The rate at which particles diffuse through a liquid is related to the viscosity, and we should expect a high diffusion coefficient to be found for fluids that have a low viscosity. That is, we can suspect that $\eta \propto 1/D$, where η is the coefficient of viscosity, and hence that

$$\eta \propto e^{E_a/RT} \tag{24}$$

(Note the change in sign of the exponent: viscosity *decreases* as the temperature is raised.) This temperature dependence is observed, at least over reasonably small temperature ranges (Fig. 10.20). One problem with the analysis of viscosity is that the density of the liquid changes as it is heated, which makes a pronounced contribution to the diffusion coefficient and to the viscosity. The temperature dependence of viscosity at constant volume, when the density is constant, is much less than that at constant pressure. The intermolecular potentials govern the magnitude of E_a, but the problem of calculating it is immensely difficult and still largely unsolved.

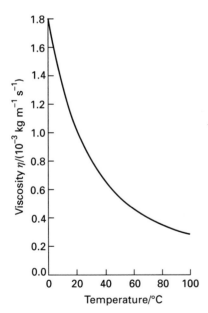

Fig. 10.20 The experimental temperature dependence of the viscosity of water. As the temperature is increased, more molecules are able to escape from the potential wells provided by their neighbours, and so the liquid becomes more fluid.

384

Exercise E10.11

Estimate the activation energy for the viscosity of water from the graph in Fig. 10.20, by using the viscosities at 40 °C and 80 °C. *Hint*: Use an equation like eqn (24) to formulate an expression for the logarithm of the ratio of the two viscosities.

$$[4\,kJ\,mol^{-1}]$$

Fig. 10.21 The arrangement of molecules in the smectic phase of a liquid crystal.

10.7 Liquid crystals

A **mesophase** is a bulk phase that is intermediate in character between a solid and a liquid. The most important mesophase is a **liquid crystal**, which is a substance having liquid-like imperfect long-range order in some directions but some crystal-like aspects of short-range order in other directions. Liquid crystals can be used as models of biological cell walls and studied to gain insight into the process of transport through membranes. They are also of considerable technological importance for their use in liquid-crystal displays on electronic equipment.

There are three important types of liquid crystal; they differ in the type of long-range order that they retain. One type of retained long-range order gives rise to a **smectic phase** (from the Greek word for soapy), in which the molecules align themselves in layers (Fig. 10.21). Other materials, and some smectic liquid crystals at higher temperatures, lack the layered structure but retain a parallel alignment (Fig. 10.22): this mesophase is the **nematic phase** (from the Greek for thread). The strongly anisotropic optical properties of nematic liquid crystals, and their response to electric fields, is the basis of their use as data displays in calculators and watches. In the **cholesteric phase**, which is so-called because some derivatives of cholesterol form them, the molecules lie in sheets at angles that change slightly between neighbouring sheets (Fig. 10.23). As a result, they form helical structures. The pitch of the helix varies with temperature, and as a result, cholesteric liquid crystals diffract light and appear to have colours that depend on the temperature. They are used for detecting temperature distributions in living material, including human patients, and have even been incorporated into fabrics.

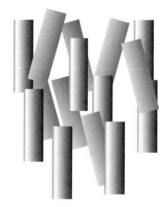

Fig. 10.22 The arrangement of molecules in the nematic phase of a liquid crystal.

10.8 Disperse systems

A **disperse system** is a dispersion of small particles of one material in another. The small particles are commonly called **colloids**. In this context, 'small' means something less than about 500 nm in diameter (about the wavelength of light). In general, they are aggregates of numerous atoms or molecules, but are too small to be seen with an ordinary optical microscope. They pass through most filter papers, but can be detected by light-scattering, sedimentation, and osmosis.

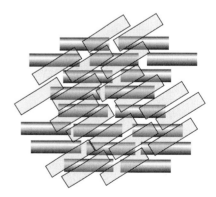

Fig. 10.23 The arrangement of molecules in the cholesteric phase of a liquid crystal. Two layers are shown; the relative orientation of these layers is repeated in successive layers, to give a helical array of molecules.

Classification

The name given to the system depends on the two physical states of the substances involved. A **sol** is a dispersion of a solid in a liquid (such as clusters of gold atoms in water) or of a solid in a solid (such as ruby glass, which is a gold-in-glass sol, and achieves its colour by scattering). An **aerosol** is a dispersion of a liquid in a gas (like fog and many sprays) and of a solid in a gas (such as smoke): the particles are often large enough to be seen with a microscope. An **emulsion** is a dispersion of a liquid in a liquid (such as milk and some paints).

A further classification of colloids is as **lyophilic** (solvent attracting) and **lyophobic** (solvent repelling); in the case of water as solvent, the terms **hydrophilic** and **hydrophobic** are used instead. Lyophobic colloids include the metal sols. Lyophilic colloids generally have some chemical similarity to the solvent, such as OH groups able to form hydrogen bonds. A **gel** is a semi-rigid mass of a lyophilic sol in which all the dispersion medium has been absorbed by the sol particles.

The preparation of aerosols can be as simple as sneezing (which produces an aerosol). Laboratory and commercial methods make use of several techniques. Material (e.g. quartz) may be ground in the presence of the dispersion medium. Passing a heavy electric current through a cell may lead to the crumbling of an electrode into colloidal particles; arcing between electrodes immersed in the support medium also produces a colloid. Chemical precipitation sometimes results in a colloid. A precipitate (for example, silver iodide) already formed may be converted to a colloid by the addition of **peptizing agent**, a substance that disperses a colloid. An example of a peptizing agent is potassium iodide, which provides ions that adhere to the colloidal particles and cause them to repel one another. Clays may be peptized by alkalis, the OH^- ion being the active agent.

Emulsions are normally prepared by shaking the two components together, although some kind of **emulsifying agent** has to be used in order to stabilize the product. This emulsifier may be a soap (the salt of a long chain fatty acid), a surfactant, or a lyophilic sol that forms a protective film around the dispersed phase. In milk, which is an emulsion of fats in water, the emulsifying agent is casein, a protein containing phosphate groups. That casein is not completely successful in stabilizing milk is apparent from the formation of cream: the dispersed fats coalesce into oily droplets which float to the surface. This separation may be prevented by ensuring that the emulsion is dispersed very finely initially: violent agitation with ultrasonics or extrusion through a very fine mesh brings this about, the product being 'homogenized' milk.

Aerosols are formed when a spray of liquid is torn apart by a jet of gas. The dispersal is aided if a charge is applied to the liquid, for then the electrostatic repulsions blast the jet apart into droplets. This procedure may also be used to produce emulsions, for the charged liquid phase may be squirted into another liquid.

Disperse systems are often purified by **dialysis**, a form of osmosis. The aim is to remove much (but not all, for reasons explained later) of the ionic material that may have accompanied their formation. A membrane (e.g. cellulose) is selected which is permeable to solvent and ions, but not to the bigger colloid particles. Dialysis is very slow, and is normally accelerated by applying an electric field and making use of the charge carried by many colloids; the technique is then called **electrodialysis**.

Surface, structure, and stability

The principal feature of colloids is the very great surface area of the dispersed phase in comparison with the same amount of ordinary material. For example, a cube of side 1 cm has a surface area of $6\,cm^2$, but when it is dispersed as 10^{18} little 10 nm cubes the total surface area is $6 \times 10^6\,cm^2$ (about the size of a tennis court). This dramatic increase in area means that surface effects are of dominating importance in the chemistry of disperse systems.

As a result of their great surface area, colloids are thermodynamically unstable with respect to the bulk: that is, colloids have a thermodynamic tendency to reduce their surface area (like a liquid). Their apparent stability must therefore be a consequence of the kinetics of collapse: disperse systems are kinetically nonlabile, not thermodynamically stable. At first sight, though, even the kinetic argument seems to fail: colloidal particles attract one another over large distances by the dispersion interaction, and so there is a long-range force tending to collapse them down into a single blob.

Several factors oppose the long-range dispersion attraction. There may be a protective film at the surface of the colloid particles that stabilizes the interface and cannot be penetrated when two particles touch. For example, the surface atoms of a platinum sol in water react chemically and are turned into $-Pt(OH)_3H_3$, and this layer encases the particle like a shell. A fat can be emulsified by a soap because the long hydrocarbon tails penetrate the oil droplet but the $-CO_2^-$ head groups (or other hydrophilic groups in detergents) surround the surface, form hydrogen bonds with water, and give rise to a shell of negative charge that repels a possible approach from another similarly charged particle.

Micelle formation and the hydrophobic interaction

By a **surfactant** we mean a species that accumulates at the interface of two fluids (one of which may be air) and modifies the properties of the surface. A typical surfactant molecule consists of a long hydrocarbon tail that dissolves in hydrocarbon and other nonpolar materials, and a hydrophilic **head group**, such as a carboxylate group, $-CO_2^-$, that dissolves in a polar solvent (typically water). Soaps, for example, consist of the alkali metal salts of long-chain carboxylic acids, and the surfactant in detergents is typically a long chain benzenesulfonic acid $(R-C_6H_4SO_3H)$. The mode of action of a surfactant in a detergent, and of soap, is to dissolve in both the aqueous phase and the hydrocarbon phase where their surfaces are in contact, and hence to solubilize the hydrocarbon phase so that it can be washed away (Fig. 10.24).

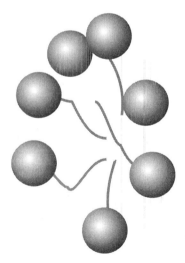

Fig. 10.24 A surfactant molecule in a detergent or soap acts by sinking its hydrophobic hydrocarbon tail into the grease, so leaving its hydrophilic head groups on the surface of the grease where they can interact attractively with the surrounding water.

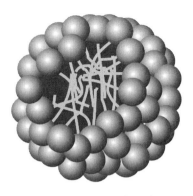

Fig. 10.25 A representation of a spherical micelle. The hydrophilic groups are represented by spheres, and the hydrophobic hydrocarbon chains are represented by the stalks: the latter are mobile.

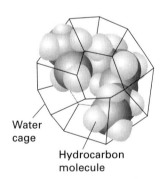

Water cage

Hydrocarbon molecule

Fig. 10.26 When a hydrocarbon molecule is surrounded by water, the water molecules form a clathrate cage. As a result of this acquisition of structure, the entropy of the water decreases, and so the dispersal of the hydrocarbon into water is entropy-opposed; the coalescence of the hydrocarbon into a single large blob is entropy-favoured.

Surfactant molecules can group together as micelles, colloid-sized clusters of molecules, even in the absence of grease droplets, for their hydrophobic tails tend to congregate, and their hydrophilic heads provide protection (Fig. 10.25). Micelles form only above the **critical micelle concentration** (CMC) and above the **Krafft temperature**. Nonionic surfactant molecules may cluster together in swarms of 1000 or more, but ionic species tend to be disrupted by the Coulomb repulsions between head groups and are normally limited to groups of between 10 and 100 molecules. The shapes of the individual micelles vary with concentration. Although spherical micelles do occur, they are more commonly flattened spheres close to the CMC, and rodlike at higher concentrations. The interior of a micelle is like a droplet of oil, and magnetic resonance shows that the hydrocarbon tails are mobile, but slightly more restricted than in the bulk.

Micelles are important in industry and biology on account of their solubilizing function: matter can be transported by water after it has been dissolved in their hydrocarbon interiors. For this reason, micellar systems are used as detergents and drug carriers, and for organic synthesis, froth flotation, and petroleum recovery.

The thermodynamics of micelle formation shows that the enthalpy of formation in aqueous systems is probably positive (that is, that they are endothermic) with $\Delta H \approx 1$–$2\,kJ$ per mole of surfactant. That they do form above the CMC indicates that the entropy change accompanying their formation must then be positive (in order for the Gibbs energy accompanying the formation process to be negative), and measurements suggest a value of about $+140\,J\,K^{-1}\,mol^{-1}$ at room temperature. That the entropy change is positive even though the molecules are clustering together shows that there must be a contribution to the entropy from the solvent and that its molecules must be more free to move once the solute molecules have herded into small clusters. This interpretation is plausible, because each individual solute molecule is held in an organized solvent cage (Fig. 10.26), but once the micelle has formed the solvent molecules need form only a single (admittedly larger) cage. The increase in energy when hydrophobic groups cluster together and reduce their structural demands on the solvent is the origin of the **hydrophobic interaction** that tends to stabilize groupings of hydrophobic groups in biological macromolecules. The hydrophobic interaction is an example of an ordering process that is stabilized by a tendency toward greater disorder of the solvent.

The electric double layer

Apart from the physical stabilization of disperse systems, a major source of kinetic stability is the existence of an electric charge on the surfaces of the colloidal particles. On account of this charge, ions of opposite charge tend to cluster nearby.

Two regions of charge must be distinguished. First, there is a fairly immobile layer of ions that stick tightly to the surface of the colloidal particle, and which may include water molecules (if that is the support medium). The radius of the sphere that captures this rigid layer is called the **radius of** shear, and is the major factor determining the mobility of the particles (Fig. 10.27). The electric potential at the radius of shear relative to its value in the distant, bulk medium is called the **zeta potential**, ζ, or the 'electrokinetic potential'. The charged unit attracts an oppositely charged ionic atmosphere. The inner shell of charge and the outer atmosphere jointly constitute the **electric double layer**.

At high concentrations of ions of high charge number, the atmosphere is dense and the potential falls to its bulk value within a short distance. In this case there is little electrostatic repulsion to hinder the close approach of two colloid particles. As a result, **flocculation**, the coalescence of the colloidal particles, occurs as a consequence of the van der Waals forces. Flocculation is often reversible, and should be distinguished from **coagulation**, which is the irreversible collapse of the colloid into a bulk phase. When river water containing colloidal clay flows into the sea, the brine induces coagulation and is a major cause of silting in estuaries.

Metal oxide sols tend to be positively charged whereas sulfur and the noble metals tend to be negatively charged. Naturally occurring macromolecules also acquire a charge when dispersed in water, and an important feature of proteins and other natural macromolecules is that their overall charge depends on the pH of the medium. For instance, in acid environments protons attach to basic groups, and the net charge of the macromolecule is positive; in basic media the net charge is negative as a result of proton loss. At the **isoelectric point**, the pH is such that there is no net charge on the macromolecule.

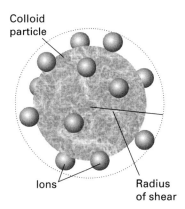

Fig. 10.27 The definition of the radius of shear for a colloidal particle. The spheres are ions attached to the surface of the particle.

Example *Determining the isoelectric point*
The speed with which bovine serum albumin (BSA) moves through water under the influence of an electric field was monitored at several values of pH, and the data are listed below. What is the isoelectric point of the protein?

pH	4.20	4.56	5.20	5.65	6.30	7.00
Velocity/(μm s^{-1})	0.50	0.18	−0.25	−0.65	−0.90	−1.25

Strategy
The macromolecule is not influenced by the electric field when it is uncharged. Therefore, the isoelectric point is the pH at which it does not migrate in an electric field. We should therefore plot speed against pH and find by interpolation the pH at which the speed is zero.

Solution
The data are plotted in Fig. 10.28. The velocity passes through zero at pH = 4.8; hence pH = 4.8 is the isoelectric point.

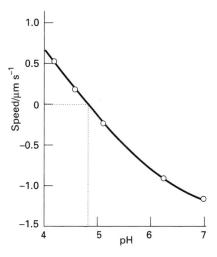

Fig. 10.28 The plot of the speed of a moving macromolecule against pH allows the isoelectric point to be detected as the pH at which the speed is zero.

Exercise E10.12

The following data were obtained for another protein:

pH	4.5	5.0	5.5	6.0
Velocity/(μm s^{-1})	−0.10	−0.20	−0.30	−0.35

Estimate the pH of the isoelectric point.

[*Answer*: 4.3]

The primary role of the electric double layer is to render the colloid kinetically nonlabile. Colliding colloidal particles break through the double layer and coalesce only if the collision is sufficiently energetic to disrupt the layers of ions and solvating molecules, or if thermal motion has stirred away the surface accumulation of charge. This kind of disruption of the double layer may occur at high temperatures, which is one reason why sols precipitate when they are heated. The protective role of the double layer is the reason why it is important not to remove all the ions when a colloid is being purified by dialysis, and why proteins coagulate most readily at their isoelectric point.

The presence of charge on colloidal particles and natural macromolecules also permits us to control their motion, as in dialysis and electrophoresis. Apart from its application to the determination of molar mass, electrophoresis has several analytical and technological applications. One analytical application is to the separation of different macromolecules, and a typical apparatus is illustrated in Fig. 10.29. Technical applications include silent ink-jet printers, the painting of objects by airborne charged paint droplets, and electrophoretic rubber forming by deposition of charged rubber molecules on anodes formed into the shape of the desired product (for example, surgical gloves).

Crystal structure

Now we turn to the structures of solids. We shall concentrate on **crystalline solids**, which consist of regular arrays of atoms, molecules, or ions. The

Fig. 10.29 The layout of a simple electrophoresis apparatus. The sample is introduced into the trough in the gel, and the different components form separated bands under the influence of the potential difference.

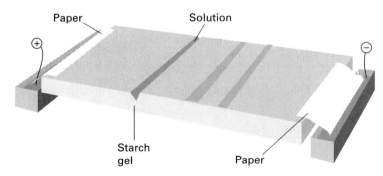

structures of crystals are of considerable practical importance, for they have implications for geology, materials, technologically advanced materials such as semiconductors and high-temperature superconductors, and in biology. The first, and often very demanding, step in an X-ray structural analysis of biological macromolecules is to form crystals in which the large molecules lie in orderly ranks. On the other hand, the crystallization of a virus particle would take it out of circulation, and one of its strategies for avoiding this kind of entombment makes unconscious use of the geometry of crystal packing.

10.9 Unit cells

Early in the history of modern science it was suggested that the regular external form of crystals implied an internal regularity. The pattern that atoms, ions, or molecules adopt in a crystal is expressed in terms of an array of points called the **lattice** that identify the locations of the individual species (Fig. 10.30). A **unit cell** of a crystal is the small three-dimensional figure obtained by joining typically eight of these points, and which may be used to construct the entire crystal lattice by purely translational displacements, much as a wall may be constructed from bricks (Fig. 10.31). An infinite number of different unit cells can describe the same structure, but it is conventional to choose the cell with sides that have the shortest lengths and which are most nearly perpendicular to one another.

Unit cells are classified into one of seven **crystal systems** according to the symmetry they possess under rotations about different axes. The **cubic system**, for example, has four threefold axes. A threefold axis is the axis of a rotation that restores the unit cell to the same appearance three times during a complete revolution, after rotations through 120°, 240°, and 360° (Fig. 10.32). The four axes make the tetrahedral angle to each other. The **monoclinic system** has one twofold axis. A twofold axis is the axis of a rotation that leaves the cell apparently unchanged twice during a complete revolution, after rotations through 180° and 360° (Fig. 10.33.) The **essential symmetries**, the properties that must be present for the unit cell to belong to a particular system, are listed in Table 10.7.

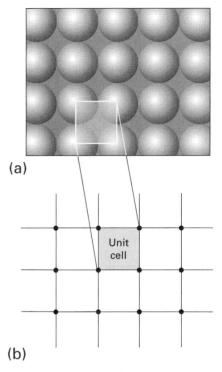

(a)

(b)

Fig. 10.30 (a) A crystal consists of a uniform array of atoms, molecules, or ions, as represented by these spheres. In many cases, the components of the crystal are far from spherical, but this diagram illustrates the general idea. (b) The location of each atom, molecule, or ion can be represented by a single point; here (for convenience only), the locations are denoted by a point at the centre of the sphere. The unit cell, which is shown shaded, is the smallest block from which the entire array of points can be constructed without rotating or otherwise modifying the block.

Table 10.7 The essential symmetries of the seven crystal systems

The systems	Essential symmetries
Triclinic	None
Monoclinic	One twofold axis
Orthorhombic	Three perpendicular twofold axes
Rhombohedral	One threefold axis
Tetragonal	One fourfold axis
Hexagonal	One sixfold axis
Cubic	Four threefold axes in a tetrahedral arrangement

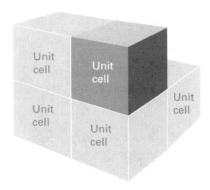

Fig. 10.31 A unit cell, here shown in three dimensions, is like a brick used to construct a wall. Once again, only pure translations are allowed in the construction of the crystal. (Some bonding patterns for actual walls use rotations of bricks, so for these a single brick is not a unit cell.)

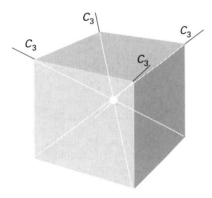

Fig. 10.32 A unit cell belonging to the cubic system has four threefold axes (denoted C_3) arranged tetrahedrally.

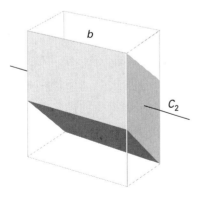

Fig. 10.33 A unit cell belonging to the monoclinic system has one twofold (denoted C_2) axis (along b).

A unit cell may have lattice points other than at the corners joined, so each crystal system can occur in a number of different varieties. For example, in some cases points may occur on the faces and in the body of the cell without destroying the cell's essential symmetry. These various possibilities give rise to fourteen distinct types of unit cell, which are called **Bravais lattices** (Fig. 10.34).

10.10 The identification of crystal planes

The spacing of the points in a lattice is an important quantitative aspect of a crystal's structure. Because two-dimensional arrays of points are easier to visualize than three-dimensional arrays, we shall introduce the concepts we need by referring to two-dimensions initially, and then extend the conclusions to three dimensions.

Consider the two-dimensional rectangular lattice formed from a unit cell of sides a and b (Fig. 10.35). We can distinguish the four sets of planes on which the points lie in the illustration by the distances along the axes at which the planes intersect the axes. One way of labelling the planes would therefore be to denote each set by the smallest intersection distances. For example, we could denote the four sets in the illustration as $(1a, 1b)$, $(3a, 2b)$, $(-1a, 1b)$, and $(\infty a, 1b)$. If, however, we agreed always to quote distances along the axes as multiples of the lengths of the unit cell, then we could omit the a and b and label the planes more simply as $(1, 1)$, $(3, 2)$, $(-1, 1)$, and $(\infty, 1)$. If the array in Fig. 10.35 is the top view of a three-dimensional rectangular lattice in which the unit cell has a length c in the z direction, then all four sets of planes intersect the z-axis at infinity, and so the full labels of the sets of planes of lattice points would be $(1, 1, \infty)$, $(3, 2, \infty)$, $(-1, 1, \infty)$, and $(\infty, 1, \infty)$.

The presence of infinity in the labels is inconvenient, and can be eliminated by taking the reciprocals of the numbers in the labels (this step turns out to have further advantages, as we shall see). The resulting **Miller indices** are the reciprocals of the numbers in the parentheses with fractions cleared. For example, the $(1, 1, \infty)$ planes in Fig. 10.35 are the (110) planes in the Miller notation. Similarly, the $(3, 2, \infty)$ planes become first $(\frac{1}{3}, \frac{1}{2}, 0)$ when reciprocals are formed, and then $(2, 3, 0)$ when fractions are cleared by multiplication through by 6, and so they are referred to as the (230) planes. Negative indices are written with a bar over the number, and Fig. 10.35(c) shows the $(\bar{1}10)$ planes. The illustration in Fig. 10.36 shows some planes in three dimensions, including an example of a lattice with axes that are not mutually perpendicular.

Exercise E10.13

A representative member of a set of planes in a crystal intersects the axes at $3a$, $2b$, and $2c$; what are the Miller indices of the planes?

[*Answer*: (233)]

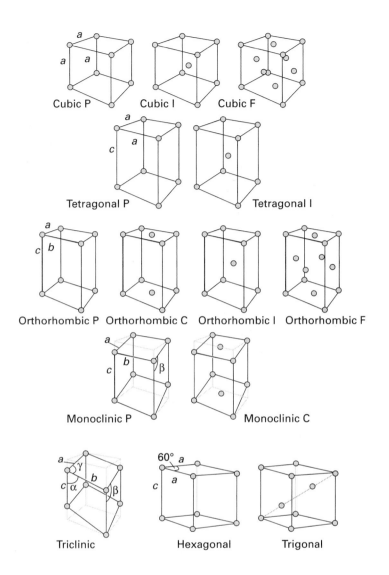

Fig. 10.34 The fourteen Bravais lattices. The letter P denotes a primitive unit cell, I a body-centred unit cell, F a face-centred unit cell, and C (or A or B) a cell with lattice points on two opposite faces.

Cubic P   Cubic I   Cubic F

Tetragonal P   Tetragonal I

Orthorhombic P   Orthorhombic C   Orthorhombic I   Orthorhombic F

Monoclinic P   Monoclinic C

Triclinic   Hexagonal   Trigonal

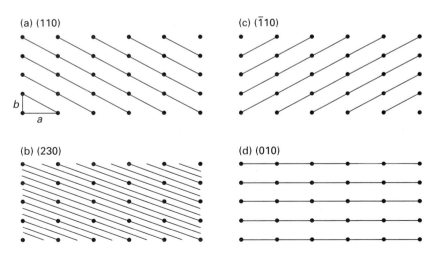

(a) ($\bar{1}$10)

(b) (230)

(c) ($\bar{1}$10)

(d) (010)

Fig. 10.35 Some of the planes that can be drawn through the points of the space lattice and their corresponding Miller indices (hkl).

Fig. 10.36 Some representative planes in three dimensions and their Miller indices. Note that a 0 indicates that a plane is parallel to the corresponding axis, and that the indexing may also be used for unit cells with nonorthogonal axes.

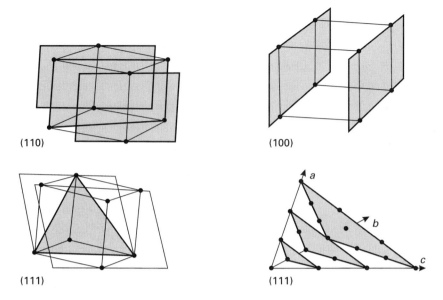

(110)   (100)

(111)   (111)

It is helpful to keep in mind the fact, as illustrated in Fig. 10.35, that the smaller the value of h in the Miller index (hkl), the more nearly parallel the plane is to the a axis. The same is true of k and the b axis and l and the c axis. When $h = 0$, the planes intersect the a axis at infinity, and so the $(0kl)$ planes are parallel to the a axis. Similarly, the $(h0l)$ planes are parallel to b and the $(hk0)$ planes are parallel to c.

The Miller indices are very useful for calculating the separation of planes. The expression for the separation, d, of the planes (hkl) in a rectangular lattice built from a unit cell of sides of lengths a, b, and c is

$$\frac{1}{d^2} = \frac{h^2}{a^2} + \frac{k^2}{b^2} + \frac{l^2}{c^2} \tag{25}$$

Example *Using the Miller indices*
Calculate the separation of (a) the (123) planes and (b) the (246) planes of an orthorhombic cell with $a = 0.82$ nm, $b = 0.94$ nm, and $c = 0.75$ nm.

Strategy
For the first part, we simply substitute the information into eqn (25). For the second part, instead of repeating the calculation, we should examine how d in eqn (25) changes when all three Miller indices are multiplied by 2 (or by a more general factor, n).

Solution
Substituting the data into eqn (25) gives

$$\frac{1}{d^2} = \frac{1^2}{(0.82\,\text{nm})^2} + \frac{2^2}{(0.94\,\text{nm})^2} + \frac{3^2}{(0.75\,\text{nm})^2} = 22\,\text{nm}^{-2}$$

It follows that $d = 0.21\,\text{nm}$. When the indices are all increased by a factor of 2, the separation becomes

$$\frac{1}{d^2} = \frac{2^2}{(0.82\,\text{nm})^2} + \frac{4^2}{(0.94\,\text{nm})^2} + \frac{6^2}{(0.75\,\text{nm})^2}$$

$$= 4 \times \left\{ \frac{1^2}{(0.82\,\text{nm})^2} + \frac{2^2}{(0.94\,\text{nm})^2} + \frac{3^2}{(0.75\,\text{nm})^2} \right\} = 4 \times 22\,\text{nm}^{-2}$$

So, for these planes $d = 0.11\,\text{nm}$. In general, increasing the indices uniformly by a factor n decreases the separation of the planes by n.

Exercise 10.14

Calculate the separation of the (133) and (399) planes in the same lattice.

[*Answer*: 0.20 nm, 0.065 nm]

10.11 The determination of structure

One of the most important techniques for the determination of the structures of crystals is **X-ray diffraction**. In its simplest form, the technique can be used to identify the lattice type and the separation of the planes of lattice points (and hence the distance between the nuclei of atoms and ions). In its most sophisticated version, X-ray diffraction provides detailed information about the location of all the atoms in molecules as complicated as proteins. The current considerable success of modern molecular biology has stemmed from X-ray diffraction techniques that have grown in sensitivity and scope as computing techniques have become more powerful. Here we shall confine attention to the principle of the technique, and illustrate how it may be used to determine the spacing of atoms in a crystal.

Diffraction

A characteristic property of waves is that they **interfere** with one another, which means that they give a greater amplitude where their displacements add and a smaller amplitude where their displacements subtract (Fig. 10.37). Because the intensity of electromagnetic radiation is proportional to the square of the amplitude of the waves, the regions of constructive and destructive interference show up as regions of enhanced and diminished intensities. The phenomenon of **diffraction** is the interference that is caused by an object in the path of waves, and the pattern of varying intensity that results is called the **diffraction pattern** (Fig. 10.38; the phenomenon was first

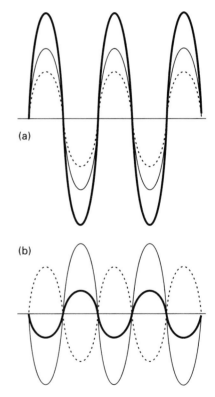

Fig. 10.37 When two waves (drawn as thin or dotted lines) are in the same region of space they interfere. Depending on their relative phase, they may interfere (a) constructively, to give an enhanced amplitude, or (b) destructively, to give a smaller amplitude.

395

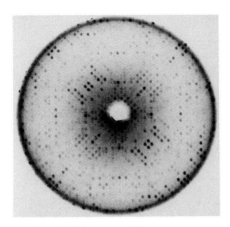

Fig. 10.38 A typical diffraction pattern obtained in a version of the X-ray diffraction technique. The black dots are the reflections, the points of maximum constructive interference, that are used to determine the structure of the crystal.

mentioned in Section 8.3 in connection with the wave properties of electrons). Diffraction occurs when the dimensions of the diffracting object are comparable to the wavelength of the radiation: sound waves (with wavelengths of the order of 1 m) are diffracted by macroscopic objects, and light waves (with wavelengths of the order of 500 nm) are diffracted by narrow slits.

X-Rays have wavelengths comparable to bond lengths in molecules and the spacing of atoms in crystals (about 100 pm), and so they are diffracted by them. By analysing the diffraction pattern, it is possible to draw up a detailed picture of the location of atoms. Electrons moving at about $2 \times 10^4 \text{ km s}^{-1}$ (after acceleration through about 4 kV) have wavelengths of about 20 pm, and may also be diffracted by molecules. Neutrons generated in a nuclear reactor, and then slowed to thermal velocities, have similar wavelengths and may also be used for diffraction studies. We shall confine attention to X-ray diffraction, the most widely used of the three techniques.

Example *Estimating the wavelength of accelerated electrons*
Calculate the wavelength of electrons that have been accelerated from rest through a potential difference of 4.0 kV.

Strategy
The key to the calculation is the de Broglie relation $\lambda = h/p$ (Section 8.3) between the wavelength λ and the linear momentum p. We first need to calculate the kinetic energy, $p^2/2m_e$, of an electron that has been accelerated from rest through a potential difference V and then to obtain an expression for the linear momentum p in terms of the accelerating potential. After acceleration from rest through a potential difference V, an electron acquires a kinetic energy eV. In the numerical calculation, we need to use the relations $1\,\text{C V} = 1\,\text{J}$ and $1\,\text{J} = 1\,\text{kg m}^2\,\text{s}^{-2}$.

Solution
By equating the expressions $p^2/2m_e$ and eV,

$$\frac{p^2}{2m_e} = eV$$

we can obtain an expression for the linear momentum:

$$p = \sqrt{2m_e eV}$$

Then, from the de Broglie relation, we can find an expression for the wavelength of the electron:

$$\lambda = \frac{h}{p} = \frac{h}{\sqrt{2m_e eV}}$$

Substitution of the data and the fundamental constants (from the tables inside the front cover) then gives

$$\lambda = \frac{6.626 \times 10^{-34}\,\text{J s}}{\sqrt{2 \times (9.109 \times 10^{-31}\,\text{kg}) \times (1.602 \times 10^{-19}\,\text{C}) \times (4.0 \times 10^{3}\,\text{V})}}$$
$$= 1.9 \times 10^{-11}\,\text{m, or 19 pm}$$

The unit cancellation was carried out as follows:

$$\frac{\text{J s}}{(\text{kg C V})^{1/2}} = \frac{\text{J s}}{(\text{kg J})^{1/2}} = \frac{\text{kg m}^2\,\text{s}^{-1}}{(\text{kg}^2\,\text{m}^2\,\text{s}^{-2})^{1/2}}$$
$$= \frac{\text{kg m}^2\,\text{s}^{-1}}{\text{kg m s}^{-1}} = \text{m}$$

Exercise E10.15

Calculate the wavelength of a neutron that is travelling with a kinetic energy kT, where k is Boltzmann's constant, in a reactor at a temperature $500\,^{\circ}\text{C}$.

[*Answer*: 110 pm]

The short wavelength electromagnetic radiation we call X-rays are produced by bombarding a metal with high-energy electrons. The electrons decelerate as they plunge into the metal and generate radiation with a continuous range of wavelengths. This radiation is called **bremsstrahlung** (*Bremse* is German for brake, *Strahlung* for ray). Superimposed on the continuum are a few high-intensity, sharp peaks. These peaks arise from the interaction of the incoming electrons with the electrons in the inner shells of the atoms. A collision expels an electron, and an electron of higher energy drops into the vacancy, emitting the excess energy as an X-ray photon. An example of the process is the expulsion of an electron from the K shell (the shell with $n = 1$) of a copper atom, followed by the transition of an outer electron into the vacancy, and the energy so released gives rise to copper's K_{α} radiation of wavelength 154 pm.

In 1923, the German physicist Max von Laue suggested that X-rays might be diffracted when passed through a crystal, for their wavelengths are comparable to the separation of atoms. Laue's suggestion was confirmed almost immediately by Walter Friedrich and Paul Knipping, and has grown since then into a technique of extraordinary power.

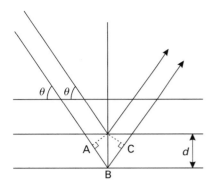

Fig. 10.39 The derivation of the Bragg law treats each lattice plane as reflecting the incident radiation. The path lengths differ by AB + BC, which depends on the glancing angle θ. Constructive interference (a 'reflection') occurs when AB + BC is equal to an integral number of wavelengths.

The Bragg law

The earliest approach to the analysis of diffraction patterns produced by crystals was to regard a plane of atoms as a semitransparent mirror, and to model a crystal as stacks of reflecting planes of separation d (Fig. 10.39). The model makes it easy to calculate the angle the crystal must make to the incoming beam of X-rays for constructive interference to occur. It has also given rise to the name **reflection** to denote an intense spot arising from constructive interference.

The path-length difference of the two rays shown in the illustration is

$$AB + BC = 2d \sin \theta$$

where θ is the **glancing angle**. When the path-length difference is equal to one wavelength (AB + BC = λ), the reflected waves are in phase and interfere constructively. It follows that a reflection should be observed when the glancing angle satisfies the **Bragg law**:

$$\lambda = 2d \sin \theta \qquad (26)$$

The primary use of the Bragg law is in the determination of the spacing between the layers of atoms, for once the angle θ corresponding to a reflection has been determined, d may readily be calculated.

Example *Using the Bragg law*
A reflection from the (111) planes of a cubic crystal was observed at a glancing angle of 11.2° when Cu K_α X-rays of wavelength 154 pm were used. What is the length of the side of the unit cell?

Strategy
The separation d of the lattice planes can be found from eqn (26) and the data. The problem then is to find the size of the unit cell by using eqn (25). Because the unit cell is cubic, $a = b = c$, so

$$\frac{1}{d^2} = \frac{h^2 + k^2 + l^2}{a^2}$$

which rearranges to

$$a = d \times \sqrt{h^2 + k^2 + l^2}$$

Solution
According to the Bragg law, the (111) planes responsible for the diffraction have separation

$$d = \frac{\lambda}{2 \sin \theta} = \frac{154 \text{ pm}}{2 \sin 11.2°}$$

It then follows that with $h = 1, k = 1, l = 1$,

$$a = \frac{154 \, \text{pm}}{2 \sin 11.2°} \times \sqrt{3} = 687 \, \text{pm}$$

Exercise E10.16

Calculate the angle at which the same lattice will give a reflection from the (123) planes.

[*Answer*: 24.8°]

Experimental techniques

Laue's original method consisted of passing a beam of X-rays of a wide range of wavelengths into a single crystal, and recording the diffraction pattern photographically. The idea behind the approach was that a crystal might not be suitably orientated to act as a diffraction grating for a single wavelength, but whatever its orientation the Bragg law would be satisfied for at least one of the wavelengths if a range of wavelengths was present in the beam. An alternative technique was developed by Peter Debye and Paul Scherrer and independently by Albert Hull. They used monochromatic radiation and a powdered sample.

When the sample is a powder, some of the crystallites will always be orientated so as to satisfy the Bragg condition. For example, some of them will be orientated so that their (111) planes, of spacing d, give rise to a diffracted intensity at the glancing angle θ (Fig. 10.40). The (111) planes

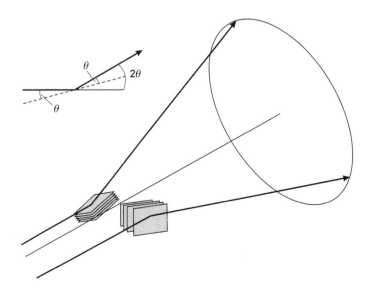

Fig. 10.40 The same set of planes in two microcrystallites with different orientations around the direction of the incident beam give diffracted rays that lie on a cone. The full powder diffraction pattern is formed by cones corresponding to reflections from all the sets of (*hkl*) planes that satisfy the Bragg law. (A reflection at a glancing angle θ gives rise to a reflection at an angle 2θ to the direction of the incident beam; see inset.)

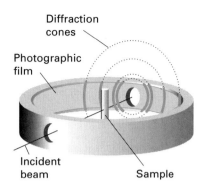

Fig. 10.41 In the Debye–Scherrer method, a monochromatic X-ray beam is diffracted by a powder sample. The crystallites give rise to cones of intensity, which are detected by a photographic film wrapped round inside the circumference of the camera.

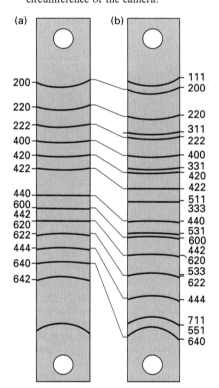

Fig. 10.42 X-ray powder photographs of (a) KCl, (b) NaCl. The numbers are the Miller indices of the sets of planes responsible for the diffraction. Note that the film covers one half of the circumference of the powder camera, the lines at the top and bottom corresponding to the smallest diffraction angles.

of some other crystallites may be at the angle θ to the beam, but at an arbitrary angle about the line of its approach. It follows that the diffracted beams will lie on a cone around the incident beam of half-angle 2θ. Other crystallites will be orientated with different planes satisfying the Bragg law and give rise to a cone of diffracted intensity with a different half-angle. In principle, each set of (hkl) planes gives rise to a different diffraction cone because some of the randomly orientated crystallites will have the correct angle to diffract the incident beam.

The **Debye–Scherrer method**, which makes use of this approach, is illustrated in Fig. 10.41. The sample is in a capillary tube, which is rotated to ensure that the crystallites are randomly orientated. The diffraction cones are photographed as arcs of circles where they cut the strip of film, and some typical patterns are shown in Fig. 10.42. In modern diffractometers the sample is spread on a flat plate and the diffraction pattern is monitored electronically. The major application is now for qualitative analysis because the diffraction pattern is a kind of fingerprint and may be recognizable. The technique is also used for the initial determination of the dimensions and symmetries of unit cells.

Modern X-ray diffraction, which utilizes an **X-ray diffractometer** (Fig. 10.43), is now a highly sophisticated technique. By far the most detailed information comes from developments of the techniques pioneered by the Braggs, in which a single crystal is employed as the diffracting grating and a monochromatic beam of X-rays is used to generate the diffraction pattern. The single crystal (which may be only a fraction of a millimetre in length) is rotated relative to the beam, and the diffraction pattern is monitored and recorded electronically for each crystal orientation. The analysis of the diffraction pattern is then carried out on a computer that is an intrinsic part of the diffractometer, and the results presented as a detailed structural map of the unit cell of the crystal, showing the relative locations of all the atoms it contains.

10.12 Information from X-ray analysis

The bonding within a solid, the origin of its cohesion, may be of various kinds. Simplest of all (in principle) is the bonding in **metallic solids**, in which electrons are delocalized over arrays of identical cations and bind the whole together into a rigid but malleable structure. The origin of this type of bonding in terms of molecular orbitals was described in Sections 9.8 and 9.9. In many cases, because the delocalized molecular orbitals can accommodate bonding patterns with very little directional character, the crystal structures of metals are determined largely by the manner in which spherical metal cations can pack together into an orderly array. In **covalent solids**, covalent bonds in a definite spatial orientation link the atoms in a network extending through the crystal. The stereochemical demands of valence now override the geometrical problem of packing spheres together, and elaborate and extensive structures may be formed. A famous example of a covalent solid is diamond (Fig. 10.44), in which each sp^3-hybridized carbon is bonded tetrahedrally by σ bonds to its four neighbours. Covalent solids are often

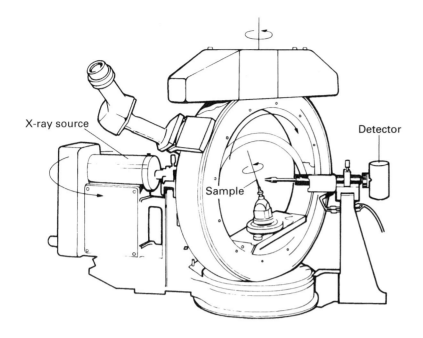

Fig. 10.43 A four-circle diffractometer. The settings of the orientations of the components is controlled by computer; each reflection is monitored in turn, and their intensities are recorded.

hard and unreactive. **Molecular solids**, which are the subject of the over-whelming majority of modern structural determinations, are bonded together by the van der Waals interactions described earlier in the chapter.

In each case, the observed crystal structure is nature's solution to the problem of condensing objects of various shapes into an aggregate of minimum energy (actually, for temperatures above zero, of minimum Gibbs energy). Exactly the same is true of the structures of proteins, nucleic acids, and other biological macromolecules, but for them the packing pattern is generally more complex because the individual units (the peptide groups in proteins) are more varied, they are held together by chemical bonds (the peptide links) as well as van der Waals forces, and the surrounding medium (typically water and the ions it contains) may play an important structural role.

10.13 The packing of identical spheres: metal crystals

Most metallic elements crystallize in one of three simple forms, two of which can be explained in terms of organizing spheres into the closest possible packing. In such **close-packed structures** the spheres representing the atoms are packed together with least waste of space and each sphere has the greatest possible number of nearest neighbours.

Close packing

A close-packed layer of identical spheres, one with maximum utilization of space, can be formed as shown in Fig. 10.45(a). A second close-packed layer

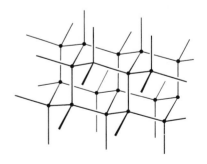

Fig. 10.44 A fragment of the structure of diamond. Each C atom is tetrahedrally bonded to four neighbours. This frameworklike structure results in a rigid crystal with a high thermal conductivity.

401

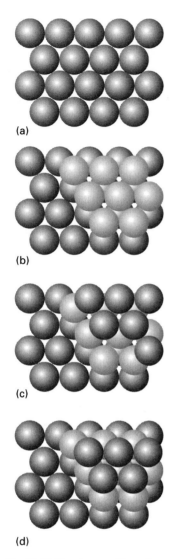

(a)

(b)

(c)

(d)

Fig. 10.45 The close-packing of identical spheres. (a) The first layer of close-packed spheres. (b) The second layer of close-packed spheres occupies the dips of the first layer. The two layers are the AB component of the structure. (c) The third layer of close-packed spheres might occupy the dips lying directly above the spheres in the first layer, resulting in an ABA structure. (d) Alternatively, the third layer might lie in the dips that are not above the spheres in the first layer, resulting in an ABC structure.

can be formed by placing spheres in the depressions of the first layer (Fig 45(b)). The third layer may be added in either of two ways, both of which result in the same degree of close packing. In one, the spheres are placed so that they reproduce the first layer (Fig 10.46(c)), to give an ABA pattern of layers. Alternatively, the spheres may be placed over the gaps in the first layer (Fig. 10.45(d)), so giving an ABC pattern.

Two structures are formed if the two stacking patterns are repeated in the vertical direction. If the ABA pattern is repeated, to give the sequence of layers ABABAB..., then the spheres are **hexagonally close-packed** (hcp); the name reflects the symmetry of the unit cell (Fig. 10.46). Metals with hcp structures include beryllium, cadmium, cobalt, manganese, titanium, and zinc. Solid helium (which forms only under pressure) also adopts this arrangement of atoms. Alternatively, if the ABC pattern is repeated to give the sequence of layers ABCABC..., the spheres are **cubic close-packed** (ccp); once again, the name reflects the symmetry of the unit cell (Fig. 10.47). Metals with this structure include silver, aluminium, gold, calcium, copper, nickel, lead, and platinum. The noble gases other than helium also adopt a ccp structure.

The compactness of the ccp and hcp structures is indicated by their **coordination number**, the number of atoms immediately surrounding any selected atom, which is 12 in both cases. Another measure of their compactness is the **packing fraction**, the fraction of space occupied by the spheres, which is 0.740. That is, in a close-packed solid of identical hard spheres, 74.0 per cent of the available space is occupied and only 26.0 per cent of the total volume is empty space. The fact that many metals are close-packed accounts for one of their common characteristics, their high densities.

Less closely packed structures

A number of common metals adopt structures that are not close-packed, which suggests that directional covalent bonding between neighbouring atoms is beginning to influence the structure and impose a specific geometrical arrangement. One such arrangement results in a **body-centred cubic** (bcc) lattice, with one sphere at the centre of a cube formed by eight others (Fig. 10.48). The bcc structure is adopted by a number of common metals, including barium, caesium, chromium, iron, potassium, and tungsten. The coordination number of a bcc lattice is 8 and its packing fraction is only 0.68, showing that only about two-thirds of the available space is actually occupied.

Exercise E10.17

What is the coordination number and the packing fraction of a primitive cubic lattice in which there is a lattice point at each corner of a cube?

[*Answer*: 6, 0.52]

402

10.14 Ionic crystals

When the structures of ionic crystals are modelled by stacks of spheres, we must allow for the fact that the two or more types of ion present in the compound have different radii (generally with the cations smaller than the anions) and different charges. The coordination number of an ionic lattice is the number of nearest neighbours of opposite charge, and cations and anions may have different environments in the same crystal.

Even if, by chance, the ions have the same size, the problem of ensuring that the unit cells are electrically neutral makes it impossible to achieve 12-coordinate close-packed structures (which is a reason why ionic solids are generally less dense than metals). The closest packing that can be achieved is the 8-coordination of the **caesium-chloride structure** (Fig. 10.49) in which each cation is surrounded by eight anions and each anion is surrounded by eight cations. In the caesium-chloride structure, an ion of one charge occupies the centre of a cubic unit cell with eight ions of opposite charge at its corners. This structure is adopted by caesium chloride itself and by calcium sulfide, caesium cyanide (with some distortion), and one type of brass (CuZn).

When the radii of the ions differ by more than in caesium chloride, even 8-coordinate packing cannot be achieved. One common structure adopted is the 6-coordinated **rock-salt structure** typified by sodium chloride (rock salt is a mineral form of sodium chloride) in which each cation is surrounded by six anions and each anion is surrounded by six cations (Fig. 10.50). The rock-salt structure is the structure of sodium chloride itself and of several other compounds of formula MX, including potassium bromide, silver chloride, and magnesium oxide.

The switch from the adoption of the caesium-chloride structure to the adoption of the rock-salt structure occurs (in a number of examples) in accord with the **radius-ratio rule**, which is based on the value of the radius ratio

$$\gamma = \frac{r_{\text{smaller}}}{r_{\text{larger}}} \qquad (27)$$

The two radii are those of the smaller and larger ions in the crystal and the rule is derived by analysing the geometrical problem of stacking together spheres of different radii. The radius-ratio rule states that the caesium-chloride structure should be expected when

$$\gamma > \sqrt{3} - 1 = 0.732$$

and that the rock-salt structure should be expected when

$$\sqrt{2} - 1 = 0.414 < \gamma < 0.732$$

For $\gamma < 0.414$, when the two types of ions have markedly different radii (like water melons and grapefruit), the most efficient packing leads to 4-coordination of the type exhibited by the sphalerite (or zinc blende) form of zinc

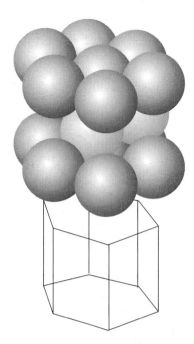

Fig. 10.46 A hexagonal close-packed structure. The tinting of the spheres (denoting the three layers of atoms) is the same as in Fig. 10.45.

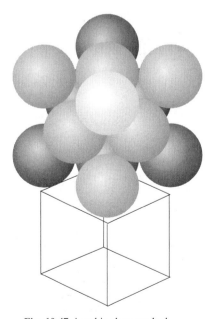

Fig. 10.47 A cubic close-packed structure. The tinting of the spheres is the same as in Fig. 10.45.

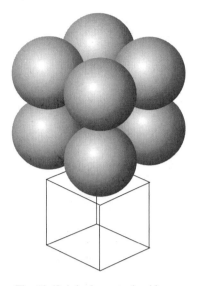

Fig. 10.48 A body-centred cubic unit cell. There is one sphere at the centre of the cube. The packing pattern leaves more empty space than in the two close-packed structures.

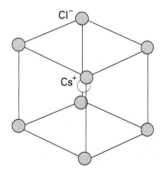

Fig. 10.49 The caesium-chloride structure consists of two interpenetrating simple cubic lattices, one of cations and the other of anions, so that each cube of ions of one kind has a counter-ion at its centre. This illustration shows a single unit cell with a Cs^+ ion at the centre. By imagining eight of these unit cells stacked together to form a bigger cube, it should be possible to imagine an alternative form of the unit cell with Cs^+ at the corners and a Cl^- ion at the centre.

sulfide, ZnS (Fig. 10.51). The radius-ratio rule, which is based on geometrical considerations of sphere packing, is moderately well supported by observation. The deviation of a structure from the prediction is often taken to be an indication of a shift from ionic towards covalent bonding.

The ionic radii that are used to calculate γ, and wherever else it is important to know the sizes of ions, are derived from the distance between the centres of adjacent ions in a crystal. However, it is necessary to apportion the total distance between the two ions by defining the radius of one ion and reporting all others on that basis. One scale that is widely used is based on the value 140 pm for the radius of the O^{2-} ion (Table 10.8). Other scales are also available (such as one based on F^- for discussing halides), and it is essential not to mix values from different scales. Because ionic radii are so arbitrary, predictions based on them (such as those made by using the radius-ratio rule) must be viewed cautiously.

Exercise E10.18

Is sodium iodide likely to have a rock-salt or caesium-chloride structure?

[*Answer*: rock salt]

Natural biopolymers

Natural macromolecules need a precisely maintained shape in order to function. The achievement of a specific shape is the major remaining problem in

Table 10.8 Ionic radii, r/pm

Li^+	Be^{2+}	B^{3+}	N^{3-}	O^{2-}	F^-
59	27	12	171	140	133
Na^+	Mg^{2+}	Al^{3+}	P^{3-}	S^{2-}	Cl^-
102	72	53	212	184	181
K^+	Ca^{2+}	Ga^{3+}	As^{3-}	Se^{2-}	Br^-
138	100	62	222	198	196
Rb^+	Sr^{2+}				I^-
149	116				220
Cs^+	Ba^{2+}				
170	136				

protein synthesis, for although polypeptides can be synthesized, the product is inactive because the correct twisting and coiling of the polypeptide chain cannot yet be produced. The overall shape of a protein molecule is sustained by a variety of intermolecular forces of the kind we have encountered in this chapter, including hydrogen bonding, the hydrophobic effect, and dipolar and dispersion interactions.

10.15 The primary and secondary structures

The **primary structure** of a biopolymer is the sequence of its monomer units. For polypeptides, which we consider here, the primary structure is an ordered list of the amino acid residues. The **secondary structure** of a polypeptide is the spatial arrangement of the polypeptide chain—its twisting into a specific shape—under the influence of hydrogen bonding between the various peptide residues (the amino acid groups).

The secondary structures of proteins can be rationalized in terms of the rules formulated by Linus Pauling and Robert Corey. The essential feature is the stabilization of structures by hydrogen bonds involving the —CO—NH— **peptide link**. The latter can act both as a donor of the H atom (the NH part of the link) and as an acceptor (the CO part). The **Corey–Pauling rules** are then as follows:

1. The atoms of the peptide link lie in a plane (Fig. 10.52).

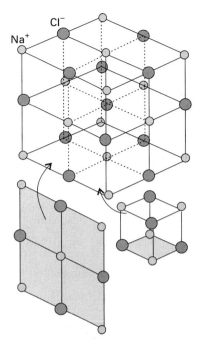

Fig. 10.50 The rock-salt (NaCl) structure consists of two mutually interpenetrating slightly expanded face-centred cubic lattices. The additional diagrams in this illustration show various details of the structure.

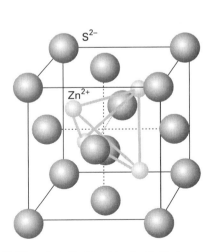

Fig. 10.51 The sphalerite (zinc-blende, ZnS) structure. This structure is typical of ions that have markedly different radii and equal but opposite charges.

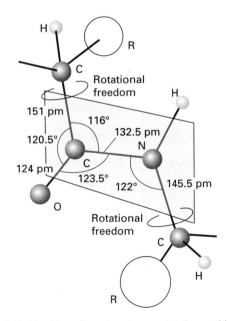

Fig. 10.52 The dimensions that characterize the peptide link. The CCONHC atoms define a plane (the CN bond has partial double-bond character), but there is rotational freedom around the CCO and NC bonds.

405

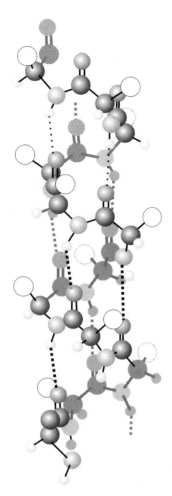

Fig. 10.53 The polypeptide α-helix. There are 3.6 residues per turn, and a translation along the helix of 150 pm per residue, giving a pitch of 544 pm. The diameter (ignoring side chains, represented here by white circles) is about 600 pm.

2. The N, H, and O atoms of a hydrogen bond lie in a straight line (with displacements of H tolerated up to not more than 30° from the N—O bond).
3. All NH and CO groups are engaged in bonding.

These rules are satisfied by two structures. One, in which hydrogen bonding occurs between peptide links of the same chain, is the α **helix**. The other, in which hydrogen bonding links different chains, is the β-**pleated sheet**; this form is the secondary structure of the protein fibroin, the constituent of silk.

The α helix is illustrated in Fig. 10.53. Each turn of the helix contains 3.6 amino acid residues, and so the period of the helix corresponds to 5 turns (18 residues). The pitch of a single turn is 544 pm. The N—H$\cdots$O bonds lie parallel to the axis and link every fifth group (so residue i is linked to residues $i-4$ and $i+4$). There is freedom for the helix to be arranged as either a right- or a left-handed screw, but the overwhelming majority of natural polypeptides are right-handed on account of the preponderance of the L-configuration of the naturally occurring amino acids. The stabilities of different polypeptide geometries can be investigated by calculating the total potential energy of all the interactions between nonbonded atoms and looking for a minimum. It turns out, in agreement with experience, that a right-handed α helix of L-amino acids has a marginally lower energy than a left-handed helix of the same acids.

10.16 Higher-order structure

Helical polypeptide chains are folded into a tertiary structure if there are other bonding influences between the residues of the chain that are strong enough to overcome the interactions responsible for the secondary structure. The folding influences include —S—S— disulfide links, ionic interactions (which depend on the pH), and strong hydrogen bonds (such as O—H$\cdots$O—), the full structure (of 2600 atoms) having been determined by X-ray diffraction. About 77 per cent of the structure is α helix, the rest being involved in the folds resulting from disulfide links.

Proteins with $M > 50\,\text{kg mol}^{-1}$ are often found to be aggregates of two or more polypeptide chains. The possibility of such **quaternary structure** often confuses the determination of their molar masses, because different techniques might give values differing by factors of 2 or more. Haemoglobin, which consists of four myoglobin-like chains, is an example of a quaternary structure.

Protein **denaturation**, or loss of structure, can be caused by several means, and different aspects of structure may be affected. The permanent waving of hair, for example, is reorganization at the quaternary level. Hair is a form of the protein keratin, and its quaternary structure is thought to be a multiple helix, with the α helices bound together by disulfide links and hydrogen bonds. The process of permanent waving consists of disrupting these links, unravelling the keratin quaternary structure, and then reforming it into a more fashionable disposition. The 'permanence' is only temporary, however, because the structure of the newly formed hair is genetically controlled. Incidentally, normal hair grows at a rate that requires at least 10 twists of the keratin helix to be produced each second, and so very close inspection of the human scalp would show it to be literally writhing with activity.

Denaturation at the secondary level is brought about by agents that destroy hydrogen bonds. Thermal motion may be sufficient, in which case denaturation is a kind of intramolecular melting. When eggs are cooked the albumin is denatured irreversibly, and the protein collapses into a structure resembling a random coil. The **helix-coil transition** is sharp, like ordinary melting, because it is a **cooperative process** in the sense that when one hydrogen bond has been broken it is easier to break its neighbours, and then even easier to break theirs, and so on. The disruption cascades down the helix, and the transition occurs sharply. Denaturation may also be brought about chemically. For instance, a solvent that forms stronger hydrogen bonds than those within the helix will compete successfully for the NH and CO groups. Acids and bases can cause denaturation by protonation or deprotonation of various groups.

EXERCISES

10.1 Estimate the lattice enthalpy of magnesium oxide from the data in Appendix 1 and the ionization and electron gain enthalpies in Chapter 2.

10.2 Estimate the ratio of the lattice enthalpies of SrO and CaO from the Born–Meyer equation by using the ionic radii in Table 10.8.

10.3 The ClF_3 molecule has five electron pairs around the central Cl atom, and so may have either three equatorial F atoms or two axial and one equatorial F atoms. The molecule is polar. Which structure does it have?

10.4 The electric dipole moment of toluene (methylbenzene) is 0.4 D. Estimate the dipole moments of the three xylenes (dimethylbenzenes). Which value can you be sure about?

10.5 From the information in the preceding problem, estimate the dipole moments of (a) 1,2,3-trimethylbenzene, (b) 1,2,4-trimethylbenzene, and (c) 1,3,5-trimethylbenzene. Which value can you be sure about?

10.6 Calculate the resultant of two dipoles of magnitude 1.5 D and 0.80 D that make an angle 109.5° to each other.

10.7 At low temperatures a substituted 1,2-dichlorethane molecule can adopt the three conformations

4, **5**, and **6** with different probabilities. Suppose that the dipole moment of each bond is 1.5 D. Calculate the mean dipole moment of the molecule when (a) all three conformations are equally likely, (b) only conformation **4** occurs, (c) the three conformations occur with probabilities in the ratio 2:1:1 and (d) 1:2:2.

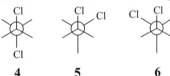

10.8 Find an expression for the potential energy of interaction of a dipole moment μ with a point charge q representing an ion that is at a distance R from the dipole and collinear with it. Base your calculation on the expression for the Coulombic potential energy of interaction of two point charges, and suppose that the two partial charges $+q'$ and $-q'$ of the dipole are separated by a distance l from one another (with $\mu = q'l$).

10.9 Modify the expression derived in the preceding exercise to the case when $l \ll R$. *Hint.* Express the denominators as $R(1 \pm l/R)$ and use the expansions

$$\frac{1}{1+x} = 1 - x + x^2 - \cdots \qquad \frac{1}{1-x} = 1 + x + x^2 - \cdots$$

Keep as few terms in the expansions as are needed to get a nonzero result.

10.10 One contribution to the shapes adopted by polypeptide chains is the energy of interaction between the dipoles of polar groups. This kind of interaction is taken into account in energy-minimization procedures. Repeat the calculation in Exercise 10.9 for two collinear, parallel, identical point dipoles separated by a distance R, and go on to specialize to the case when $l \ll R$.

10.11 Another contribution to shape is the attraction and repulsion between nonbonded atoms in a polypeptide. This contribution can be represented by the Lennard-Jones potential. Given that the force is the negative slope of the potential, calculate the distance-dependence of the force acting between the atoms. What is the separation at which the force is zero? *Hint* Calculate the slope by considering the potential energy at R and $R + \delta R$, with $\delta R \ll R$, and evaluating $\{V(R + \delta R) - V(R)\}/\delta R$. You should use the expansions in Exercise 10.9 together with

$$(1 \pm x + \cdots)^6 = 1 \pm 6x + \cdots$$
$$(1 \pm x + \cdots)^{12} = 1 \pm 12x + \cdots$$

At the end of the calculation, let δR become vanishingly small.

10.12 Acetic acid vapour contains a proportion of planar, hydrogen bonded dimers. The apparent dipole moment of molecules in pure gaseous acetic acid increases with increasing temperature. Suggest an interpretation of the latter observation.

10.13 Suppose you distrusted the Lennard-Jones (12,6) potential for assessing a particular polypeptide conformation, and replaced the repulsive term by an exponential function of the form $e^{-r/\sigma}$. Sketch the form of the potential energy and locate the distance at which it is a minimum.

10.14 Suppose that at a particular temperature it was found that the third virial coefficient C was approximately equal to B^2. Show that under these circumstances, $V_m = RT/p + B$.

10.15 Many liquids preserve some local features of the structure of the solid from which they form or to which they freeze. Sketch the form of the radial distribution function for a liquid that locally resembles (a) a cubic close-packed structure and (b) a body-centred cubic structure. In each case, show only the first two spheres of neighbours (the nearest and the next nearest).

10.16 The mobility of species through fluids is of the greatest importance for nutritional processes. (a) Estimate the diffusion coefficient for a molecule that leaps 150 pm each 1.8 ps. (b) What would be the diffusion coefficient if the molecule travelled only half as far on each step?

10.17 Pollutants spread through the environment by convection (winds and currents) and by diffusion. How many steps must a molecule take to be 1000 step lengths away from its origin if it undergoes (a) a one-dimensional, (b) a three-dimensional random walk?

10.18 Calculate the packing fraction of (a) a stack of cylinders and (b) a primitive cubic lattice.

10.19 How many (a) nearest neighbours, (b) next-nearest neighbours are there in a body-centred cubic structure. What are their distances if the side of the cube is 500 nm?

10.20 How many (a) nearest neighbours, (b) next-nearest neighbours are there in a cubic close-packed structure. What are their distances if the side of the cube is 500 nm?

10.21 The thermal and mechanical processing of materials is an important step in ensuring that they have the appropriate physical properties for their intended application. Suppose a metallic element underwent a phase transition in which its crystal structure changed from cubic close-packed to body-centred cubic. (a) Would it become more or less dense? (b) By what factor would its density change?

10.22 Draw a set of points as a rectangular array based on unit cells of side a and b, and mark the planes with Miller indices (10), (01), (11), (12), (23), (41), (4$\bar{1}$).

10.23 Repeat Exercise 10.22 for an array of points in which the a and b axes make $60°$ to each other.

10.24 In a certain unit cell, planes cut through the crystal axes at $(2a, 3b, c)$, (a, b, c), $(6a, 3b, 3c)$, $(2a, -3b, -3c)$. Identify the Miller indices of the planes.

10.25 Draw an orthorhombic unit cell and mark on it the (100), (010), (001), (011), (101), and (101) planes.

10.26 Draw a triclinic unit cell and mark on it the (100), (010), (001), (011), (101), and (101) planes.

10.27 Calculate the separations of the planes (111), (211), and (100) in a crystal in which the cubic unit cell has sides of length 532 pm.

10.28 The glancing angle of a Bragg reflection from a set of crystal planes separated by 97.3 pm is 19.85°. Calculate the wavelength of the X-rays.

10.29 Copper K_α radiation consists of two components of wavelengths 154.433 pm and 154.051 pm. Calculate the separation of the diffraction lines arising from the two components in a powder diffraction pattern recorded in a camera of radius 5.74 cm from planes of separation 73.2 pm.

10.30 The compound Rb_3TlF_6 has a tetragonal unit cell with dimensions $a = 651$ pm and $c = 934$ pm. Calculate the volume of the unit cell and the density of the solid.

10.31 The orthorhombic unit cell of $NiSO_4$ has the dimensions $a = 634$ pm, $b = 784$ pm, and $c = 516$ pm, and the density of the solid is estimated as $3.9\,\mathrm{g\,cm^{-3}}$. Determine the number of formula units per unit cell and calculate a more precise value of the density.

10.32 The unit cells of $SbCl_3$ are orthorhombic with dimensions $a = 812$ pm, $b = 947$ pm, and $c = 637$ pm. Calculate the spacing of (a) the (321) planes, (b) the (642) planes.

10.33 The separation of (100) planes of lithium metal is 350 pm and its density is $0.53\,\mathrm{g\,cm^{-3}}$. Is the structure of lithium fcc or bcc?

10.34 Copper crystallizes in an fcc structure with unit cells of side 361 pm. Predict the appearance of the powder diffraction pattern using 154 pm radiation. What is the density of copper?

Molecular spectroscopy

Spectroscopy is the analysis of the electromagnetic radiation emitted, absorbed, or scattered by molecules. We have already seen that photons act as messengers from inside atoms, and that atomic spectra can be used to obtain detailed information about electronic structure. Photons of radiation ranging from long radio waves to very short X-rays, or even γ rays, also bring information to us about molecules. The difference between molecular and atomic spectroscopy is that the energy of a molecule can change not only as a result of electronic transitions but also because it can make transitions between its rotational and vibrational states. The spectra are correspondingly more complicated, but they also contain more information. As well as its role in structural investigation—the determination of electronic energy levels, bond lengths, bond angles, bond strengths, and other features—molecular spectroscopy is used to monitor changing concentrations in kinetic studies (as was touched on in Section 7.1). Molecular spectroscopy is also the basis of our understanding of colour in the world around us, and how we can contribute to that colour by the formulation of new materials.

General features of spectroscopy

In **emission spectroscopy**, a molecule undergoes a transition from a state of high energy, E_1, to a state of lower energy, E_2, and emits the excess energy as a photon (Fig. 11.1). In **absorption spectroscopy**, the absorption of nearly monochromatic (single-frequency) incident radiation is monitored as it is swept over a range of frequencies. In **Raman spectroscopy**, a monochromatic incident beam—typically in the visible region of the spectrum and generated by a laser—is passed through the sample and the scattered radiation is analysed. Some of the incident photons collide with the molecules, giving up some of their energy or gaining energy from the molecule, and emerge with a different frequency (Fig. 11.2).

The energy $h\nu$ of the photon emitted or absorbed, and therefore the frequency, ν, of the radiation emitted or absorbed, is given by the Bohr frequency condition (Section 8.7):

$$h\nu = |E_1 - E_2| \tag{1}$$

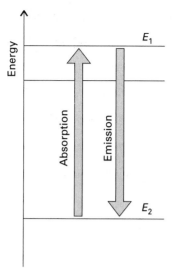

Fig. 11.1 In emission spectroscopy, a molecule returns to a lower state (typically the ground state) from an excited state, and emits the excess energy as a photon. The same transition may be observed in absorption, when the incident radiation supplies a photon that can excite the molecule from its ground state to an excited state.

where E_1 and E_2 are the energies of the two states between which the transition occurs. (Raman scattering is a special case, and we deal with it later.) This relation is often expressed in terms of the **wavelength**, λ, of the radiation (see *Further information 6*) by using the relation

$$\lambda = \frac{c}{\nu} \tag{2}$$

or the **wavenumber**, $\tilde{\nu}$:

$$\tilde{\nu} = \frac{1}{\lambda} = \frac{\nu}{c} \tag{3}$$

The units of wavenumber are almost always chosen as reciprocal centimetres, cm^{-1}. The chart in Fig. 11.3 summarizes the frequencies, wavelengths, and wavenumbers of the various regions of the electromagnetic spectrum.

11.1 Experimental techniques

Emission, absorption, and Raman spectroscopy give the same information about energy level separations, but practical considerations generally determine which technique is employed. In practice, emission spectroscopy, if it is used at all, is used mainly for the study of energetically excited reaction products. Absorption spectroscopy is much more widely used, and we shall concentrate on it. Raman spectroscopy became a feasible technique once intense, monochromatic laser radiation became readily available.

The apparatus

The radiation source generally produces radiation spanning a range of frequencies, but in a few cases (including lasers) it generates nearly monochromatic radiation. For the far infrared, the source is a mercury arc inside a quartz envelope, most of the radiation being generated by the hot quartz. A **Nernst filament** is used to generate radiation in the near infrared: it consists of a heated ceramic filament containing rare-earth oxides, and emits radiation closely resembling that of a hot black body (of the kind considered in Section 8.1). For the visible region of the spectrum, a tungsten–iodine lamp is used, which gives out intense white light. A discharge through deuterium gas or xenon in quartz is still widely used for the near ultraviolet. A device called a **klystron** (which is also used in radar installations) is used to generate microwaves. Radio-frequency radiation (of the kind used in nuclear magnetic resonance) is generated by causing electric currents to oscillate in coils of wire.

In all but specialized techniques using monochromatic microwave radiation and certain types of laser radiation, spectrometers include a component for separating the frequencies of the radiation so that the variation of the absorption with frequency can be monitored. In conventional spectrometers, this component is a **dispersing element** that separates different frequencies into different spatial directions. The simplest dispersing element is a glass or quartz prism (Fig. 11.4), but a diffraction grating is also widely used. A **diffraction grating** consists of a glass or ceramic plate into which fine grooves have been cut about 1000 nm apart (a spacing comparable to the wavelength of visible light) and covered with a reflective aluminium coating. The grating causes interference between waves reflected from its surface, and constructive interference occurs at specific angles that depend on the frequency of the radiation being used. That is, each wavelength of light is directed into a specific direction.

The third component of a spectrometer is the **detector**, the device that converts incident radiation into an electric current for the appropriate signal processing or plotting. Radiation-sensitive semiconductor devices are increasingly dominating this role in the spectrometer. However, in the optical and ultraviolet region, photographic recording or a **photomultiplier** are widely used. In the latter device, an incident photon ejects an electron from a photosensitive surface, the electron is accelerated by a potential difference, and ejects a shower of electrons where it strikes a screen. The latter electrons are accelerated, and each one releases a further shower on impact with another screen. Thus the impact of the initial photon is converted into a cascade of electrons, which is converted into a current in an external circuit.

Although semiconductor detectors are increasingly being used in the infrared, thermocouples are still widely used. A thermocouple detector consists typically of a blackened gold foil to which are attached thermoelectric alloys which generate an electric current when the temperature is raised. A **thermistor bolometer** is essentially a resistance thermometer, and is typically formed from a mixture of oxides deposited on quartz. In each case the radiation is chopped by a shutter that rotates in the beam so that an alternating signal is obtained from the detector (an oscillating signal is easier to

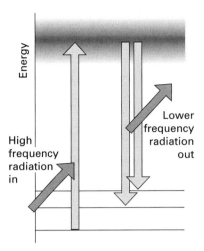

Fig. 11.2 In Raman spectroscopy, an incident photon is scattered from a molecule with either an increase in frequency (if the radiation collects energy from the molecule) or—as shown here—with a lower frequency if it loses energy to the molecule. The process can be regarded as taking place by an excitation of the molecule to a wide range of states (represented by the shaded band), and the subsequent return of the molecule to a lower state; the net energy change is then carried away by the photon.

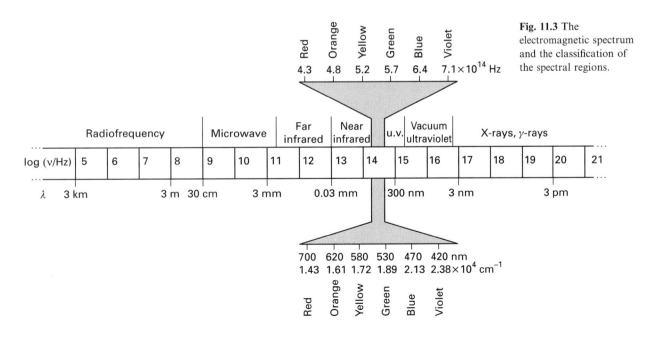

Fig. 11.3 The electromagnetic spectrum and the classification of the spectral regions.

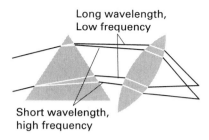

Fig. 11.4 One simple dispersing element is a prism, which separates frequencies spatially by making use of the higher refractive index for high-frequency radiation. The shortest wavelengths for which a glass prism can be used is about 400 nm, but quartz can be used down to 180 nm.

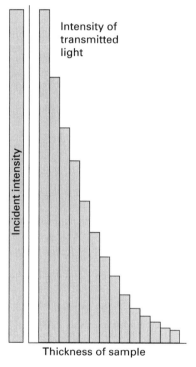

Fig. 11.5 The intensity of light transmitted by an absorbing sample decreases exponentially with the path length through the sample. The decrease in intensity for various sample thicknesses of a particular sample is depicted in this illustration for an incident intensity depicted by the height of the bar on the left.

amplify than a steady signal). A microwave detector is typically a crystal diode consisting of a tungsten tip in contact with a semiconductor, such as germanium, silicon, or gallium arsenide.

The highest resolution is obtained when the sample is gaseous and of such low pressure that collisions between the molecules are infrequent. Gaseous samples are essential for rotational (microwave) spectroscopy, for only in that phase can molecules rotate freely. To achieve sufficient absorption, the path lengths of gaseous samples must be very long, of the order of metres. Long path lengths are achieved by multiple passage of the beam between two parallel mirrors at each end of the sample cavity.

For infrared spectroscopy, the sample is typically a liquid held between windows of sodium chloride (which is transparent down to $700\,cm^{-1}$) or potassium bromide (down to $400\,cm^{-1}$). Other ways of preparing the sample include grinding into a paste with 'Nujol', a hydrocarbon oil, or pressing it into a solid disk, perhaps with powdered potassium bromide.

Measures of intensity

The intensity with which radiation is absorbed depends on the identity of the absorbing species (which we denote J), the frequency ν of the radiation, the molar concentration [J] of the species in the sample, and the path length l of the radiation through the sample. It is found experimentally that the **transmittance**, T, the ratio of the emerging intensity I to the incident intensity I_0, is given by the **Beer–Lambert law**:

$$\log T = -\varepsilon[\text{J}]l \qquad T = \frac{I}{I_0} \qquad (4)$$

The coefficient ε is the **molar absorption coefficient** (formerly the 'extinction coefficient') of the species, and it depends on the frequency of the incident light. Its units are those of $1/(\text{concentration} \times \text{length})$, and it is normally convenient[†] to express it in $mol^{-1}\,L\,cm^{-1}$. The dimensionless product

$$\mathcal{A} = \varepsilon[\text{J}]l \qquad (5)$$

is called the **absorbance** (formerly the 'optical density') of the sample. The Beer–Lambert law is sometimes written in the form

$$I = I_0 \times 10^{-\varepsilon[\text{J}]l} \qquad (6)$$

This expression shows that the transmitted intensity decreases exponentially with the length of the sample (Fig. 11.5).

[†]The alternative units $cm^2\,mol^{-1}$ bring out the point that ε is a molar cross-section for absorption, and the greater the cross-section of the molecule for absorption, the greater the reduction in the intensity of the beam for a given pathlength, concentration, and frequency.

Justification

The Beer–Lambert law can be derived by noting that the change in intensity, dI, that occurs when electromagnetic radiation passes through a layer of thickness dx is proportional to the thickness of the layer, the concentration, and the incident intensity in the region of the absorbing species. Because dI is negative (the intensity is reduced by absorption), we can write

$$dI = -\kappa[J]I \, dx$$

where κ is the proportionality coefficient, or equivalently

$$\frac{dI}{I} = -\kappa[J] \, dx$$

These expressions apply to each successive layer into which the sample can be regarded as being divided (Fig. 11.6). Therefore, to obtain the intensity, I, that emerges from a sample of thickness l when the incident intensity is I_0, we add (that is, integrate) all the successive changes:

$$\int_{I_0}^{I} \frac{dI}{I} = -\kappa \int_0^l [J] \, dx$$

When the concentration is uniform, $[J]$ is independent of x, and the expression integrates to

$$\ln \frac{I}{I_0} = -\kappa[J]l$$

The relation between natural and common logarithms is

$$\ln x = \ln 10 \times \log x$$

Therefore, by writing $\kappa = \varepsilon \ln 10$, we obtain

$$\log \frac{I}{I_0} = -\varepsilon[J]l$$

which, on substituting $T = I/I_0$, is the Beer–Lambert law.

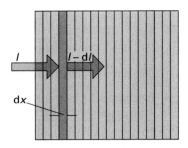

Fig. 11.6 To establish the Beer–Lambert law, the sample is supposed to be sliced into a large number of planes. The reduction in intensity caused by one plane is proportional to the intensity incident on it (after passing through the preceding planes), the thickness of the plane, and the concentration of absorbing species.

Example *Calculating the molar absorption coefficient*
Radiation of wavelength 256 nm passed through 1.0 mm of a solution that contained C_6H_6 at a concentration of $0.050 \, mol \, L^{-1}$. The light intensity is reduced to 16 per cent of its initial value (so $T = 0.16$). Calculate the absorbance and the molar absorption coefficient of the sample. What would be the transmittance through a 2.0 mm cell?

Strategy:
For ε we use eqn (4) rearranged into

$$\varepsilon = -\frac{\log T}{[\mathrm{X}]l}$$

and then combine this expression with the definition of A in eqn (5):

$$A = \varepsilon[\mathrm{X}]l = -\log T$$

For the transmittance through the thicker cell, we use eqn (6) and the value of ε calculated here.

Solution
The molar absorption coefficient is

$$\varepsilon = -\frac{\log 0.16}{(0.050\,\mathrm{mol\,L^{-1}}) \times (1.0\,\mathrm{mm})} = 16\,\mathrm{mol^{-1}\,L\,mm^{-1}}$$

The units could be tidied up (into $\mathrm{m^2\,mol^{-1}}$), but those shown are convenient in practice. The absorbance is

$$A = -\log 0.16 = 0.80$$

The absorbance of a sample of length 2.0 mm is

$$A = (16\,\mathrm{mol^{-1}\,L\,mm^{-1}}) \times (0.050\,\mathrm{mol\,L^{-1}}) \times (2.0\,\mathrm{mm}) = 1.6$$

It follows that the transmittance is

$$T = 10^{-A} = 10^{-1.6} = 0.025$$

That is, the emergent light is reduced to 2.5 per cent of its incident intensity.

Exercise E11.1

The transmittance of an aqueous solution that contained Cu^{2+} ions at a molar concentration of $0.10\,\mathrm{mol\,L^{-1}}$ at 600 nm was measured as 0.30 in a cell of length 5.0 mm. Calculate the molar absorption coefficient of $Cu^{2+}(\mathrm{aq})$ at that wavelength, and the absorbance of the solution. What would be the transmittance through a cell of length 1.0 mm?

[*Answer*: $10\,\mathrm{mol^{-1}\,L\,cm^{-1}}$, $A = 0.50$, $T = 0.79$]

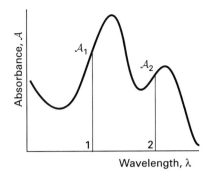

The maximum value of the molar absorption coefficient, ε_{max}, is an indication of the intensity of a transition. Typical values for strong transitions are of the order of 10^4–$10^5\,\mathrm{mol^{-1}\,L\,cm^{-1}}$, indicating that in a solution of molar concentration $0.01\,\mathrm{mol\,L^{-1}}$ the intensity of light (of the appropriate frequency) falls to 10 per cent of its initial value after passing through about 0.1 mm of solution.

Beer's law (as it is normally called) is used to determine the concentrations of species of known molar absorption coefficients. To do so, we measure the absorbance of a sample and rearrange eqn (5) into

$$[J] = \frac{A}{\varepsilon l} \tag{7}$$

Fig. 11.7 The concentrations of two absorbing species in a mixture can be determined from their molar absorption coefficients and the measurement of their absorbances at two different wavelengths lying within their joint absorption region.

More importantly, particularly in biological applications, it is possible to make measurements at two wavelengths and use them to find the concentrations of two components A and B in a mixture. For this analysis, we write the total absorbance as

$$A = A_A + A_B = \varepsilon_A[A]l + \varepsilon_B[B]l$$
$$= (\varepsilon_A[A] + \varepsilon_B[B])l$$

Then, for two measurements at wavelengths at which the molar absorption coefficients are ε_1 and ε_2 (Fig. 11.7), we have

$$A_1 = (\varepsilon_{A1}[A] + \varepsilon_{B1}[B])l$$
$$A_2 = (\varepsilon_{A2}[A] + \varepsilon_{B2}[B])l$$

These two simultaneous equations can be solved for the two unknowns (the molar concentrations of A and B), and we find

$$[A] = \frac{\varepsilon_{B2}A_1 - \varepsilon_{B1}A_2}{l\{\varepsilon_{A1}\varepsilon_{B2} - \varepsilon_{A2}\varepsilon_{B1}\}}$$
$$[B] = \frac{\varepsilon_{A1}A_2 - \varepsilon_{A2}A_1}{l\{\varepsilon_{A1}\varepsilon_{B2} - \varepsilon_{A2}\varepsilon_{B1}\}}$$

There may be a wavelength at which the molar extinction coefficients of the two species are equal. This wavelength is called the **isosbestic wavelength**, λ°. At this wavelength, the total absorbance of the mixture is

$$A^\circ = \varepsilon^\circ([A] + [B])l \tag{8}$$

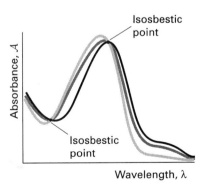

Fig. 11.8 One or more isosbestic points are formed when there are two interrelated absorbing species in solution.

Even if A and B are interconverted (as a result of changing the pH of a solution containing an indicator, for instance), because the total concentration remains constant, so does the absorbance at the isosbestic wavelength. As a result, one or more **isosbestic points**, invariant points in the absorption spectrum, may be observed (Fig. 11.8). Because it is very unlikely that three or more species would have the same molar extinction coefficients at a single wavelength, the observation of an isosbestic point, or at least not more than one such point, is compelling evidence that a solution consists of only two solutes in equilibrium with each other with no intermediates.

II.2 Intensities and linewidths

The intensity of a spectral line depends on the number of molecules that are in the initial state and the strength with which individual molecules are able to interact with the electromagnetic field and generate or absorb photons. If we confine our attention to vibrational and electronic spectroscopy, then the situation is very simple: almost all vibrational absorptions and all electronic absorptions occur from the ground state of a molecule, because that is the only state populated at room temperature. However, molecules can be prepared in short-lived excited states as a result of chemical reaction, electric discharge, or photolysis. In these cases the populations may be quite different from those at thermal equilibrium, and absorption and emission spectra—if they can be recorded quickly enough—then arise from transitions from all the populated levels.

Spectroscopic lines are not infinitely narrow, and in condensed media may spread over several thousand reciprocal centimetres. One important broadening process in gaseous samples is the **Doppler effect**, in which radiation is shifted in frequency when the source is moving towards or away from the observer. When a source emitting radiation of frequency ν recedes with a speed s, the observer detects radiation of frequency

$$\nu' = \left(\frac{1 - s/c}{1 + s/c}\right)^{1/2} \nu \tag{9}$$

where c is the speed of the radiation (the speed of light for electromagnetic radiation, the speed of sound for sound waves). A source approaching the observer appears to be emitting radiation of frequency

$$\nu' = \left(\frac{1 + s/c}{1 - s/c}\right)^{1/2} \nu \tag{10}$$

Exercise E11.2

A laser line occurs at 628.443 cm^{-1}. What wavenumber will an observer detect when approaching the laser at (a) 1 m s^{-1}, (b) 1000 m s^{-1}?

[*Answer*: (a) 628.443 cm^{-1}, (b) 628.445 cm^{-1}]

Molecules reach high speeds in all directions in a gas, and a stationary observer detects the corresponding Doppler shifted range of frequencies. Some molecules approach the observer, some move away; some move quickly, others slowly. The detected spectroscopic 'line' is the absorption or emission profile arising from all the resulting Doppler shifts. The profile reflects the Maxwell distribution of molecular speeds (Section 1.7), and we observe a bell-shaped Gaussian curve (a shape of the form e$^{-x^2}$). The

Doppler line shape is therefore also a Gaussian curve (Fig. 11.9), and calculation shows that when the temperature is T and the molar mass of the molecule is M, the width of the line at half its maximum height is

$$\delta\lambda = \frac{2\lambda}{c}\sqrt{\frac{2RT\ln 2}{M}} \qquad (11)$$

The Doppler width increases with temperature because the molecules acquire a wider range of speeds. Therefore, to obtain spectra of maximum sharpness, it is best to work with cold gaseous samples.

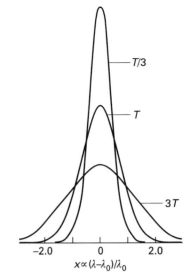

Fig. 11.9 The shape of a Doppler-broadened spectral line reflects the Maxwell distribution of speeds in the sample at the temperature of the experiment. Notice that the line broadens as the temperature is increased. The width at half height is given by eqn (11).

Exercise E11.3

The Sun emits a spectral line at 677.4 nm which has been identified as arising from a transition in highly ionized ^{57}Fe. Its width at half-height is 5.3 pm. What is the temperature of the Sun's surface?

[*Answer*: 6.8×10^3 K]

Another source of line broadening is related to the lifetimes of the states involved in the transition. When the Schrödinger equation is solved for a system that is changing with time, it is found that its states do not have precisely defined energies. If on average a system survives in a state for a time τ, the lifetime of the state, then its energy levels are blurred to an extent of order δE, where

$$\delta E \approx \frac{\hbar}{\tau} \qquad (12)$$

The energy spread inherent to the states of systems that have finite lifetimes is called **lifetime broadening**. When we express the energy spread in wavenumbers through $\delta E = hc\delta\tilde{\nu}$ and use the values of the fundamental constants, the practical form of this relation becomes

$$\delta\tilde{\nu} \approx \frac{5.3\,\text{cm}^{-1}}{\tau/\text{ps}} \qquad (13)$$

Only if τ is infinite can the energy of a state be specified exactly (with $\delta E = 0$). However, no excited state has an infinite lifetime; therefore, all states are subject to some lifetime broadening, and the shorter the lifetimes of the states involved in a transition, the broader the spectral lines.

419

Exercise E11.4

What is the width (in wavenumbers) of a transition from a state with a lifetime of 5.0 ps?

[*Answer*: $1.0 \, cm^{-1}$]

Two processes are principally responsible for the finite lifetimes of excited states, and hence for the widths of transitions to or from them. The dominant one is **collisional deactivation**, which arises from collisions between molecules or with the walls of the container. If the **collisional lifetime** is τ_{col}, then the resulting collisional linewidth is $\delta E_{col} \approx \hbar/\tau_{col}$. In gases, the collisional life-time can be lengthened, and the broadening minimized, by working at low pressures. The second contribution is **spontaneous emission**, the emission of radiation when an excited state collapses into a lower state. The rate of spontaneous emission depends on details of the wavefunctions of the excited and lower states. Because its rate cannot be changed (without changing the molecule), it is a natural limit to the lifetime of an excited state. The resulting lifetime broadening is the **natural linewidth** of the transition. The natural linewidth cannot be changed by modifying the temperature or pressure. Natural linewidths depend strongly on the transition frequency ν (they increase as ν^3), so low-frequency transitions (such as the microwave transi-tions of rotational spectroscopy) have very small natural linewidths, and collisional and Doppler line-broadening processes are dominant. The natural lifetimes of electronic transitions are very much shorter than for vibrational transitions, so the natural linewidths of electronic transitions are much greater than those of vibrational and rotational transitions. For example, a typical electronic excited state natural lifetime is about $10^{-8} \, s$ (10^4 ps), corre-sponding to a natural width of about $5 \times 10^{-4} \, cm^{-1}$ (equivalent to 15 MHz).

Rotational spectroscopy

Molecules of substances that can be vaporized and form a low pressure gas undergo free rotation. The energy levels of rotating molecules are quantized, and transitions between these levels gives rise to the **rotational spectrum** of a molecule. Very little energy is needed to change the state of rotation of a molecule, and the electromagnetic radiation emitted or absorbed lies in the microwave region, with wavelengths of the order of 1 cm.

11.3 The rotational energy levels of molecules

To a first approximation, the rotational states of molecules are based on a model system called a **rigid rotor**, a body that can rotate but is not distorted by the rotation in any way. The simplest type of rigid rotor is called a **linear rotor**, and corresponds to a linear molecule, such as HCl, CO_2, or $HC{\equiv}CH$ that is supposed not to be able to bend or stretch. When the Schrödinger equation is solved for a linear rotor (in much the same way as was indicated in Section 8.6), the energy levels turn out to be

$$E_J = hcBJ(J+1) \qquad J = 0, 1, 2, \cdots \qquad (14)$$

These energy levels are illustrated in Fig. 11.10: note that the separation increases with J. Note also that, because J may be 0, the lowest possible energy is 0: there is no zero-point rotational energy for molecules. The constant B is called the **rotational constant** of the molecule, and is defined as

$$B = \frac{\hbar}{4\pi cI} \qquad (15)$$

where I is the **moment of inertia** of the molecule. The moment of inertia plays a role in rotation analogous to mass in translation. A body with a high moment of inertia (like that of a flywheel or a heavy molecule) undergoes only a small rotational acceleration when a twisting force (a torque) is applied, but a body with a small moment of inertia undergoes a large acceleration for the same torque. The moment of inertia of a molecule depends on the masses of the atoms and their distances from the centre of mass of the molecule (the point about which rotation occurs), and for a diatomic molecule of bond length R and atomic masses m_A and m_B it is

$$I = \mu R^2 \qquad \mu = \frac{m_A m_B}{m_A + m_B} \qquad (16)$$

Note that, because B is inversely proportional to I, the larger the moment of inertia (corresponding to a long bond and heavy atoms), the smaller the rotational constant.

A number of nonlinear molecules can be modelled as a **symmetric rotor**, a rigid rotor in which the moments of inertia about two axes are the same but different from a third (and all three are nonzero). An example is ammonia, NH_3, and another is phosphorus pentachloride, PCl_5 (Fig. 11.11). The energy levels of a symmetric rotor are determined by two quantum numbers, J and K, and are

$$E_{J,K} = hcBJ(J+1) + hc(A-B)K^2 \quad \begin{cases} J = 0, 1, 2, \cdots \\ K = J, J-1, \cdots, -J \end{cases} \qquad (17)$$

The rotational constants A and B correspond to the moments of inertia parallel and perpendicular to the axis of the molecule (Fig. 11.12):

$$A = \frac{\hbar}{4\pi cI_{\parallel}} \qquad B = \frac{\hbar}{4\pi cI_{\perp}} \qquad (18)$$

The quantum number K tells us the extent to which the molecule is rotating about its axis: when $K = 0$, the molecule is rotating end-over-end; when

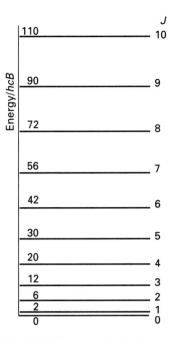

Fig. 11.10 The energy levels of a linear rigid rotor as multiples of hcB.

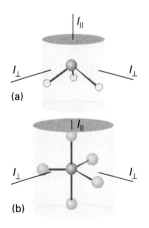

Fig. 11.11 The two different moments of inertia of (a) a trigonal pyramidal molecule and (b) a trigonal bipyramidal molecule.

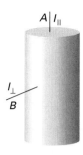

Fig. 11.12 The two rotational constants of a symmetric rotor, which are inversely proportional to the moments of inertia parallel and perpendicular to the axis of the molecule.

$K = \pm J$, it is rotating mainly about its symmetry axis. Intermediate values of K correspond to a mixture of the two modes of rotation (Fig. 11.13).

A special case of a symmetric rotor is a **spherical rotor**, a rigid body with three equal moments of inertia (like a sphere). Methane, CH_4, is one example; the octahedral SF_6 molecule is another. When $I_{\parallel} = I_{\perp}$, $A = B$, and then the energy levels simplify to the same expression as in eqn (14).

11.4 Rotational transitions: microwave spectroscopy

We met the concept of a **selection rule** in Section 8.10 as a statement about which transitions are forbidden and which are allowed. Selection rules also apply to molecular spectra, and the form they take depends on the type of transition. The idea to keep in mind is that, for the molecule to interact with the electromagnetic field and absorb or create a photon of frequency ν, it must possess, at least transiently, a dipole oscillating at that frequency. The dipole moment associated with the transition is called a **transition moment**. A large transition moment gives a strong jolt to the electromagnetic field and results in an intense transition.

A **gross selection rule** specifies the general features a molecule must have if it is to have a spectrum of a given kind. For a rotation of a molecule to give rise to an absorption or emission spectrum, *the molecule must be polar*. The classical basis of this rule is that a stationary observer watching the rotating polar molecule sees its partial charges moving backwards and forwards (Fig. 11.14), and their motion shakes the electromagnetic field into oscillation. Because the molecule must be polar, it follows that tetrahedral (CH_4, for instance), octahedral (SF_6), symmetric linear (CO_2), and homonuclear diatomic (H_2) molecules do *not* have rotational spectra. On the other hand, heteronuclear diatomic (HCl) and less symmetrical polar polyatomic molecules (NH_3) do have rotational spectra.

A **specific selection rule** tells us which quantum states transitions may occur between, provided that transitions are allowed by the gross selection rule. For rotational transitions, the specific selection rules are

$$\Delta J = \pm 1 \qquad \Delta K = 0$$

The first of these selection rules can be traced, like the $\Delta l = \pm 1$ rule for atoms (Section 8.10), to the preservation of the total angular momentum when a photon is absorbed or created. A photon is a spin-1 particle, and when one is absorbed or created the angular momentum of the molecule must change by a compensating amount. The rule about K can be traced to the fact that the partial charges of a molecule do not move when a molecule rotates around its symmetry axis, so there can be no acceleration or deceleration about that axis by the absorption or emission of electromagnetic radiation.

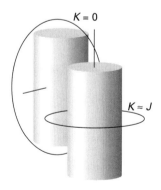

Fig. 11.13 When $K = 0$ for a symmetric rotor, the entire motion of the molecule is around an axis perpendicular to the symmetry axis of the rotor. When the value of K is close to J, almost all the motion is around the symmetry axis.

When a molecule changes its rotational quantum number from J to $J + 1$ in an absorption, the change in energy is

$$\Delta E = E_{J+1} - E_J = hcB\{(J+1)(J+2) - J(J+1)\}$$
$$= 2hcB(J+1) \qquad (19)$$

The energies of the allowed transitions are $2hcB, 4hcB, 6hcB, \cdots$, and the wavenumbers of the radiation absorbed are therefore $2B, 4B, 6B, \cdots$. (To obtain the latter, we have identified ΔE with the energy of a photon written in terms of its wavenumber as $hc\tilde{\nu}$, and then cancelled the hc.) A rotational spectrum of a polar linear molecule (HCl) and of a polar symmetric rotor (NH_3), therefore consists of a series of lines separated by $2B$ (Fig. 11.15). Substitution of typical values of B for molecules shows that rotational absorption occurs in the microwave region of the spectrum, so rotational spectroscopy is also called **microwave spectroscopy**.

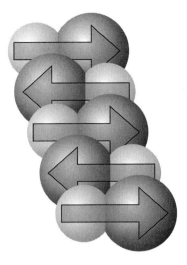

Fig. 11.14 To an external observer, a rotating polar molecule has an electric dipole (the arrow) that oscillates in direction. This oscillating dipole can interact with the electromagnetic field.

Example *Estimating the wavenumber of a rotational transition*
Estimate the wavenumber and frequency of the $J = 0 \rightarrow 1$ transition of the ${}^1H^{35}Cl$ molecule. The masses of the two atoms are 1.673×10^{-27} kg and 5.807×10^{-26} kg, respectively, and the equilibrium bond length is 127.4 pm.

Strategy
The calculation depends on the value of B, which we can obtain by substituting the data into eqn (15). The wavenumber of the transition is $2B$, and to convert this wavenumber to a frequency, multiply by c.

Solution
The moment of inertia of the molecule is

$$I = \mu R^2$$
$$= \frac{(1.673 \times 10^{-27}\,\text{kg}) \times (5.807 \times 10^{-26}\,\text{kg})}{(1.673 \times 10^{-27}\,\text{kg}) + (5.807 \times 10^{-26}\,\text{kg})} \times (1.274 \times 10^{-10}\,\text{m})^2$$
$$= 2.639 \times 10^{-47}\,\text{kg m}^2$$

Therefore, the rotational constant is

$$B = \frac{\hbar}{4\pi cI}$$
$$= \frac{(1.054\,57 \times 10^{-34}\,\text{J s})}{4\pi \times (2.99795 \times 10^8\,\text{m s}^{-1}) \times (2.639 \times 10^{-47}\,\text{kg m}^2)}$$
$$= 1.061 \times 10^3\,\text{m}^{-1}$$

which corresponds to $10.61\,\text{cm}^{-1}$. It follows that the wavenumber of the transition is

$$\tilde{\nu} = 2B = 21.22\,\text{cm}^{-1}$$

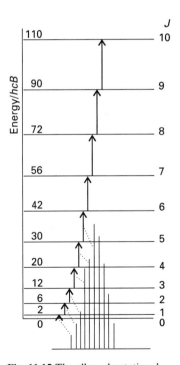

Fig. 11.15 The allowed rotational transitions (shown as absorptions) for a polar linear molecule.

This wavenumber corresponds to the frequency

$$\nu = c\tilde{\nu} = (2.99795 \times 10^8 \, \text{m s}^{-1}) \times (2.122 \times 10^3 \, \text{m}^{-1}) = 6.362 \times 10^{11} \, \text{s}^{-1}$$

or 636.2 GHz.

Exercise E11.5

What is the wavenumber and frequency of the same transition in the $^2\text{H}^{35}\text{Cl}$ molecule? The mass of ^2H is 3.344×10^{-27} kg. Before commencing the calculation, decide whether the frequency should be higher or lower than for $^1\text{H}^{35}\text{Cl}$.

[*Answer*: $10.91 \, \text{cm}^{-1}$, 327.0 GHz]

Once we measure the separation between neighbouring lines, we can use the value of B to obtain a value for the moment of inertia $I_\perp$. For a diatomic molecule, we can convert that value to a value of the bond length, R, by using eqn (16). Highly accurate bond lengths can be obtained in this way. More complicated procedures can be used to obtain the bond lengths in simple polyatomic molecules and to measure the electric dipole moments of polar molecules from the modification of the appearance of their rotational spectra caused by applied electric fields.

11.5 Rotational Raman spectra

In a Raman spectrum, the incident radiation—which is typically monochromatic blue light or ultraviolet radiation from a laser—is scattered by the molecules in the sample and is detected by a detector placed at 90° to the incident beam (Fig. 11.16). It is found that the scattered radiation is shifted in frequency and that it consists of several frequencies. Lines shifted to lower frequency are called **Stokes lines** and lines shifted to higher frequency are called **anti-Stokes lines**. The Stokes lines arise because the collision of a photon with a molecule results in the excitation of rotation, and so the photon responsible for the excitation loses some energy and hence travels to the detector with a lower energy and hence a lower frequency. The anti-Stokes radiation arises from photon–molecule collisions in which the incoming photon captures some rotational energy from the molecule, and so travels to the detector with higher energy and hence higher frequency.

The gross selection rule for rotational Raman spectra is that *the polarizability of the molecule must be anisotropic*. There are two terms to appreciate here, the polarizability and its anisotropy. The **polarizability** of a molecule is a measure of the extent to which an applied electric field can induce an electric dipole moment (recall Section 10.2). The **anisotropy** of the

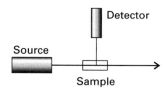

Fig. 11.16 The experimental arrangement for the observation of Raman spectra. The source is a laser, and the scattered radiation is detected electronically.

polarizability is the variation of this polarizability with the orientation of the molecule relative to the applied field. Tetrahedral (CH_4), octahedral (SF_6), and icosohedral (C_{60}) molecules have the same polarizability regardless of their orientations, so these molecules are rotationally Raman inactive: they do not have rotational Raman spectra. All other molecules, including homonuclear diatomic molecules such as H_2, are rotationally Raman active.

The specific selection rules for the rotational Raman transitions of linear molecules (the only ones we consider) are

$$\Delta J = +2 \,(\text{Stokes lines}) \qquad \Delta J = -2 \,(\text{anti-Stokes lines})$$

It follows that the change in energy when a molecule makes the transition $J \rightarrow J + 2$ is

$$\begin{aligned} \Delta E &= hcB(J + 2)(J + 3) - hcBJ(J + 1) \\ &= 2hcB(2J + 3) \end{aligned} \qquad (20)$$

Therefore, when a photon scatters from molecules in the rotational states $J = 0, 1, 2, 3, \cdots$, and transfers energy to them, its energy is decreased by $6hcB, 10hcB, 14hcB, \cdots$ and its wavenumber is reduced by $6B, 10B, 14B, \cdots$ from the wavenumber of the incident radiation. If the photon acquires energy during the collision, then a similar argument shows that the anti-Stokes lines occur with wavenumbers $6B, 10B, 14B, \cdots$ higher than the incident radiation (Fig. 11.17). It follows that from a measurement of the separation of the lines, the value of B can be determined and from that the bond length can be found. Because homonuclear diatomic species are rotationally Raman active, this technique can be applied to them as well as to heteronuclear species. The incident radiation must be highly monochromatic, for otherwise it would mask the Raman lines.

Vibrational spectra

All molecules are capable of vibrating, and complicated molecules may do so in a large number of different modes. Even benzene, with 12 atoms, can vibrate in 30 different modes, some of which involve the periodic enlargement and contraction of the ring and others its buckling into various distorted shapes. A molecule as big as a protein can vibrate in tens of thousands of different ways, twisting, stretching, and buckling in different regions and in different manners. However, according to quantum mechanics, no vibration can be excited unless the molecule has been provided with a certain minimum energy. Vibrations can be excited by the absorption of electromagnetic radiation, and observing the frequencies at which absorption occurs gives very valuable information about the identity of the molecule and provides quantitative information about the flexibility of its bonds.

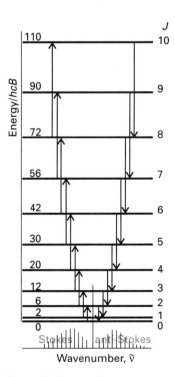

Fig. 11.17 The transitions responsible for the Stokes and anti-Stokes lines of a rotational Raman spectrum of a linear molecule.

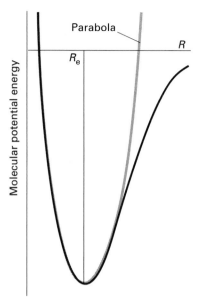

Fig. 11.18 The molecular potential energy curve can be approximated by a parabola near the bottom of the well. (A parabola is described by the equation $y = ax^2$, with a a constant; here $x = R - R_e$.) A parabolic potential results in harmonic oscillation. At high vibrational excitation energies the parabolic approximation is poor.

11.6 The vibrations of molecules

We shall base our discussion on Fig. 11.18, which shows a typical potential energy curve (it is a reproduction of Fig. 9.1) of a diatomic molecule as its bond is lengthened by pulling one atom away from the other or shortened by pushing the atoms together. In regions close to the equilibrium bond length R_e (at the minimum of the curve) the potential energy can be approximated by a parabola (a curve of the form $y = ax^2$), and we can write

$$V = \tfrac{1}{2}k(R - R_e)^2 \qquad (21)$$

where k is the **force constant** of the bond. The name derives from the relation between the restoring force experienced by the atoms when a bond is stretched and the extent of the stretching:

$$\text{restoring force} = -k(R - R_e) \qquad (22)$$

That is, k is the constant of proportionality between the displacement from equilibrium and the strength of the resulting restoring force. The steeper the walls of the potential (the stiffer the bond), the greater the force constant (Fig. 11.19). Bending modes of molecules generally have weaker force constants than stretching modes.

Exercise E11.6

What are the SI units of k? Confirm that V is expressed in joules.

[*Answer*: $N\,m^{-1}$; $(1\,N\,m^{-1}) \times (1\,m)^2 = 1\,J$]

Vibrational energy levels

The permitted energy levels of a vibrating diatomic molecule are obtained by solving the Schrödinger equation for the motion of the two atoms of masses m_1 and m_2 with the potential energy in eqn (21), and are

$$E_v = (v + \tfrac{1}{2})\hbar\omega, \qquad v = 0, 1, 2, \cdots \qquad (23)$$

where

$$\omega = \sqrt{\frac{k}{\mu}} \qquad \mu = \frac{m_1 m_2}{m_1 + m_2} \qquad (24)$$

The quantity ω is the *circular* frequency of the oscillator in radians per second. It is related to the frequency ν (in hertz, cycles per second) by $\omega = 2\pi\nu$. These energy levels are illustrated in Fig. 11.20, which shows that they form a uniformly spaced ladder, with separation $\hbar\omega$ between neighbours. It is important to note that the vibrational energy levels depend on the **effective mass**, μ, of the molecule, not its total mass. If atom 1 were as heavy as a brick wall, we could neglect m_2 in the denominator of μ, and so would find $\mu \approx m_2$, the mass of the lighter atom. The vibration would then be that of a light atom relative to a stationary wall (this is approximately the case in HI, for example, where the I atom barely moves and $\mu \approx m_H$). In the case of a homonuclear diatomic molecule, for which $m_1 = m_2 = m$, the effective mass is half the mass of one atom: $\mu = \frac{1}{2}m$.

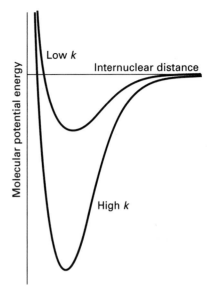

Exercise E11.7

An HCl molecule has a force constant of $516\,\text{N m}^{-1}$, a reasonably typical value. Calculate the vibrational frequency, ω, of the molecule and the energy separation between any two neighbouring vibrational energy levels.

[*Answer*: $5.63 \times 10^{14}\,\text{s}^{-1}$, $5.94 \times 10^{-20}\,\text{J}$]

Fig. 11.19 A low value of k indicates a loose bond; a high value indicates a stiff bond. Although the value of k is not directly related to the strength of the bond, this illustration indicates that it is likely that a strong bond (one with a deep minimum) has a large force constant.

Incident electromagnetic radiation can excite a molecular vibration if its photons carry enough energy. Because a typical vibrational excitation energy is of the order of 10^{-20}–10^{-19} J (see Exercise E11.7), the frequency of the radiation should be of the order of 10^{13}–10^{14} Hz (from $\Delta E = h\nu$). This frequency corresponds to infrared radiation, so vibrational transitions are observed by **infrared spectroscopy**.

A significant feature of eqn (23) is that, because the lowest vibrational state has $v = 0$, corresponding to an energy of $\frac{1}{2}\hbar\omega$, a molecular vibration cannot be stopped completely. That is, each vibrational mode of a molecule has an irremovable zero-point energy. We can think of the molecule as perpetually quivering about its equilibrium bond lengths and angles (this quivering would occur even at $T = 0$). For a diatomic molecule, with only one mode of vibration, the energy locked up as zero-point vibration is quite small, but for a macromolecule with tens of thousands of vibrations, the energy stored in this way may be considerable.

The allowed transitions

For a vibration of a molecule to give rise to an absorption spectrum, *the electric dipole moment of the molecule must change during the vibration*. The basis of this rule is that the molecule can shake the electromagnetic field into oscillation only if it has an electric dipole moment that oscillates as the molecule vibrates (Fig. 11.21). The molecule need not have a permanent dipole: the rule requires only a *change* in dipole moment, possibly from zero. Some vibrations do not affect the molecule's dipole moment (for example, the stretching motion of a homonuclear diatomic molecule), and so they neither absorb nor generate radiation. Homonuclear diatomic molecules are

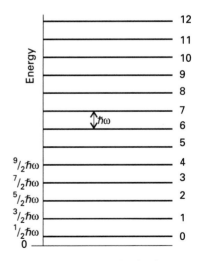

Fig. 11.20 The energy levels of an harmonic oscillator. The quantum number v ranges from 0 to infinity, and the permitted energy levels form a uniform ladder with spacing $\hbar\omega$. The molecule has a zero-point energy of $\frac{1}{2}\hbar\omega$, which can never be removed.

427

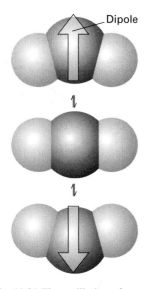

Fig. 11.21 The oscillation of a molecule, even if it is nonpolar, may result in an oscillating dipole that can interact with the electromagnetic field. Here we see a representation of a bending mode of CO_2.

therefore **infrared inactive**, because their dipole moments remain zero however long the bond, but heteronuclear diatomic molecules—which have a dipole moment that changes as the bond lengthens and contracts—are **infrared active**.

Example *Using the gross selection rules*
State which of the following molecules have vibrational absorption spectra: N_2, CO_2, OCS, H_2O, $CH_2 = CH_2$, C_6H_6.

Strategy
Molecules that give rise to vibrational spectra have dipole moments that change during the course of a vibration. Therefore, judge whether a distortion of the molecule can change its dipole moment (including changing it from zero).

Solution
All the molecules except N_2 possess at least one vibrational mode that results in a change of dipole moment, and so all except N_2 can show a vibrational absorption spectrum. It should be noted that not all the modes of complicated molecules are vibrationally active. For example, a vibration of CO_2 in which the O—C—O bonds stretch and contract symmetrically is inactive because it leaves the dipole moment unchanged (at zero).

Exercise E11.8

Repeat the example for H_2, NO, N_2O, CH_4.

[*Answer*: NO, N_2O, CH_4]

A more detailed study of the vibrational wavefunctions leads to the specific selection rule:

$$\Delta v = \pm 1 \qquad (25)$$

(In this expression, $\Delta v = +1$ corresponds to absorption and $\Delta v = -1$ corresponds to emission.) It follows that the change in energy for the transition from a state with quantum number v to one with quantum number $v + 1$ is

$$\Delta E = (v + \tfrac{3}{2})\hbar\omega - (v + \tfrac{1}{2})\hbar\omega = \hbar\omega \qquad (26)$$

The frequency ν (in hertz) of the photon absorbed is $\omega/2\pi$ and its wavenumber is $\omega/2\pi c$. That is, absorption occurs at the frequency

$$\nu = \frac{1}{2\pi}\sqrt{\frac{k}{\mu}} \tag{27}$$

Molecules with stiff bonds (large k) joining atoms with low masses (small μ) have high vibrational frequencies. As we have seen, bending modes are usually less stiff than stretching modes, so bends tend to occur at lower frequencies than stretches.

At room temperature, almost all the molecules will be in their vibrational ground states initially. Hence, the dominant spectral transition will be from $v = 0$ to $v = 1$. It follows from the calculation of ω for HCl (in Exercise E11.7), that $\nu = 8.95 \times 10^{13}$ Hz, so the infrared spectrum of the molecule will be an absorption at that frequency. The corresponding wavenumber and wavelength are $2990\,\text{cm}^{-1}$ and $3.35\,\mu\text{m}$, respectively.

Exercise E11.9

The force constant of the CO group in a peptide link is approximately $1.2\,\text{kN}\,\text{m}^{-1}$. At what wavenumber would you expect it to absorb? [*Hint*: For the effective mass, treat the group as a CO molecule.]

[*Answer*: at approximately $1700\,\text{cm}^{-1}$]

Vibrational Raman spectra of diatomic molecules

The gross selection rule for vibrational Raman transitions is that *the molecular polarizability should change as the molecule vibrates*. The polarizability plays a role in vibrational Raman spectroscopy because the molecule must be squeezed and stretched by the incident radiation in order that a vibrational excitation may occur during the photon–molecule collision. Both homonuclear and heteronuclear diatomic molecules swell and contract during a vibration, and the control of the nuclei over the electrons, and hence the molecular polarizability, changes too. Both types of diatomic molecule are therefore vibrationally Raman active.

The specific selection rule for vibrational Raman transitions is the same as for infrared transitions ($\Delta v = \pm 1$). The photons that are scattered with a lower wavenumber than that of the incident light, the Stokes lines, are those for which $\Delta v = +1$. They are stronger than the anti-Stokes lines, the radiation scattered with a higher wavenumber and arising from transitions in which $\Delta v = -1$, because very few molecules are in an excited vibrational state initially.

The information available from vibrational Raman spectra adds to that from infrared spectroscopy because homonuclear diatomic molecules

Table 11.1 Properties of diatomic molecules

	$\tilde{\nu}/cm^{-1}$	r/pm	$k/(N\ m^{-1})$	$D/(kJ\ mol^{-1})$
$^1H_2^+$	2333	106	160	256
$^1H_2$	4400	74	575	432
$^2H_2$	3118	74	577	440
$^1H^{19}F$	4138	92	955	564
$^1H^{35}Cl$	2991	127	516	428
$^1H^{81}Br$	2649	141	412	363
$^1H^{127}I$	2308	161	314	295
$^{14}N_2$	2358	110	2294	942
$^{16}O_2$	1580	121	1177	494
$^{19}F_2$	892	142	445	154
$^{35}Cl_2$	560	199	323	239

can also be studied. The spectra can be interpreted in terms of the force constants, dissociation energies, and bond lengths, and some of the information obtained is included in Table 11.1

11.7 The vibrations of polyatomic molecules

There is only one mode of vibration in a diatomic molecule, the stretching and compression of the bond. In a nonlinear polyatomic molecule of N atoms there are $3N - 6$ modes because bonds may stretch and angles may bend. For example, H_2O is a triatomic nonlinear molecule, and has three modes of vibration. Naphthalene, $C_{10}H_8$, has 48 distinct modes of vibration. In a linear molecule of N atoms, there are $3N - 5$ vibrational modes. Any diatomic molecule ($N = 2$) has one mode; carbon dioxide ($N = 3$) has four modes.

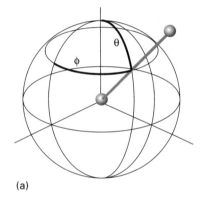

(a)

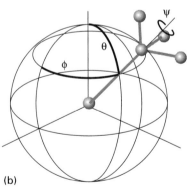

(b)

Fig. 11.22 (a) The orientation of a linear molecule requires the specification of two angles (the latitude and longitude of its axis). (b) The orientation of a nonlinear molecule requires the specification of three angles (the latitude and longitude of its axis and the angle of twist—the azimuthal angle—around that axis).

Justification

Each atom may change its location by varying its position along any of three perpendicular axes, and so the total number of such displacements that are available in a molecule in which there are N atoms is $3N$. Three of these coordinates are needed to specify the location of the centre of mass of the molecule, and so three of these displacements correspond to the translational motion of the molecule as a whole. The remaining $3N - 3$ displacements are 'internal' modes of the molecule that leave its centre of mass unchanged. Three angles are needed to specify the orientation of a nonlinear molecule in space (Fig. 11.22). Therefore three of the $3N - 3$ internal displacements are rotational, which leaves $3N - 6$ displacements that change neither the centre of mass of the molecule nor its orientation in space. These $3N - 6$ displacements are the vibrational modes. A similar calculation for linear molecules, which require only two angles to specify their orientation in space, shows that such molecules have $3N - 5$ vibrational modes.

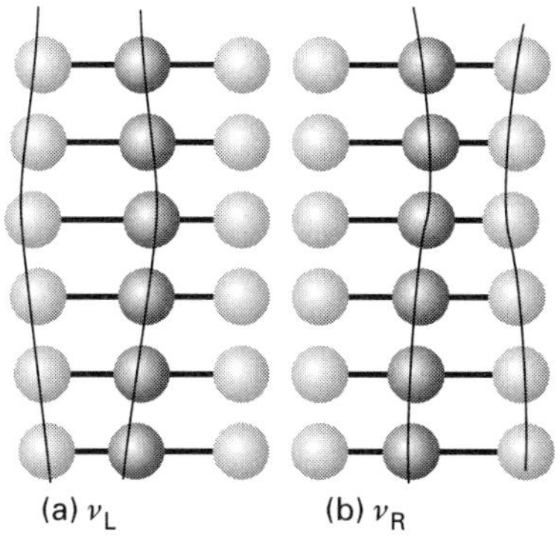

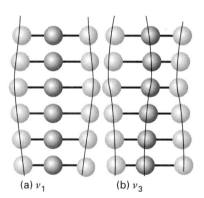

(a) ν_L (b) ν_R

Fig. 11.23 The stretching vibrations of a CO_2 molecule can be represented in a number of ways, In this representation, (a) one OC bond vibrates and the remaining O atom is stationary, and (b) the CO bond vibrates while the other O atom is stationary. Because the stationary atom is linked to the C atom, it does not remain stationary for long. That is, if one vibration begins, it rapidly stimulates the other to occur.

The description of the vibrational motion of a polyatomic molecule is much simpler if combinations of the stretching and bending motions of individual bonds are taken. For example, although we could describe two of the four vibrations of a CO_2 molecule as individual carbon–oxygen bond stretches, ν_L and ν_R in Fig. 11.23, the description of the motion becomes much simpler if two combinations of these vibrations are used instead. One combination is ν_1 in Fig. 11.24: this combination is the **symmetric stretch**. Another combination is ν_3, the **antisymmetric stretch**, in which the two O atoms always move in the same directions and opposite to the C atom. The two modes are independent in the sense that if one is excited, then its motion does not excite the other. They are two of the **normal modes** of the molecule, its independent, collective vibrational displacements. The two other normal modes are the **bending modes**, ν_2. In general, a normal mode is an independent, synchronous motion of atoms or groups of atoms that may be excited without leading to the excitation of any other normal mode.

Exercise E11.10

How many normal modes of vibration are there in (a) ethyne (HC≡CH) and (b) a protein molecules of 4000 atoms?

[*Answer*: (a) 7, (b) 11994]

The four normal modes of CO_2, and the N_{vib} normal modes of polyatomic molecules in general, are the key to the description of molecular vibrations. Each normal mode behaves like an independent harmonic oscillator and the energies of the vibrational levels are given by the same expression as in eqn (23), but with an effective mass, μ, that depends on the extent to which each of the atoms contributes to the vibration. Atoms that do not move, such as the C atom in the symmetric stretch of CO_2, do not contribute to the effective mass. The force constant also depends in a complicated way on the extent to which bonds bend and stretch during a vibration. Typically, a normal mode that is largely a bending motion has a lower force constant (and hence a lower frequency) than a normal mode that is largely a stretching motion.

(a) ν_1 (b) ν_3

Fig. 11.24 Alternatively, linear combinations of the two modes can be taken to give these two normal modes of the molecule. The mode in (a) is the symmetric stretch and that in (b) is the antisymmetric stretch. The two modes are independent, and if either of them is stimulated, the other remains unexcited. A normal mode description greatly simplifies the description of the vibrations of the molecule.

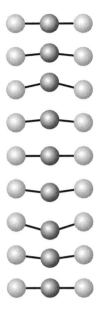

Fig. 11.25 The bending mode of CO_2 results in the formation of an electric dipole, which oscillates as the molecule vibrates.

The gross selection rule for infrared activity is that *the motion corresponding to a normal mode should give rise to a changing dipole moment.* Deciding whether this is so can sometimes be done by inspection. For example, the symmetric stretch of CO_2 leaves the dipole moment unchanged (at zero), and so this mode is infrared inactive and makes no contribution to the molecule's infrared spectrum. The antisymmetric stretch, however, changes the dipole moment because the molecule becomes unsymmetrical as it vibrates, and so this mode is infrared active. The fact that the mode does absorb infrared radiation enables carbon dioxide to act as a 'greenhouse gas' by absorbing infrared radiation emitted from the surface of the Earth. Because the dipole moment change is parallel to the molecular axis in the antisymmetric stretching mode, the transitions arising from this mode are classified as **parallel bands** in the spectrum. Both bending modes are also infrared active: they are accompanied by a changing dipole perpendicular to the molecular axis (Fig. 11.25), and so transitions involving them lead to a **perpendicular band** in the spectrum.

Exercise E11.11

State the ways in which the infrared spectrum of dinitrogen oxide (nitrous oxide, N_2O) will differ from that of carbon dioxide.
[*Answer*: different frequencies on account of different atomic masses and force constants; all four modes infrared active]

Some of the normal modes of organic molecules can be regarded as motions of individual functional groups. Others cannot be regarded as localized in this way, and are internal, collective motions of the molecule as a whole. The latter are generally of relatively low frequency, and occur below about $1500 \, cm^{-1}$ in the spectrum. The resulting whole-molecule region of the absorption spectrum is called the **fingerprint region** of the spectrum, for it is characteristic of the molecule. The matching of the fingerprint region with a spectrum of a known compound in a library of infrared spectra is a very powerful way of confirming the presence of a particular substance.

The characteristic vibrations of functional groups that occur outside the fingerprint region are very useful for the identification of an unknown compound. Most of these vibrations can be regarded as stretching modes, for the lower frequency bending modes usually occur in the fingerprint region and so are less readily identified. The characteristic wavenumbers of some functional groups are listed in Table 11.2.

Table 11.2 Typical vibrational wavenumbers

Vibration type	$\tilde{\nu}/cm^{-1}$
C—H stretch	2850-2960
C—H bend	1340-1465
C—C stretch, bend	700-1250
C=C stretch	1620-1680
C≡C stretch	2100-2260
O—H stretch	3590-3650
C=O stretch	1640-1780
C≡N stretch	2215-2275
N—H stretch	3200-3500
Hydrogen bonds	3200-3570

Example *Interpreting an infrared spectrum*
The infrared spectrum of an organic compound is shown in Fig. 11.26. Suggest an identification.

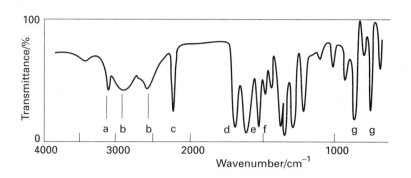

Fig. 11.26 A typical infrared absorption spectrum taken by forming a sample into a disk with potassium bromide. As explained in the example, the substance can be identified as $O_2NC_6H_4CCCOOH$.

Strategy

Some of the features at wavenumbers above $1500\,cm^{-1}$ can be identified by comparison with the data in Table 11.2.

Solution

(a) C—H stretch of a benzene ring, indicating a substituted benzene; (b) carboxylic acid O—H stretch, indicating a carboxylic acid; (c) the strong absorption of a conjugated $C\equiv C$ group, indicating a substituted alkyne; (d) this strong absorption is also characteristic of a carboxylic acid that is conjugated to a carbon–carbon multiple bond; (e) a characteristic vibration of a benzene ring, confirming the deduction drawn from (a); (f) a characteristic absorption of a nitro group ($-NO_2$) connected to a multiply bonded carbon–carbon system, suggesting a nitro-substituted benzene. The molecule contains as components a benzene ring, an aromatic carbon–carbon bond, a —COOH group, and a —NO_2 group. The molecule is in fact $O_2N-C_6H_4-C\equiv C-COOH$ (and more detailed analysis and comparison of the fingerprint region shows it to be the 1,4-isomer).

Exercise E11.12

Suggest an identification of the organic compound responsible for the spectrum shown in Fig. 11.27. [*Hint*: The molecular formula of the compound is C_3H_5ClO.]

[*Answer*: $CH_2=CClCH_2OH$]

11.8 Vibrational Raman spectra of polyatomic molecules

When a photon strikes a polyatomic molecule it may deposit energy in it as a vibrational motion, or acquire additional energy from it. As a result, the

Fig. 11.27 The spectrum considered in Exercise E11.12.

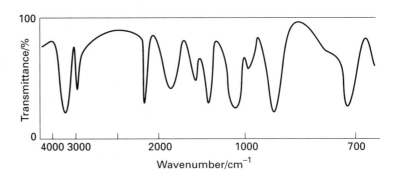

Fig. 11.27 The spectrum considered in Exercise E11.12.

incident photon may undergo Raman scattering and travel away from the molecule with a reduced or an increased frequency. The normal modes of vibration of molecules are Raman active and participate in this kind of scattering if they are accompanied by a changing polarizability. However, it is often quite difficult to judge by inspection when this is so. The symmetric stretch of CO_2, for example, alternately swells and contracts the molecule: this motion changes its polarizability, and so the mode is Raman active. The other modes of CO_2 leave the polarizability unchanged (although that is hard to justify pictorially), and so they are Raman inactive.

In some cases it is possible to make use of a very general rule about the infrared and Raman activity of vibrational modes:

> The **exclusion rule** states that if the molecule is symmetric under inversion, then no modes can be both infrared and Raman active.

(A mode may be inactive in both.) A molecule is symmetric under inversion if it looks unchanged when each atom is projected through a single point, the **centre of inversion**, and out an equal distance on the other side (Fig. 11.28). Because we can often judge intuitively when a mode changes the molecular dipole moment, we can use this rule to identify modes that are not Raman active. The rule applies to CO_2 but to neither H_2O nor CH_4 because they have no centre of symmetry.

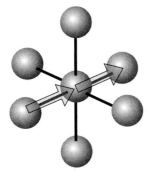

Fig. 11.28 In an inversion operation, we consider every point in a molecule, and project them all through the centre of the molecule out to an equal distance on the other side.

Exercise E11.13

One vibrational mode of benzene is a 'breathing mode' in which the ring alternately expands and contracts. Is it vibrationally Raman active?

[*Answer*: yes]

One application of vibrational Raman spectroscopy is to the determination of the structures of nonpolar molecules such as XeF_4 and SF_6. Another application makes use of the fact that the intensity characteristics of

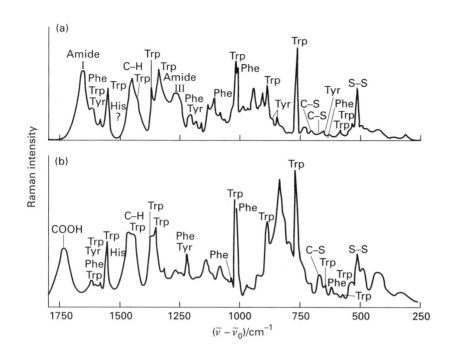

Fig. 11.29 The vibrational Raman spectrum of lysozyme in water and the superposition of theRaman spectra of the constituent amino acids. (From *Raman spectroscopy*, D.A. Long. Copyright 1977, McGraw-Hill Inc. Used with the permission of the McGraw-Hill Book Company.)

Raman transitions, which depend on molecular polarizabilities, are more readily transferred from molecule to molecule than the intensities of infrared spectra, which depend on dipole moments and are more sensitive to the other groups present in a molecule and to the solvent. Hence, Raman spectra are useful in the identification of organic and inorganic species in solution. An example of the technique is shown in Fig. 11.29, which shows the vibrational Raman spectrum of an aqueous solution of lysozyme and, for comparison, a superposition of the Raman spectra of the constituent amino acids. The differences are indications of the effects of conformation, environment, and specific interactions (such as disulfide links, S—S) in the enzyme molecule.

A modification of the basic Raman effect involves using incident radiation that coincides with the frequency of an electronic transition of the sample (Fig. 11.30; compare Fig. 11.2, where the incident radiation did not coincide with an actual electronic transition of the sample). The technique is then called **resonance Raman spectroscopy**. It is characterized by a much greater intensity in the scattered radiation and, because only a few vibrational modes contribute to the scattering, to a greatly simplified Raman spectrum. The resonance Raman spectrum shown in Fig. 11.31, for example, is of solid potassium chromate, K_2CrO_4. The nine peaks that are identified are the Stokes lines that correspond to the excitation of the symmetric breathing mode of the tetrahedral CrO_4^{2-} ion and the transfer of up to nine vibrational quanta during the photon–ion collision.The high intensity of the resonance Raman transitions is employed to examine the metal ions in biological macromolecules (such as the iron in haemoglobin and cytochromes or the cobalt in vitamin B), which are present in such low abundances that conventional Raman spectroscopy cannot detect them.

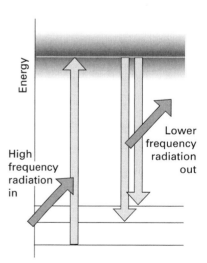

Fig. 11.30 In the resonance Raman effect, the incident radiation has a frequency corresponding to an actual electronic excitation of the molecule. A photon is emitted when the excited state returns to a state close to the ground state.

Fig. 11.31 The resonance Raman spectrum of solid potassium chromate. The peaks represent transitions in which the CrO_4^{2-} ion, initially in its vibrational ground state, is left in a vibrational state with $v = 1, 2, \cdots$. The vibrational mode concerned is the symmetric breathing mode of the tetrahedral ion.

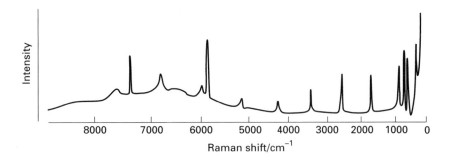

Electronic transitions: ultraviolet and visible spectra

The energies needed to change the distributions of electrons in molecules are of the order of several electronvolts (1 eV is about $8000\,\text{cm}^{-1}$). Consequently, the photons emitted or absorbed when such changes occur lie in the visible and ultraviolet regions of the spectrum, which spread from about $14\,000\,\text{cm}^{-1}$ for red light to $21\,000\,\text{cm}^{-1}$ for blue, and on to $50\,000\,\text{cm}^{-1}$ for ultraviolet radiation (Table 11.3). Indeed, many of the colours of the objects in the world around us, including the green of vegetation, the colours of flowers and of synthetic dyes, and the colours of pigments and minerals, stem from transitions in which electrons are shifted from one orbital of a molecule or ion into another. The migration of electrons that takes place when chlorophyll absorbs red and blue light (leaving green to be reflected) is the primary energy harvesting step by which our planet captures energy from the Sun and uses it to drive the non-spontaneous reactions of photosynthesis. In some cases the relocation of electrons may be so extensive that it results in the breaking of a bond and the dissociation of the molecule: such

Table 11.3 Colour, frequency, and energy of light

Colour	λ/nm	$\nu/10^{14}\ \text{Hz}$	$\tilde{\nu}/10^4\ \text{cm}^{-1}$	E/eV	$E/\text{kJ mol}^{-1}$
Infrared	1000	3.00	1.00	1.24	120
Red	700	4.28	1.43	1.77	171
Orange	620	4.84	1.61	2.00	193
Yellow	580	5.17	1.72	2.14	206
Green	530	5.66	1.89	2.34	226
Blue	470	6.38	2.13	2.64	254
Violet	420	7.14	2.38	2.95	285
Near ultraviolet	300	10.0	3.33	4.15	400
Far ultraviolet	200	15.0	5.00	6.20	598

processes give rise to the numerous reactions of photochemistry, including the reactions that sustain or damage the atmosphere.

To understand the contribution of electronic transitions to the colours of objects, we need to know that white light is a mixture of light of all different colours. The removal, by absorption, of any one of these colours from white light results in the **complementary colour** being observed. For instance, the absorption of red light from white light by an object results in that object appearing green, the complementary colour of red. Conversely, the absorption of green results in the object appearing red. The pairs of complementary colours are neatly summarized by the artist's colour wheel shown in Fig. 11.32, where complementary colours lie opposite one another along a diameter.

It should be stressed, however, that the perception of colour is a very subtle phenomenon. For example, although an object may appear green because it absorbs red light, it may also appear green because it absorbs all colours from the incident light except green. This is the origin of the colour of vegetation, because chlorophyll absorbs in two regions of the spectrum (Fig. 11.33). Moreover, an absorption band may be very broad, and although it may be a maximum at one particular wavelength, it may have a long tail that spreads into other regions (Fig. 11.34). In such cases, it is very difficult to predict the perceived colour from the location of the absorption maximum.

11.9 The Franck–Condon principle

In the electronic ground state of a molecule, the nuclei take up locations in response to the Coulombic forces acting on them, which arise from the electrons and the other nuclei. After an electronic transition, the nuclei are subjected to different Coulomb forces from the surrounding electrons, and the molecule may respond by bursting into vibration. As a result, some of the energy used to redistribute an electron is in fact used to stimulate the vibrations of the absorbing molecules. Therefore, instead of a single, sharp, and purely electronic absorption line being observed, the absorption spectrum consists of many lines. This **vibrational structure** of an electronic transition can be resolved if the sample is gaseous, but in a liquid or solid the lines usually merge together and result in a broad, almost featureless band (Fig. 11.35).

The details of the appearance of the vibrational structure of a band are explained by the **Franck–Condon principle**:

Because nuclei are so much more massive than the electrons, an electronic transition takes place faster than the nuclei can respond.

As a result of the electronic transition, electron density builds up rapidly in new regions of the molecule and is removed from others, and the initially stationary nuclei suddenly experience a new force field. They respond by beginning to vibrate, and (in classical terms) swing backwards and forwards from their original separation (which was maintained during the rapid

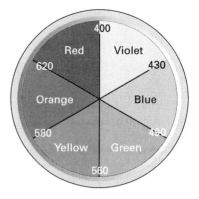

Fig. 11.32 An artist's colour wheel: complementary colours are opposite one another on a diameter.

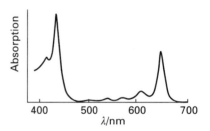

Fig. 11.33 The absorption spectrum of chlorophyll in the visible region. Note that it absorbs in the red and blue regions, and that green light is not absorbed.

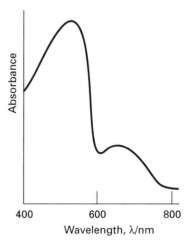

Fig. 11.34 An electronic absorption of a species in solution is typically very broad and consists of several broad bands.

437

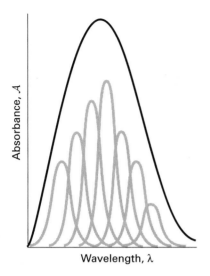

Fig. 11.35 An electronic absorption band consists of many superimposed bands which merge together to give a single broad band with unresolved vibrational structure.

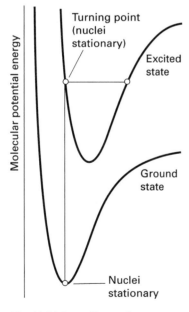

Fig. 11.36 According to the Franck–Condon principle, the most intense vibronic transition is from the ground vibrational state to the vibrational state lying vertically above it. Transitions to other vibrational levels also occur, but with lower intensity.

electronic excitation). The stationary equilibrium separation of the nuclei in the initial electronic state therefore becomes a **turning point**, the point of a vibration at which the nuclei are at the end points of their swing, in the final electronic state (Fig. 11.36). In practice, the electronically excited molecule may be formed in one of several excited vibrational states, so the absorption occurs at several different wavenumbers. As remarked above, in a condensed medium, the individual transitions are so broad that they merge together to give a broad band of absorption.

11.10 Specific types of transitions

The absorption of a photon can often be traced to the excitation of electrons that belong to a small group of atoms. For example, when a carbonyl group is present, an absorption at about 290 nm is normally observed. Groups with characteristic optical absorptions are called **chromophores** (from the Greek for 'colour bringer'), and their presence often accounts for the colours of substances.

A *d*-metal complex may absorb light as a result of the transfer of an electron from the ligands into the *d* orbitals of the central atom, or vice versa (Fig. 11.37). In such **charge-transfer transitions** the electron moves through a considerable distance, which means that the redistribution of charge may be large and the absorption correspondingly intense. This mode of chromophore activity is shown by the permanganate ion, MnO_4^-: the charge redistribution that accompanies the migration of an electron from the O atoms to the central Mn atom accounts for its intense violet colour (resulting from absorption in the range 420 to 700 nm).

The transition responsible for absorption in carbonyl compounds can be traced to the lone pairs of electrons on the O atom. One of these electrons may be excited into an empty π^* orbital of the carbonyl group (Fig. 11.38), which gives rise to an ***n*-to-π^* transition**. Typical absorption energies are about 4 eV.

Exercise E11.14

Estimate the wavelength of maximum absorption for a transition of energy 4.3 eV.

[*Answer*: 290 nm]

A C=C double bond acts as a chromophore because the absorption of a photon excites a π electron into an antibonding π^* orbital (Fig. 11.39). The chromophore activity is therefore due to a **π-to-π^* transition**. Its energy is around 7 eV for an unconjugated double bond, which corresponds to an absorption at 180 nm (in the ultraviolet). When the double bond is part of a conjugated chain, the energies of the molecular orbitals lie closer together and the transition shifts into the visible region of the spectrum. Many of the reds and yellows of vegetation are a consequence of transitions of this kind. For example, the carotenes that are present in green leaves (but are concealed by the intense absorption of the chlorophyll until the latter decays in the autumn) collect some of the solar radiation incident on the leaf by a π-to-π^* transition in their long conjugated hydrocarbon chains.

An important example of a π-to-π^* transition is provided by the photochemical mechanism of vision. The retina of the eye contains rhodopsin, which is a protein in combination with 11-*cis*-retinal (**1**). The 11-*cis*-retinal acts as a chromophore, and is the primary receptor for photons entering the eye. A solution of 11-*cis*-retinal absorbs at about 380 nm, but in combination with the protein (a link which might involve the elimination of the terminal carbonyl) the absorption maximum shifts to about 500 nm and tails into the blue. The conjugated double bonds are responsible for the ability of the molecule to absorb over the entire visible region, but they also play another important role. In its electronically excited state the conjugated chain can isomerize, one half of the chain being able to twist about an excited $C{=}C$ bond, which is now no longer torsionally rigid, and forming all-*trans*-retinal (**2**). On account of its different shape, the new isomer cannot fit into the protein. The primary step in vision therefore appears to be photon absorption followed by isomerization: the uncoiling of the molecule then triggers a nerve impulse to the brain.

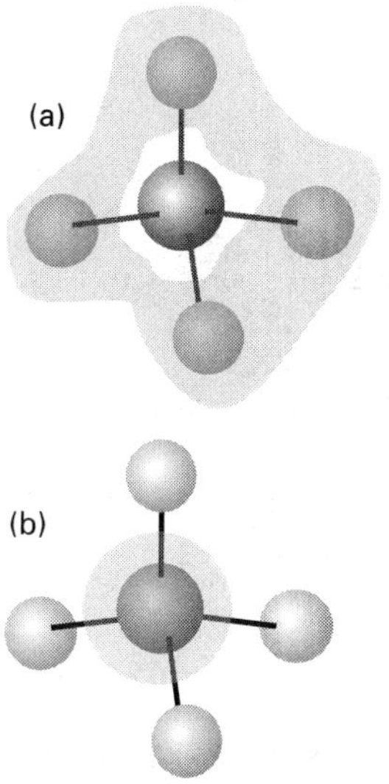

(a)

(b)

Fig. 11.37 A schematic representation of a charge transfer transition in which an electron initially on the outer atoms, represented by the grey tint in (a), migrates to the central ion in (b).

11.11 Radiative decay

In most cases, the excitation energy of a molecule that has absorbed a photon is degraded into the disordered thermal motion of its surroundings. However, one process by which an electronically excited molecule can discard its excess energy is by **radiative decay**, in which an electron relaxes back into a lower energy orbital and in the process generates a photon. As a result, an observer sees the sample glowing (if the emitted radiation is in the visible region of the spectrum).

There are two principal modes of radiative decay, fluorescence and phosphorescence. In **fluorescence**, the spontaneously emitted radiation ceases immediately after the exciting radiation is extinguished. In **phosphorescence**, the spontaneous emission may persist for long periods (even hours, but characteristically seconds or fractions of seconds). The difference suggests that fluorescence is an immediate conversion of absorbed light into re-emitted energy and that phosphorescence involves the storage of energy in a reservoir from which it slowly leaks.

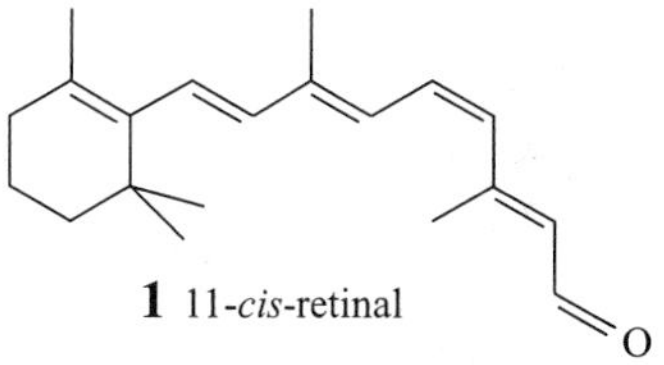

Fluorescence and phosphorescence

The diagram in Fig. 11.40 is a simple example of a **Jablonski diagram**, a schematic portrayal of molecular electronic and vibrational energy levels, which shows the sequence of steps involved in fluorescence. The initial absorption takes the molecule to an excited electronic state, and if the

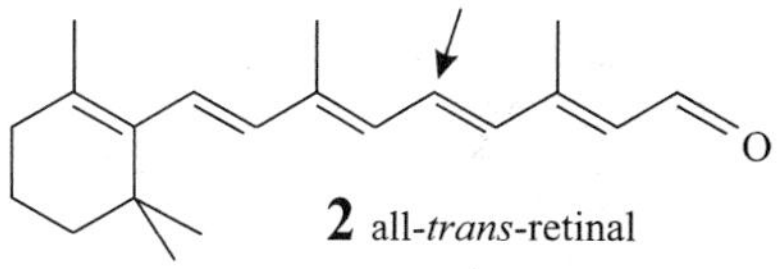

1 11-*cis*-retinal

2 all-*trans*-retinal

439

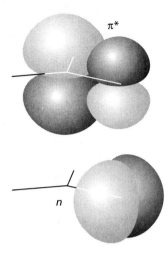

Fig. 11.38 A carbonyl group acts as a chromophore primarily on account of the excitation of a nonbonding O lone-pair electron to an antibonding CO π^* orbital.

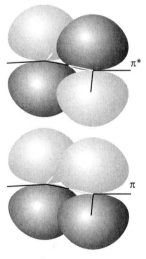

Fig. 11.39 A carbon–carbon double bond acts as a chromophore. One of its important transitions is the $\pi \rightarrow \pi^*$ transition illustrated here, in which an electron is promoted from a π orbital to the corresponding antibonding orbital.

absorption spectrum were monitored it would look like the one shown in Fig. 11.41(a). The excited molecule is subjected to collisions with the surrounding molecules, and as it gives up energy it steps down the ladder of vibrational levels. The surrounding molecules, however, might be unable to accept the larger energy needed to lower the molecule to the ground electronic state. It might therefore survive long enough to generate a photon and emit the remaining excess energy as radiation. The downward electronic transition is 'vertical' (in accord with the Franck–Condon principle) and the **fluorescence spectrum** (Fig. 11.41(b)) has a vibrational structure characteristic of the lower electronic state.

Fluorescence occurs at a lower frequency than that of the incident radiation because the fluorescence radiation is emitted after some vibrational energy has been discarded into the surroundings. The vivid oranges and greens of fluorescent dyes are an everyday manifestation of this effect: they absorb in the ultraviolet and blue, and fluoresce in the visible. The mechanism also suggests that the intensity of the fluorescence ought to depend on the ability of the solvent molecules to accept the electronic and vibrational quanta. It is indeed found that a solvent composed of molecules with widely spaced vibrational levels (such as water) may be able to accept the large quantum of electronic energy and so quench the fluorescence.

The diagram in Fig. 11.42 is a Jablonski diagram that shows the sequence of events leading to phosphorescence. The first steps are the same as in fluorescence, but the presence of a **triplet state** plays a decisive role. A triplet state is one in which two electrons in different orbitals have parallel spins: the ground state of O_2 which was discussed in Section 9.5 is an example. The name 'triplet' reflects the (quantum mechanical) fact that two parallel spins (↑↑) can adopt only three orientations with respect to an external magnetic field. An ordinary spin-paired state (↑↓) is called a **singlet state** because there is only one orientation in space for such a pair of spins.

The ground state of a typical phosphorescent molecule is a singlet because its electrons are all paired, and the excited state to which the absorption excites the molecule is also a singlet. The peculiar feature of a phosphorescent molecule, however, is that it possesses an excited triplet state of an energy similar to that of the excited singlet state, and into which the excited singlet state may convert. Hence, if there is a mechanism for unpairing two electron spins (and so converting ↑↓ into ↑↑), then the molecule may undergo **intersystem crossing** and become a triplet state. The unpairing of electron spins is possible if the molecule contains atoms of a heavy element (such as sulfur), because their nuclei can exert such strong magnetic fields on a nearby electron that the spin may be reversed.[†] After an excited singlet molecule crosses into a triplet state, it continues to discard energy into the surroundings and to step down the

[†]This argument is greatly abbreviated. The phenomenon responsible for the reversal of an electron spin is called 'spin–orbit coupling', in which the magnetic field arising from an electron's orbital motion around the nucleus interacts with the spin magnetic moment of the electron, and flips it into a new orientation. The strength of the orbitally generated magnetic field increases as the nuclear charge increases.

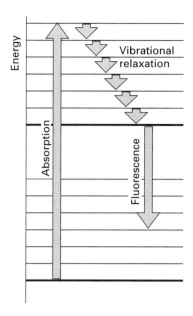

ladder of vibrational states. However, it is now stepping down the triplet's ladder, and at the lowest vibrational energy level it is trapped. The solvent cannot extract the final, large quantum of electronic excitation energy, and the molecule cannot radiate its energy because return to the ground state is forbidden: a triplet state cannot convert into a singlet state because an electron spin cannot reverse in a transition. The radiative transition, however, is not totally forbidden because the mechanism that was responsible for the intersystem crossing also breaks the selection rule. The molecules are therefore able to emit weakly, and the emission may continue long after the original excited state was formed.

The mechanism of phosphorescence summarized in Fig. 11.42 accounts for the observation that the excitation energy seems to become trapped in a slowly leaking reservoir. It also suggests (as is confirmed experimentally) that phosphorescence should be most intense from solid samples: energy transfer is then less efficient and the intersystem crossing has time to occur as the singlet excited state loses vibrational energy. The mechanism also suggests that the phosphorescence efficiency should depend on the presence of a moderately heavy atom (with its ability to flip electron spins, see the footnote), which is in fact the case.

Another fate for an electronically excited molecule is **dissociation**, or fragmentation (Fig. 11.43). The onset of dissociation can be detected in an absorption spectrum by seeing that the vibrational structure of a band terminates at a certain energy. Absorption occurs in a continuous band above this dissociation limit (the highest frequency before the onset of continuous absorption) because the final state is unquantized translational motion of the fragments. Locating the dissociation limit is a valuable way of determining the bond dissociation energy.

Lasers

Lasers have transformed experimental chemistry as much as they have the everyday world. The word **laser** is an acronym formed from light amplification by stimulated emission of radiation. As this name suggests, it is a process that depends on stimulated emission as distinct from the spontaneous emission processes characteristic of fluorescence and phosphorescence. In **stimulated emission**, an excited state is stimulated to emit a photon by the presence of radiation of the same frequency, and the more photons there are present, the greater the probability of the emission. To picture the process, we can think of the oscillations of the electromagnetic field as periodically (at the frequency at which the field oscillates) distorting the excited molecule at the frequency of the transition and hence encouraging the molecule to generate a photon of the same frequency. The essential feature of laser action is the strong **gain**, or growth of intensity, that results: the more photons present of the appropriate frequency, the more photons of that frequency the excited molecules will be stimulated to form, and so the laser medium fills with photons.

One requirement of laser action is the existence of an excited state that has a long enough lifetime for it to participate in stimulated emission.

Fig. 11.40 A Jablonski diagram showing the sequence of steps leading to fluorescence. After the initial absorption the upper vibrational states undergo radiationless decay—the process of vibrational relaxation—by giving up energy to the surroundings. A radiative transition then occurs from the ground state of the upper electronic state. In practice, the separation of the ground states of the electronic states (the heavy horizontal lines) is 10 to 100 times greater than the separation of the vibrational levels.

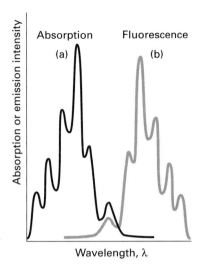

Fig. 11.41 The absorption spectrum (a) shows a vibrational structure characteristic of the upper state. The fluorescence spectrum (b) shows a structure characteristic of the lower state; it is also displaced to lower frequencies and resembles a mirror image of the absorption.

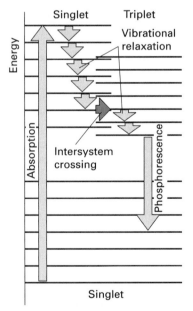

Fig. 11.42 The sequence of steps leading to phosphorescence. The important step is the intersystem crossing from an excited singlet to an excited triplet state. The triplet state acts as a slowly radiating reservoir because the return to the ground state is very slow.

Another requirement is the existence of a greater population in the upper state than in the lower state where the transition terminates, for otherwise there would be a net absorption rather than the net emission required. Because at thermal equilibrium the population is greater in the lower energy state, it is necessary to achieve a **population inversion** in which there are more molecules in the upper state than in the lower. One way of achieving population inversion is illustrated in Fig. 11.44. The inversion is achieved indirectly through an intermediate state I. Thus, the molecule is excited to I, which then gives up some of its energy nonradiatively (by passing energy on to vibrations of the surroundings) and changes into a lower state B; the laser transition is the return of B to a lower state A. Because four levels are involved overall, this arrangement leads to a **four-level laser**. The transition from X to I is caused by an intense flash of light in the process called **pumping**. In some cases the pumping flash is achieved with an electric discharge through xenon or with the radiation from another laser.

Laser radiation has a number of advantages for applications in chemistry. One advantage is its highly monochromatic character, which enables very precise spectroscopic observations to be made. Another advantage is the ability of laser radiation to be produced in very short pulses (currently, as brief as about $1\,\text{fs}$, $10^{-15}\,\text{s}$): as a result, very fast chemical events, such as the individual transfers of atoms during a chemical reaction, can be followed. Laser radiation is also very intense, which reduces the time needed for spectroscopic observations: this characteristic is particularly useful in Raman spectroscopy, where the scattered radiation is of very low intensity.

11.12 Circular dichroism

When light is passed through a suitable medium, it can be **plane polarized** in the sense that its electric field lies in a plane (Fig. 11.45). When this plane-polarized radiation passes through a sample, it is observed in certain cases that the plane of rotation is rotated. This rotation is the phenomenon of **optical activity**. It is observed when the molecules in the sample are **chiral**, which means distinguishable from their mirror image (Fig. 11.46). In many cases, organic chiral compounds are easy to detect, because they contain a carbon atom to which are bonded four different groups (the amino acid alanine, $NH_2CH(CH_3)COOH$, is an example). Mirror image pairs of molecules, so called **enantiomers**, rotate light of a given frequency through exactly the same angle but in opposite directions.

Optically active molecules have a second characteristic: they absorb left and right **circularly polarized light** to different extents. In a circularly polarized ray of light, the electric field describes a helical path as the wave travels through space (Fig. 11.47), and the rotation may be either clockwise or counterclockwise. The differential absorption of left- and right-circularly polarized light is called **circular dichroism**. In terms of the absorbances for the two components, $\mathcal{A}_L$ and $\mathcal{A}_R$, the circular dichroism of a sample of molar concentration [J] is reported as

$$\Delta\varepsilon = \varepsilon_L - \varepsilon_R = \frac{\mathcal{A}_L - \mathcal{A}_R}{l[J]} \qquad (28)$$

where l is the path length of the sample.

Circular dichroism is a useful adjunct to ordinary visible and UV spectroscopy. For example, the CD spectra of chiral d-metal complexes are distinctly different, whereas there is little difference between their absorption spectra (Fig. 11.48). Moreover, CD spectra can be used to assign the absolute configuration of complexes by comparing the observed spectrum with the CD spectrum of a similar complex of known handedness. The CD spectra of polypeptides and nucleic acids give similar structural information. In these cases the spectrum of the polymer chain arises from the chirality of individual monomer units and, in addition, a contribution from the helical structure of the polymer itself. By subtracting the CD spectra of a mixture of monomers, the remaining structure is due largely to the secondary structure of the polymer, and in this way its conformation may be investigated.

11.13 Photoelectron spectroscopy

The exposure of a molecule to high frequency radiation can result in the ejection of an electron. This **photoejection** is the basis of another type of spectroscopy in which the energies of the ejected **photoelectrons** are monitored. The principle of the technique is that an incoming photon of frequency ν has an energy $h\nu$; that energy may be transferred to an electron to remove it from its orbital (which requires an energy I, where I is the ionization energy of the electron from the orbital it occupies), leaving the remainder of the original energy of the photon to appear as kinetic energy of the electron, $\frac{1}{2}m_e v^2$, where v is the speed of the photoelectron after it has been ejected. From the conservation of energy, we can write

$$h\nu = I + \tfrac{1}{2}m_e v^2 \qquad (29)$$

Therefore, by monitoring the velocity of the photoelectron, and knowing the frequency of the incident radiation, it is possible to deduce the ionization energy of the electron and hence the strength with which the electron was bound in the molecule (Fig. 11.49, p. 446): the slower the ejected electron, the lower in energy the orbital from which it was ejected. The apparatus is a modification of a mass spectrometer (Fig. 11.50, p. 447), in which the velocity of the photoelectrons is measured by determining the strength of the electric field required to bend their paths on to the detector.

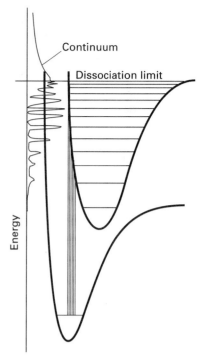

Fig. 11.43 When absorption occurs to unbound states of the upper electronic state, the molecule dissociates and the absorption is a continuum. Below the dissociation limit the electronic spectrum has a normal vibrational structure.

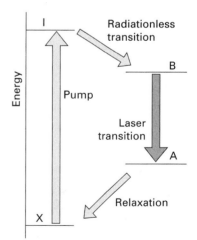

Fig. 11.44 The transitions involved in a four-level laser. Because the laser transition terminates in an excited state (A), the population inversion between A and B is much easier to achieve than when the lower state of the laser transition is the ground state.

443

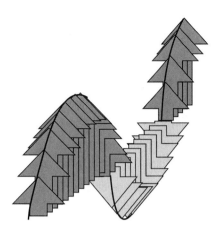

Fig. 11.45 In plane-polarized light, the electric field (represented by the arrows) oscillates in a plane.

A typical **photoelectron spectrum** (of HBr) is shown in Fig. 11.51. If we disregard the fine structure, we see that the HBr lines fall into two main groups. The least tightly bound electrons (with the lowest ionization energies and hence highest kinetic energies when ejected) are those in the lone pairs of the Br atom. The next ionization energy lies at 15.2 eV, and corresponds to the removal of an electron from the HBr σ bond.

The HBr spectrum shows that ejection of a σ electron is accompanied by a considerable amount of vibrational excitation. The Franck–Condon principle would account for this observation if ejection were accompanied by an appreciable change of equilibrium bond length between HBr and HBr$^+$: if that is so, then the ion is formed in a bond-compressed state, which is consistent with the important bonding effect of the σ electrons. The lack of much vibrational structure in the other band is consistent with the nonbonding role of the Br$4p_x$ and Br$4p_y$ lone-pair electrons, for the equilibrium bond length is little changed when one is removed.

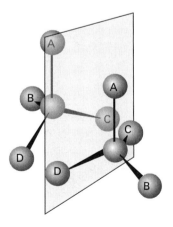

Fig. 11.46 A chiral molecule is one that is not identical to its mirror image. A carbon atom attached to four different groups is an example of a chiral centre in a molecule. Such molecules are optically active.

Example *Interpreting a UV photoelectron spectrum*
The highest kinetic energy electrons in the spectrum of H_2O using 21.22 eV He radiation are at about 9 eV and show a large vibrational spacing of 0.41 eV. The symmetric stretching mode of the neutral H_2O molecule lies at 3652 cm^{-1}. What conclusions can be drawn from the nature of the orbital from which the electron is ejected?

Strategy
To convert from electronvolts to reciprocal centimetres, use 1 eV = 8065.5 cm^{-1}. If the vibrational separation in the ion is similar to that in the molecule, then the ejected electron had little influence on bonding in the molecule. A lot of vibrational structure would suggest that the electron had been heavily involved in bonding.

Solution
Because 0.41 eV corresponds to 3.3×10^3 cm^{-1}, which is similar to the 3652 cm^{-1} of the nonionized molecule, we can suspect that the electron is ejected from an orbital that has little influence on the bonding in the molecule. That is, photoejection is from a largely nonbonding orbital.

Exercise E11.16

In the same spectrum of H_2O, the band near 7.0 eV shows a long vibrational series with spacing 0.125 eV. The bending mode of H_2O lies at 1596 cm^{-1}. What conclusions can you draw about the characteristics of the orbital occupied by the photoelectron?

[*Answer*: the electron contributes to long-distance HH bonding across the molecule]

Nuclear magnetic resonance

One of the most widely used and helpful forms of spectroscopy, and a technique that has transformed the practice of chemistry and its dependent disciplines, makes use of an effect that is familiar from classical physics. When two pendulums are joined by the same slightly flexible support and one is set in motion, the other is forced into oscillation by the motion of the common axle, and energy flows between the two. The energy transfer occurs most efficiently when the frequencies of the two oscillators are identical (Fig. 11.52). The condition of strong effective coupling when the frequencies are identical is called **resonance**, and the excitation energy is said to **resonate** between the coupled oscillators.

Resonance is the basis of a number of everyday phenomena, including the response of radios to the weak oscillations of the electromagnetic field generated by a distant transmitter. In this section we explore a spectroscopic

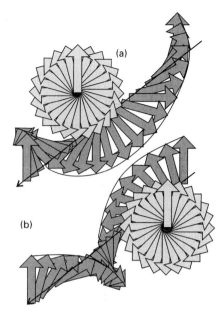

Fig. 11.47 In circularly polarized light, the electric field at different points along the direction of propagation rotates. The rosette-like arrays of arrows in these illustrations show the view of the electric field when looking toward the oncoming ray: (a) left-circularly polarized, (b) right-circularly polarized light.

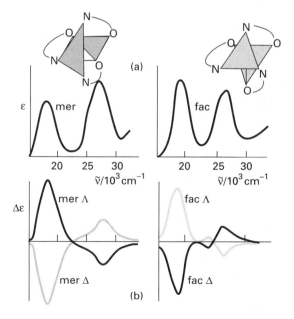

Fig. 11.48 (a) The absorption spectra of two isomers of $[Co(ala)_3]$, where ala is the conjugate base of alanine, and (b) the corresponding CD spectra. The left- and right-handed forms of these isomers give identical absorption spectra. However, the CD spectra are distinctly different, and the absolute configurations have been assigned by comparison with the CD spectra of a complex of known absolute configuration.

Molecular spectroscopy

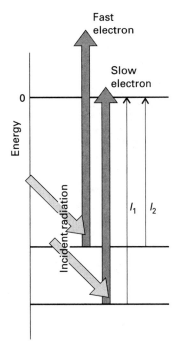

Fig. 11.49 The basic principle of photoelectron spectroscopy. An incoming photon of known energy collides with an electron in one of the orbitals and expels it with a kinetic energy that is equal to the difference between the energy supplied by the photon and the ionization energy from the occupied orbital. An electron from an orbital with a low ionization energy will emerge with a high kinetic energy (and high speed) whereas an electron from an orbital with a high ionization energy will be ejected with a low kinetic energy (and low speed).

application that when originally developed (and in some cases still) depends on matching a set of energy levels to a source of monochromatic radiation and observing the strong absorption that occurs at resonance.

11.14 Principles of magnetic resonance

The application of resonance that we describe here depends on the fact that many nuclei possess spin angular momentum (just as an electron does, but nuclei possess a wider range of values). A nucleus with spin quantum number I (the analogue of s for electrons, and which may be an integer or a half-integer) may take $2I + 1$ different orientations relative to an arbitrary axis that are distinguished by the quantum number m_I:

$$m_I = I, I - 1, \cdots, -I$$

A proton has $I = \frac{1}{2}$ (the same as an electron) and may adopt either of two orientations ($m_I = +\frac{1}{2}$ and $-\frac{1}{2}$). A ^{14}N nucleus has $I = 1$ and may adopt any of three orientations ($m_I = +1, 0, -1$). In this section we shall consider only **spin-$\frac{1}{2}$ nuclei**, those with $I = \frac{1}{2}$. Spin-$\frac{1}{2}$ nuclei include protons (^{1}H), ^{13}C, ^{19}F, and ^{31}P nuclei (Table 11.4). As for electrons, the state with $m_I = +\frac{1}{2}$ (↑) is denoted α, and that with $m_I = -\frac{1}{2}$ (↓) is denoted β. It is worth bearing in mind that two very common nuclei, ^{12}C and ^{16}O, have zero spin and hence are invisible in magnetic resonance.

The energies of nuclei in magnetic fields

A nucleus with nonzero spin behaves like a tiny magnet. The orientation of this magnet is determined by the value of m_I, and in a magnetic field B the $2I + 1$ orientations of the nucleus have different energies, which are given by

$$E_{m_I} = -g_I \mu_N B m_I \tag{30}$$

where g_I is the **nuclear g-factor**, a characteristic of the nucleus, and μ_N is the **nuclear magneton**:

$$\mu_N = \frac{e\hbar}{2m_p} = 5.051 \times 10^{-27} \, \text{J T}^{-1} \tag{31}$$

In this expression, m_p is the mass of the proton and T denotes the unit tesla, which is used to measure the intensity of a magnetic field ($1 \, \text{T} = 1 \, \text{kg s}^{-2} \text{A}^{-1}$). Nuclear g-factors are experimentally determined dimensionless quantities of the order of 1: for protons, $g_I = 5.5857$. Positive values of g_I indicate that the nuclear magnet lies in the same direction as the nuclear spin (this is the case for protons) whereas negative values of g_I indicate that the magnet points in the opposite direction. The strength of a nuclear magnet is about 2000 times weaker than that of the magnet associated with electron spin.

The energy separation of the two states of spin-$\frac{1}{2}$ nuclei (Fig. 11.53) is

$$\Delta E = E_\beta - E_\alpha = \tfrac{1}{2}g_I\mu_N\mathcal{B} - (-\tfrac{1}{2}g_I\mu_N\mathcal{B}) = g_I\mu_N\mathcal{B} \tag{32}$$

For most nuclei, g_I is positive, so the β state lies above the α state and there are slightly more α spins than β spins. If the sample is bathed in radiation of frequency ν, then the energy separations come into resonance with the radiation when the frequency satisfies the **resonance condition**:

$$h\nu = g_I\mu_N\mathcal{B} \tag{33}$$

At resonance there is strong coupling between the nuclear spins and the radiation, and strong absorption occurs as the spins flip from ↑ to ↓.

Exercise E11.17

Calculate the frequency at which radiation comes into resonance with proton spins in a 12 T magnetic field.

[*Answer*: 510 MHz]

The technique

In its simplest form, **nuclear magnetic resonance** (NMR) is the study of the properties of molecules containing magnetic nuclei by means of the

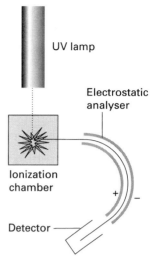

Fig. 11.50 A photoelectron spectrometer consists of a source of ionizing radiation (such as a helium discharge lamp for UPS and an X-ray source for XPS), an electrostatic analyser, and an electron detector. The deflection of the electron paths caused by the analyser depends on their speed.

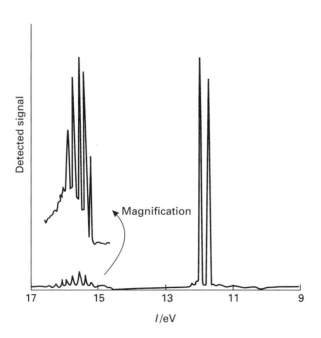

Fig. 11.51 The photoelectron spectrum of HBr. The lowest ionization energy band corresponds to the ionization of a Br lone-pair electron. The higher ionization energy band corresponds to the ionization of a bonding electron. The structure on the latter is due to the vibrational excitation of HBr$^+$ that results from the ionization.

Molecular spectroscopy

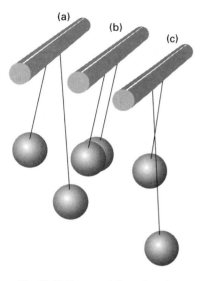

Fig. 11.52 Two pendulums hanging from the same support (which helps to convey the motion of one pendulum to the other, and so weakly couples them together) show the phenomenon of resonance when their two natural frequencies (their lengths) are the same. When their natural frequencies are different (a and c), the motion of one has only a slight effect on the other. When their natural frequencies are the same (b), if one is set in motion the other begins to swing in time with it.

Table 11.4 Nuclear spin properties

Nucleus	Natural abundance/ per cent	Spin, I
^{1}H	99.98	$\frac{1}{2}$
^{1}H (D)	0.0156	1
^{12}C	98.99	0
^{13}C	1.11	$\frac{1}{2}$
^{14}N	99.64	1
^{16}O	99.96	0
^{17}O	0.037	$\frac{5}{2}$
^{19}F	100	$\frac{1}{2}$
^{31}P	100	$\frac{1}{2}$
^{35}Cl	75.4	$\frac{3}{2}$
^{37}Cl	24.6	$\frac{3}{2}$

application of a magnetic field and the observation of the frequency at which they come into resonance with a radiofrequency electromagnetic field. When applied to proton spins, the technique is occasionally called **proton magnetic resonance** (^{1}H-NMR). In the early days of the technique the only nuclei that could be studied were protons (which behave like relatively strong magnets because g_I is large), but now a wide variety of nuclei (especially ^{13}C and ^{31}P) are investigated routinely.

An NMR spectrometer consists of a magnet that can produce a uniform, intense field and the appropriate sources of radiofrequency radiation. In simple instruments the magnetic field is provided by an electromagnet; for serious work, a superconducting magnet capable of producing fields of the order of 10 T and more is used. (A magnetic field of 10 T is very strong: a small magnet, for example, gives a magnetic field of only a few millitesla.) The use of high magnetic fields has two advantages. One is that the field increases the energy separation and therefore the population difference between the two spin states, and hence a stronger net absorption is obtained. Secondly, a high field simplifies the appearance of certain spectra.

Exercise E11.18

Calculate the ratio of numbers of ↑ and ↓ protons in a sample that is exposed to (a) a 1.0 T magnetic field, (b) a 10 T magnetic field at 20 °C. [*Hint*: Use the Boltzmann distribution (*Further information 10*) and the energy separation calculated from eqn (32).]

[*Answer*: (a) 1.000 007 0, (b) 1.000 070]

All modern work is done using a modification of this basic technique. In **Fourier transform NMR** (FT-NMR), the sample is held in a strong magnetic field (generated by a superconducting magnet) and exposed to one or more brief bursts of radiofrequency radiation. This radiation changes the orientations of the nuclear spins in a controlled way, and the radiofrequency radiation they emit as they return to equilibrium is monitored and analysed mathematically (the latter is the 'Fourier transform' part of the technique). The detected radiation contains all the information in the spectrum obtained by the earlier technique, but it is a much more efficient way of obtaining the spectrum (and hence more sensitive). Moreover, by choosing different sequences of exciting pulses, the data can be analysed much more closely.

11.15 The information in NMR spectra

Nuclear spins interact with the local magnetic field. The local field may differ from the applied field either on account of the local electronic structure of the molecule or because there is another magnetic nucleus nearby.

The chemical shift

The applied magnetic field can induce a circulating motion of the electrons in the molecule, and that motion gives rise to a small additional magnetic field, $\delta\mathcal{B}$. This additional field is proportional to the applied field, and it is conventional to express it as

$$\delta\mathcal{B} = -\sigma\mathcal{B} \tag{34}$$

where σ is the **shielding constant**. The constant σ may be positive or negative according to whether the induced field lies in the opposite or same direction as the applied field. The ability of the applied field to induce the circulation of electrons through the nuclear framework of the molecule depends on the details of the electronic structure near the magnetic nucleus of interest, and so nuclei in different chemical groups have different shielding constants.

Because the total local field is

$$\mathcal{B}_{\mathrm{loc}} = \mathcal{B} + \delta\mathcal{B} = (1-\sigma)\mathcal{B}$$

the resonance condition is

$$h\nu = g_I\mu_\mathrm{N}\mathcal{B}_{\mathrm{loc}} = g_I\mu_\mathrm{N}(1-\sigma)\mathcal{B} \tag{35}$$

and is different for nuclei in different environments (because σ varies with the environment). Hence, different nuclei, even of the same element so long as they are in different parts of a molecule, come into resonance at different frequencies.

The **chemical shift** of a nucleus is the difference between its resonance frequency and that of a reference standard. The standard for protons is the proton resonance in tetramethylsilane, $Si(CH_3)_4$, commonly referred to as TMS, which bristles with protons and dissolves without reaction in many solutions. Other references are used for other nuclei. For ^{13}C, the reference frequency is the ^{13}C resonance in TMS, and for ^{31}P it is the ^{31}P resonance in 85 per cent $H_3PO_4(aq)$. The separation of the resonance of a particular group of nuclei from the standard increases with the strength of the applied magnetic field because the induced field is proportional to the applied field, and the stronger the latter, the greater the shift.

Chemical shifts are reported on the δ **scale**, which is defined as

$$\delta = \frac{\nu - \nu^{\ominus}}{\nu^{\ominus}} \times 10^6 \tag{36}$$

where $\nu^{\ominus}$ is the resonance frequency of the standard. The advantage of the δ scale is that shifts reported on it are independent of the applied field (because both numerator and denominator are proportional to the applied field). The resonance frequencies themselves, however, do depend on the applied field through

$$\nu - \nu^{\ominus} = \delta \times \nu^{\ominus} \times 10^{-6} \tag{37}$$

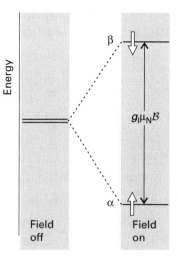

Fig. 11.53 The energy levels of a spin-$\frac{1}{2}$ nucleus (e.g. ^1H or ^{13}C) in a magnetic field. Resonance occurs when the energy separation of the levels matches the energy of the photons in the electromagnetic field.

449

Fig. 11.54 The range of typical chemical shifts for ^1H resonances.

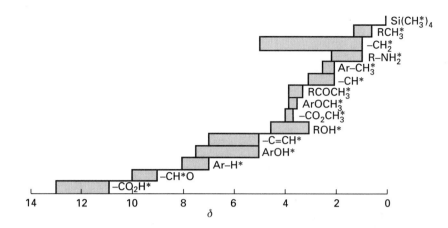

If $\delta > 0$, we say that the nucleus is **deshielded**; if $\delta < 0$, then it is **shielded**. A positive δ indicates that the resonance frequency of the group of nuclei in question is higher than that of the standard. Hence $\delta > 0$ indicates that the local magnetic field is stronger than that experienced by the nuclei in the standard under the same conditions. Some typical chemical shifts are given in Fig. 11.54.

Exercise E11.19

What is the shift of the resonance from TMS of a group of nuclei with $\delta = 3.50$ and an operating frequency of 350 MHz?

[*Answer*: 1.23 kHz]

The existence of a chemical shift explains the general features of the spectrum of ethanol shown in Fig. 11.55. The CH_3 protons form one group of nuclei with $\delta = 1$. The two CH_2 are in a different part of the molecule, experience a different local magnetic field, and hence resonate at $\delta = 3$. Finally, the OH proton is in another environment, and has a chemical shift of $\delta = 4$.

We can use the relative intensities of the signal (the areas under the absorption lines) to help distinguish which group of lines corresponds to which chemical group, and spectrometers can integrate the absorption—that is, determine the areas under the absorption signal—automatically (as is shown in Fig. 11.55). In ethanol the group intensities are in the ratio 3:2:1 because there are three CH_3 protons, two CH_2 protons, and one OH proton in each molecule. Counting the number of magnetic nuclei as well as noting their chemical shifts is valuable analytically because it helps us identify the compound present in a sample.

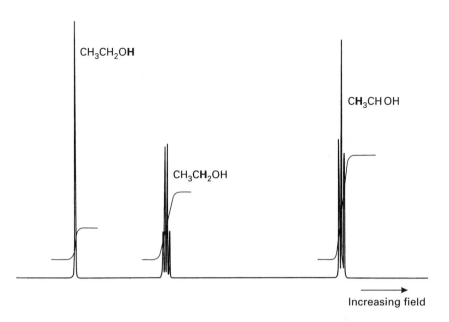

For some purposes it is convenient to regard σ as the sum of a positive **diamagnetic contribution**, σ_d, and a negative **paramagnetic contribution**, σ_p, and to write

$$\sigma = \sigma_d + \sigma_p \qquad (38)$$

Then σ is positive if the diamagnetic contribution dominates and is negative if the paramagnetic contribution dominates.

The diamagnetic contribution arises from the ability of the applied field to generate orbital motion of the electrons in the molecule. The resulting circulation of charge generates a magnetic field that opposes the applied field and reduces the frequency needed for resonance. The paramagnetic contribution, σ_p, arises from the ability of the applied field to force the electrons to circulate through the molecule by taking advantage of the availability of orbitals that are unoccupied in the ground state. It is zero in free atoms and around the axes of linear molecules (such as $H-C\equiv C-H$) where the electrons can circulate freely. Proton shielding constants are often dominated by the diamagnetic contribution, but this is not the case for the nuclei of other elements.

The fine structure

The splitting of resonances into individual lines in Fig. 11.55 is called the **fine structure** of the spectrum. It arises because each magnetic nucleus contributes to the local field experienced by the other nuclei and modifies their resonance frequencies. The strength of the interaction is expressed in terms of the **spin–spin coupling constant**, J, and reported in hertz (Hz). Spin

Fig. 11.56 The effect of spin–spin coupling on an NMR spectrum of two spin-$\frac{1}{2}$ nuclei with widely different chemical shifts. Each resonance is split into two lines separated by J. Full circles indicate α spins, open circles indicate β spins.

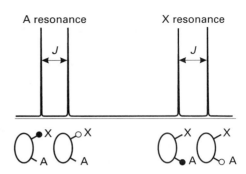

coupling constants are an intrinsic property of the molecule and independent of the strength of the applied field.

We shall consider first a molecule that contains two spin-$\frac{1}{2}$ nuclei A and X. Suppose that the spin of X is α; then A will resonate at a certain frequency as a result of the combined effect of the external field, the shielding constant, and the spin–spin interaction of the nucleus A with X. The spin–spin coupling will result in one line in the spectrum of A being shifted by $\frac{1}{2}J$ from the frequency it would have in the absence of coupling. When the spin of X is β, A will resonate at a frequency shifted by $-\frac{1}{2}J$. Therefore, instead of a single line from A, we get a doublet of lines separated by a frequency J (Fig. 11.56). The same splitting occurs in the X resonance: instead of a single line it is a doublet with splitting J (the same value as for the splitting of A).

If there is another X nucleus in the molecule with the same chemical shift as the first X (giving an AX_2 species), the resonance of A is split into a doublet by one X, and each line of the doublet is split again by the same amount (Fig. 11.57) by the second X. This splitting results in three lines in the intensity ratio 1:2:1 (because the central frequency can be obtained in two ways). As in the AX case discussed above, the X resonance of the AX_2 species is split into a doublet by A.

Three equivalent X nuclei (an AX_3 species) split the resonance of A into four lines of intensity ratio 1:3:3:1 (Fig. 11.58). The X resonance remains a doublet as a result of the splitting caused by A. In general, N equivalent spin-$\frac{1}{2}$ nuclei split the resonance of a nearby spin or group of equivalent spins into $N+1$ lines with an intensity distribution given by Pascal's triangle (**3**). Subsequent rows of the triangle are formed by adding together the two adjacent numbers in the line above.

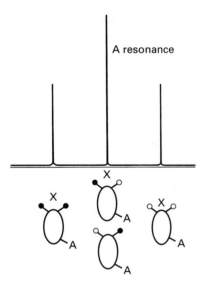

Fig. 11.57 The origin of the 1:2:1 triplet in the A resonance of an AX_2 species. The two X nuclei may have the $2^2 = 4$ spin arrangements (↑↑); (↑↓), (↑↓); (↓↓). The middle two arrangements are responsible for the coincident resonances of A.

```
            1
         1     1
      1     2     1
   1     3     3     1
1     4     6     4     1
```

3 Pascal's triangle

Exercise E11.20

Complete the next line of the triangle, the pattern arising from five equivalent protons.

[*Answer*: 1:5:10:10:5:1]

Example *Accounting for the fine structure in a spectrum*
Account for the fine structure in the ^{1}H-NMR spectrum of the C—H protons of ethanol.

Strategy
Refer to Pascal's triangle to determine the effect of a group of N equivalent protons on a proton, or (equivalently) a group of protons, of interest.

Solution
The three protons of the CH_3 group split the single resonance of the CH_2 protons into a 1:3:3:1 quartet with a splitting J. Likewise, the two protons of the CH_2 group split the single resonance of the CH_3 protons into a 1:2:1 triplet. Each of these lines is split into a doublet to a small extent by the OH proton.

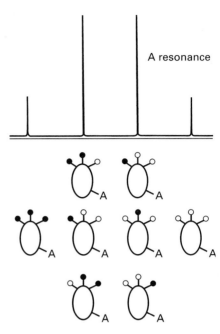

Fig. 11.58 The origin of the 1:3:3:1 quartet in the A resonance of an AX_3 species where A and X are spin-$\frac{1}{2}$ nuclei with widely different chemical shifts. There are $2^3 = 8$ arrangements of the spins of the three X nuclei, and their effects on the A nucleus give rise to four groups of resonances.

Exercise E11.23

What fine-structure can be expected for the protons in NH_4^+?

[*Answer*: a 1:1:1 triplet from N]

The spin–spin coupling constant of two nuclei joined by N bonds is normally denoted ^NJ, with subscripts for the types of nuclei involved (the N in this expression is not the same as the N in the expansion of Pascal's triangle!). Thus, $^1J_{CH}$ is the coupling constant for a proton joined directly to a ^{13}C atom, and $^2J_{CH}$ is the coupling constant when the same two nuclei are separated by two bonds (as in ^{13}C—C—H). A typical value of $^1J_{CH}$ is between 10^2 to 10^3 Hz; the value of $^2J_{CH}$ is about 10 times less, between about 10 and 10^2 Hz. Both ^3J and ^4J give detectable effects in a spectrum, but couplings over larger numbers of bonds can generally be ignored.

Example *Interpreting an NMR spectrum*
Suggest an interpretation of the ^{1}H-NMR spectrum in Fig. 11.59.

Strategy
We need to look for groups with characteristic chemical shifts (Fig. 11.54) and account for the fine structure as was done for ethanol.

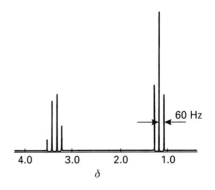

Fig. 11.59 The NMR spectrum considered in the example.

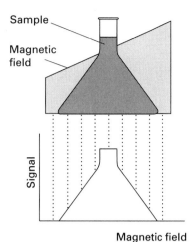

Fig. 11.60 A simple illustration of the principle of magnetic resonance imaging (MRI). The sample containing protons (in water, in this instance) is subjected to a strong uniform field on which is superimposed a linearly increasing field (represented by the tinted shape). Different numbers of protons come into resonance at different applied fields, and the resonance signal reproduces the shape of the distribution of protons (the shape of the sample in the flask).

Solution

The resonance at $\delta = 3.4$ corresponds to CH_2 in an ether; that at $\delta = 1.2$ corresponds to CH_3 in CH_3CH_2. The fine structure of the CH_2 group (a 1:3:3:1 quartet) is characteristic of splitting caused by CH_3; the fine structure of the CH_3 resonance is characteristic of splitting caused by CH_2. The fine structure constant is $J = 60\,Hz$ (the same for each group). The compound is probably $(CH_3CH_2)_2O$.

Exercise E11.22

What changes in the spectrum would be observed on recording it in a spectrometer that operates at a magnetic field that is five times stronger?

[*Answer*: groups of lines 5 times further apart in frequency (but the same δ values); no change in spin–spin splitting]

11.16 Magnetic resonance imaging

One of the most striking applications of nuclear magnetic resonance is in medicine. The technique of **magnetic resonance imaging** (MRI) is a simple portrayal of the concentrations of protons in a solid object. If an object containing hydrogen nuclei (a flask of water or a human body) is placed in an NMR spectrometer and exposed to a uniform magnetic field, then a single resonance signal will be detected. However, if the magnetic field varies linearly across the object, then the protons in different regions will be resonant at different frequencies, and the intensity of the signal will be proportional to the numbers of protons at each magnetic field. The intensity of the signal will therefore be a map of the distribution of protons in the sample. More precisely, it will be a projection of the numbers of protons on a line parallel to the field gradient (Fig. 11.60). If the body is rotated into different orientations, another projection can be determined. Then, it is only a matter of analysing the data on a computer to reconstruct the distribution of protons (for example, as water) in the living tissue, to detect abnormalities, and to observe metabolic processes.

It is fitting to conclude this text with an example of a technique that is in many respects a symbol of the role of physical chemistry. Magnetic resonance imaging has evolved from a recondite experiment in physics (as the original demonstration of NMR was) to an application that has helped to save countless lives.

EXERCISES

11.1 The kinetic energy of a bicycle wheel rotating once per second is about 0.2 J. To what rotational quantum number does that correspond?

11.2 To a first approximation, the rotation of an HI molecule can be pictured as an orbiting of the hydrogen atom in a plane at a radius of 160 pm about a stationary I atom. What wavelength radiation would be emitted in the transition $m_l = +1 \rightarrow m_l = 0$? *Hint.* The energy levels for rotation in a plane are $E = hcBm_J^2$ with $m_J = 0, \pm1, \pm2, \cdots$ and $B = \hbar/4\pi I$ with $I = m_H R^2$.

11.3 Which of the following molecules can have a pure rotational spectrum? (a) HCl, (b) N_2O, (c) O_3, (d) SF_4, (e) XeF_4.

11.4 Which of the molecules in the preceding exercise can have a rotational Raman spectrum?

11.5 A methane molecule free to rotate in three dimensions is described by the quantum numbers J, M_J, and K. How many rotational states have an energy equal to $hcBJ(J + 1)$ with $J = 10$?

11.6 Suppose the methane molecule in the previous exercise is replaced by chloromethane. How many rotational states now have an energy equal to $hcBJ(J + 1)$ with $J = 10$?

11.7 Express the moment of inertia of an octahedral AB_6 molecule in terms of its bond lengths and the masses of the B atoms.

11.8 Derive expressions for the moments of interia of a square-planar AB_4 molecule in terms of its bond lengths and the masses of the B atoms.

11.9 The rotational constant of $^1H^{35}Cl$ is $10.6\,cm^{-1}$. What is the separation of the line in its pure rotational spectrum (a) in reciprocal centimetres, (b) in gigahertz?

11.10 The rotational constant of $^1H^{35}Cl$ is $10.6\,cm^{-1}$. What is the separation of the lines in its rotational Raman spectrum (a) in reciprocal centimetres, (b) in gigahertz?

11.11 Suppose that hydrogen is replaced by deuterium in $^1H^{35}Cl$. Would you expect the $J \rightarrow 0$ transition to move to higher or lower wavenumber?

11.12 The rotational constant of $^{12}C^{16}O_2$ (from Raman spectroscopy) is $0.3904\,cm^{-1}$. What is the

CO bond length in the molecule? Take $m(^{16}O) = 15.9949\,u$.

11.13 The microwave spectrum of $^1H^{127}I$ consists of a series of lines separated by $12.8\,cm^{-1}$. Compute its bond length. What would be the separation of $^2H^{127}I$? $(m(^{127}I) = 126.9045\,u.)$

11.14 Which of the following molecules may show infrared absorption spectra: (a) H_2, (b) HCl, (c) CO_2, (d) H_2O, (e) CH_3CH_3, (f) CH_4, (g) CH_3Cl, (h) N_2?

11.15 The wavenumber of the fundamental vibrational transition of Cl_2 is $565\,cm^{-1}$. Calculate the force constant of the bond.

11.16 What is the Doppler-shifted wavelength of a red (660 nm) traffic light approached at 55 m.p.h.? At what speed would it appear green (520 nm)?

11.17 A spectral line of $^{48}Ti^{8+}$ in a distant star was found to be shifted from 654.2 nm to 706.5 nm and to be broadened to 61.8 pm. What is the speed of recession and the surface temperature of the star?

11.18 Estimate the lifetime of a state that gives rise to a line of width (a) $0.1\,cm^{-1}$, (b) $1\,cm^{-1}$, (c) 1.0 GHz.

11.19 A molecule in a liquid undergoes about 1×10^{13} collisions in each second. Suppose that (a) every collision is effective in deactivating the molecule vibrationally and (b) that one collision in 200 is effective. Calculate the width (in cm^{-1}) of vibrational transitions in the molecule.

11.20 The hydrogen halides have the following fundamental vibrational wavenumbers:

	HF	HCl	HBr	HI
$\tilde{\nu}/cm^{-1}$	4141.3	2988.9	2649.7	2309.5

Calculate the force constants of the hydrogen–halogen bonds.

11.21 From the data in Exercise 11.20, predict the fundamental vibrational wavenumbers of the deuterium halides.

11.22 Consider the vibrational mode that corresponds to the uniform expansion of the benzene ring. Is it (a) Raman, (b) infrared active?

11.23 The molar absorption coefficient of a substance dissolved in hexane is known to be $743\,mol^{-1}\,L\,cm^{-1}$

at 285 nm. Calculate the percentage reduction in intensity when light of that wavelength passes through 2.5 mm of a solution of concentration $3.25 \times 10^{-3} \, \text{mol L}^{-1}$.

11.24 When light of wavelength 410 nm passes through 2.5 mm of a solution of the dye responsible for the yellow of daffodils at a concentration $4.33 \times 10^{-4} \, \text{mol L}^{-1}$, the transmission is 71.5 per cent. Calculate the molar absorption coefficient of the colouring matter at this wavelength and express the answer in $\text{cm}^2 \, \text{mol}^{-1}$.

11.25 The molar absorption coefficient of cytochrome P450, one of the compounds involved in electron transport in cells, at 522 nm is $291 \, \text{L mol}^{-1} \, \text{cm}^{-1}$. When light of that wavelength passes through a cell of length 6.5 mm containing a solution of the solute, 39.8 per cent of the light was absorbed. What is the molar concentration of the solution?

11.26 The compound $CH_3CH{=}CHCHO$ has a strong absorption in the ultraviolet at $46\,950 \, \text{cm}^{-1}$ and a weak absorption at $30\,000 \, \text{cm}^{-1}$. Justify these features in terms of the structure of the compound.

11.27 The following data were obtained for the absorption by Br_2 in carbon tetrachloride using a cell of length 2.0 mm. Calculate the molar absorption coefficient (ε) of bromine at the wavelength employed:

$[Br_2]/(\text{mol L}^{-1})$	0.0010	0.0050	0.0100	0.0500
$T/(\text{per cent})$	81.4	35.6	12.7	3.0×10^{-3}

11.28 A cell of length 2.0 mm was filled with a solution of benzene in a nonabsorbing solvent. The concentration of the benzene was $0.010 \, \text{mol L}^{-1}$ and the wavelength of the radiation was 256 nm (where there is a maximum in the absorption). Calculate the molar absorption coefficient of benzene at this wavelength given that the transmission was 48 per cent. What will the transmittance be in a cell of length 4.0 mm at the same wavelength?

11.29 A swimmer enters a gloomier world (in one sense) on diving to greater depths. Given that the mean molar absorption coefficient of sea water in the visible region is $6.2 \times 10^{-5} \, \text{mol}^{-1} \, \text{L cm}^{-1}$, calculate the depth at which a diver will experience (a) half the surface intensity of light, (b) one tenth that intensity.

11.30 In a particular photoelectron spectrum using 21.21 eV photons, electrons were ejected with kinetic energies of 11.01 eV, 8.23 eV, and 5.22 eV. Sketch the molecular orbital energy level diagram for the species, showing the ionization energies of the three identifiable orbitals.

11.31 The molar absorption coefficients of tryptophan and tyrosine at 240 nm are $2.00 \times 10^3 \, \text{L mol}^{-1} \, \text{cm}^{-1}$ and $1.12 \times 10^4 \, \text{L mol}^{-1} \, \text{cm}^{-1}$, respectively, and at 280 nm they are $5.40 \times 10^3 \, \text{L mol}^{-1} \, \text{cm}^{-1}$ and $1.50 \times 10^3 \, \text{L mol}^{-1} \, \text{cm}^{-1}$. The absorbance of a sample obtained by hydrolysis of a protein was measured in a cell of thickness 1.00 cm, and was found to be 0.660 at 240 nm and 0.221 at 280 nm. What are the concentrations of the two amino acids?

11.32 What would be the nuclear magnetic resonance spectrum for a proton resonance line that was split by interaction with seven identical protons?

11.33 What would be the nuclear magnetic resonance spectrum for a proton resonance line that was split by interaction with (a) two, (b) three equivalent nitrogen nuclei?

11.34 The nucleus ^{32}S has a spin of $\frac{3}{2}$ and a nuclear g factor of 0.4289. Calculate the energies of the nuclear spin states in a magnetic field of 7.500 T.

11.35 Calculate the resonance frequency of a ^{14}N nucleus ($I = 1$, $g_I = 0.4036$) in a 15.00 T magnetic field.

11.36 Calculate the magnetic field needed to satisfy the resonance condition for unshielded protons in a 550.0 MHz radiofrequency field.

11.37 The chemical shift of the CH_3 protons in acetaldehyde (ethanal) is $\delta = 2.20$ and that of the CHO proton is 9.80. What is the difference in local magnetic field between the two regions of the molecule when the applied field is (a) 1.5 T, (b) 6.0 T?

11.38 Using the information in Fig. 11.54, state the splitting (in Hz) between the methyl and aldehydic proton resonances in a spectrometer operating at (a) 300 MHz, (b) 550 MHz.

11.39 Sketch the appearance of the ^{1}H-NMR spectrum of acetaldehyde using $J = 2.90 \, \text{Hz}$ and the data in Fig. 11.54 in a spectrometer operating at (a) 300 MHz, (b) 550 MHz.

11.40 Sketch the form of the ^{19}F-NMR spectra of a natural sample of $^{10}BF_4^-$ and $^{11}BF_4^-$.

11.41 Sketch the form of an $A_3M_2X_4$ spectrum, where A, M, and X are protons with distinctly different chemical shifts and $J_{AM} > J_{AX} > J_{MX}$.

FURTHER INFORMATION 1: Mathematical techniques

The art of doing mathematics correctly is to do nothing at each step of a calculation. That is, it is permissible to develop an equation by ensuring that the left-hand side of an expression remains equal to the right-hand side. There are several ways of modifying the *appearance* of an expression without upsetting its balance.

Algebraic equations and graphs

The simplest types of equation we have to deal with have the form

$$y = ax + b$$

It may be modified by subtracting b from both sides, to give

$$y - b = ax$$

It may be modified further by dividing both sides by a, to give

$$\frac{y - b}{a} = x$$

This series of manipulations is called *rearranging* the expression for y in terms of x to give an expression for x in terms of y. A short cut, as can be seen by inspecting these two steps, is that an added term can be moved through the equals sign provided that as it passes $=$ it changes sign (that happened to b in the example). Similarly, a multiplying factor becomes a divisor (and vice versa) when it passes through the $=$ sign.

There are several more complicated manipulations that are required in certain cases. The only one that we make use of in this text is to find the values of x that satisfy an equation of the form

$$ax^2 + bx + c = 0$$

or any equation that can be rearranged into this form by the steps we have already illustrated. An equation in which x occurs as its square is called a **quadratic equation**. Its solutions are found by inserting the values of the constants a, b, and c into the expression

$$x = \frac{-b \pm \sqrt{b^2 - 4ac}}{2a}$$

where the two values of x given by this expression (one by using the $+$ sign and the other by using the $-$ sign) are called the two **roots** of the original quadratic equation.

A **function**, f, tells us how something changes as a variable is changed. For example, we might write

$$f(x) = ax + b$$

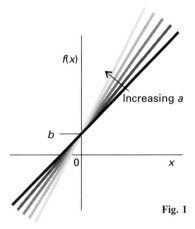

Fig. 1

to show how a property f changes as x is changed. The variation of f with x is best shown by drawing a graph in which $f(x)$ is plotted on the vertical axis and x is plotted horizontally. The graph of the function we have just written is shown in Fig. 1. The important point about this graph is that it is linear (that is, it is a straight line); its intercept with the vertical axis (the value of f when $x = 0$) is b, and its slope is a. That is, a straight line has the form

$$f = \text{slope} \times x + \text{intercept}$$

A positive value of a indicates an upward slope from left to right (increasing x); a change of sign of a results in a negative slope, down from left to right.

The solutions of the equation $f(x) = 0$ can be visualized graphically: they are the values of x for which f cuts through the horizontal axis (the axis corresponding to $f = 0$). For example, the solution of the quadratic equation given earlier is depicted in Fig. 2. In general, a quadratic equation has a graph that cuts through the horizontal axis at two points (the equation has two roots), a cubic equation (an equation in which x^3 is the highest power of x) cuts through it three times (the equation has three roots), and so on.

Logarithms, exponentials, and powers

Some equations are most readily solved by using logarithms and related functions. The **natural logarithm** of a number x is denoted $\ln x$, and is defined as the power to which a certain number designated e must be raised for the result to be equal to x. The number e, which is equal to $2.718\cdots$ may seem to be decidedly unnatural; however, it falls out naturally from various manipulations in mathematics and its use greatly simplifies calculations. On a calculator, $\ln x$ is obtained simply by entering

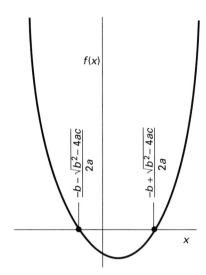

Fig. 2.

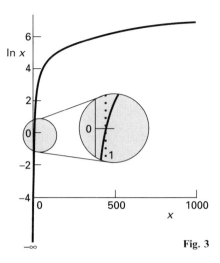

Fig. 3

the number x and pressing the 'ln' key or its equivalent. It follows from the definition of logarithms that

$$\ln x + \ln y = \ln xy \qquad \ln x - \ln y = \ln \frac{x}{y} \qquad a \ln x = \ln x^a$$

Thus, $\ln 5 + \ln 3$ is the same as $\ln 15$ and $\ln 6 - \ln 2$ is the same as $\ln 3$, as may readily be checked with a calculator. The last of these three relations is very useful for finding an awkward root of a number. For example, suppose we wanted the fifth root of 28. We write the required root as x, with $x^5 = 28$. We take logarithms of both sides, which gives $\ln x^5 = \ln 28$, and then rewrite the left-hand side of this equation as $5 \ln x$. At this stage we see that we have to solve

$$5 \ln x = \ln 28$$

To do so, we divide both sides by 5, which gives

$$\ln x = \frac{\ln 28}{5} = 0.6664 \cdots$$

All we need do at this stage is find the antilogarithm of the number on the right, the value of x for which the natural logarithm is the number quoted. The natural antilogarithm of a number is obtained by pressing the 'exp' key on a calculator (where 'exp' is an abbreviation for exponential), and in this case the answer is $1.947 \cdots$.

There are a number of useful points to remember about logarithms, and they are summarized in Fig. 3. You can see how logarithms increase only very slowly as x increases. For instance, when x increases from 1 to 1000, $\ln x$ increases from 0 to only 6.9. Another point is that the logarithm of 1 is 0: $\ln 1 = 0$. The logarithms of numbers less than 1 are negative, and in elementary mathematics the logarithms of negative numbers are not defined.

You will also encounter **common logarithms**, which are logarithms compiled with 10 in place of e; they are denoted $\log x$. For example, $\log 5$ is the power to which 10 must be raised to obtain 5, and is $0.69897 \cdots$. Common logarithms follow the same rules of addition and subtraction as natural logarithms. They are largely of historical interest now that calculators are so readily available, but they survive in the context of acid–base chemistry and pH. Common and natural logarithms (log and ln, respectively) are related by

$$\ln x = \ln 10 \times \log x = 2.303 \cdots \times \log x$$

The **exponential function**, e^x, plays a very special role in the mathematics of chemistry. It is evaluated by entering x and pressing the 'exp' key on a calculator. The following properties are important:

$$e^x \times e^y = e^{x+y} \qquad \frac{e^x}{e^y} = e^{x-y} \qquad (e^x)^a = e^{ax}$$

(These relations are the analogues of the relations for logarithms.) A graph of e^x is shown in Fig. 4. As you can see, it

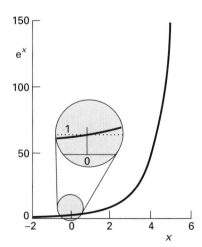

Fig. 4

is positive for all values of x. It is less than 1 for all negative values of x, is equal to 1 when $x = 0$, and rises ever more rapidly towards infinity as x increases. This sharply rising character of e^x is the origin of the popular expression 'exponentially increasing' that is widely but loosely used in the media. (Strictly, a function increases exponentially if its rate of change is proportional to its current value.)

Differentiation and integration

Rates of change of functions—slopes—are best discussed in terms of the infinitesimal calculus. The slope of a function, like the slope of a hill, is obtained by dividing the rise of the hill by the horizontal distance (Fig. 5). However, because the slope may vary from point to point, we should take the horizontal distance between the points as small as possible. In fact, we let it become infinitesimally small—hence the name *infinitesimal* calculus. The values of a function f at two locations x and $x + \delta x$ are $f(x)$ and $f(x + \delta x)$, respectively. Therefore, the slope of the function f at x is the vertical distance, which we write δf divided by the horizontal distance, which we write δx:

$$\text{slope} = \frac{\delta f}{\delta x} = \frac{\text{rise in value}}{\text{horizontal distance}} = \frac{f(x + \delta x) - f(x)}{\delta x}$$

The slope exactly *at* x itself is obtained by letting the horizontal distance become zero, which we write $\lim \delta x \to 0$. In this limit, the δ is replaced by a d, and we write

$$\text{slope at } x = \frac{\mathrm{d}f}{\mathrm{d}x} = \lim_{\delta x \to 0} \frac{f(x + \delta x) - f(x)}{\delta x}$$

To work out the slope of any function, we work out the expression on the right: this process is called **differentiation**. It leads to the following three important expressions:

$$\frac{\mathrm{d}x^n}{\mathrm{d}x} = nx^{n-1} \qquad \frac{\mathrm{d}e^{ax}}{\mathrm{d}x} = ae^a \qquad \frac{\mathrm{d}\sin ax}{\mathrm{d}x} = a\cos ax$$

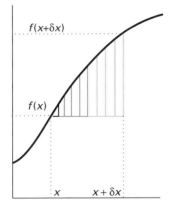

Fig. 5

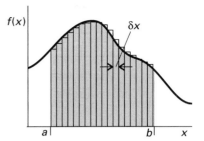

Fig. 6

The area under a graph of any function f is found by the techniques of **integration**. For instance, the area under the graph of the function f drawn in Fig. 6 can be written as the value of f evaluated at a point multiplied by the width of the region, δx, and then all those products $f(x)\delta x$ summed (denoted by the symbol $\sum$) over all the regions:

$$\text{area between } a \text{ and } b = \sum f(x)\delta x$$

When we allow δx to become infinitesimally small, written $\mathrm{d}x$, and sum an infinite number of strips, we write

$$\text{area between } a \text{ and } b = \int_a^b f(x)\,\mathrm{d}x$$

The elongated S symbol on the right is called the **integral** of the function f. Some important integrals are

$$\int x^n\,\mathrm{d}x = \frac{x^{n+1}}{n+1} \qquad \int e^{ax}\,\mathrm{d}x = \frac{e^{ax}}{a} \qquad \int \cos ax\,\mathrm{d}x = \frac{\sin ax}{a}$$

It may be verified from these examples—and this is a very deep result of infinitesimal calculus—that integration is the inverse of differentiation. That is, if we integrate a function and then differentiate the result, we get back the original function.

A **differential equation** is an equation that tells us how the slope of a function varies from place to place. For example, if the slope increased as x increased, we would write

$$\frac{\mathrm{d}f}{\mathrm{d}x} = ax$$

where a is a constant. To solve a differential equation, we have to look for the function f that satisfies it. In chemical kinetics, for instance, we may know that the reaction rate is proportional to the concentration of a reactant, and look for a solution of the rate equation (a differential equation) which tells us how the concentration varies with time as the reaction proceeds. The solution of differential equations is a very powerful technique in the physical sciences, but is often very difficult.

FURTHER INFORMATION 2: Quantities and units

The result of a measurement is a **physical quantity** (such as mass or density) that is reported as a numerical multiple of agreed **units:**

$$\text{physical quantity} = \text{numerical value} \times \text{unit}$$

For example, the mass of an object may be reported as 2.5 g and its density as $1.01\,\text{g cm}^{-3}$ where the units are respectively 1 gram (1 g) and 1 gram per centimetre cubed ($1\,\text{g cm}^{-3}$). Units are treated like algebraic quantities, and may be multiplied, divided, and cancelled. Thus, the expression (physical quantity)/unit is simply the numerical value of the measurement in the specified units, and hence is a dimensionless quantity. For instance, the mass reported above could be denoted $m/\text{g} = 2.5$ and the density as $d/(\text{g cm}^{-3}) = 1.01$.

In the **International System** of units (SI, from the French *Système International*), the units are formed from seven **base units** lited in Table 1. All other physical quantities may be expressed as combinations of these physical quantities and reported in terms of **derived units.** Thus, volume is (length)^3 and may be reported as a multiple of 1 metre cubed ($1\,\text{m}^3$), and density, which is mass/volume, may be reported as a multiple of 1 kilogram per metre cubed ($1\,\text{kg m}^{-3}$). A number of derived units have special names and symbols. Among the most important for our purposes are listed in Table 2.

In all cases (both for base and derived quantities), the units may be modified by a prefix that denotes a factor of a power of 10. Among the most common prefixes are those listed in Table 3. Examples of the use of these prefixes are

$$1\,\text{nm} = 10^{-9}\,\text{m} \qquad 1\,\text{ps} = 10^{-12}\,\text{s} \qquad 1\,\mu\text{mol} = 10^{-6}\,\text{mol}$$

The kilogram (kg) is anomalous: although it is a base unit, it is interpreted as 10^3 g, and prefixes are attached to the gram (as in

Table 1 The SI base units

Physical quantity	Symbol for quantity	Base unit
length	l	metre, m
mass	m	kilogram, kg
time	t	second, s
electric current	I	ampere, A
thermodynamic temperature	T	kelvin, K
amount of substance	n	mole, mol
luminous intensity	I	candela, cd

Table 2 A selection of derived units

Physical quantity	Derived unit	Name of derived unit
Force	$1\,\text{kg m s}^{-2}$	newton, N
Pressure	$1\,\text{kg m}^{-2}\,\text{s}^{-2}$	pascal, P
Energy	$1\,\text{kg m}^2\,\text{s}^{-2}$	joule, J
Power	$1\,\text{kg m}^{-2}\,\text{s}^{-3}$	watt, W

Table 3 Common SI prefixes

Prefix	f	p	n	μ	m	c	d
Name	femto	pico	nano	micro	milli	centi	deci
Factor	10^{-15}	10^{-12}	10^{-9}	10^{-6}	10^{-3}	10^{-2}	10^{-1}

Prefix	k	M	G	T
Name	kilo	mega	giga	tera
Factor	10^3	10^6	10^9	10^{12}

$1\,\text{mg} = 10^{-3}\,\text{g}$). Powers of units apply to the prefix as well as the unit they modify:

$$1\,\text{cm}^3 = 1\,(\text{cm})^3 = 1\,(10^{-2}\,\text{m})^3 = 10^{-6}\,\text{m}^3$$

Note that $1\,\text{cm}^3$ does *not* mean $\text{c}(\text{m}^3)$. When carrying out numerical calculations, it is usually safest to write out the numerical value of an observable as powers of 10.

Table 4 Some common units

Physical quantity	Name of unit	Symbol for unit	Value
Time	minute	min	60 s
	hour	h	3600 s
	day	d	86 400 s
Length	ångström	Å	10^{-10} m
Volume	litre	L	$1\,\text{dm}^3$
Mass	tonne	t	10^3 kg
Pressure	bar	bar	10^5 Pa
	atmosphere	atm	101.325 kPa
Energy	electronvolt	eV	$1.602\,177\,33 \times 10^{-19}$ J
			$96.485\,31\,\text{kJ mol}^{-1}$

All values in the final column are exact, except for the definition of 1 eV.

There are a number of units that are in wide use but are not a part of the International System. Some are *exactly* equal to multiples of SI units. These include the **litre** (L), which is exactly $10^3 \, cm^3$ (or $1 \, dm^3$) and the **atmosphere** (atm), which is exactly $101.325 \, kPa$. Others rely on the values of fundamental constants, and hence are liable to change when the values of the fundamental constants are modified by more accurate or more precise measurements. Thus, the size of the energy unit **electronvolt** (eV), the energy acquired by an electron that is accelerated through a potential difference of exactly $1 \, V$, depends on the value of the charge of the electron, and the present (1996) conversion factor is $1 \, eV = 1.602\,177\,33 \times 10^{-19} \, J$. Table 4 gives the conversion factors for a number of these convenient units.

FURTHER INFORMATION 3: Energy and force

Matter can store energy in two ways, as kinetic energy and as potential energy.

Kinetic energy is the energy that a body (a block of matter, an atom, or an electron) possesses by virtue of its motion. The formula for calculating the kinetic energy of a body of mass m that is travelling at a speed v is

$$\text{kinetic energy} = \tfrac{1}{2}mv^2$$

This expression shows that a body may have a high kinetic energy if it is heavy (m large) and is travelling rapidly (v large). A stationary body ($v = 0$) has zero kinetic energy, whatever its mass. The energy of a sample of perfect gas is entirely due to the kinetic energy of its molecules: they travel more rapidly (on average) at high temperatures than at low, and so raising the temperature of a gas increases the kinetic energy of its molecules.

Potential energy is the energy that a body has by virtue of its position. A body on the surface of the Earth has a potential energy on account of the gravitational force it experiences: if the body is raised, then its potential energy is increased. There is no general formula for calculating the potential energy of a body because there are several kinds of force. For a body of mass m at a height h above (but close to) the surface of the Earth, the gravitational potential energy is

$$\text{potential energy} = mgh$$

where g is the acceleration of free fall ($g = 9.81 \, m\,s^{-2}$). A heavy object at a certain height has a greater potential energy than a light object at the same height. One very important contribution to the potential energy is encountered when a charged particle is brought up to another charge. In this case the potential energy is inversely proportional to the distance between the charges (see *Further information 5*):

$$\text{potential energy} \propto \frac{1}{r}$$

This **Coulomb potential energy** decreases with distance, and two infinitely widely separated charged particles have zero potential energy of interaction. The Coulomb potential energy plays a central role in the structures of atoms, molecules, and solids.

The **total energy** of a body is the sum of its kinetic and potential energies. It is a central feature of physics that the total energy of a body that is free from external influences is constant. Thus, a stationary ball at a height h above the surface of the Earth has a potential energy of magnitude mgh; if it is released and begins to fall to the ground, it loses potential energy (as it loses height), but gains the same amount of kinetic energy (and therefore accelerates). Just before it hits the surface, it has lost all its potential energy, and all its energy is kinetic.

The state of motion of a body is changed by a **force**. According to Newton's second law of motion, a force changes the momentum of a body such that the acceleration of the body (its rate of change of velocity) is proportional to the strength of the force:

$$\text{force} = \text{mass} \times \text{acceleration}$$

Thus, to accelerate a heavy particle by a given amount requires a stronger force than to accelerate a light particle by the same amount. A force can be used to change the kinetic energy of a body, by accelerating the body to a higher speed. It may also be used to change the potential energy of a body by moving it to another position (for example, by raising it near the surface of the Earth).

The units of energy and force are given in *Further information 2*.

FURTHER INFORMATION 4: The kinetic theory of gases

Consider the system in Fig. 1. When a particle of mass m collides with the wall on the right, its component of linear momentum (the product of its mass and its velocity) parallel to the x axis changes from mv_x (when it is travelling to the right) to $-mv_x$ (when it is travelling to the left). Its momentum therefore changes by $2mv_x$ on each collision. The number of collisions in an interval Δt is equal to the number of particles able to reach the wall in that interval. Because a particle with speed v_x can travel a distance $v_x\Delta t$ in an interval Δt, all the particles within a distance $v_x\Delta t$ of the wall will strike it if they are travelling towards it. Therefore, if the wall has area A, then all the particles in a volume $Av_x\Delta t$ will reach the wall (if they are travelling towards it). If the number density, the number of particles divided by the total volume, is $\mathcal{N}$, the number in the volume $Av_x\Delta t$ is $\mathcal{N}Av_x\Delta t$.

On average, half the particles are moving to the right, and half are moving to the left. Therefore, the average number of collisions with the wall during the interval Δt is $\frac{1}{2}\mathcal{N}Av_x\Delta t$. The total momentum change in that interval is the product of this number and the change $2mv_x$ that an individual molecule experiences:

$$\text{momentum change} = \tfrac{1}{2}\mathcal{N}Av_x\Delta t \times 2mv_x = m\mathcal{N}Av_x^2\Delta t$$

The rate of change of momentum is this change of momentum divided by the interval Δt during which it occurs:

$$\text{rate of change of momentum} = m\mathcal{N}Av_x^2$$

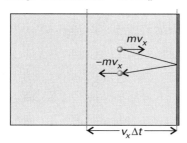

Fig. 1

The rate of change of momentum is equal to the force (by Newton's second law of motion), and so the force exerted by the gas on the wall is also $m\mathcal{N}Av_x^2$. It follows that the pressure, the force divided by the area A on which the force acts, is

$$\text{pressure} = m\mathcal{N}v_x^2$$

The detected pressure, p, is the average (denoted $\langle\cdots\rangle$) of the quantity just calculated:

$$p = m\mathcal{N}\langle v_x^2\rangle$$

The root mean square speed, c, of the particles is

$$c = \langle v^2\rangle^{1/2} = \left(\langle v_x^2\rangle + \langle v_y^2\rangle + \langle v_z^2\rangle\right)^{1/2}$$

However, because the particles are moving randomly, the average of v_x^2 is the same as the average of the analogous quantities in the y and z directions. Because $\langle v_x^2\rangle$, $\langle v_y^2\rangle$, and $\langle v_z^2\rangle$ are all equal,

$$c = \sqrt{3\langle v_x^2\rangle}$$

which implies that

$$\langle v_x^2\rangle = \tfrac{1}{3}c^2$$

Therefore,

$$p = \tfrac{1}{3}\mathcal{N}mc^2$$

The value of $\mathcal{N}$ is the product of the amount (the number of moles, n) and the Avogadro constant, N_A, divided by the volume, V, so the last equation becomes

$$pV = \tfrac{1}{3}nN_A mc^2 = \tfrac{1}{3}nMc^2$$

where $M = m \times N_A$ is the molar mass of the molecules. This expression is the equation used in the text.

FURTHER INFORMATION 5: Concepts of electrostatics

The fundamental expression in **electrostatics,** the interactions of stationary electric charges, is the **Coulomb potential energy** of one charge of magnitude q at a distance r from another charge q':

$$V = \frac{1}{4\pi\varepsilon_0} \times \frac{qq'}{r}$$

That is, the potential energy is inversely proportional to the separation of the charges. The fundamental constant ε_0 is the **vacuum permittivity**; its value is

$$\varepsilon_0 = 8.854\,187\,816 \times 10^{-12}\ \text{J}^{-1}\,\text{C}^2\,\text{m}^{-1}$$

With r in metres, m, and the charges in coulombs, C, the potential energy is in joules, J. The potential energy is equal to the work that must be done to bring up a charge q from infinity to a distance r from a charge q'. The potential energy of a charge q in the presence of another charge q' can be expressed in terms of the **Coulomb potential**, ϕ,

$$V = q \times \phi \qquad \phi = \frac{1}{4\pi\varepsilon_0} \times \frac{1}{r}$$

The units of potential are joules per coulomb, JC^{-1}, so that when ϕ is multiplied by a charge in coulombs, the result is in joules. The combination joules per coulomb occurs widely in electrostatics, and is called a **volt**, V:

$$1\,V = 1\,J\,C^{-1}$$

(which implies that $1\,V\,C = 1\,J$). If there are several charges $q_1, q_2, \cdots$ present in the system, then the total potential experienced by the charge q is the sum of the potential generated by each charge:

$$\phi = \phi_1 + \phi_2 + \cdots$$

For example, the potential generated by a dipole is the sum of the potentials of the two equal and opposite charges: these potentials do not in general cancel because the point of interest is at different distances from the two charges (Fig. 1). The motion of charge gives rise to an **electric current**, I. Electric current is measured in **amperes**, A, where

$$1\,A = 1\,C\,s^{-1}$$

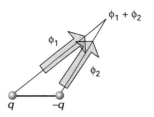

Fig. 1

If the electric charge is that of electrons (as it is through metals and semiconductors), then a current of 1 A represents the flow of 6×10^{18} electrons per second. If the current flows from a region of potential ϕ_i to ϕ_f, through a **potential difference** $\Delta\phi = \phi_f - \phi_i$ (for which we have sometimes used the symbol $\mathcal{V}$), then the rate of doing work is the current (the rate of transfer of charge) multiplied by the potential difference, $I \times \Delta\phi$. The rate of doing work is called **power**, P, so

$$P = I \times \Delta\phi$$

With current in amperes and the potential difference in volts, the power works out in joules per second, or **watts**, W:

$$1\,W = 1\,J\,s^{-1}$$

The total energy supplied in a time t is the power (the energy per second) multiplied by the time:

$$E = P \times t = I \times \Delta\phi \times t$$

The energy is obtained in joules with the current in amperes, the potential difference in volts, and the time in seconds.

FURTHER INFORMATION 6: Electromagnetic radiation and photons

Electromagnetic radiation, which includes γ radiation, ultraviolet radiation, visible light, infrared radiation, microwave radiation, and radio waves, is a wavelike, oscillating electric and magnetic field that propagates through space with a constant speed c, the 'speed of light'. The radiation is depicted in Fig. 1: the electric field and magnetic fields are perpendicular to each other and vary sinusoidally with a **wavelength**, λ (lambda), and **frequency**, ν (nu), that are related by

$$\lambda\nu = c$$

Therefore, the shorter the wavelength, the higher the frequency of the radiation. The rate of change of the fields is also commonly reported as the **wavenumber**, $\tilde{\nu}$ (nu tilde), which is defined as

$$\tilde{\nu} = \frac{\nu}{c}$$

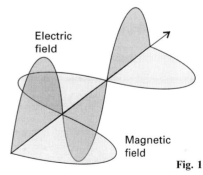

Fig. 1

The wavenumber is the number of complete wavelengths in a given region divided by the length of the region (typically, 1 cm). According to classical physics, the **intensity** of a ray is proportional to the square of the amplitude of the wave, so an

Table 1 The regions of the electromagnetic spectrum[†]

Regions	Wavelength	Frequency/Hz
Radiofrequency	> 30 cm	$< 10^9$
Microwave	3 mm to 30 cm	10^9 to 10^{11}
Infrared	1000 nm to 3 mm	10^{11} to 3×10^{14}
Visible	400 nm to 800 nm	4×10^{14} to 8×10^{14}
Ultraviolet	3 nm to 300 nm	10^{15} to 10^{17}
X-Rays, γ-rays	< 3 nm	$> 10^{17}$

[†]The boundaries of the regions are only approximate.

Table 2 Colour, frequency, and wavelength of light[†]

	Frequency/ 10^{14} Hz	Wavelength/ nm	Energy of photon/ 10^{-19} J
X-Rays and γ-rays	10^3 and above	3 and below	660 and above
Ultraviolet	10	300	6.6
Visible light			
Violet	7.1	420	4.7
Blue	6.4	470	4.2
Green	5.7	530	3.7
Yellow	5.2	580	3.4
Orange	4.8	620	3.2
Red	4.3	700	2.8
Infrared	3.0	1000	1.9
Microwaves and radiowaves	3×10^{-11} Hz and below	3×10^6 and above	2.0×10^{-22} J and below

[†] The values given are approximate but typical.

intense ray would be a wave of electromagnetic field that oscillated with a large amplitude. The classification of the electromagnetic spectrum into different regions according to the frequency and wavelength of the radiation is summarized in Table 1.

The wave shown in the illustration is **plane polarized**: it is so called because the electric field oscillates in a single plane. The plane may be orientated in any direction around the direction of propagation (with the electric field perpendicular to that direction). An alternative mode of polarization is **circular polarization**, in which the electric field rotates around the direction of propagation in either a clockwise or a counterclockwise sense.

According to quantum theory, a ray of frequency ν consists of a stream of **photons**, each one of which has energy

$$E = h\nu$$

where h is the Planck constant (Section 8.1). Thus, a photon of high frequency radiation has more energy than a photon of low frequency radiation. The greater the intensity of the ray, the greater the number of photons in it. In a vacuum, each photon travels with the speed of light. The frequency of the radiation determines the colour of visible light because different visual receptors in the eye respond to photons of different energy. The relation between colour and frequency is shown in Table 2, which also gives the energy carried by each type of photon.

Photons may also be polarized. A plane polarized ray of light consists of plane polarized photons and a circularly polarized ray consists of circularly polarized photons. The latter can be regarded as spinning either clockwise or counterclockwise about their direction of propagation.

FURTHER INFORMATION 7: Oxidation numbers

A simple way of judging whether a monatomic species has undergone oxidation or reduction is to note if the charge number of the species has changed. For example, an increase in the charge number of a monatomic ion (which corresponds to electron loss), as in the conversion of Fe^{2+} to Fe^{3+}, is an oxidation. A decrease in charge number (to a less positive or more negative value, as a result of electron gain), as in the conversion of Br to Br^-, is a reduction.

It is possible to assign to an atom in a polyatomic species an effective charge number, called the **oxidation number**, ω. (There is no standard symbol for this quantity.) The oxidation number is defined so that an increase in its value ($\Delta\omega > 0$) corresponds to oxidation, and a decrease ($\Delta\omega < 0$) corresponds to reduction.

An oxidation number is assigned to an element in a compound by supposing that it is present as an ion with a

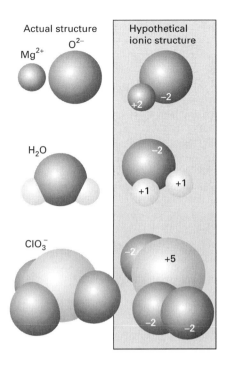

Actual structure | Hypothetical ionic structure

Mg^{2+} O^{2-}

H_2O

ClO_3^-

Fig. 1

characteristic charge; for instance, oxygen is present as O^{2-} in most of its compounds, and fluorine is present as F^- (Fig. 1). The more electronegative element is supposed to be present as the anion. This procedure implies that:

1. The oxidation number of an elemental substance is zero, $\omega(\text{element}) = 0$.
2. The oxidation number of a monatomic ion is equal to the charge number of that ion: $\omega(E^{z\pm}) = \pm z$.
3. The sum of the oxidation numbers of all the atoms in a species is equal to the overall charge number of the species.

Thus, hydrogen, oxygen, iron, and all the elements have $\omega = 0$ in their elemental forms; $\omega(Fe^{3+}) = +3$ and $\omega(Br^-) = -1$. It follows that the conversion of Fe to Fe^{3+} is an oxidation (because $\Delta\omega > 0$) and the conversion of Br to Br^- is a reduction (because $\Delta\omega < 0$). The definition of oxidation number and its relation to oxidation and reduction are consistent with the definitions in terms of electron loss and gain.

As an illustration, consider the oxidation numbers of the elements in SO_2 and SO_4^{2-}. The sum of oxidation numbers of the atoms in SO_2 must be 0, so we can write

$$\omega(S) + 2\omega(O) = 0$$

Each O atom has $\omega = -2$. Hence,

$$\omega(S) + 2 \times (-2) = 0$$

which solves to $\omega(S) = +4$. Now consider SO_4^{2-}. The sum of oxidation numbers of the atoms in the ion is -2, so we can write

$$\omega(S) + 4\omega(O) = -2$$

Because $\omega(O) = -2$,

$$\omega(S) + 4 \times (-2) = -2$$

which solves to $\omega(S) = +6$. The sulfur is more highly oxidized in the sulfate ion than in sulfur dioxide.

Exercise

Calculate the oxidation numbers of the elements in (a) H_2S, (b) PO_4^{3-}, (c) NO_3^-.

[*Answer*: (a) $\omega(H) = +1$, $\omega(S) = -2$; (b) $\omega(P) = +5$, $\omega(O) = -2$; (c) $\omega(N) = +5$, $\omega(O) = -2$]

FURTHER INFORMATION 8: The Lewis theory of covalent bonding

In his original formulation of a theory of the covalent bond, G.N. Lewis proposed that each bond consisted of one electron pair. Each atom in a molecule shared electrons until it had acquired an octet characteristic of a noble gas atom near it in the periodic table. (Hydrogen is an exception: it acquires a duplet of electrons.) Thus, to write down a Lewis structure:

1. Arrange the atoms as they are found in the molecule.
2. Add one electron pair (represented by dots, :) between each bonded atom.
3. Use the remaining electron pairs to complete the octets of all the atoms present either by forming lone pairs or by forming multiple bonds.

4. Replace bonding electron pairs by bond lines (—) but leave lone pairs as dots (:).

A Lewis structure does not (except in very simple cases), portray the actual geometrical structure of the molecule; it is a topological map of the arrangement of bonds.

As an example, consider the Lewis structure of methanol, CH_3OH, in which there are $4 \times 1 + 4 + 6 = 14$ electrons (and hence seven electron pairs) to accommodate. The first step is to write the atoms in the arrangement (1); the pale rectangles have been included to indicate which atoms are linked. The next step is to add electron pairs to denote bonds (2). The C atom now has a complete octet and all four H atoms have complete duplets. There are two unused electron pairs, which are used as lone pairs to complete the octet of the O atom (3). Finally, replace the bonding pairs by lines to indicate bonds (4). An example of a species with a multiple bond is acetic acid (5).

1

2

3

4

5

In some cases, more than one structure can be written in which the only difference is the location of multiple bonds or lone pairs. In such cases, the molecule's structure is interpreted as a **resonance hybrid** (a quantum mechanical blend) of the individual structures. Resonance is depicted by a double headed arrow. For example, the ozone molecule, O_3, is a resonance hybrid of two structures (6). Resonance distributes multiple-bond character over the participating atoms.

6

Many molecules cannot be written in a way that conforms to the octet rule. Those classified as **hypervalent molecules** require an expansion of the octet. Although it is often stated that octet expansion requires the involvement of d-orbitals, and is therefore confined to Period 3 and subsequent elements, there is good evidence to suggest that octet expansion is a conse-

quence of an atom's size, not its intrinsic orbital structure. Whatever the reason, octet expansion is need to account for the structures of PCl_5 with expansion to ten electrons (7), SF_6, expansion to 12 electrons (8), and XeO_4, expansion to 16 electrons (9). Octet expansion is also encountered in species that do not necessarily require it, but which, if it is permitted, may acquire a lower energy. Thus, of the structures (10a) and (10b) of the SO_4^{2-} ion, the second has a lower energy than the first. The actual structure of the ion is a resonance hybrid of both structures (together with analogous structures with double bonds in different locations), but the latter structure makes the dominant contribution.

7

8

9

10a

10b

Octet completion is not always energetically appropriate. Such is the case with boron trifluoride, BF_3. Two of the possible Lewis structures for this molecule are (11a) and (11b). In the former, the B atom has an **incomplete octet**. Nevertheless, it has a lower energy than the other structure,

11a

11b

for to form the latter structure one F atom has had partially to relinquish an electron pair, which is energetically demanding for such an electronegative element. The actual molecule is a resonance hybrid of the two structures (and of others with the double bond in different locations), but the overwhelming contribution is from the former structure. Consequently, we regard BF_3 as a molecule with an incomplete octet. This feature is responsible for its ability to act as a Lewis acid (an electron pair acceptor).

The Lewis approach fails for a class of **electron-deficient compounds**, which are molecules that have too few electrons for a Lewis structure to be written. The most famous example is diborane, B_2H_6, which requires at least seven pairs of electrons to bind the eight atoms together, but it has only twelve valence electrons in all. The structures of such molecules can be explained in terms of molecular orbital theory and the concept of delocalized electron pairs, in which the influence of an electron pair is distributed over several atoms.

FURTHER INFORMATION 9: The VSEPR model

In the **valence shell electron pair repulsion model** (VSEPR) we focus on a single, central atom and consider the local arrangement of atoms that are linked to it. For example, in considering the H_2O molecule, we concentrate on the electron pairs in the valence shell of the central O atom. This procedure can be extended to molecules in which there is no obvious central atom, such as in benzene, C_6H_6, or hydrogen peroxide, H_2O_2, by focusing attention on a group of atoms, such as a C—CH—C fragment of benzene or an H—O—O fragment of hydrogen peroxide, and considering the arrangement of electron pairs around the central atom of the fragment.

The basic assumption of the VSEPR model is that *the valence-shell electron pairs of the central atom adopt positions that maximize their separations*. Thus, if the atom has four electron pairs in its valence shell, then the pairs adopt a tetrahedral arrangement around the atom; if the atom has five pairs, then the arrangement is trigonal bipyramidal. The maximum-separation arrangements of two to seven electron pairs are summarized in Table 1.

Once the basic shape of the arrangement of electron pairs has been identified, the pairs are identified as bonding or nonbonding pairs. For instance, in the H_2O molecule, two of the tetrahedrally arranged pairs are bonding pairs and two are nonbonding pairs. Then the shape of the molecule is classified by noting the arrangement of the atoms around the central atom. The H_2O molecule, for instance, has an underlying tetrahedral arrangement of lone pairs, but as only two of the pairs are bonding pairs, the molecule is classified as angular (Fig. 1) It is important to keep in mind the distinction between the arrangement of electron pairs and the shape of the resulting

Table I Electron pair arrangements

Number of electron pairs	Arrangement
2	Linear
3	Trigonal planar
4	Tetrahedral
5	Trigonal bipyramidal
6	Octahedral
7	Pentagonal bipyramidal

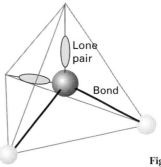

Lone pair

Bond

Fig. 1

molecule: the latter is determined by the relative locations of the atoms, not the lone pairs (Fig. 2).

For example, to predict the shape of an ethane molecule we concentrate on one of the C atoms initially. That atom has four electron pairs in its valence shell (in the molecule), and they adopt a tetrahedral arrangement. All four electron pairs are bonding: three bond H atoms and the fourth bonds the second C atom. Therefore, the arrangement of atoms is tetrahedral around the C atom. The second C atom has the same environment, so we conclude that the ethane molecule consists of two tetrahedral CH_3 groups (**1**).

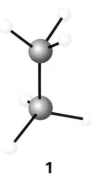

1

The next stage in the application of the VSEPR model is to accommodate the greater repelling effect of lone pairs compared with that of bonding pairs. That is, *bonding pairs tend to move away from lone pairs even though that might reduce their separation from other bonding pairs*. The NH_3 molecule provides a simple example. The N atom has four electron pairs in its valence shell and they adopt a tetrahedral arrangement. Three

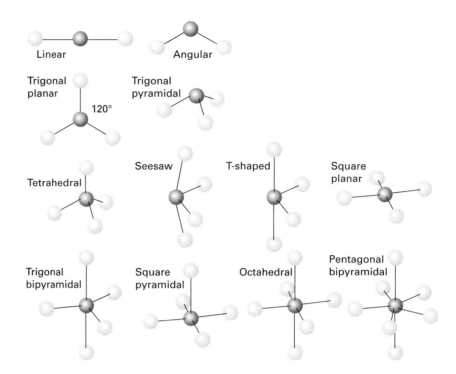

Fig. 2

of the pairs are bonding pairs, and the fourth is a lone pair. The basic shape of the molecule is therefore trigonal pyramidal. However, a lower energy is achieved if the three bonding pairs move away from the lone pair, even though they are brought slightly closer together (**2**). We therefore predict an HNH bond angle of slightly less than the tetrahedral angle of 109.5°, which is consistent with the observed angle of 107°.

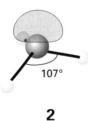

2

As an example, consider the shape of an SF_4 molecule. The first step is to write a Lewis (electron dot) structure for the molecule to identify the number of lone pairs in the valence shell of the S atom (**3**). This structure shows that there are five electron pairs on the S atom. Reference to Table 1 shows that the five pairs are arranged as a trigonal bipyramid. Four of the pairs are bonding pairs and one is a lone pair. The repulsions stemming from the lone pair are minimized if the lone pair is placed in an equatorial position: then it is close to the axial pairs (**4**), whereas if it had adopted an axial position it would

have been close to three equatorial pairs (**5**). Finally, the four bonding pairs are allowed to relax away from the single lone pair, to give a distorted seesaw arrangement (**6**).

$$:\ddot{F}\text{---}\underset{\ddot{F}}{\overset{..}{S}}\text{---}\ddot{F}: $$

3

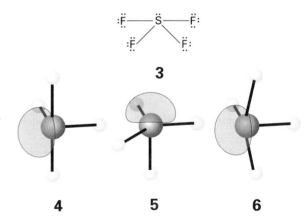

4   **5**   **6**

So far, we have not considered how to take into account multiple bonding. It is supposed that the two or three electron pairs, respectively, of double and triple bonds are pinned together in their respective bonding regions, and hence act like a single 'superpair'. For example, each C atom in an ethene (ethylene) molecule, $CH_2\text{=}CH_2$, is regarded as having three pairs (one of them the superpair of two electrons pairs of the double bond); they adopt a trigonal planar arrangement around each atom, so the shape of the molecule is trigonal planar at each C atom (**7**). Another example is the SO_3^{2-} ion: if we adopt the Lewis structure in (**8**), then we see that there are four pairs (one of them a superpair) around the S atom, indicating a tetrahedral arrangement of pairs. One pair is a lone pair, so overall the ion is trigonal pyramidal (**9**). We would reach the same conclusion if we adopted the alternative Lewis structure (**10**) in which there are four electron pairs (none of them a superpair).

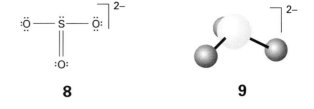

8   **9**

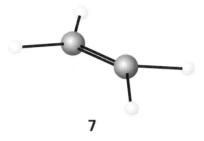

7

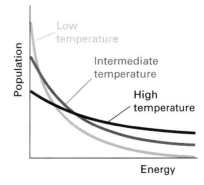

10

FURTHER INFORMATION 10: The Boltzmann distribution

The Boltzmann distribution expresses the probability, p, that a molecule will be found in a state with energy E:

$$p = \frac{e^{-E/kT}}{q} \qquad q = \sum_i e^{-E_i/kT}$$

where the E_i are the energies of all the states of the system. The constant k is the **Boltzmann constant**, with the value

$$k = 1.380\,658 \times 10^{-23}\,\mathrm{J\,K^{-1}}$$

This constant is a fundamental constant of nature, and when multiplied by the Avogadro constant yields the gas constant:

$$R = N_A \times k$$

The Boltzmann distribution shows that the population decreases exponentially with increasing energy (Fig. 1). Specifically, it follows from the distribution that the ratio of the probabilities p' and p of finding a system in states with energies E' and E at a temperature T is

$$\frac{p'}{p} = e^{-(E'-E)/kT}$$

which shows the exponential dependence explicitly. The spread of populations to high energies increases with temperature (Fig. 2). No simple expression can be given in this case because the denominator q varies with temperature, and so does not cancel in the expression for the ratio of probabilities. The energy kT (in molar terms, RT) is the dividing line between states that are

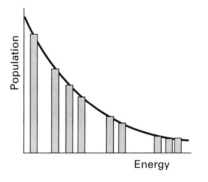

Fig. 1

significantly populated and those that are not. At 25 °C, $RT = 2.5\,\mathrm{kJ\,mol^{-1}}$, and kT corresponds to $207\,\mathrm{cm^{-1}}$ or $0.026\,\mathrm{eV}$, so these three values are the frontier between substantial occupation and substantial emptiness at about room temperature.

Fig. 2

The quantity q is called the **partition function** for the molecule. Broadly speaking, q is a measure of the number of states of the system that are significantly populated at the temperature of interest. Thus, q varies from 1 at $T = 0$ (when only the state of lowest energy can be populated) to infinity at infinite temperature (when all states of the system are accessible).

Under certain circumstances, the Boltzmann distribution can be used to calculate the fraction of molecules that have an energy of at least a certain minimum (such as the activation energy E_a in the collision theory of reactions). Thus, suppose that the system can have any energy E between zero and infinity. The partition function is then given by the integral

$$q = \text{constant} \times \int_0^\infty e^{-E/kT}\, dE = \text{constant} \times kT$$

The constant of proportionality will cancel shortly. The fraction of molecules with an energy of at least E_a is then the sum of the probabilities of having an energy between E_a and infinity:

fraction with energy greater than E_a

$$= \frac{\text{constant}}{q} \times \int_{E_a}^\infty e^{-E/kT} = \frac{\text{constant}}{q} \times kT e^{-E_a/kT}$$

The constant cancels the same term in the expression for q, and we are left with

$$\text{fraction with energy greater than } E_a = e^{-E_a/kT}$$

That is, the fraction decreases exponentially with the energy E_a and increases sharply as the temperature is raised.

Appendix I

Table AI.I Thermodynamic data for organic compounds (all values relate to 298 K)

	$M/$ g mol^{-1}	$\Delta_f H^{\ominus}/$ kJ mol^{-1}	$\Delta_f G^{\ominus}/$ kJ mol^{-1}	$S^{\ominus}/$ J K^{-1} mol^{-1}	$C_{p,m}/$ J K^{-1} mol^{-1}	$\Delta_{com} H^{\ominus}/$ kJ mol^{-1}
C(s) (graphite)	12.011	0	0	5.740	8.527	−393.51
C(s) (diamond)	12.011	+1.895	+2.900	2.377	6.113	−395.40
CO_2(g)	44.010	−393.51	−394.36	213.74	37.11	
Hydrocarbons						
CH_4(g), methane	16.04	−74.81	−50.72	186.26	35.31	−890
CH_3(g), methyl	15.04	+145.69	+147.92	194.2	38.70	
C_2H_2(g), ethyne	26.04	+226.73	+209.20	200.94	43.93	−1300
C_2H_4(g), ethene	28.05	+52.26	+68.15	219.56	43.56	−1411
C_2H_6(g), ethane	30.07	−84.68	−32.82	229.60	52.63	−1560
C_3H_6(g), propene	42.08	+20.42	+62.78	267.05	63.89	−2058
C_3H_6(g), cyclopropane	42.08	+53.30	+104.45	237.55	55.94	−2091
C_3H_8(g), propane	42.10	−103.85	−23.49	269.91	73.5	−2220
C_4H_8(g), 1-butene	56.11	−0.13	+71.39	305.71	85.65	−2717
C_4H_8(g), *cis*-2-butene	56.11	−6.99	+65.95	300.94	78.91	−2710
C_4H_8(g), *trans*-2-butene	56.11	−11.17	+63.06	296.59	87.82	−2707
C_4H_{10}(g), butane	58.13	−126.15	−17.03	310.23	97.45	−2878
C_5H_{12}(g), pentane	72.15	−146.44	−8.20	348.40	120.2	−3537
C_5H_{12}(l)	72.15	−173.1				
C_6H_6(l), benzene	78.12	+49.0	+124.3	173.3	136.1	−3268
C_6H_6(g)	78.12	+82.93	+129.72	269.31	81.67	−3320
C_6H_{12}(l), cyclohexane	84.16	−156	+26.8		156.5	−3902
C_6H_{14}(l), hexane	86.18	−198.7		204.3		−4163
$C_6H_5CH_3$(g), methylbenzene						
(toluene)	92.14	+50.0	+122.0	320.7	103.6	−3953
C_7H_{16}(l), heptane	100.21	−224.4	+1.0	328.6	224.3	
C_8H_{18}(l), octane	114.23	−249.9	+6.4	361.1		−5471
C_8H_{18}(l), iso-octane	114.23	−255.1				−5461
$C_{10}H_8$(s), naphthalene	128.18	+78.53				−5157
Alcohols and phenols						
CH_3OH(l), methanol	32.04	−238.66	−166.27	126.8	81.6	−726

Table AI.I — continued

	$M/$ g mol^{-1}	$\Delta_f H^{\ominus}/$ kJ mol^{-1}	$\Delta_f G^{\ominus}/$ kJ mol^{-1}	$S^{\ominus}/$ $\text{J K}^{-1}\text{mol}^{-1}$	$C_{p,m}/$ $\text{J K}^{-1}\text{mol}^{-1}$	$\Delta_{com} H^{\ominus}/$ kJ mol^{-1}
Alcohols and phenols (continued)						
$CH_3OH(g)$	32.04	−200.66	−161.96	239.81	43.89	−764
$C_2H_5OH(l)$, ethanol	46.07	−277.69	−174.78	160.7	111.46	−1368
$C_2H_5OH(g)$	46.07	−235.10	−168.49	282.70	65.44	−1409
$C_6H_5OH(s)$, phenol	94.12	−165.0	−50.9	146.0		−3054
Carboxylic acids, hydroxy acids, and esters						
$HCOOH(l)$, formic	46.03	−424.72	−361.35	128.95	99.04	−255
$CH_3COOH(l)$, acetic	60.05	−484.5	−389.9	159.8	124.3	−875
$CH_3COOH(aq)$	60.05	−485.76	−396.46	178.7		
$CH_3CO_2^-(aq)$	59.05	−486.01	−369.31	86.6	−6.3	
$(COOH)_2(s)$, oxalic	90.04	−827.2			117	−254
$C_6H_5COOH(s)$, benzoic	122.13	−385.1	−245.3	167.6	146.8	−3227
$CH_3CH(OH)COOH(s)$, lactic	90.08	−694.0				−1344
$CH_3COOC_2H_5(l)$, ethyl acetate	88.11	−479.0	−332.7	259.4	170.1	−2231
Alkanals and alkanones						
$HCHO(g)$, methanal	30.03	−108.57	−102.53	218.77	35.40	−571
$CH_3CHO(l)$, ethanal	44.05	−192.30	−128.12	160.2		−1166
$CH_3CHO(g)$	44.05	−166.19	−128.86	250.3	57.3	−1192
$CH_3COCH_3(l)$, propanone	58.08	−248.1	−155.4	200.4	124.7	−1790
Sugars						
$C_6H_{12}O_6(s)$, α-D-glucose	180.16	−1274				−2808
$C_6H_{12}O_6(s)$, β-D-glucose	180.16	−1268	−910	212		
$C_6H_{12}O_6(s)$, β-D-fructose	180.16	−1266				−2810
$C_{12}H_{22}O_{11}(s)$, sucrose	342.30	−2222	−1543	360.2		−5645
Nitrogen compounds						
$CO(NH_2)_2(s)$, urea	60.06	−333.51	−197.33	104.60	93.14	−632
$CH_3NH_2(g)$, methylamine	31.06	−22.97	+32.16	243.41	53.1	−1085
$C_6NH_2(l)$, aniline	93.13	+31.1				−3393
$CH_2(NH_2)COOH(s)$, glycine	75.07	−532.9	−373.4	103.5	99.2	−969

Table AI.2 Thermodynamic data (all values relate to 298 K)

	$M/$ $\mathrm{g\,mol^{-1}}$	$\Delta_f H^{\ominus}/$ $\mathrm{kJ\,mol^{-1}}$	$\Delta_f G^{\ominus}/$ $\mathrm{kJ\,mol^{-1}}$	$S^{\ominus}/$ $\mathrm{J\,K^{-1}\,mol^{-1}}$	$C_{p,m}/$ $\mathrm{J\,K^{-1}\,mol^{-1}}$
Aluminium (aluminum)					
Al(s)	26.98	0	0	28.33	24.35
Al(l)	26.98	+10.56	+7.20	39.55	24.21
Al(g)	26.98	+326.4	+285.7	164.54	21.38
Al^{3+}(g)	26.98	+5483.17			
Al^{3+}(aq)	26.98	−531	−485	−321.7	
Al_2O_3(s, α)	101.96	−1675.7	−1582.3	50.92	79.04
$AlCl_3$(s)	133.24	−704.2	−628.8	110.67	91.84
Argon					
Ar(g)	39.95	0	0	154.84	20.786
Antimony					
Sb(s)	121.75	0	0	45.69	25.23
SbH_3(g)	153.24	+145.11	+147.75	232.78	41.05
Arsenic					
As(s, α)	74.92	0	0	35.1	24.64
As(g)	74.92	+302.5	261.0	174.21	20.79
As_4(g)	299.69	+143.9	+92.4	314	
AsH_3(g)	77.95	+66.44	+68.93	222.78	38.07
Barium					
Ba(s)	137.34	0	0	62.8	28.07
Ba(g)	137.34	+180	+146	170.24	20.79
Ba^{2+}(aq)	137.34	−537.64	−560.77	9.6	
BaO(s)	153.34	−553.5	−525.1	70.43	47.78
$BaCl_2$(s)	208.25	−858.6	−810.4	123.68	75.14
Beryllium					
Be(s)	9.01	0	0	9.50	16.44
Be(g)	9.01	+324.3	+286.6	136.27	20.79
Bismuth					
Bi(s)	208.98	0	0	56.74	25.52
Bi(g)	208.98	+207.1	+168.2	187.00	20.79
Bromine					
Br_2(l)	159.82	0	0	152.23	75.689
Br_2(g)	159.82	+30.907	+3.110	245.46	36.02
Br(g)	79.91	+111.88	+82.396	175.02	20.786
Br^-(g)	79.91	−219.07			
Br^-(aq)	79.91	−121.55	−103.96	82.4	−141.8
HBr(g)	90.92	−36.40	−53.45	198.70	29.142

Table AI.2 — Continued

	$M/$ g mol^{-1}	$\Delta_f H^{\ominus}/$ kJ mol^{-1}	$\Delta_f G^{\ominus}/$ kJ mol^{-1}	$S^{\ominus}/$ J K^{-1} mol^{-1}	$C_{p,m}/$ J K^{-1} mol^{-1}
Cadium					
Cd(s, γ)	112.40	0	0	51.76	25.98
Cd(g)	112.40	+112.01	+77.41	167.75	20.79
Cd^{2+}(aq)	112.40	−75.90	−77.612	−73.2	
CdO(s)	128.40	−258.2	−228.4	54.8	43.43
CdCO$_3$(s)	172.41	−750.6	−669.4	92.5	
Caesium (cesium)					
Cs(s)	132.91	0	0	85.23	32.17
Cs(g)	132.91	+76.06	+49.12	175.60	20.79
Cs$^+$(aq)	132.91	−258.28	−292.02	133.05	−10.5
Calcium					
Ca(s)	40.08	0	0	41.42	25.31
Ca(g)	40.08	+178.2	+144.3	154.88	20.786
Ca^{2+}(aq)	40.08	−542.83	−553.58	−53.1	
CaO(s)	56.08	−635.09	−604.03	39.75	42.80
CaCO$_3$(s) (calcite)	100.09	−1206.9	−1128.8	92.9	81.88
CaCO$_3$(s) (aragonite)	100.09	−1207.1	−1127.8	88.7	81.25
CaF$_2$(s)	78.08	1219.6	−1167.3	68.87	67.03
CaCl$_2$(s)	110.99	−795.8	−748.1	104.6	72.59
CaBr$_2$(s)	199.90	−682.8	−663.6	130	
Carbon (for 'organic' compounds of carbon, see Table A1.1)					
C(s) (graphite)	12.011	0	0	5.740	8.527
C(s) (diamond)	12.011	+1.895	+2.900	2.377	6.133
C(g)	12.011	+716.68	+671.26	158.10	20.838
C$_2$(g)	24.022	+831.90	+775.89	199.42	43.21
CO(g)	28.011	−110.53	−137.17	197.67	29.14
CO$_2$(g)	44.010	−393.51	−394.36	213.74	37.11
CO$_2$(aq)	44.010	−413.80	−385.98	117.6	
H$_2$CO$_3$(aq)	62.03	−699.65	−623.08	187.4	
HCO$_3^-$(aq)	61.02	−691.99	−586.77	91.2	
CO$_3^{2-}$(aq)	60.01	−677.14	−527.81	−56.9	
CCl$_4$(l)	153.82	−135.44	−65.21	216.40	131.75
CS$_2$(l)	76.14	+89.70	+65.27	151.34	75.7
HCN(g)	27.03	+135.1	+124.7	201.78	35.86
HCN(l)	27.03	+108.87	+124.97	112.84	70.63
CN$^-$(aq)	26.02	+150.6	+172.4	94.1	
Chlorine					
Cl$_2$(g)	70.91	0	0	223.07	33.91
Cl(g)	35.45	+121.68	+105.68	165.20	21.840
Cl$^-$(g)	35.45	−233.13			
Cl$^-$(aq)	35.45	−167.16	−131.23	56.5	−136.4

Table AI.2 — Continued

	$M/$ g mol^{-1}	$\Delta_f H^{\ominus}/$ kJ mol^{-1}	$\Delta_f G^{\ominus}/$ kJ mol^{-1}	$S^{\ominus}/$ $\text{J K}^{-1}\text{mol}^{-1}$	$C_{p,m}/$ $\text{J K}^{-1}\text{mol}^{-1}$
Chlorine (continued)					
HCl(g)	36.46	−92.31	−95.30	186.91	29.12
HCl(aq)	36.46	−167.16	−131.23	56.5	−136.4
Chromium					
Cr(s)	52.00	0	0	23.77	23.35
Cr(g)	52.00	+396.6	+351.8	174.50	20.79
CrO_4^{2-}(aq)	115.99	−881.15	−727.75	50.21	
$Cr_2O_7^{2-}$(aq)	215.99	−1490.3	−1301.1	261.9	
Copper					
Cu(s)	63.54	0	0	33.150	24.44
Cu(g)	63.54	+338.32	+298.58	166.38	20.79
Cu^+(aq)	63.54	+71.67	+49.98	40.6	
Cu^{2+}(aq)	63.54	+64.77	+65.49	−99.6	
Cu_2O(s)	143.08	−168.6	−146.0	93.14	63.64
CuO(s)	79.54	−157.3	−129.7	42.63	42.30
$CuSO_4$(s)	159.60	−771.36	−661.8	109	100.0
$CuSO_4 \cdot H_2O$(s)	177.62	−1085.8	−918.11	146.0	134
$CuSO_4 \cdot 5H_2O$(s)	249.68	−2279.7	−1879.7	300.4	280
Deuterium					
D_2(g)	4.028	0	0	144.96	29.20
HD(g)	3.022	+0.318	−1.464	143.80	29.196
D_2O(g)	20.028	−249.20	−234.54	198.34	34.27
D_2O(l)	20.028	−294.60	−243.44	75.94	84.35
HDO(g)	19.022	−245.30	233.11	199.51	33.81
HDO(l)	19.022	−289.89	−241.86	79.29	
Fluorine					
F_2(g)	38.00	0	0	202.78	31.30
F(g)	19.00	+78.99	+61.91	158.75	22.74
F^-(aq)	19.00	−332.63	−278.79	−13.8	−106.7
HF(g)	20.01	−271.1	−273.2	173.78	29.13
Gold					
Au(s)	196.97	0	0	47.40	25.42
Au(g)	196.97	+366.1	+326.3	180.50	20.79
Helium					
He(g)	4.003	0	0	126.15	20.786
Hydrogen (see also deuterium)					
H_2(g)	2.016	0	0	130.684	28.824
H(g)	1.008	+217.97	+203.25	114.71	20.784
H^+(aq)	1.008	0	0	0	0

Table AI.2 — Continued

	$M/$ g mol^{-1}	$\Delta_f H^{\ominus}/$ kJ mol^{-1}	$\Delta_f G^{\ominus}/$ kJ mol^{-1}	$S^{\ominus}/$ J K^{-1} mol^{-1}	$C_{p,m}/$ J K^{-1} mol^{-1}
Hydrogen (continued)					
$H_2O(l)$	18.015	−285.83	−237.13	69.91	75.291
$H_2O(g)$	18.015	−241.82	−228.57	188.83	33.58
$H_2O_2(l)$	34.015	−187.78	−120.35	109.6	89.1
Iodine					
$I_2(s)$	253.81	0	0	116.135	54.44
$I_2(g)$	253.81	+62.44	+19.33	260.69	36.90
$I(g)$	126.90	+106.84	+70.25	180.79	20.786
$I^-(aq)$	126.90	−55.19	−51.57	111.3	−142.3
$HI(g)$	127.91	+26.48	+1.70	206.59	29.158
Iron					
$Fe(s)$	55.85	0	0	27.28	25.10
$Fe(g)$	55.85	+416.3	+370.7	180.49	25.68
$Fe^{2+}(aq)$	55.85	−89.1	−78.90	−137.7	
$Fe^{3+}(aq)$	55.85	−48.5	−4.7	−315.9	
$Fe_3O_4(s)$ (magnetite)	231.54	−1184.4	−1015.4	146.4	143.43
$Fe_2O_3(s)$ (haematite)	159.69	−824.2	−742.2	87.40	103.85
$FeS(s, \alpha)$	87.91	−100.0	−100.4	60.29	50.54
$FeS_2(s)$	119.98	−178.2	−166.9	52.93	62.17
Krypton					
$Kr(g)$	83.80	0	0	164.08	20.786
Lead					
$Pb(s)$	207.19	0	0	64.81	26.44
$Pb(g)$	207.19	+195.0	+161.9	175.37	20.79
$Pb^{2+}(aq)$	207.19	−1.7	−24.43	10.5	
$PbO(s, yellow)$	223.19	−217.32	−187.89	68.70	45.77
$PbO(s, red)$	223.19	−218.99	−188.93	66.5	45.81
$PbO_2(s)$	239.19	−277.4	−217.33	68.6	64.64
Lithium					
$Li(s)$	6.94	0	0	29.12	24.77
$Li(g)$	6.94	+159.37	+126.66	138.77	20.79
$Li^+(aq)$	6.94	−278.49	−293.31	13.4	68.6
Magnesium					
$Mg(s)$	24.31	0	0	32.68	24.89
$Mg(g)$	24.31	+147.70	+113.10	148.65	20.786
$Mg^{2+}(aq)$	24.31	−466.85	−454.8	−138.1	
$MgO(s)$	40.31	−601.70	−569.43	26.94	37.15

	$M/$ $\mathrm{g\,mol^{-1}}$	$\Delta_f H^{\ominus}/$ $\mathrm{kJ\,mol^{-1}}$	$\Delta_f G^{\ominus}/$ $\mathrm{kJ\,mol^{-1}}$	$S^{\ominus}/$ $\mathrm{J\,K^{-1}\,mol^{-1}}$	$C_{p,m}/$ $\mathrm{J\,K^{-1}\,mol^{-1}}$
Magnesium (continued)					
$MgCO_3(s)$	84.32	−1095.8	−1012.1	65.7	75.52
$MgCl_2(s)$	95.22	−641.32	−591.79	89.62	71.38
Mercury					
$Hg(l)$	200.59	0	0	76.02	27.983
$Hg(g)$	200.59	+61.32	+31.82	174.96	20.786
$Hg^{2+}(aq)$	200.59	+171.1	+164.40	−32.2	
$Hg_2^{2+}(aq)$	401.18	+172.4	+153.52	84.5	
$HgO(s)$	216.59	−90.83	−58.54	70.29	44.06
$Hg_2Cl_2(s)$	472.09	−265.22	−210.75	192.5	102
$HgCl_2(s)$	271.50	−224.3	−178.6	146.0	
$HgS(s, black)$	232.65	−53.6	−47.7	88.3	
Neon					
$Ne(g)$	20.18	0	0	146.33	20.786
Nitrogen					
$N_2(g)$	28.013	0	0	191.61	29.125
$N(g)$	14.007	+472.70	+455.56	153.30	20.786
$NO(g)$	30.01	+90.25	+86.55	210.76	29.844
$N_2O(g)$	44.01	+82.05	+104.20	219.85	38.45
$NO_2(g)$	46.01	+33.18	+51.31	240.06	37.20
$N_2O_4(g)$	92.01	+9.16	+97.89	304.29	77.28
$N_2O_5(s)$	108.01	−43.1	+113.9	178.2	143.1
$N_2O_5(g)$	108.01	+11.3	+115.1	355.7	84.5
$HNO_3(l)$	63.01	−174.10	−80.71	155.60	109.87
$HNO_3(aq)$	63.01	−207.36	−111.25	146.4	−86.6
$NO_3^-(aq)$	62.01	−205.0	−108.74	146.4	−86.6
$NH_3(g)$	17.03	−46.11	−16.45	192.45	35.06
$NH_3(aq)$	17.03	−80.29	−26.50	113.3	
$NH_4^+(aq)$	18.04	−132.51	−79.31	113.4	79.9
$NH_2OH(s)$	33.03	−114.2			
$HN_3(l)$	43.03	+264.0	+327.3	140.6	43.68
$HN_3(g)$	43.03	+294.1	+328.1	238.97	98.87
$N_2H_4(l)$	32.05	+50.63	+149.43	121.21	139.3
$NH_4NO_3(s)$	80.04	−365.56	−183.87	151.08	84.1
$NH_4Cl(s)$	53.49	−314.43	−202.87	94.6	
Oxygen					
$O_2(g)$	31.999	0	0	205.138	29.355
$O(g)$	15.999	+249.17	+231.73	161.06	21.912
$O_3(g)$	47.998	+142.7	+163.2	238.93	39.20
$OH^-(aq)$	17.007	−229.99	−157.24	−10.75	−148.5

	$M/$ g mol^{-1}	$\Delta_f H^{\ominus}/$ kJ mol^{-1}	$\Delta_f G^{\ominus}/$ kJ mol^{-1}	$S^{\ominus}/$ J K^{-1} mol^{-1}	$C_{p,m}/$ J K^{-1} mol^{-1}
Phosphorus					
P(s, wh)	30.97	0	0	41.09	23.840
P(g)	30.97	+314.64	+278.25	163.19	20.786
P_2(g)	61.95	+144.3	+103.7	218.13	32.05
P_4(g)	123.90	+58.91	+24.44	279.98	67.15
PH_3(g)	34.00	+5.4	+13.4	210.23	37.11
PCl_3(g)	137.33	−287.0	−267.8	311.78	71.84
PCl_3(l)	137.33	−319.7	−272.3	217.1	
PCl_5(g)	208.24	−374.9	−305.0	364.6	112.8
PCl_5(s)	208.24	−443.5			
H_3PO_3(s)	82.00	−964.4			
H_3PO_3(aq)	82.00	−964.8			
H_3PO_4(s)	94.97	−1279.0	−1119.1	110.50	106.06
H_3PO_4(l)	94.97	−1266.9			
H_3PO_4(aq)	94.97	−1277.4	−1018.7	−222	
PO_4^{3-}(aq)	94.97	−1277.4	−1018.7	−222	
P_4O_{10}(s)	283.89	−2984.0	−2697.0	228.86	211.71
P_4O_6(s)	219.89	−1640.1			
Potassium					
K(s)	39.10	0	0	64.18	29.58
K(g)	39.10	+89.24	+60.59	160.336	20.786
K^+(g)	39.10	+514.26			
K^+(aq)	39.10	−252.38	−283.27	102.5	21.8
KOH(s)	56.11	−424.76	−379.08	78.9	64.9
KF(s)	58.10	−576.27	−537.75	66.57	49.04
KCl(s)	74.56	−436.75	−409.14	82.59	51.30
KBr(s)	119.01	−393.80	−380.66	95.90	52.30
KI(s)	166.01	−327.90	−324.89	106.32	52.93
Silicon					
Si(s)	28.09	0	0	18.83	20.00
Si(g)	28.09	+455.6	+411.3	167.97	22.25
SiO_2(s, α)	60.09	−910.93	−856.64	41.84	44.43
Silver					
Ag(s)	107.87	0	0	42.55	25.351
Ag(g)	107.87	+284.55	+245.65	173.00	20.79
Ag^+(aq)	107.87	+105.58	+77.11	72.68	21.8
AgBr(s)	187.78	−100.37	−96.90	107.1	52.38
AgCl(s)	143.32	−127.07	−109.79	96.2	50.79
Ag_2O(s)	231.74	−31.05	−11.20	121.3	65.86
$AgNO_3$(s)	169.88	−124.39	−33.41	140.92	93.05

Table AI.2 — Continued

	$M/$ $\mathrm{g\,mol^{-1}}$	$\Delta_f H^{\ominus}/$ $\mathrm{kJ\,mol^{-1}}$	$\Delta_f G^{\ominus}/$ $\mathrm{kJ\,mol^{-1}}$	$S^{\ominus}/$ $\mathrm{J\,K^{-1}\,mol^{-1}}$	$C_{p,m}/$ $\mathrm{J\,K^{-1}\,mol^{-1}}$
Sodium					
Na(s)	22.99	0	0	51.21	28.24
Na(g)	22.99	+107.32	+76.76	153.71	20.79
Na^+(aq)	22.99	−240.12	−261.91	59.0	46.4
NaOH(s)	40.00	−425.61	−379.49	64.46	59.54
NaCI(s)	58.44	−411.15	−384.14	72.13	50.50
NaBr(s)	102.90	−361.06	−348.98	86.82	51.38
NaI(s)	149.89	−287.78	−286.06	98.53	52.09
Sulfur					
S(s, α) (rhombic)	32.06	0	0	31.80	22.64
S(s, β) (monoclinic)	32.06	+0.33	+0.1	32.6	23.6
S(g)	32.06	+278.81	+238.25	167.82	23.673
S_2(g)	64.13	+128.37	+79.30	228.18	32.47
S^{2-}(aq)	32.06	+33.1	+85.8	−14.6	
SO_2(g)	64.06	−296.83	−300.19	248.22	39.87
SO_3(g)	80.06	−395.72	−371.06	256.76	50.67
H_2SO_4(l)	98.08	−813.99	−690.00	156.90	138.9
H_2SO_4(aq)	98.08	−909.27	−744.53	20.1	−293
SO_4^{2-}(aq)	96.06	−909.27	−744.53	20.1	−293
HSO_4^-(aq)	97.07	−887.34	−755.91	131.8	−84
H_2S(g)	34.08	−20.63	−33.56	205.79	34.23
H_2S(aq)	34.08	−39.7	−27.83	121	
HS^-(aq)	33.072	−17.6	+12.08	62.08	
SF_6(g)	146.05	−1209	−1105.3	291.82	97.28
Tin					
Sn(s, β)	118.69	0	0	51.55	26.99
Sn(g)	118.69	+302.1	+267.3	168.49	20.26
Sn^{2+}(aq)	118.69	−8.8	−27.2	−17	
SnO(s)	134.69	−285.8	−256.8	56.5	44.31
SnO_2(s)	150.69	−580.7	519.6	52.3	52.59
Xenon					
Xe(g)	131.30	0	0	169.68	20.786
Zinc					
Zn(s)	65.37	0	0	41.63	25.40
Zn(g)	65.37	+130.73	+95.14	160.98	20.79
Zn^{2+}(aq)	65.37	−153.89	−147.06	−112.1	46
ZnO(s)	81.37	−348.28	−318.30	43.64	40.25

Appendix 2

Table A2.I Standard potentials at 298 K in In electrochemical order

Reduction half-reaction	$E^{\ominus}/\text{V}$	Reduction half-reaction	$E^{\ominus}/\text{V}$
Strongly oxidizing		$Cu^{2+} + e^- \rightarrow Cu^+$	$+0.16$
$H_4XeO_6 + 2H^+ + 2e^- \rightarrow XeO_3 + 3H_2O$	$+3.0$	$Sn^{4+} + 2e^- \rightarrow Sn^{2+}$	$+0.15$
$F_2 + 2e^- \rightarrow 2F^-$	$+2.87$	$AgBr + e^- \rightarrow Ag + Br^-$	$+0.07$
$O_3 + 2H^+ + 2e^- \rightarrow O_2 + H_2O$	$+2.07$	$Ti^{4+} + e^- \rightarrow Ti^{3+}$	0.00
$S_2O_8^{2-} + 2e^- \rightarrow 2SO_4^{2-}$	$+2.05$	$2H^+ + 2e^- \rightarrow H$	$0, \text{by definition}$
$Ag^{2+} + e^- \rightarrow Ag^+$	$+1.98$	$Fe^{3+} + 3e^- \rightarrow Fe$	-0.04
$Co^{3+} + e^- \rightarrow Co^{2+}$	$+1.81$	$O_2 + H_2O + 2e^- \rightarrow HO_2^- + OH^-$	-0.08
$HO_2 + 2H^+ + 2e^- \rightarrow 2H_2O$	$+1.78$	$Pb^{2+} + 2e^- \rightarrow Pb$	-0.13
$Au^+ + e^- \rightarrow Au$	$+1.69$	$In^+ + e^- \rightarrow In$	-0.14
$Pb^{4+} + 2e^- \rightarrow Pb^{2+}$	$+1.67$	$Sn^{2+} + 2e^- \rightarrow Sn$	-0.14
$2HClO + 2H^+ + 2e^- \rightarrow Cl_2 + 2H_2O$	$+1.63$	$AgI + e^- \rightarrow Ag + I^-$	-0.15
$Ce^{4+} + e^- \rightarrow Ce^{3+}$	$+1.61$	$Ni^{2+} + 2e^- \rightarrow Ni$	-0.23
$2HBrO + 2H^+ + 2e^- \rightarrow Br_2 + 2H$	$+1.60$	$Co^{2+} + 2e^- \rightarrow Co$	-0.28
$MnO_4^- + 8H^+ + 5e^- \rightarrow Mn^{2+} + 4H_2O$	$+1.51$	$In^{3+} + 3e^- \rightarrow In$	-0.34
$Mn^{3+} + e^- \rightarrow Mn^{2+}$	$+1.51$	$Tl^+ + e^- \rightarrow Tl$	-0.34
$Au^{3+} + 3e^- \rightarrow Au$	$+1.40$	$PbSO_4 + 2e^- \rightarrow Pb + SO_4^{2-}$	-0.36
$Cl_2 + 2e^- \rightarrow 2Cl^-$	$+1.36$	$Ti^{3+} + e^- \rightarrow Ti^{2+}$	-0.37
$Cr_2O_7^{2-} + 14H^+ + 6e^- \rightarrow 2Cr^{3+} + 7H_2O$	$+1.33$	$Cd^{2+} + 2e^- \rightarrow Cd$	-0.40
$O_3 + H_2O + 2e^- \rightarrow O_2 + 2OH^-$	$+1.24$	$In^{2+} + e^- \rightarrow In^+$	-0.40
$O_2 + 4H^+ + 4e^- \rightarrow 2H_2O$	$+1.23$	$Cr^{3+} + e^- \rightarrow Cr^{2+}$	-0.41
$ClO_4^- + 2H^+ + 2e^- \rightarrow ClO_3^- + H_2O$	$+1.23$	$Fe^{2+} + 2e^- \rightarrow Fe$	-0.44
$MnO_2 + 4H^+ + 2e^- \rightarrow Mn^{2+} + 2H_2O$	$+1.23$	$In^{3+} + 2e^- \rightarrow In^+$	-0.44
$Br_2 + 2e^- \rightarrow 2Br^-$	$+1.09$	$S + 2e^- \rightarrow S^{2-}$	-0.48
$Pu^{4+} + e^- \rightarrow Pu^{3+}$	$+0.97$	$In^{3+} + e^- \rightarrow In^{2+}$	-0.49
$NO_3^- + 4H^+ + 3e^- \rightarrow NO + 2H_2O$	$+0.96$	$U^{4+} + e^- \rightarrow U^{3+}$	-0.61
$2Hg^{2+} + 2e^- \rightarrow Hg_2^{2+}$	$+0.92$	$Cr^{3+} + 3e^- \rightarrow Cr$	-0.74
$ClO^- + H_2O + 2e^- \rightarrow Cl^- + 2OH^-$	$+0.89$	$Zn^{2+} + 2e^- \rightarrow Zn$	-0.76
$Hg^{2+} + 2e^- \rightarrow Hg$	$+0.86$	$Cd(OH)_2 + 2e^- \rightarrow Cd + 2OH^-$	-0.81
$NO_3^- + 2H^+ + e^- \rightarrow NO_2 + H_2O$	$+0.80$	$2H_2O + 2e^- \rightarrow H_2 + 2OH^-$	-0.83
$Ag^+ + e^- \rightarrow Ag$	$+0.80$	$Cr^{2+} + 2e^- \rightarrow Cr$	-0.91
$Hg_2^{2+} + 2e^- \rightarrow 2Hg$	$+0.79$	$Mn^{2+} + 2e^- \rightarrow Mn$	-1.18
$Fe^{3+} + e^- \rightarrow Fe^{2+}$	$+0.77$	$V^{2+} + 2e^- \rightarrow V$	-1.19

Table A2.I Standard potentials at 298 K in electrochemical order

Reduction half-reaction	$E^{\ominus}/V$	Reduction half-reaction	$E^{\ominus}/V$
$BrO^- + H_2O + 2e^- \rightarrow Br^- + 2OH^-$	+0.76	$Ti^{2+} + 2e^- \rightarrow Ti$	−1.63
$Hg_2SO_4 + 2e^- \rightarrow 2Hg + SO_4^{2-}$	+0.62	$Al^{3+} + 3e^- \rightarrow Al$	−1.66
$MnO_4^{2-} + 2H_2O + 2e^- \rightarrow MnO_2 + 4OH^-$	+0.60	$U^{3+} + 3e^- \rightarrow U$	−1.79
$MnO_4^- + e^- \rightarrow MnO_4^{2-}$	+0.56	$Mg^{2+} + 2e^- \rightarrow Mg$	−2.36
$I_2 + 2e^- \rightarrow 2I^-$	+0.54	$Ce^{3+} + 3e^- \rightarrow Ce$	−2.48
$Cu^+ + e^- \rightarrow Cu$	+0.52	$La^{3+} + 3e^- \rightarrow La$	−2.52
$I_3^- + 2e^- \rightarrow 3I^-$	+0.53	$Na^+ + e^- \rightarrow Na$	−2.71
$NiOOH + H_2O + e^- \rightarrow Ni(OH)_2 + OH^-$	+0.49	$Ca^{2+} + 2e^- \rightarrow Ca$	−2.87
$Ag_2CrO_4 + 2e^- \rightarrow 2Ag + CrO_4^{2-}$	+0.45	$Sr^{2+} + 2e^- \rightarrow Sr$	−2.89
$O_2 + 2H_2O + 4e^- \rightarrow 4OH^-$	+0.40	$Ba^{2+} + 2e^- \rightarrow Ba$	−2.91
$ClO_4^- + H_2O + 2e^- \rightarrow ClO_3^- + 2OH^-$	+0.36	$Ra^{2+} + 2e^- \rightarrow Ra$	−2.92
$[Fe(CN)_6]^{3-} + e^- \rightarrow [Fe(CN)_6]^{4-}$	+0.36	$Cs^+ + e^- \rightarrow Cs$	−2.92
$Cu^{2+} + 2e^- \rightarrow Cu$	+0.34	$Rb^+ + e^- \rightarrow Rb$	−2.93
$Hg_2Cl_2 + 2e^- \rightarrow 2Hg + 2Cl^-$	+0.27	$K^+ + e^- \rightarrow K$	−2.93
$AgCl + e^- \rightarrow Ag + Cl^-$	+0.22	$Li^+ + e^- \rightarrow Li$	−3.05
$Bi^{3+} + 3e^- \rightarrow Bi$	+0.20	**Strongly reducing**	

A2.I Standard potentials at 298 K in alphabetical order

Reduction half-reaction	$E^{\ominus}/V$	Reduction half-reaction	$E^{\ominus}/V$
$Ag^+ + e^- \rightarrow Ag$	+0.80	$I_2 + 2e^- \rightarrow 2I^-$	+0.54
$Ag^{2+} + e^- \rightarrow Ag^+$	+1.98	$I_3^- + 2e^- \rightarrow 3I^-$	+0.53
$AgBr + e^- \rightarrow Ag + Br^-$	+0.0713	$In^+ + e^- \rightarrow In$	−0.14
$AgCl + e^- \rightarrow Ag + Cl^-$	+0.22	$In^{2+} + e^- \rightarrow In^+$	−0.40
$Ag_2CrO_4 + 2e^- \rightarrow 2Ag + CrO_4^{2-}$	+0.45	$In^{3+} + 2e^- \rightarrow In^+$	−0.44
$AgF + e^- \rightarrow Ag + F^-$	+0.78	$In^{3+} + 3e^- \rightarrow In$	−0.34
$AgI + e^- \rightarrow Ag + I^-$	−0.15	$In^{3+} + e^- \rightarrow In^{2+}$	−0.49
$Al^{3+} + 3e^- \rightarrow Al$	−1.66	$K^+ + e^- \rightarrow K$	−2.93
$Au^+ + e^- \rightarrow Au$	+1.69	$La^{3+} + 3e^- \rightarrow La$	−2.52
$Au^{3+} + 3e^- \rightarrow Au$	+1.40	$Li^+ + e^- \rightarrow Li$	−3.05
$Ba^{2+} + 2e^- \rightarrow Ba$	−2.91	$Mg^{2+} + 2e^- \rightarrow Mg$	−2.36
$Be^{2+} + 2e^- \rightarrow Be$	−1.85	$Mn^{2+} + 2e^- \rightarrow Mn$	−1.18
$Bi^{3+} + 3e^- \rightarrow Bi$	+0.20	$Mn^{3+} + e^- \rightarrow Mn^{2+}$	+1.51
$Br_2 + 2e^- \rightarrow 2Br^-$	+1.09	$MnO_2 + 4H^+ + 2e^- \rightarrow Mn^{2+} \rightarrow 2H_2O$	+1.23
$BrO^- + H_2O + 2e^- \rightarrow Br^- + 2OH^-$	+0.76	$MnO_4^- + 8H^+ + 5e^- \rightarrow Mn^{2+} + 4H_2O$	+1.51
$Ca^{2+} + 2e^- \rightarrow Ca$	−2.87	$MnO_4^- + e^- \rightarrow MnO_4^{2-}$	+0.56
$Cd(OH)_2 + 2e^- \rightarrow Cd + 2OH^-$	−0.81	$MnO_4^{2-} + 2H_2O + 2e^- \rightarrow MnO_2 + 4OH^-$	+0.60
$Cd^{2+} + 2e^- \rightarrow Cd$	−0.40	$Na^+ + e^- \rightarrow Na$	−2.71
$Ce^{3+} + 3e^- \rightarrow Ce$	−2.48	$Ni^{2+} + 2e^- \rightarrow Ni$	−0.23
$Ce^{4+} + e^- \rightarrow Ce^{3+}$	+1.61	$NiOOH + H_2O + e^- \rightarrow Ni(OH)_2 + OH^-$	+0.49
$Cl_2 + 2e^- \rightarrow 2Cl^-$	+1.36	$NO_3^- + 2H^+ + e^- \rightarrow NO_2 + H_2O$	+0.80
$ClO^- + H_2O + 2e^- \rightarrow Cl^- + 2OH^-$	+0.89	$NO_3^- + 4H^+ + 3e^- \rightarrow NO + 2H_2O$	+0.96
$ClO_4^- + 2H^+ + 2e^- \rightarrow ClO_3^- + H_2O$	+1.23	$NO_3^- + H_2O + 2e^- \rightarrow NO_2^- + 2OH^-$	+0.10
$ClO_4^- + H_2O + 2e^- \rightarrow ClO_3^- + 2OH^-$	+0.36	$O_2 + 2H_2O + 4e^- \rightarrow 4OH^-$	+0.40

Table A2.1 Continued

Reduction half-reaction	$E^{\ominus}/V$	Reduction half-reaction	$E^{\ominus}/V$
$Co^{2+} + 2e^- \rightarrow Co$	-0.28	$O_2 + 4H^+ + 4e^- \rightarrow 2H_2O$	$+1.23$
$Co^{3+} + e^- \rightarrow Co^{2+}$	$+1.81$	$O_2 + e^- \rightarrow O_2^-$	-0.56
$Cr^{2+} + 2e^- \rightarrow Cr$	-0.91	$O_2 + H_2O + 2e^- \rightarrow HO_2^- + OH^-$	-0.08
$Cr_2O_7^{2-} + 14H^+ + 6e^- \rightarrow 2Cr^{3+} + 7H_2O$	$+1.33$	$O_3 + 2H^+ + 2e^- \rightarrow O_2 + H_2O$	$+2.07$
$Cr^{3+} + 3e^- \rightarrow Cr$	-0.74	$O_3 + H_2O + 2e^- \rightarrow O_2 + 2OH^-$	$+1.24$
$Cr^{3+} + e^- \rightarrow Cr^{2+}$	-0.41	$Pb^{2+} + 2e^- \rightarrow Pb$	-0.13
$Cs^+ + e^- \rightarrow Cs$	-2.92	$Pb^{4+} + 2e^- \rightarrow Pb^{2+}$	$+1.67$
$Cu^+ + e^- \rightarrow Cu$	$+0.52$	$PbSO_4 + 2e^- \rightarrow Pb + SO_4^{2-}$	-0.36
$Cu^{2+} + 2e^- \rightarrow Cu$	$+0.34$	$Pt^{2+} + 2e^- \rightarrow Pt$	$+1.20$
$Cu^{2+} + e^- \rightarrow Cu^+$	$+0.16$	$Pu^{4+} + e^- \rightarrow Pu^{3+}$	$+0.97$
$F_2 + 2e^- \rightarrow 2F^-$	$+2.87$	$Ra^{2+} + 2e^- \rightarrow Ra$	-2.92
$Fe^{2+} + 2e^- \rightarrow Fe$	-0.44	$Rb^+ + e^- \rightarrow Rb$	-2.93
$Fe^{3+} + 3e^- \rightarrow Fe$	-0.04	$S + 2e^- \rightarrow S^{2-}$	-0.48
$Fe^{3+} + e^- \rightarrow Fe^{2+}$	$+0.77$	$S_2O_8^{2-} + 2e^- \rightarrow 2SO_4^{2-}$	$+2.05$
$[Fe(CN)_6]^{3-} + e^- \rightarrow [Fe(CN)_6]^{4-}$	$+0.36$	$Sn^{2+} + 2e^- \rightarrow Sn$	-0.14
$2H^+ + 2e^- \rightarrow H_2$	0, by definition	$Sn^{4+} + 2e^- \rightarrow Sn^{2+}$	$+0.15$
$2H_2O + 2e^- \rightarrow H_2 + 2OH^-$	-0.83	$Sr^{2+} + 2e^- \rightarrow Sr$	-2.89
$2HBrO + 2H^+ + 2e^- \rightarrow Br_2 + 2H_2O$	$+1.60$	$Ti^{2+} + 2e^- \rightarrow Ti$	-1.63
$2HClO + 2H^+ + 2e^- \rightarrow Cl_2 + 2H_2O$	$+1.63$	$Ti^{3+} + e^- \rightarrow Ti^{2+}$	-0.37
$H_2O_2 + 2H^+ + 2e^- \rightarrow 2H_2O$	$+1.78$	$Ti^{4+} + e^- \rightarrow Ti^{3+}$	0.00
$H_4XeO_6 + 2H^+ + 2e^- \rightarrow XeO_3 + 3H_2O$	$+3.0$	$Tl^+ + e^- \rightarrow Tl$	-0.34
$Hg_2^{2+} + 2e^- \rightarrow 2Hg$	$+0.79$	$U^{3+} + 3e^- \rightarrow U$	-1.79
$Hg_2Cl_2 + 2e^- \rightarrow 2Hg + 2Cl^-$	$+0.27$	$U^{4+} + e^- \rightarrow U^{3+}$	-0.61
$Hg^{2+} + 2e^- \rightarrow Hg$	$+0.86$	$V^{2+} + 2e^- \rightarrow V$	-1.19
$2Hg^{2+} + 2e^- \rightarrow Hg_2^{2+}$	$+0.92$	$V^{3+} + e^- \rightarrow V^{2+}$	-0.26
$Hg_2SO_4 + 2e^- \rightarrow 2Hg + SO_4^{2-}$	$+0.62$	$Zn^{2+} + 2e^- \rightarrow Zn$	-0.76

Answers to Exercises

1.1 (a) 825 Torr; (b) 0.984 atm; (c) 0.212 atm; (d) 9.64×10^4 Pa.
1.2 89.20 kPa.
1.3 4.27 kPa.
1.4 2.48×10^{-3} mol.
1.5 66.4 kPa.
1.6 10.9 atm.
1.7 418 kPa.
1.8 173 kPa.
1.9 29.5 K.
1.10 388 K.
1.11 (a) 3.6 m^3; (b) 178 m^3.
1.12 1.32 L.
1.13 132 g mol^{-1}.
1.14 16.4 g mol^{-1}.
1.15 (a) (i) 1.0 atm, (ii) 8.2×10^2 atm;
      (b) (i) 0.99 atm, (ii) 1.7×10^3 atm.
1.16 (a) 2.0 atm H_2, 1.0 atm N_2; (b) 3.0 atm.
1.17 0.5 m^3.
1.18 1.5 kPa.
1.19 3.2×10^{-2} atm.
1.20 (a) Perfect gas, 5631 kPa; (b) van der Waals' gas, 4363 kPa.
1.21 $B = b - a/RT$, $C = b^2$;
      $a = 1.26$ L^2 atm mol^{-2},
      $b = 34.6$ cm^3 mol^{-1}.
1.22 1017 K.
1.23 (a) (i) 693, (ii) 1363, (iii) 2497 m s^{-1};
      (b) (i) 346, (ii) 682, (iii) 1249 m s^{-1}.
1.24 81 mPa.
1.25 13 MPa.
1.26 1 µm.
1.27 (a) 5×10^{10} s^{-1}; (b) 5×10^9 s^{-1}; (c) 5×10^4 s^{-1}.
1.28 (a) 6.1×10^{33} s^{-1}; (b) 6×10^{31} s^{-1}; (c) 6×10^{21} s^{-1}.
1.29 4×10^8 s^{-1}.
1.30 (a) 6.8 nm; (b) 68 nm; 0.7 cm.
1.31 9.1×10^{-3}.
1.32 λ independent of T.

2.1 (a) 98 J; (b) 16 J.
2.2 2.6 kJ.
2.3 3.03 J.
2.4 –0.10 kJ.
2.5 (a) horizontally 895 J; (b) vertically 899 J.
2.6 (a) –88 J; (b) –167 J.

2.7 +123 J.
2.8 –2.25 kJ.
2.9 –1.25 kJ.
2.10 25.3×10^3 kJ.
2.11 (a) 2.4×10^3 kJ; (b) 2.26×10^3 kJ.
2.12 23.7 J K^{-1}.
2.13 42 kJ.
2.14 301 kJ.
2.15 30 J K^{-1} mol^{-1}, 22 J K^{-1} mol^{-1}.
2.16 640 kJ, 640 s.
2.17 80 J K^{-1}, –1.2 kJ, –1.2 kJ.
2.18 +2.2 kJ, +2.2 kJ, +1.6 kJ.
2.19 –1.0 kJ, +13 kJ, +12 kJ.
2.20 (a) –92.22 kJ; (b) –46.11 kJ.
2.21 (a) 1560 kJ (b) 2340 kJ.
2.22 –4.56 MJ mol^{-1}.
2.23 –85 kJ mol^{-1}.
2.24 –432 MJ mol^{-1}.
2.25 +79 kJ mol^{-1}.
2.26 641 J K^{-1}.
2.27 4.21 kJ K^{-1}, 0.770 K.
2.28 (a) –2.80 MJ mol^{-1}; (b) –2.80 MJ mol^{-1}; (c) –1.28 MJ mol^{-1}.
2.29 84.4 kJ mol^{-1}.
2.30 –383 kJ mol^{-1}.
2.31 1.90 kJ mol^{-1}.
2.32 2.19 kJ mol^{-1}.
2.33 –25 kJ, 9.8 m.
2.34 (a) –2205 kJ mol^{-1}; (b) –2200 kJ mol^{-1}.
2.35 (a) exothermic, negative ΔH;
      (b) endothermic; (c) endothermic
      (d) endothermic; (e) endothermic
2.36 –138.2 kJ mol^{-1}.
2.37 +11.3 kJ mol^{-1}.
2.38 –56.98 kJ mol^{-1}.
2.39 (a) decrease; (b) decrease; (c) increase
2.40 (a) increase; (b) increase
2.41 higher.

3.1 0.410 J K^{-1}.
3.2 (a) 0.12 kJ K^{-1}; (b) –0.12 kJ K^{-1}.
3.3 –45.1 kJ.
3.5 55.9 J K^{-1}.
3.6 2.90 L.
3.7 –7.9 J.

3.8 $T_f = 0.630\ T_i$

3.10 5.3 J K^{-1}.

3.11 +87.8 J K^{-1} mol^{-1}, –87.8 J K^{-1} mol^{-1}.

3.12 (a) –386.1 J K^{-1}; (b) +92.6 J K^{-1}; (c) –153.1 J K^{-1}; (d) –21.0 J K^{-1}; (e) +512.0 J K^{-1}.

3.13 (a) 1.20 J; (b) 0.449 J.

3.14 –5.03 kJ K^{-1}.

3.15 (a) –521.5 kJ; (b) +25.8 kJ; (c) –178.7 kJ; (d) –212.40 kJ; (e) –5798 kJ.

3.16 (a) –522.1 kJ; (b) +25.78 kJ; (c) –178.6 kJ; (d) –212.55 kJ; (e) –5798 kJ.

3.17 (a) 5.5×10^4 kJ; (b) 5.1×10^4 kJ.

3.18 (a) -1.41×10^4 kJ; (b) $+1.7 \times 10^4$ kJ.

3.19 (a) non-exp = -1.67×10^4 kJ, water vapor exp = –79.0 kJ, total = -1.68×10^4 kJ; (b) non-exp = total work = -1.69×10^4 kJ liquid water.

3.20 –49.7 kJ mol^{-1}.

3.21 817.9 kJ mol^{-1}.

3.22 +0.95 J K^{-1} mol^{-1}.

3.23 –2.42 kJ mol^{-1}.

3.24 3.01.

3.25 $K_2^2 = K_1$.

3.26 –294 kJ mol^{-1}.

3.27 $K_{eq} = 1$.

3.28 1500 K.

3.29 –53.4 kJ mol^{-1}.

3.30 Expect $\Delta G^{\ominus}$ to be more negative for a gaseous reaction because $\Delta S^{\ominus}$ would be larger and positive.

3.31 Since the $\Delta G^{\ominus}$ values do not depend on the presence of a catalyst, $\Delta G^{\ominus}$ values should be the same.

3.32 $\Delta H^{\ominus} = +2.77$ kJ; $\Delta S^{\ominus}$ –16.5 J K^{-1} mol^{-1}.

3.33 +12.3 kJ mol^{-1}.

3.34 0.90, 0.10.

3.35 –41.0 kJ mol^{-1}.

3.36 (a, c, and e).

3.37 (b and d).

3.38 (a) +53 kJ mol^{-1}; (b) –53 kJ mol^{-1}.

3.39 –14.38 kJ mol^{-1}.

3.40 –118 kJ work produced.

3.41 (a) 9.24; (b) –12.9 kJ mol^{-1}; (c) +161 kJ mol^{-1}; (d) +248 J K^{-1} mol^{-1}.

3.42 (a) 1110 K; (b) 397 K.

3.43 –53.36 kJ mol^{-1}.

4.1 (a) 2.37 kg; (b) 41.9 kg; (c) 1.87 kg.

4.2 Yes; 3 Torr or more.

4.6 886.8 cm^3.

4.7 6.4 MPa.

4.8 $x_A = 0.920$, $x_B = 0.080$; $y_A = 0.968$, $y_B = 0.032$.

4.9 53.8 g mol^{-1}.

4.10 207 g mol^{-1}.

4.11 $K = \{1 - \rho s \Delta p / cp\}/c\{1 - 2\rho s \Delta p / cp\}^2$, $c = [A]_{nominal}$

4.12 –0.09°C.

4.13 (a) 1.32 mmol kg^{-1}; (b) 3.3 mmol kg^{-1}.

4.14 0.51 mmol kg^{-1} N$_2$, 0.27 mmol kg^{-1} O$_2$.

4.15 0.100 mol L^{-1}

4.16 –0.27°C.

4.17 87 kg mol^{-1}.

4.18 14 kg mol^{-1}.

4.19 (a) $y_T = 0.36$; (b) $y_T = 0.82$.

4.24 (a) 80 per cent by mass.

5.1 (a) $a(COCl)a(Cl)/a(CO)a(Cl_2)$; (b) $a(SO_3)^2/a(SO_2)^2a(O_2)$; (c) $a(HBr)^2/a(H_2)a(Br_2)$; (d) $a(O_2)^3/a(O_3)^2$.

5.2 4.46

5.3 (a) 1.15×10^9; (b) 1.8×10^2.

5.4 5.4×10^{-4}.

5.5 2.6×10^{-4} bar.

5.6 (a) 2.25×10^{-2} mol PCl$_5$ L^{-1}; (b) 42 per cent.

5.7 0.02 bar, 7.6×10^{-5} bar.

5.8 $K = 4\alpha^2 p/(1 - \alpha^2) \approx 4\alpha^2 p$.

5.9 (a) $H_2SO_4(A_1) + H_2O(B_2) \rightleftharpoons H_3O^+(A_2) + HSO_4^-(B_1)$; (b) $HF(A_1) + H_2O(B_2) \rightleftharpoons H_3O^+(A_2) + F^-(B_1)$; (c) $C_6H_5NH_3^+(A_1) + H_2O(B_2) \rightleftharpoons H_3O^+(A_2) + C_6H_5NH_2 (B_1)$; (d) $H_2PO_4^-(A_1) + H_2O(B_2) \rightleftharpoons H_3O^+(A_2) + HPO_4^{2-}(B_1)$; (e) $HCOOH(A_2) + H_2O(B_2) \rightleftharpoons H_3O^+(A_2) + HCO_2^-(B_1)$; (f) $NH_2NH_3^+(A_1) + H_2O(B_2) \rightleftharpoons H_3O^+(A_2) + N_2H_4 (B_1)$.

5.10 (a) $CH_3CHOHCOOH + H_2O \rightarrow CH_3CHOHCO_2^- + H_3O^+$
(b) $(H_2COOH)C(NH_2)COOH + H_2O \rightarrow$
$^-O_2CH_2C(NH_3^+)COOH + H_2O \rightarrow$
$^-O_2CH_2C(NH_3^+)COOH + H_2O \rightarrow$
$^-O_2CH_2C(NH_3^+)CO_2^- + H_3O^+$
(c) $NH_2CH_2COOH + H_2O \rightarrow (NH_3^+)CH_2CO_2^-$
(d) $HOOCCOOH + H_2O \rightarrow HOOCCO_2^- + H_3O^+$
$HOOCCO_2^- + H_2O \rightarrow {}^-O_2CCO_2^- + H_3O^+$

5.11 (a) 1.6×10^{-7} mol L^{-1}, 6.80; (b) 1.6×10^{-7} mol L^{-1}, 6.80.

5.12 (a) $2D_2O\ (l) \rightleftharpoons D_3O^+\ (aq) + OD^-\ (aq)$; (b) p$K_w$ =14.87; (c) $K_w = (D_3O^+)\ (OD^-) = X^2$, $X = 3.67 \times 10^{-8}$; (d) pD = pOH = 7.43; (e) pK_w = pD + pOD = 14.87.

5.13 (a) 4.82, 9.18; (b) 2.32, 11.18; (c) 7.0, 7.0; (d) 4.30, 9.70.

5.14 (a) 9.5×10^{-3} M, 2.02; (b) 0.025 M, 12.40; (c) 5.35×10^{-2} M, 1.27.

5.15 (a) <7, $NH_4^+ + H_2O \rightleftharpoons H_3O^+ + NH_3$; (b) >7, $CO_3^{2-} + H_2O \rightleftharpoons HCO_3^- + OH^-$; (c) >7, $H_2O + F^- \rightleftharpoons HF + OH^-$; (d) 7; (e) <7, $[Al(OH_2)_6]^{3+} + H_2O \rightleftharpoons H_3O^+ + [Al(OH_2)_5OH]^{2+}$; (f) <7, $[Co(OH_2)_6]^{2+} + H_2O \rightleftharpoons H_3O^+ + [Co(OH_2)_5OH]^+$.

5.16 (a) 9.14; (b) 4.83; (c) strong acid, 0.

5.17 (a) 3.08, 8.3×10^{-4}; (c) 2.8.

5.18 (a) 13.48; (b) 34 mL HCl

5.19 (a) 1.6 per cent; (b) 0.33 per cent; (c) 2.4 per cent.

5.20 (a) 2.00, 12.00, 0.083; (b) 4.85, 9.15, 0.10; (c) 1.14, 12.86, 0.73.

5.21 (a) 6.5; (b) 2.1; (c) 1.5.

5.22 8.3.

5.23 (a) 1.58×10^{-5}; (b) 1; (c) 5.0.

5.24 (a) 0.09 M, 0.06 M; (b) 0.065, 9.2×10^{-5} M, 7.1×10^{-15} M, 9.2×10^{-5} M, 1.1×10^{-10} M.

5.25 (a) 2.87; (b) 4.56; (c) 12.5 mL; (d) 4.74; (e) 25.0 mL; (f) 8.72.

5.26 (a) 4.74; (b) 5.04, 0.3; (c) 4.14, –0.60.

5.27 (a) 2 to 4; (b) 3 to 5; (c) 6 to 8.

5.28 3.24.

5.29 (a) 5.04; (b) 8.96; (c) 2.78.

5.30 7.94.

5.32 (a) H$_3$PO$_4$/NaH$_2$PO$_4$; (b) NaH$_2$PO$_4$/Na$_2$HPO$_4$.

5.33 (a) $K_s = [Ag^+]\ [I^-]$; (b) $K_s = [Hg_2^{2+}]\ [S^{2-}]$; (c) $K_s = [Fe^{3+}]\ [OH^-]^3$; (d) $K_s = [Ag^+]^2\ [CrO_4^{2-}]$.

5.34 (a) 1.0×10^{-5}; (b) 1.1×10^{-4}; (c) 9.28×10^{-11}; (d) 2.4×10^{-4}.

5.35 (a) 5.5×10^{-10} M; (b) 0.91 M;
(c) 1.6×10^{-7} M; (d) 2.45×10^{-7} M.
5.36 380.
5.37 -8.3×10^2 kJ mol^{-1}.

6.1 1.25×10^{-5} mol L^{-1}.
6.2 41 mV.
6.3 -0.021 mV.
6.4 Mn $\mid$ MnCl$_2$ (aq) $\mid$ Cl$_2$ (g) $\mid$ Pt,
-1.18 V.
6.5 (a) Ag$^+$(aq) + e$^-$ → Ag(s), Ag(s) → Ag$^+$(aq) + e$^-$,
cell: Ag$^+$(R) → Ag$^+$(L) concentration cell;
(b) H$_2$(g) (R) → H$_2$(g) (L);
(c) MnO$_2$(s) + 4H$^+$(aq) + 2e$^-$ → Mn^{2+}(aq) + 2H$_2$O(l),
3[Fe(CN)$_6$]$^{3-}$(aq) + 3e$^-$ → 3[Fe(CN)$_6$]$^{4-}$(aq),
cell: 3MnO$_2$(s) + 12H$^+$(aq) + 6[Fe(CN)$_6$]$^{4-}$(aq)→
3Mn^{2+}(aq) + 6[Fe(CN)$_6$]$^{3-}$(aq);
(d) Br$_2$(l) + 2e$^-$ → 2Br$^-$(aq), Cl$_2$(g) + 2e$^-$ → 2Cl$^-$(aq),
cell: Br$_2$(l) + 2Cl$^-$(aq) →2Br$^-$(aq)+ Cl$_2$(g);
(e) Sn^{4+}(aq)+2e$^-$ → Sn^{2+}(aq), 2Fe^{3+}(aq) + 2e$^-$ → 2Fe^{2+}(aq),
cell: Sn^{4+}(aq) + 2Fe^{2+}(aq) → Sn^{2+}(aq) + 2Fe^{3+}(aq);
(f) MnO$_2$(s) + 4H$^+$(aq) + 2e$^-$ → Mn^{2+}(aq) + 2H$_2$O(l),
Fe^{2+}(aq) + 2e$^-$ → Fe(s),
Fe(s) + MnO$_2$(s) + 4H$^+$(aq) → Fe^{2+}(aq) + Mn^{2+}(aq) +
2H$_2$O(l).
6.6 (a) $-RT/F \ln m_L/m_R$;
(b) $-RT/F \ln p_{H_2,L}/p_{H_2,R}$;
(c) $-RT/6F \ln [Mn^{2+}]^3[Fe(CN)_6^{3-}]^6/[H^+]^{12}[Fe(CN)_6^{4-}]^6$;
(d) $-RT/2F \ln [Br^-]^2 p_{Cl_2}/[Cl^-]^2$;
(e) $-RT/2F \ln [Fe^{3+}]^2 [Sn^{2+}]/[Fe^{2+}]^2 [Sn^{4+}]$;
(f) $-RT/2F \ln [Mn^{2+}] [Fe^{2+}]/[H^+]^4$.
6.7 (a) 2; (b) 2; (c) 4; (d) 2; (e) 2; (f) 1.
6.8 (a) & (b) see Nernst equation; (c) +0.87 V; (d) −0.27 V;
(e) −0.62 V; (f) 1.67 V.
6.9 (a) +0.08 V; (b) +0.27 V; (c) +1.23 V; (d) +0.695 V;
(e) +0.54 V; (f) +0.366 V.
6.10 (a) $|E|$ increase; (b) $|E|$ would increase;
(c) $E^{\ominus}$ becomes smaller in absolute value;
(d) $|E^{\ominus}|$ becomes smaller; (e) $|E|$ increases.
6.11 (a) $|E|$ increases; (b) $|E|$ increases; (c) $|E|$ decreases;
(d) $|E|$ decreases; (e) $|E|$ increases; (f) no change.
6.12 (a) −1.20 V; (b) −1.18 V.
6.13 (a) −394 kJ mol^{-1}; (b) −394 kJ mol^{-1}; (c) +75 kJ mol^{-1};
(d) −291 kJ mol^{-1}; (e) −291 kJ mol^{-1}; (f) +5.0 × 10^2 kJ mol^{-1}.
6.14 (a) +0.324 V; (b) +0.45 V.
6.15 −0.023 V.
6.16 87.9 kJ mol^{-1}.
6.17 (a) 0.3108 V; (b) −0.519 V;
(c) $E = E^{\ominus} − RT/2F \ln [HCO_3^-] [OH]/[CO_3^{2-}] [H_2O]$;
(d) $E = E' − 2.303 \, RT/2F$ pH $= −0.933$V;
(e) $K_a = 4.7 \times 10^{-11}$.
6.18 (a) 1.6×10^{-8} mol L^{-1}; (b) +0.12 V.
6.19 (a) 6.5×10^9; (b) 1.2×10^7; (c) 4.4×10^{69}; (d) 1.0×10^{25};
(e) 5.6×10^{-7}.
6.20 1.8×10^{-10}, 9.04×10^{-7}.
6.21 $E = E^{\ominus} − (RT/6F) \ln Q$,
$Q = a(Cr^{3+})^2/a(Cr_2O_7^{2-})a(H^+)^{14}$.
6.22 0.78.
6.23 0.
6.24 (a) 9.19×10^{-9} mol L^{-1}; (b) 8.5×10^{-17}.

6.25 −21.4 V.
6.26 5.53 V.
6.27 (a) Cathode will be on left; (b) 0.67 V.
6.28 0.94 V.
6.29 3.6×10^{-8}.
6.30 +0.22 V.

7.1 0.73 mol L^{-1} s^{-1}, 1.47 mol L^{-1} s^{-1}, 1.47 mol L^{-1} s^{-1}.
7.2 L^2 mol^{-2} s^{-1}.
7.3 (a) (molecules m^{-3})$^{-1}$ s^{-1}, (molecules m^{-3})$^{-2}$ s^{-1};
(b) kPa^{-1} s^{-1}, kPa^{-2} s^{-1}.
7.4 1.03×10^4 s; (a) 501 Torr; (b) 530 Torr.
7.5 1.12×10^{-4} s^{-1}.
7.6 8.8×10^{-3} s^{-1}.
7.7 1.09×10^{-6} mmol^{-1} s^{-1}.
7.8 1.12×10^{-4} s^{-1}.
7.9 1.9×10^{-5} Pa^{-1} s^{-1}.
7.10 (a) 0.014 kPa s^{-1}; (b) 1500 s.
7.11 1330 s.
7.12 3067 y.
7.13 (a) 0.63 µg; (b) 0.16 µg.
7.14 (a) 0.142 M; (b) 0.095 M.
7.15 2.79×10^4 s.
7.16 first order, 2.8×10^{-4} s^{-1}.
7.17 3.66×10^{11} mol L^{-1} s^{-1}.
7.18 $T = 299.2$ K.
7.19 one with larger E_a, 52 kJ mol^{-1}.
7.20 21.6 kJ mol^{-1}.
7.21 120 kJ mol^{-1}.
7.22 $E_a = −21.6$ kJ mol^{-1}.
7.23 48 kJ mol^{-1}.
7.24 5.42×10^4 s.
7.25 H$_2$O(1) , Br$^-$(1), overall (2).
7.26 rate $= k_2K^{1/2}[A_2]^{1/2}[B]$.
7.27 rate $= k_1K[A][B]$, $k = k_1k_2/k_2'$.
7.28 1.62×10^{-3} mol L^{-1} s^{-1}.
7.29 [S] $= K_M$.
7.30 rate $= −k_1[R_2] −k_2(k_1/k_4)^{1/2} [R_2]^{3/2}$.
7.31 0.28 to 2.2 kPa, 0.14 to 8.9 kPa;
> 0.11 kPa.
7.32 3.1×10^{18} .
7.33 0.412.
7.34 rate $= k_1k_3[A][AH][B]/\{k_2[BH^+] +k_3[A]\}$.
7.35 rate $= k_3KK_a[HA][BH^+]$.
7.36 rate $= k' [AH]$,
$k' = k_a + k_ak_c/2kk_d$, with
$$k = \left(\frac{k_a}{4k_b}\right)\left\{1 + \sqrt{1 + \frac{8k_bk_c}{k_ak_d}}\right\}$$
7.37 0.40 s^{-1}.
7.38 non-competitive.

8.1 4.6×10^3 W.
8.2 8.41×10^{11} .
8.3 1.32×10^6 m s^{-1}.
8.4 (a) 9.14×10^{-28} kg m s^{-1}; (b) 8.8×10^{-24} kg m s^{-1};
(c) 3.3×10^{-35} kg m s^{-1}.
8.5 (a) 1.02×10^{-27} kg m s^{-1}; (b) 1.02×10^{-33} kg m^{-1} s^{-1};
(c) 9.8×10^{26} s.
8.6 50.6 nm.
8.7 90 nm.

485

8.8 9.85×10^{-32} J.

8.9 $x = L/4$ and $x = 3L/4$.

8.10 (a) 2.17×10^{-20} J;
(b) 9.16×10^{-6} m.

8.11 (a) 3.31×10^{-19} J, 199 kJ mol^{-1};
(b) 3.61×10^{-19} J, 218 kJ mol^{-1};
(c) 4.97×10^{-19} J, 299 kJ mol^{-1};
(d) 9.93×10^{-19} J, 598 kJ mol^{-1};
(e) 1.32×10^{-15} J, 7.98×10^{5} kJ mol^{-1};
(f) 1.99×10^{-23} J, 0.012 kJ mol^{-1}.

8.12 2.2×10^{-24} m s^{-1}.

8.13 (a) 1.7×10^{18} s^{-1}; (b) 1.7×10^{20} s^{-1}.

8.14 6.90×10^{29} s^{-1}.

8.15 6000 K.

8.16 (a) no ejection; (b) 9.96×10^{5} m s^{-1}.

8.17 (a) 400 kJ mol^{-1}; (b) 20 kJ mol^{-1}; (c) 8×10^{-13} kJ mol^{-1}.

8.18 (a) 6.6×10^{-31} m; (b) 6.6×10^{-39} m; (c) 99.7 pm.

8.19 (a) 1.23 nm; (b) 39 pm; (c) 3.88 pm.

8.20 $\Delta v_{min} = 2.2 \times 10^{-29}$ m s^{-1}.

8.21 $\Delta q = 1 \times 10^{-26}$ m.

8.22 5×10^{-25} kg m s^{-1}, 5×10^{5} m s^{-1}.

8.23 1.12 fJ.

8.24 (a) 1.6×10^{-33} J m^{-3}; (b) 0.25 mJ m^{-3}.

8.25 6.52×10^{-34} J s.

8.26 434 nm.

8.27 6.

8.28 (a) 6842 cm^{-1}; (b) 1.36×10^{-19} J.

8.29 14.0 eV.

8.30 101 pm, 376 pm.

8.31 0, 90°, 180°, or 270°.

8.32 (a) 0; (b) 0; (c) $\sqrt{6}\hbar$; (d) $\sqrt{2}\hbar$;
(e) $\sqrt{2}\hbar$; angular nodes $= l$; radial nodes $= n - l - 1$.

8.33 (a) $g = 1$ (1 s); (b) $g = 9$ (3 s);
(c) $g = 49$, $n = 7$ (7s, three 7p, five 7d, seven 7f, nine 7g).

8.34 37 pm.

8.35 (a) 40 or 13 pm; (b) 29 or 24 pm.

8.36 1/6.

8.37 (a) F; (b) A; (c) A; (d) F; (e) A.

8.38 (a) 0, 2 e$^-$; (b) 3, 14 e$^-$; (c) 5, 22 e$^-$.

8.39 H $1s^1$; He $1s^2$; Li [He]$2s^1$;
Be [He]$2s^2$; B [He]$2s^2 2p^1$;
C [He]$2s^2 2p^2$; N [He]$2s^2 2p^3$;
O [He]$2s^2 2p^4$; F [He]$2s^2 2p^5$;
Ne [He]$2s^2 2p^6$; Na [Ne]$3s^1$;
Mg [Ne]$3s^2$; Al [Ne]$3s^2 3p^1$;
Si [Ne]$3s^2 3p^2$; P [Ne]$3s^2 3p^3$;
Si [Ne]$3s^2 3p^4$; Cl [Ne]$3s^2 3p^5$;
Ar [Ne]$3s^2 3p^6$.

8.40 $n_2 \rightarrow 6$.

8.41 3092 nm.

8.42 397.13 nm, 3.40 eV.

8.43 987 663 cm^{-1}; lines at
137 175 cm^{-1}, 185 187 cm^{-1}, . . . ;
122.5 eV.

8.44 (b) Mg and Ti.

9.1 (a) $1\sigma_g^2$, $b = 1$; (b) $1\sigma_g^2 2\sigma_u^2$, $b = 0$; (c) $1\sigma_g^2 2\sigma_u^2 2\pi_u^4$, $b = 2$.

9.2 (a) $1\sigma_g^2 2\sigma_u^1$; (b) $\ldots 1\pi_u^4 2\sigma_g^2$; (c) $1\pi_g^1 1\pi_g^1$.

9.3 (a) CO (14e$^-$) $1s\sigma^2 \, 1s\sigma^{*2} \, 2s\sigma^2 \, 2s\sigma^{*2} \, 2p\sigma^2 \, 2p\pi^4$;
(b) NO (15e$^-$) $1s\sigma^2 \, 1s\sigma^{*2} \, 2s\sigma^2 \, 2s\sigma^{*2} \, 2p\sigma^2 \, 2p\pi^4 \, 2p\pi^{*1}$;
(c) CN$^-$ (14e$^-$) $1s\sigma^2 \, 1s\sigma^{*2} \, 2s\sigma^2 \, 2s\sigma^{*2} \, 2p\sigma^2 \, 2p\pi^4$.

9.4 C_2.

9.5 (a) C_2, CN; (b) NO, O_2, F_2.

9.6 $R(XeF^+) < R(XeF)$.

9.7 (a) g; (b) not relevant; (c) g; (d) u.

9.8 $n = 1$ g; $n = 2$ u; $n = 3$ g; $n = 4$ u.

9.9 a_2 (g), e_1 (g), e_2 (u), b_2 (g).

9.10 N_2.

9.11 $F_2^+ < F_2 < F_2^-$.

9.13 $\phi_B = s - (\sqrt{2}s - p)/\sqrt{3}$

9.14 $N = 1/(1 + 2\lambda S + \lambda^2)^{1/2}$.

9.15 (a, c).

9.16 (a) angular; (b) octahedral; (c) square planar; (d) see-saw.

9.18 (a) $6\alpha + 8\beta$; (b) $5\alpha + 7b$.

9.19 4.3×10^{-19} J (460 nm).

10.1 3.9×10^{3} kJ mol^{-1}.

10.2 1.07

10.3 2 ax + 1 eq.

10.4 (a) ortho: 0.7 D; (b) meta: 0.4 D; (c) para: 0 (certain).

10.5 (a) m-xylene + 0.4 D = 1.1 D;
(b) p-xylene + 0.4 cos 60° = 0.2 D;
(c) m-xylene − 0.4 D = 0 (certain).

10.6 1.4 D.

10.7 (a) 1.87 D; (b) 3.0 D; (c) 2.15 D; (d) 1.64.

10.8 $V = -\mu q/4\pi\epsilon_0 R^2$.

10.9 $4\pi\epsilon_0 V = -q_1 q_2/(r - \frac{1}{2}l) + q_1 q_2/(r + \frac{1}{2}l)$

10.10 $V = 2\mu^2/4\pi\epsilon_0 R^3$.

10.11 $2^{1/6} \sigma$

10.16 (a) 6.25×10^{-9} m^2 s^{-1}; (b) 1.56×10^{-9} m^2 s^{-1}.

10.17 (a) $N = 10^6$; (b) same.

10.18 (a) 0.9069; (b) 0.5236.

10.19 (a) 8; (b) ($d = 4$, 433nm) 6, next nearest ($d = 500$ nm).

10.20 (a) 12; 354 nm; (b) 20, 613 nm.

10.21 (a) less dense; (b) 2.

10.24 (326), (111), (122), (3$\bar{2}\bar{2}$).

10.25 249 pm, 176 pm, 432 pm.

10.27 307 pm, 217 pm, 532 pm.

10.28 70.7 pm.

10.30 2.40×10^6 g m^{-3}.

10.31 4, 4.01 g cm^{-3}.

10.32 (a) 220 pm; (b) 110 pm.

10.33 bcc.

10.34 8.97 g cm^{-3}.

11.1 3.0×10^{32}.

11.2 $\lambda = 5.1$ pm.

11.3 (a) yes; (b) yes; (c) yes; (d) yes; (e) no.

11.4 all.

11.5 one.

11.6 10 rotational E states.

11.7 $4m_B r^2$.

11.8 $I = \frac{8}{3} m_B r^2$.

11.9 (a) 21.2 cm^{-1}; (b) 63.6×10^{10} Hz.

11.10 1.91×10^{12} Hz.

11.11 I would be higher; B smaller, ν smaller.

11.12 251 pm.

11.13 (a) 168 pm; (b) 6.44 cm^{-1}.
11.14 (b, c, d, e, f, g).
11.15 328.7 N m^{-1}.
11.16 6.36×10^7 m s^{-1} or 1.4×10^8 mph.
11.17 2.4×10^4 km s^{-1}; 8.4×10^5 K.
11.18 (a) 53 ps; (b) same; (c) 159 ps.
11.19 (a) 53 cm^{-1}; (b) 0.27 cm^{-1}.
11.20 HF: 967.1 N m^{-1},
HCl: 515.6 N m^{-1},
HBr: 411.8 N m^{-1},
HI: 314.2 N m^{-1}.
11.21 DF: 3002 cm^{-1}, DI: 2144 cm^{-1},
DBr: 1886 cm^{-1}, DI: 1640 cm^{-1}.
11.22 (a) yes; (b) no.

11.23 75 per cent reduction.
11.24 1.35×10^6 mol^{-1} cm^2.
11.25 1.16×10^{-3} mol L^{-1}.
11.27 450 mol L^{-1} cm^{-1}.
11.28 160 mol L^{-1} cm^{-1}, 23 per cent.
11.29 (a) 0.9 m; (b) 3 m.
11.31 2.11×10^{-5} M, 7.14×10^{-5} M.
11.32 1: 7: 21: 35: 35: 21: 7: 1.
11.33 1 : 6 : 15 : 20 : 15 : 6 : 1 heptet.
11.34 -1.625×10^{-26} J $\times m_I$.
11.35 92.3 MHz.
11.36 12.92 T.
11.37 (a) 11 µT; (b) 46 µT
11.38 (a) 2.28 kHz; (b) 4.18 kHz.

Index

T after a page number refers to a Table in the text.

Index